Fourier Transforms in NMR, Optical, and Mass Spectrometry

A User's Handbook

Fourier Transforms in NMR, Optical, and Mass Spectrometry

A User's Handbook

Alan G. Marshall and Francis R. Verdun

Departments of Chemistry and Biochemistry, The Ohio State University, Columbus, OH 43210-1173, U.S.A.

ELSEVIER
Amsterdam — Oxford — New York — Tokyo 1990

ELSEVIER SCIENCE PUBLISHERS B.V.
Sara Burgerhartstraat 25
P.O. Box 211, 1000 AE Amsterdam, The Netherlands

Distributors for the United States and Canada:

ELSEVIER SCIENCE PUBLISHING COMPANY INC.
655, Avenue of the Americas
New York, NY 10010, U.S.A.

Library of Congress Cataloging-in-Publication Data

Marshall, Alan G., 1944-
Fourier transforms in NMR, optical, and mass spectrometry : a user's handbook / by Alan G. Marshall and Francis R. Verdun.
p. cm.
Includes index.
ISBN 0-444-87360-0
1. Fourier transform spectroscopy. 2. Nuclear magnetic resonance spectroscopy. 3. Spectrum analysis. 4. Mass spectrometry.
I. Verdun, Francis R. II. Title.
QD96.F68M37 1989
543'.0877--dc20 89-1518
CIP

ISBN 0-444-87360-0 (hard bound)
ISBN 0-444-87412-7 (paperback)

This book is printed on acid-free paper.

Printed in The Netherlands

CONTENTS

CHAPTER 6 Non-FT methods for proceeding from time- to frequency-domain

CHAPTER 7 Fourier transform ion cyclotron resonance mass spectrometry

CHAPTER 9 FT/interferometry

Dedicated to Devon W. Meek

for helping to bring one of the authors (A.G.M.) to The Ohio State University in 1980, and more generally for his unstinting and selfless efforts on behalf of faculty, staff, and students during his terms as Chairman of the OSU Chemistry Department.

PREFACE

This book is offered as a teaching and reference text for Fourier transform methods as they are applied in spectroscopy. Whereas several other books treat either the mathematics or one type of spectroscopy, or offer a collection of chapters by several authors, the present monograph offers a unified treatment of the three most popular types of FT/spectroscopy, with uniform notation and complete indexing of specialized terms. All mathematics is self-contained, and requires only a knowledge of simple calculus. The main emphasis is on pictures (~200 original illustrations) and physical analogs, with sufficient supporting algebra to enable the reader to enter the literature. Instructive problems offer extensions from the basic treatment. Solutions are given or outlined for all problems.

Because the book aims to inform practicing spectroscopists, non-ideal effects are treated in detail: noise (source- and detector-limited); non-linear response; limits to spectrometer performance based on finite detection period, finite data size, mis-phasing, etc. Common puzzles and paradoxes are explained: e.g., use of mathematically complex variables to represent physically real quantities; interpretation of negative-frequency signals; on-resonance *vs.* off-resonance response; interpolation (when it helps and when it doesn't); ultimate accuracy of discrete representation of an analog signal; differences between linearly- and circularly-polarized radiation; multiplex advantage or disadvantage, etc.

Chapter 1 introduces the fundamental line shapes encountered in spectroscopy, from a simple classical mass-on-a-spring model. The Fourier transform relationship between the time-domain response to a sudden impulse and the steady-state frequency-domain response (absorption and dispersion spectra) to a continuous oscillation are established and illustrated. Chapters 2 and 3 summarize the basic mathematics (definitions, formulas, theorems, and examples) for continuous (analog) and discrete (digital) Fourier transforms, and their practical implications. Experimental aspects which are common to the signal (Chapter 4) and noise (Chapter 5) in all forms of Fourier transform spectrometry are followed by separate treatments of those features which are unique to FT/MS (Chapter 7), FT/NMR (Chapter 8), FT/optical (Chapter 9), other types (Chapter 10) of FT/spectrometry. In Chapter 6, non-FT methods (e.g., autoregression, maximum entropy) are presented and critically compared to FT methods.

The list of references includes both historical and comprehensive reviews and monographs, along with articles describing several key developments. The Appendices include Fast Fourier and Fast Hartley Transform algorithms in Fortran and Basic, a look-up table of useful integrals (definite and indefinite), and a pictorial atlas of the Fourier transform time/frequency functions most commonly encountered in FT spectroscopy. The comprehensive Index is designed to enable the reader to locate particular key words, including those with more than one name—e.g., foldover (aliasing), throughput (étendue or Jacquinot) advantage; apodization (windowing); Hilbert transform (Kramers-Kronig transform, Bode relation), etc.

ACKNOWLEDGMENTS

The authors acknowledge, with deep appreciation, the inspiration and assistance provided by past and present collaborators, particularly M. B. Comisarow, T. L. Ricca, D. C. Roe, T.-C. L. Wang, L. Chen, A. T. Hsu, C. E. Cottrell, J. E. Meier, C. P. Williams, M. Wang, P. B. Grosshans, Z. Liang, G. M. Alber, E. C. Craig, J. Skilling, and S. Goodman. We also thank several individuals for their critical and helpful comments about the manuscript: M. B. Comisarow, P. K. Dutta, T. Gäumann, P. R. Griffiths, T. L. Gustafson, J. A. de Haseth, M. Wang, G. M. Alber, P. B. Grosshans, N. M. M. Nibbering, and E. Williams. We thank G. M. Alber, E. J. Behrman, R. B. Cody, C. E. Cottrell, R. Doskotch, W. G. Fateley, B. S. Freiser, T. Gäumann, G. Horlick, D. Horton, D. Hunt, F. W. McLafferty, M. D. Morris, P. Schmalbrock, M. Wang, and J. Wu for providing illustrations. F.R.V. especially thanks T. Gäumann for his help and indulgence during the preparation of the manuscript. Finally, A.G.M. wishes to thank his family for their patience and support.

CHAPTER 1

Spectral line shape derived from the motion of a damped mass on a spring

1.1 Terminology

Virtually everything we need to know about line shapes in Fourier transform spectra can be understood from the simple physical model of a mass on a spring. It is easy to see why. In general, if the (time-independent) force, F, acting on a particle of mass, m, varies with distance, x, we can express Newton's second law as a Taylor series in x, in which x is measured from some arbitrary origin (e.g., the center of an atom).

$$\text{Net force} = m\frac{d^2x}{dt^2} = a_0 + a_1 x + a_2 x^2 + \cdots \tag{1.1}$$

However, the representation of even a relatively simple force, such as the Coulomb force, $m\,d^2x/dt^2 = q^2/x^2$, between two particles each of charge, q, may require an infinite number of terms in Eq. 1.1. Fortunately, two simplifications render the problem tractable and useful.

First, we can dispense with the constant term, a_0, simply by choosing a suitable "reference" frame. For example, we can neglect the (constant) force of gravity in most problems. Second, if the force is so weak that the observed displacement, x, is very small (i.e., $x >> x^2 >> x^3$ ...), then we may conveniently neglect all of the higher order terms, leaving

$$\text{Net force} = m\frac{d^2x}{dt^2} \cong a_1 x \tag{1.2}$$

If we then rename the constant, a_1, as a new constant, $-k$, then Eq. 1.2 becomes identical to the equation for the (restoring) force for a weight (of mass, m) on a spring of force constant, k.

$$\text{Net force} = m\frac{d^2x}{dt^2} = -k\,x \tag{1.3}$$

For example, even though we know that the force binding an electron to an atom or molecule is a Coulomb force, we may still treat the electron as if it were bound by springs, provided that whatever forces we apply do not displace the electron very far from its equilibrium position. Thus, because the electric and magnetic forces from electromagnetic waves (light, x-rays, radiofrequencies, microwaves, infrared, etc.) on atoms and molecules are indeed weak, we can successfully describe many effects of electromagnetic radiation on ions, atoms, and molecules using the same mathematics that describe the response of a simple mechanical weight on a spring to some externally applied "jiggling".

Before we discuss the motion of a "jiggled" spring, it is useful to review the vocabulary of wave motion, since electromagnetic radiation can be described by the same language as that for ordinary water waves.

A *transverse* wave is a disturbance whose displacement oscillates in a direction *perpendicular* to the direction of propagation of the wave. For example, a cork floating on water moves up and down vertically as the water wave moves along horizontally, as shown in Figure 1.1.

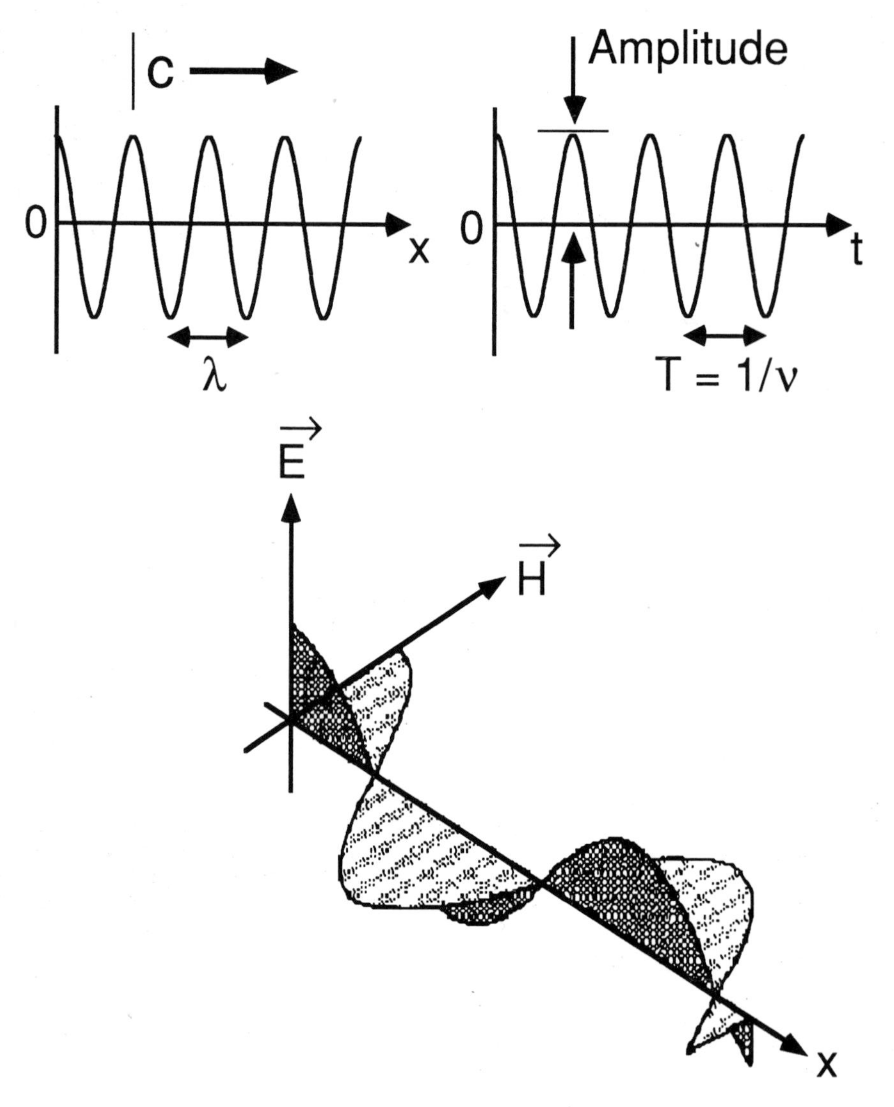

Figure 1.1 Transverse monochromatic waves. Top: water wave, viewed according either to its displacement as a function of distance at a given time instant (left), or to its displacement as a function of time at a particular point in space. Bottom: electromagnetic wave (plane-polarized for simplest display). Both waves are generated continuously at the left of the figure, so that propagation is from left to right. See text for definitions of terms.

The *amplitude* of the wave is the maximum displacement from the equilibrium position. The *wavelength*, λ, for a monochromatic (see below) wave is the distance (along the direction of propagation) between two successive points of maximal displacement at a given instant in time. The *velocity*, c, of a monochromatic wave is defined as the distance a given wave crest moves per unit time. The *frequency*, ν (in cycles per second, or Hz), of a monochromatic wave is the number of times per second that the wave displacement at a given point in space passes through its maximum value. The *period*, T, is the time (in seconds) required to complete one cycle of the oscillation of a monochromatic wave. Finally, the *phase* (or "phase angle"), φ, of a wave at a given time instant, t, can be defined as the number of radians of oscillation accumulated since time zero (there are 2π radians per oscillation cycle). Beginning at an (arbitrary) zero time, phase angle accumulates at an "angular" velocity (or "angular" frequency), $\omega = 2\pi\nu$ radians per second, for a wave of frequency, ν oscillations/second. The relations between the previous definitions are given by the following equations:

$$c = \lambda \nu \qquad \text{meter/second = (meter/cycle)·(cycles/second)} \tag{1.4}$$

$$T = \frac{1}{\nu} \qquad \text{seconds/cycle} \tag{1.5}$$

$$\varphi = \varphi_0 + \omega t \quad \text{radians accumulated since time zero} \tag{1.6}$$

$$\omega = 2 \pi \nu \qquad \text{radians/second = (radians/cycle)·(cycles/second)} \tag{1.7}$$

The *principle of superposition* states that when two waves travel through the same region of space, their displacements add, as, for example, when two light waves of different frequency (color) travel together. The ways (coherent, incoherent) in which two or more waves combine (interference, diffraction) will be discussed in Chapter 9.2.1. A wave is said to be *monochromatic* if all of its components have the same frequency (and thus the same wavelength). The *intensity* (joule m^{-2} s^{-1}) of a wave is the energy flow per unit time across unit area perpendicular to the direction of propagation. Radiation *power* (Watts) is the product of intensity and cross-sectional area across which the wave passes. Intensity is proportional to the *square* of the wave amplitude, as may be seen by analogy to the water wave. If the amplitude of a water wave doubles, then the water molecules must travel twice as far in a given length of time, and therefore have on the average twice as much velocity. Since kinetic energy is proportional to the square of velocity, intensity must be proportional to the square of wave amplitude.

$$\text{Intensity} \propto (\text{amplitude})^2 \tag{1.8}$$

Returning to Eq. 1.3, we recall that the equation of motion of a mass, m, attached to a spring of stiffness (spring constant), k, in the *absence* of any damping or driving force is

$$m \frac{d^2 x}{dt^2} + k x = 0 \tag{1.3}$$

The reader can quickly verify by substitution that Eq. 1.9 is a solution of Eq. 1.3 for the undamped, undriven mass on a spring.

$$x = x_0 \cos \omega_0 t \tag{1.9}$$

in which

$$\omega_0 = \sqrt{\frac{k}{m}} = \text{natural angular frequency for mass on a spring} \qquad (1.10)$$

for a maximum initial displacement, $x = x_0$ at time, $t = 0$. Since there is no frictional "damping", the displacement (position) of the mass continues to oscillate sinusoidally indefinitely at the *natural* frequency, $\nu_0 = \omega_0/2\pi$ Hz. Eq. 1.9 also shows why angular frequency is almost always used in the mathematical description of spring motion, since the wave frequency, ν_0, appears in the equations in the form, $\omega_0 = 2\pi\nu_0$. Armed with the necessary vocabulary, we are now prepared to discuss the motion of a mass on a spring in the presence of additional driving and damping forces.

There are two ways to discover the natural frequency of a weight on a spring. One could simply strike the weight to displace it suddenly from its equilibrium position, and then count the oscillations per unit time as the mass moves up and down. Alternatively, one could jiggle the spring steadily at each of various "driving" frequencies and wait until the system settles into a steady-state motion; when the jiggling frequency matches the natural frequency, the system will oscillate with maximal amplitude (just as a tuning fork is set into motion by sound applied at its natural frequency). These two approaches are shown schematically in Figure 1.2.

If the driven spring were completely free to move, then when the driving frequency is set equal to (in "resonance" with) the natural spring frequency, ν_0, we would expect the spring to absorb energy continuously and execute larger and larger amplitude motion without limit. However, any real spring (including an electron bound to a molecule) is hindered in its motion by frictional resistance, or drag, expressed by a "damping" force (characterized by frictional coefficient, f) proportional to the velocity of the moving mass.

$$\text{Damping force} = -f\,\frac{dx}{dt} \qquad (1.11)$$

In the spectroscopic examples to be discussed later, the natural frequencies of various springs correspond to the frequencies of power absorption (peaks) in the spectrum and the damping constant of a given spring is related to the width of a spectral line. The practical interest in these parameters is that the natural frequencies indicate the strength of the spring of interest, and are related to color, chemical bond strength or type, optical activity, and ionic mass-to-charge ratio; spectral line widths reflect rates of molecular collisions, chemical reactions and molecular motions.

1.2 Transient response to a sudden impulse: relaxation

The transient experiment consists of monitoring the displacement as a function of time for a damped mass on a spring (damped harmonic oscillator), just after the mass has been displaced by an initial distance, x_0, and then released. Mathematically, the equation of motion becomes

$$\text{Net force} = m\,\frac{d^2x}{dt^2} = -kx - f\,\frac{dx}{dt} \qquad (1.12)$$

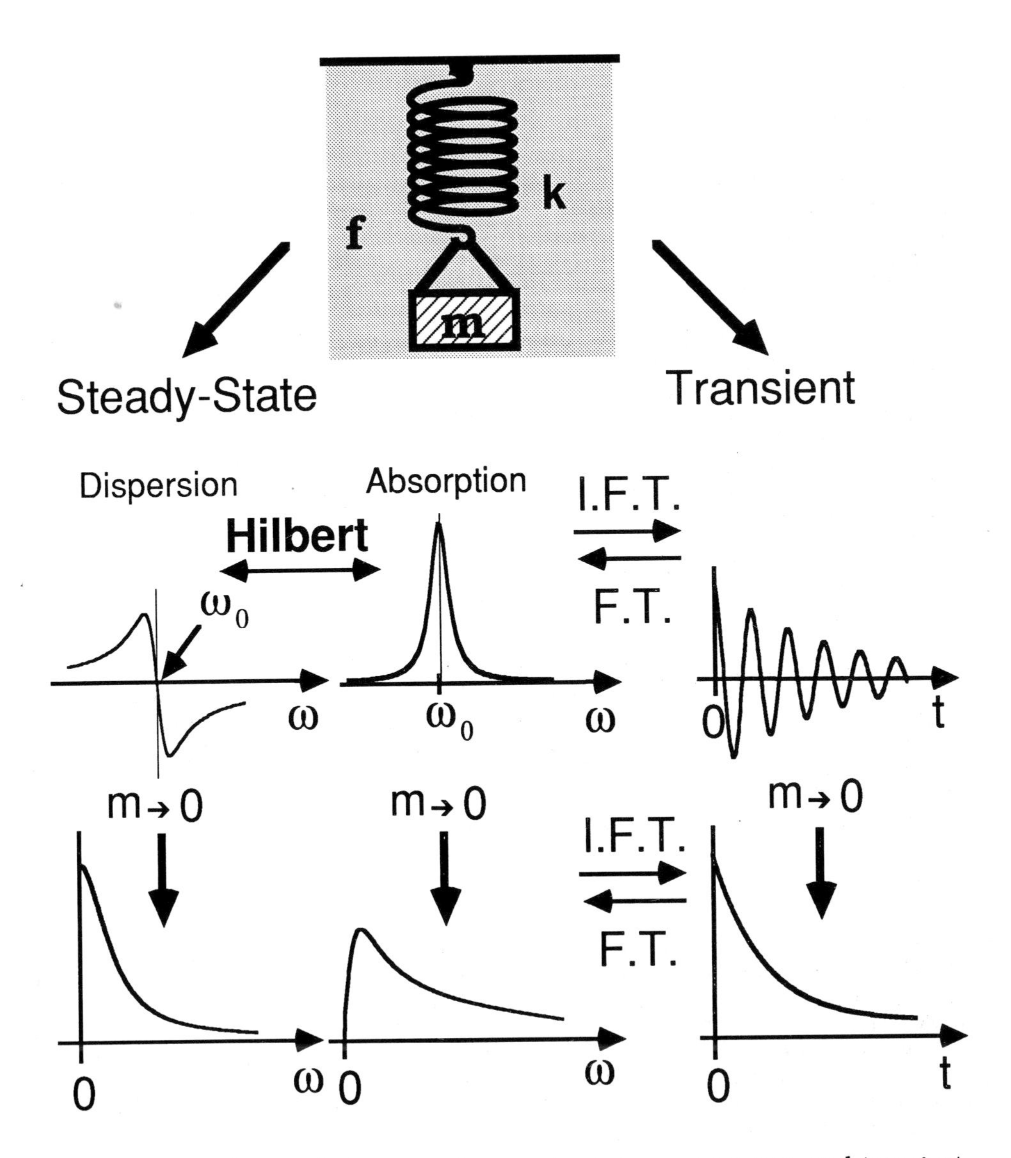

Figure 1.2 Interrelations between various scattering, spectroscopy, and transient experiments, by analogy to the motion of either a sinusoidally driven or a suddenly displaced damped weight on a spring. The weight has mass, m, bound to a spring of force constant, k, immersed in a medium whose effect is represented by frictional coefficient, f, and is subjected either to a sinusoidally time-varying continuous force, $F(t) = F_0 \cos \omega t$ (steady-state response, left), or to a sudden displacement (transient response, right) The massless limit corresponds to scattering (steady-state) or relaxation (transient) experiments. The various results are related by Fourier and Hilbert transforms as shown.

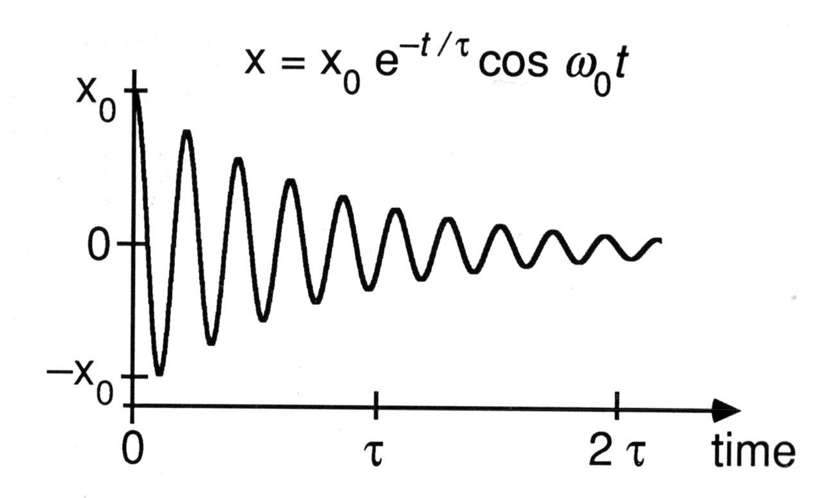

Figure 1.3 Position, $x(t)$, of a damped mass on a spring as a function of time, just after the mass has been released from an initial displacement, x_0, at time zero. (See Eq. 1.16.)

It is left as an exercise (see Problems) for the reader to verify that Eq. 1.13 is a solution of Eq. 1.12.

$$x = x_0 \exp(-t/\tau) \cos Wt \tag{1.13}$$

in which

$$\frac{1}{\tau} = \frac{f}{2m} \tag{1.14a}$$

and

$$W = \sqrt{\omega_0^2 - (1/\tau)^2} \tag{1.14b}$$

This result becomes especially simple in the usual physical limit that damping is weak:

$$(1/\tau) << \omega_0 \quad \text{(slight damping)} \tag{1.15}$$

for which

$$\lim_{(1/\tau)<<\omega_0} (x) = x_0 \exp(-t/\tau) \cos \omega_0 t \tag{1.16}$$

Equation 1.16 indicates that when the mass is initially displaced by x_0 at time zero, it will then *oscillate at approximately its natural angular frequency*, $\omega_0 = \sqrt{k/m}$, with an amplitude (i.e., envelope of the oscillations) that decreases exponentially with exponential time constant, $\tau = 2m/f$, as shown in Figure 1.3. In ordinary language, striking a tuning fork will cause it to vibrate at its natural pitch, with a loudness that decreases exponentially with time. Although the transient experiment is conceptually simple, its interpretation becomes more complicated when several natural frequencies are present (i.e., a multi-peak spectrum). We therefore next turn to the steady-state experiment, in which each spectral peak can be observed separately.

1.3 Steady-state response to a sinusoidal driving force

1.3.1 Absorption and dispersion spectra

Figure 1.4 shows that a mass on a spring subjected to a continuous, sinusoidally time-varying driving force, $F = F_0 \cos \omega t$, will eventually settle into a steady vibration *at the same (angular) frequency* (ω) *as the driver*, in contrast to the undriven mass on a spring (Eq. 1.13) which vibrates at its own *natural* (angular) frequency, ω_0. In the absence of friction, the driven spring would oscillate exactly *in-phase* with the driver. However, in the presence of friction, the driven spring displacement will in general lag behind the driver as shown in Figure 1.4. It is physically meaningful and mathematically convenient (see Problems) to analyze the steady-state displacement, x, into a sum of two components (each of which oscillates at the driver frequency) which are respectively either exactly *in-phase* (amplitude = x') or exactly *90°-out-of-phase* (amplitude = x'') with the driver, as shown in Figure 1.4.

The spectrum (i.e., variation with frequency) of x' is called the *dispersion*, and the spectrum of x'' is called the *absorption*, These terms can be understood from the special case that the weight-on-a-spring is an electron bound to an atom or molecule, and the driving force is the oscillating electric field of a light wave acting on the electronic charge. In that experiment, x'' describes the normalized power *absorption* as a function of the light frequency, and (x' + 1) describes the refractive index as a function of frequency. Since "dispersion" of light into its component colors (frequencies) by a prism results from a refractive index that varies with frequency, x' has come to be known as the *dispersion* spectrum.

The great advantage of using the mechanical model of Figure 1.4 to generate the mathematical forms of x' and x'' is that we can then describe various types of spectra (e.g., visible/ultraviolet, infrared, magnetic resonance, optical rotatory dispersion, etc.) without any additional effort—only the names and scaling of the physical quantities vary. With this motivation, we proceed to trace the mathematical steps and assumptions that correspond to the graphs in Figure 1.4.

The equation of motion for the position, x, of a weight of mass, m, on a spring of force constant, k, subject to friction, and driven by a (sinusoidally oscillating) force, F, is given by

$$\text{Net force} = m\, d^2x/dt^2 = -kx - f\, dx/dt + F_0 \cos \omega t \tag{1.17}$$

in which $-kx$ represents the restoring force from the spring, $-f\, dx/dt$ is the frictional force opposing the motion, and $F_0 \cos \omega t$ is the applied driving force. It is left to the reader to confirm that Eq. 1.18 is a solution of Eq. 1.17, in which Eq. 1.10 [$\omega_0 = \sqrt{k/m}$] has been used to simplify the notation.

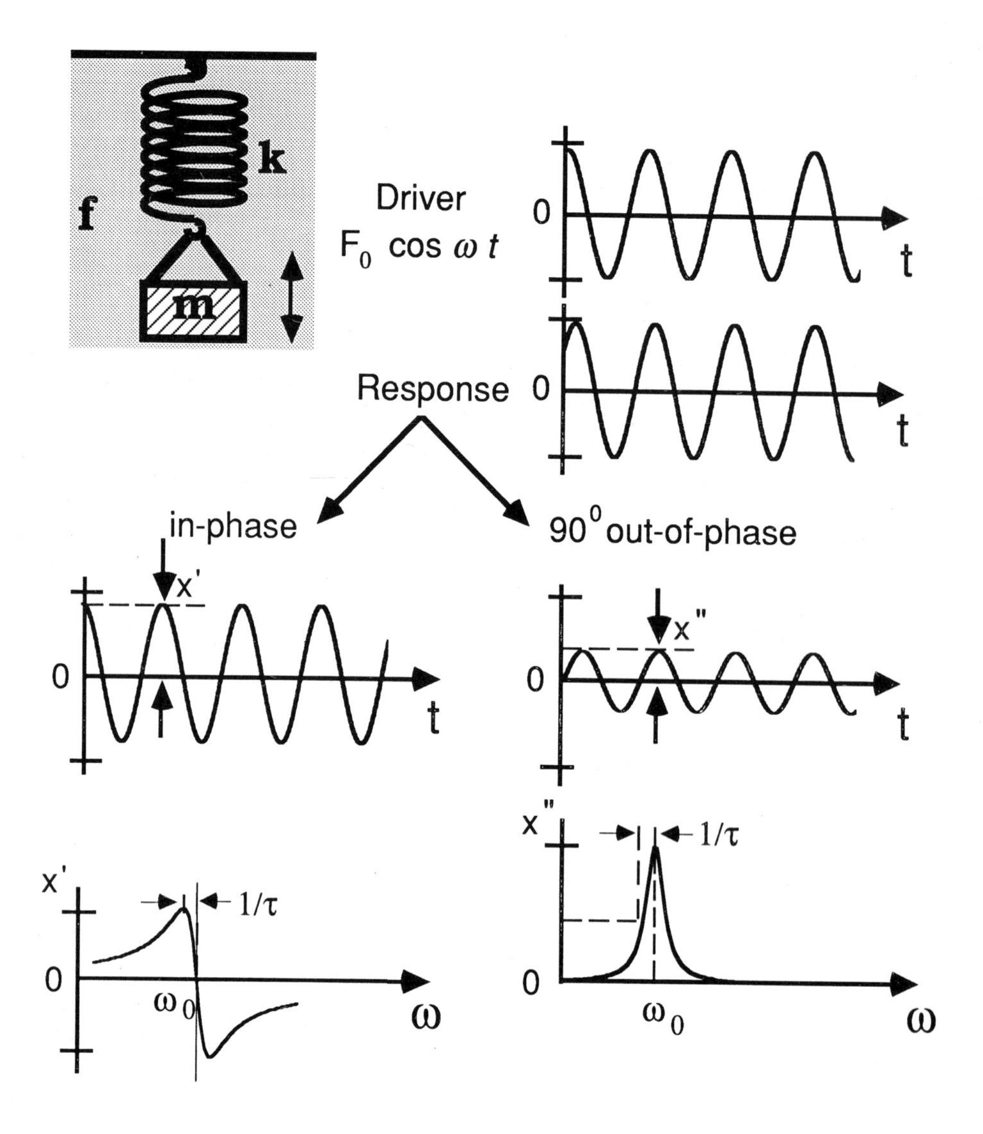

Figure 1.4 Origin of dispersion and absorption spectra. When subjected to a sinusoidally time-varying driving force (upper right), the mass on a damped spring (upper left) will eventually settle into a steady-state oscillation ("response") at the *same frequency* as the driver, but with different phase. The displacement, x, may be analyzed as the sum of one component that is exactly *in-phase* with the driver with amplitude, x', and a second component exactly *90°-out-of-phase* with the driver with amplitude, x''. If the experiment is repeated at many different driving (angular) frequencies, ω, then the resultant variations of x' and x'' with ω are the dispersion and absorption spectra shown at the bottom of the Figure.

$$x = x' \cos \omega t + x'' \sin \omega t \tag{1.18a}$$

in which

$$x' = F_0 \left(\frac{m(\omega_0^2 - \omega^2)}{m^2(\omega_0^2 - \omega^2)^2 + f^2\omega^2} \right) \tag{1.18b}$$

$$x'' = F_0 \left(\frac{f\omega}{m^2(\omega_0^2 - \omega^2)^2 + f^2\omega^2} \right) \tag{1.18c}$$

Equations 1.18 state mathematically what is shown graphically in Figure 1.4, namely that the steady-state displacement, x, may be analyzed into two components, $x' \cos \omega t$ and $x'' \sin \omega t$, which represent the response components which are either exactly *in-phase* with the driver (i.e., vary with time as $\cos \omega t$) or exactly *90°-out-of-phase* with the driver (i.e., vary with time as $\sin \omega t$). [Note that $\cos(\omega t - \pi/2) = \sin \omega t$.] x' and x'', the amplitudes of the in-phase and 90°-out-of-phase components of the response, are the quantities of physical interest, and will henceforth be referred to as the dispersion and absorption spectra.

Although the result displayed in Eqs. 1.18 is rigorously correct, we can simplify the expressions significantly in the (usual physical) limit that the observing (driving) angular frequency, ω, is very close to the natural angular frequency, ω_0. Specifically, if

$$|\omega_0 - \omega| << (\omega_0 + \omega) \tag{1.19}$$

then $\omega \cong \omega_0$. Substitution of Eq. 1.19 in Eqs. 1.18b,c then yields

$$x' \cong F_0 \left(\frac{2m\omega_0(\omega_0 - \omega)}{4m^2\omega_0^2(\omega_0 - \omega)^2 + f^2\omega^2} \right) \tag{1.20a}$$

$$x'' \cong F_0 \left(\frac{f\omega}{4m^2\omega_0^2(\omega_0 - \omega)^2 + f^2\omega^2} \right) \tag{1.20b}$$

Finally, since the frictional (damping) coefficient, f, has units of mass/time, we may further simplify Eqs. 1.20 by defining a characteristic "relaxation" time, τ, according to

$$\tau = \frac{2m}{f} \tag{1-21}$$

to give the final dispersion and absorption expressions of Eqs. 1.22. The bracketed expressions in Eqs. 1.22 are plotted at the bottom of Figure 1.4.

$$x' \cong \frac{F_0}{2m\omega_0} \left(\frac{(\omega_0 - \omega)\tau^2}{1 + (\omega_0 - \omega)^2\tau^2} \right) \tag{1.22a}$$

$$x'' \cong \frac{F_0}{2m\omega_0} \left(\frac{\tau}{1 + (\omega_0 - \omega)^2\tau^2} \right) \tag{1.22b}$$

The bracketed expressions in Eqs. 1.22 are known as the normalized *Lorentziar* dispersion and absorption [$D(\omega)$ and $A(\omega)$ in Eqs. 1.23], and occur universally in spectroscopy. The parameters of physical interest in Eqs. 1.23 (or 1.16) are the natural angular frequency, ω_0, and the relaxation time, τ. For example, in infrared spectroscopy, ω_0 represents one of the vibrational normal modes of a molecule modeled as a set of point masses connected by springs (chemical bonds). The relaxation time, τ, is inversely related to the friction which connects the spring to its environment. Thus, relaxation time measurements inform us about the surroundings near the system (e.g., a neighboring magnetic nucleus, in the case of NMR) of interest.

$$D(\omega) = \frac{(\omega_0 - \omega)\tau^2}{1 + (\omega_0 - \omega)^2 \tau^2} \tag{1.23a}$$

$$A(\omega) = \frac{\tau}{1 + (\omega_0 - \omega)^2 \tau^2} \tag{1.23b}$$

It is readily shown (see Problems) that the full width at half-maximum height of the absorption spectrum is $2/\tau$ rad s^{-1}, and that the peak-to-peak separation for the dispersion spectrum is also $2/\tau$ rad s^{-1}. Finally, it is important to recognize (see Problems) that the *dispersion is different* (both in physical origin and mathematical form) *from the derivative of the absorption spectrum*, even though the two functions are similar in shape.

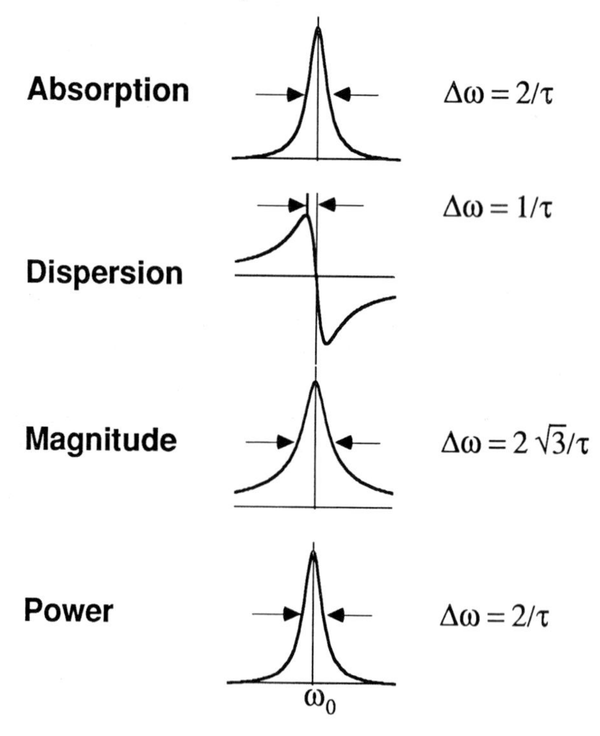

Figure 1.5 Lorentzian spectral line shapes.

1.3.2 Magnitude (absolute-value) and power spectra

As we shall later learn, detection of the absorption or dispersion spectrum requires determination of the phase spectrum (i.e., the phase difference, as a function of frequency, between the driver and the detected response). When the phase spectrum is not available, the usual display mode is either the *magnitude* mode [also known as *absolute-value* mode, $M(\omega)$], or the *power* spectrum, $P(\omega)$.

$$M(\omega) = \sqrt{[A(\omega)]^2 + [D(\omega)]^2} \tag{1.24}$$

$$P(\omega) = [M(\omega)]^2 = [A(\omega)]^2 + [D(\omega)]^2 \tag{1.25}$$

Lorentzian magnitude and power spectra are plotted in Figure 1.5. It is clear that the magnitude-mode spectrum is broader than the absorption-mode spectrum [by a factor of $\sqrt{3}$ for a Lorentzian line (see Problems)].

1.3.3 Hilbert transform and DISPA

The reader may have observed that either the dispersion or absorption spectrum offers the same information: ω_0 (from the absorption-mode peak center or the zero-crossing of the dispersion-mode) and τ (from the reciprocal of either the half-width at half-maximum height of the absorption-mode or half of the peak-to-peak separation of the dispersion-mode). When any two functions are expressed in terms of the same parameters, one expects to find a mathematical relationship between the two functions. In this case, the relationship is called a Hilbert transform, also known as a Kramers-Kronig or Bode relationship, Eq. 1.26.

$$D(\omega) = \frac{1}{\pi} \int_{-\infty}^{+\infty} \frac{A(\omega')}{\omega - \omega'} \, d\omega' \tag{1.26}$$

A remarkable property (see Problems) of the Lorentzian spectrum is that a plot of dispersion vs. absorption ("DISPA") for an isolated Lorentzian line yields the circle shown in Figure 1.6. This reference circle is tangent to the origin, and the circle's diameter is equal to the absorption-mode peak height. What makes the DISPA plot useful is that it can discriminate between various line-broadening mechanisms. For example, two or more overlapping Lorentzians of the same width but different peak *position* will produce a DISPA curve displaced *outside* its reference circle, whereas two or more overlapping Lorentzians of the same position but different *width* will displace a DISPA curve *inside* its reference circle (see Figure 1.6). Perhaps the most useful feature of the DISPA plot is for spectral phasing (see Further Reading).

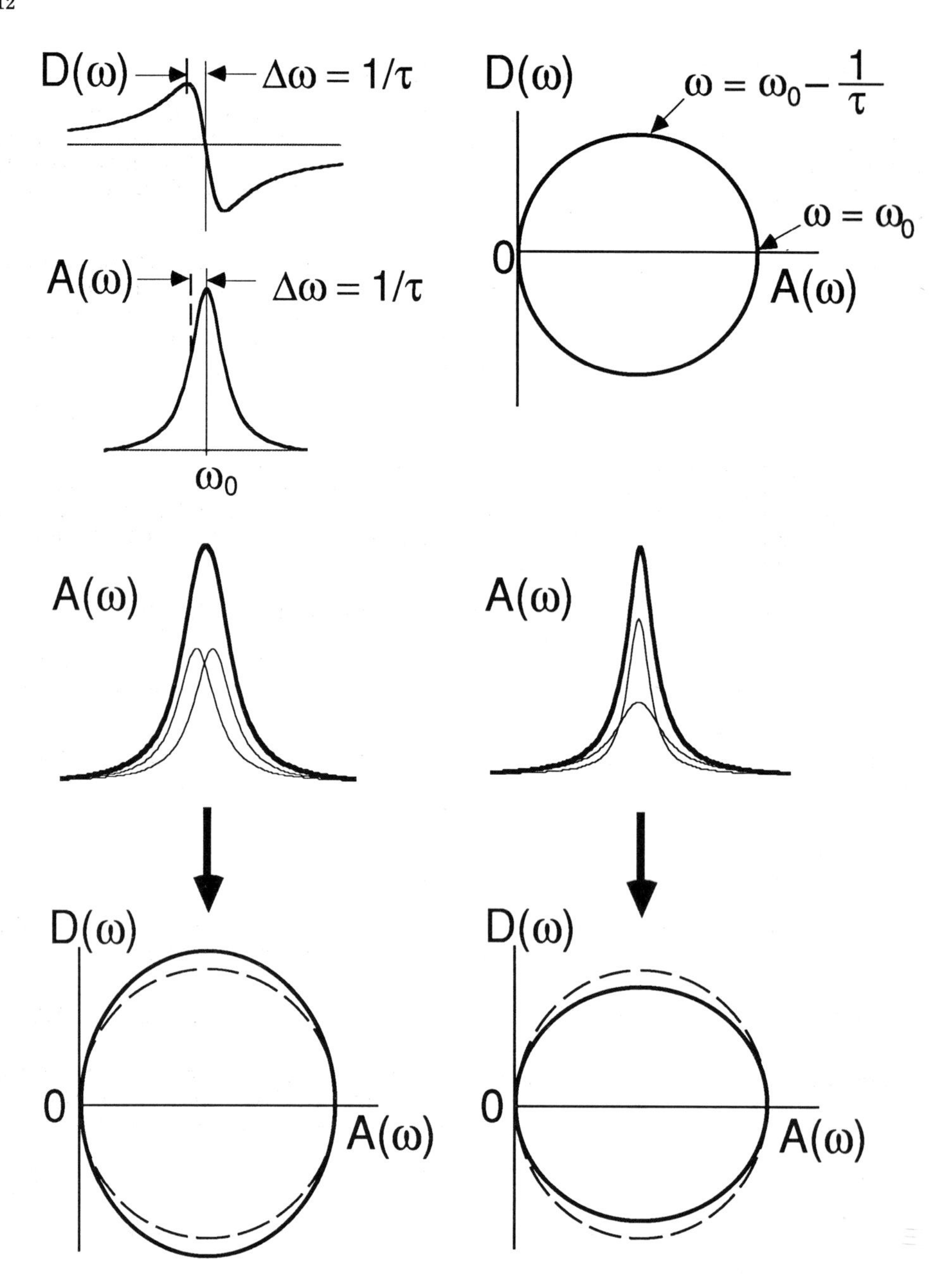

Figure 1.6 Dispersion *vs.* Absorption (DISPA) plots for: (top) a single Lorentzian line; (bottom left) superposition of two Lorentzians of equal width but different position; (bottom right) superposition of two Lorentzians of the same position but different width.

1.3.4 Physical meaning of mathematically complex quantities

To this stage, our development has been conducted wholly with mathematically *real* (as opposed to mathematically *complex*) variables. Although real variables suffice to account for all of the phenomena we will want to discuss, most of the Fourier transform spectroscopic literature is couched in complex notation for two excellent reasons.

First, Fourier analysis is based on the sinusoidal functions, $\cos \omega t$ and $\sin \omega t$. Unfortunately, the time derivative of $\cos \omega t$ is $-\omega \sin \omega t$ and of $\sin \omega t$ is $\omega \cos \omega t$. The algebra for problems such as the driven, damped, harmonic oscillator (damped mass on a spring) involves time derivatives and thus expands rapidly during calculations. However, the complex function,

$$\exp(\pm i \omega t) = \cos \omega t \pm i \sin \omega t \tag{1.27}$$

has the very useful property that its first derivative is simply $\pm i \omega \exp(\pm i \omega t)$, so that the first derivative and the function itself differ only by a constant. Therefore, it is often actually easier to solve a mathematically *real* problem by replacing $\cos \omega t$ by $\exp(i\omega t)$ in the calculations, and then throwing away the mathematically *imaginary* part when the calculation is completed!

The second reason that complex variables are useful is that they provide a convenient and automatic way to keep two physically distinct parts of a problem separated as the real and imaginary parts of a complex number. Complex numbers thus serve the same sort of purpose as two perpendicular axes for keeping two components of a vector separated in plane geometry.

Both of the above advantages emerge when we start over again to solve the driven, damped harmonic oscillator problem, this time with complex variables (see Chapter 6.3). The problem we actually want to solve is

$$m \frac{d^2 x}{dt^2} + f \frac{dx}{dt} + k x = F_0 \cos \omega t \, , \quad \text{in which } x \text{ is real} \tag{1.17}$$

Suppose, however, that we add the (imaginary) term, $i F_0 \sin \omega t$, to the right-hand side of Eq. 1.17, and then solve for the (mathematically complex) $\boldsymbol{x}$:

$$m \frac{d^2 \boldsymbol{x}}{dt^2} + f \frac{d\boldsymbol{x}}{dt} + k \boldsymbol{x} = F_0 \exp(i \omega t), \quad \text{in which } \boldsymbol{x} \text{ is complex} \tag{1.28}$$

In Equation 1.28, it is understood that $\boldsymbol{x}$ is a complex number, but that we will only ever be interested in the real part of $\boldsymbol{x}$ when we have finished solving the problem. Next, suppose that there exists a steady-state solution of the form,

$$\boldsymbol{x} = \chi \exp(i \omega t), \quad \text{in which } \chi \text{ is complex} \tag{1.29}$$

Substitution of Eq. 1.29 into Eq. 1.28 readily yields an expression for χ:

$$\chi = F_0 \frac{1}{(k - m\omega^2) + i f \omega} = F_0 \frac{1}{m(\omega_0^2 - \omega^2) + i f \omega} \tag{1.30}$$

In order to rid the denominator of imaginary terms, we apply the useful identity,

$$\frac{1}{a + ib} = \frac{a - ib}{a^2 + b^2} \tag{1.31}$$

We may then rewrite Eq. 1.30 as

$$\boldsymbol{x} = F_0 \, \frac{m(\omega_0{}^2 - \omega^2) - i f \omega}{m^2(\omega_0{}^2 - \omega^2)^2 + f^2 \omega^2} \tag{1.32}$$

Now let us examine just the real part of the complex $\boldsymbol{x}$ solution:

$$\mathrm{Re}(\boldsymbol{x}) = \mathrm{Re}[\, \chi . \exp(i \omega t)] = \mathrm{Re}[\chi \, (\cos \omega t + i \sin \omega t)]$$

$$= F_0 \left(\frac{m(\omega_0{}^2 - \omega^2)}{m^2(\omega_0{}^2 - \omega^2)^2 + f^2 \omega^2} \right) \cos \omega t + F_0 \left(\frac{f \omega}{m^2(\omega_0{}^2 - \omega^2)^2 + f^2 \omega^2} \right) \sin \omega t \tag{1.33}$$

$$= x' \cos \omega t + x'' \sin \omega t \tag{1.34}$$

In other words, the real part of the solution of the (complex) Eq. 1.28 is identical to the (real) solution of the original real Eq. 1.17 that we wanted in the first place. Furthermore, it is now apparent that the complex quantity, χ, can be expressed as

$$\chi = x' - i x'' \tag{1.35}$$

in which x' and x'' appear as mathematically real and imaginary components of the complex amplitude, χ, of the complex solution, $\boldsymbol{x}$, of Eq. 1.28. Thus, spectroscopists sometimes refer to x' and x'' as the "real" and "imaginary" parts of a "complex" susceptibility. However, it is clear from the above treatment that the quantities of physical interest, x' and x'' are both physically and mathematically real—the complex notation was simply a way to shorten the algebra. Moreover, the mathematically complex χ provides the advertised advantage of algebraically separating x' and x'' as the real and imaginary parts of a complex number. The reader can confirm (see Problems) that the algebraic difficulty in proceeding from Eq. 1.29 to Eq. 1.30 in complex notation is significantly less than that for proceeding from Eq. 1.18 to Eq. 1.17 in real notation.

1.4 Transient versus steady-state experiments

1.4.1 Fourier transform relation between the two experiments

We previously observed that the same parameters (ω_0 and τ) could be extracted from either the Lorentzian absorption or dispersion spectrum from a steady-state experiment. We then found a mathematical relationship (Hilbert transform, Chapter 1.3.3) between those two functions.

Similarly, it is clear that ω_0 and τ can also be recovered from the transient experiment (Eq. 1.16): $\omega_0/2\pi = \nu_0$ is the number of oscillations per second in the time-domain transient, and τ is the time constant for exponential decay of the envelope of the transient signal. Therefore, we should not be surprised to discover that there is a mathematical relation—namely, the Fourier transform—between the time-domain response to a sudden (impulse) excitation and the frequency-domain steady-state response to a sinusoidal excitation.

Eqs. 1.36a, 1.36b, and 1.37 define the "cosine Fourier transform", $C(\omega)$, "sine Fourier transform", $S(\omega)$, and "complex Fourier transform" $\boldsymbol{F}(\omega)$ of a time-domain real function, $f(t)$, or a time-domain complex function, $\boldsymbol{f}(t)$. Magnitude and power spectra are still defined by Eqs. 1.24 and 1.25.

$$\left.\begin{aligned} C(\omega) &= \int_{-\infty}^{+\infty} f(t)\cos\omega t\; dt \\ C(\nu) &= \int_{-\infty}^{+\infty} f(t)\cos 2\pi\nu t\; dt \end{aligned}\right\} \quad \text{for real } f(t) \qquad (1.36a)$$

$$\left.\begin{aligned} S(\omega) &= \int_{-\infty}^{+\infty} f(t)\sin\omega t\; dt \\ S(\nu) &= \int_{-\infty}^{+\infty} f(t)\sin 2\pi\nu t\; dt \end{aligned}\right\} \quad \text{for real } f(t) \qquad (1.36b)$$

$$\left.\begin{aligned} \boldsymbol{F}(\omega) &= \int_{-\infty}^{+\infty} \boldsymbol{f}(t)\exp(-i\omega t)\, dt \\ \boldsymbol{F}(\nu) &= \int_{-\infty}^{+\infty} \boldsymbol{f}(t)\exp(-i2\pi\nu t)\, dt \end{aligned}\right\} \quad \text{for complex } \boldsymbol{f}(t) \qquad (1.37)$$

For a real time-domain signal in the absence of phase shifts (see Chapter 2.4), the frequency-domain functions of Eqs. 1.36 and 1.37 are related by Eqs. 1.38 to the steady-state absorption and dispersion spectra, $A(\omega)$ and $D(\omega)$, as the reader can confirm (See Appendix D) for the particular (real) time-domain signal, $f(t) = \exp(-t/\tau)\cos\omega_0 t$.

$$A(\omega) = \mathrm{Re}\,[\boldsymbol{F}(\omega)] \qquad (1.38a)$$

$$D(\omega) = \mathrm{Im}\,[\boldsymbol{F}(\omega)] = -S(\omega) \qquad (1.38b)$$

If the same information (namely, ω_0 and τ) is contained in both $\boldsymbol{f}(t)$ and $\boldsymbol{F}(\omega)$, then there must also be an "inverse" Fourier transform for proceeding from the frequency-domain to the time-domain, as shown in Eqs. 1.39.

$$\boldsymbol{f}(t) = \frac{1}{2\pi}\int_{-\infty}^{+\infty} \boldsymbol{F}(\omega)\exp(i\omega t)\, dt \quad \text{for complex } \boldsymbol{F}(\omega) \qquad (1.39a)$$

$$\boldsymbol{f}(t) = \int_{-\infty}^{+\infty} \boldsymbol{F}(\nu)\exp(i2\pi\nu t)\, dt \quad \text{for complex } \boldsymbol{F}(\nu) \qquad (1.39b)$$

Eqs. 1.39 yields the identical (complex) $f(t)$ we started with in Eq. 1.37. Although there are several different conventions for the coefficient preceding the integrals in Eqs. 1.36 to 1.39, the basic requirement is that a forward FT followed by an inverse FT gives back the original $f(t)$ function. We have used the physical convention (for time/frequency problems), in which a factor of $(1/2\pi)$ appears in the inverse FT integral for angular frequency (Eq. 1.39, top), because $\omega = 2\pi\nu$ (see Chapter 2 Problems). Some mathematicians prefer to include a factor of $\sqrt{1/2\pi}$ in each of the forward and inverse transforms, for symmetry. The reader should always check to be sure which convention is used by a particular author.

With recourse to the following definite integrals and trigonometric identities (see Appendix A):

$$\int_0^\infty \exp(-at)\cos bt \; dt = \frac{a}{a^2+b^2} \tag{A.75}$$

$$\int_0^\infty \exp(-at)\sin bt \; dt = \frac{b}{a^2+b^2} \tag{A.76}$$

$$\sin a \cos b = \frac{1}{2}\left[\sin(a+b) + \sin(a-b)\right] \tag{A.25a}$$

$$\cos a \cos b = \frac{1}{2}\left[\cos(a+b) + \cos(a-b)\right] \tag{A.26b}$$

the reader can use Eqs. 1.36 to show (see Problems) that the cosine and sine Fourier transforms of the (real) normalized $f(t)$ for the damped mass on a spring,

$$\begin{aligned} f(t) &= \exp(-t/\tau)\cos\omega_0 t \qquad \text{for } t \geq 0 \\ f(t) &= 0 \qquad\qquad\qquad\qquad\; \text{for } t < 0 \end{aligned} \tag{1.16}$$

yield (see Eqs. 1.38) the same dispersion and absorption spectra, $D(\omega)$ and $A(\omega)$, which would have been obtained from a steady-state experiment.

$$D(\omega) = \frac{1}{2}\left(\frac{(\omega_0-\omega)\tau^2}{1+(\omega_0-\omega)^2\tau^2} + \frac{(\omega_0+\omega)\tau^2}{1+(\omega_0+\omega)^2\tau^2}\right) \tag{1.40a}$$

$$A(\omega) = \frac{1}{2}\left(\frac{\tau}{1+(\omega_0-\omega)^2\tau^2} + \frac{\tau}{1+(\omega_0+\omega)^2\tau^2}\right) \tag{1.40b}$$

Eqs. 1.40 contain the expected Lorentzian peaks centered at $\omega = \omega_0$, but also contain Lorentzians centered at $\omega = -\omega_0$. These negative-frequency peaks arise from our assumption that $f(t)$ is real, corresponding to linearly-polarized radiation. In any case, provided that the Lorentzian line width ($2/\tau$ rad s^{-1} at half-maximum absorption peak height) is small compared to ω_0, we can neglect the second terms in Eqs. 1.40, as will be discussed in Chapter 4.3.

1.4.2 Zero-friction limit: Thomson and Rayleigh scattering

A good way to improve one's understanding of any equation is to examine its limiting behavior as one of its parameters becomes very large or very small. For the general steady-state solution, Eqs. 1.18, to the driven, damped mass on a spring, two such mathematical limits have direct physical consequences.

First, consider the limit that friction is small and/or that the driving (angular) frequency, ω, is far from the system's resonant (angular) frequency, ω_0 :

$$\lim_{f \ll \left| \frac{m(\omega_0^2 - \omega^2)}{\omega} \right|} (x') = \frac{F_0}{m(\omega_0^2 - \omega^2)} \tag{1.41a}$$

$$\lim_{f \ll \left| \frac{m(\omega_0^2 - \omega^2)}{\omega} \right|} (x'') = 0 \tag{1.41b}$$

In other words, when the driving frequency is far from resonance, there is no energy absorption, and the driven mass oscillates completely *in-phase* ($\omega \ll \omega_0$) or exactly 180°-*out-of-phase* ($\omega \gg \omega_0$) with the driver. When the driver is the oscillating electric field of an electromagnetic (e.g., light) wave, then the power of the incident wave is re-radiated ("scattered") by the oscillating driven charge. From Maxwell's laws, one finds that the scattered electric field amplitude, E_{scatt}, is proportional to the acceleration (i.e., the second derivative with respect to time) of $x' \cos \omega t$. From Eq. 1.41a, one can quickly show that

$$E_{\text{scatt}} \propto \frac{-F_0\,\omega^2}{m(\omega_0^2 - \omega^2)} \cos \omega t = \frac{-F_0}{m[(\omega_0/\omega)^2 - 1]} \cos \omega t \tag{1.42}$$

Since (see Eq. 1.8) the intensity, I_{scatt}, of the scattered wave is proportional to the square of the scattered amplitude, we infer that

$$I_{\text{scatt}} = \frac{\text{constant}}{[(\omega_0/\omega)^2 - 1]^2} \tag{1.43}$$

Electrons bound to molecules can be modeled as if the two were bound by springs. The natural frequency of the electrons (as the reader may recall from solving the hydrogen atom energy levels in elementary quantum mechanics) is in the ultraviolet wavelength range. What happens next depends on whether Eq. 1.43 arises from $\omega \gg \omega_0$ (Thomson scattering) or $\omega \ll \omega_0$ (Rayleigh scattering).

Thomson scattering ($\omega \gg \omega_0$)

For $\omega \gg \omega_0$ (as when ω is in the x-ray range and ω_0 is in the ultraviolet),

$$\lim_{(\omega \gg \omega_0)} I_{\text{scatt}} = \text{constant} \quad \text{(i.e., independent of } \omega\text{)} \tag{1.44}$$

Thus, Thomson scattering (e.g., scattering of x-rays by electrons bound to atoms) is independent of x-ray wavelength, which is why x-ray images are inherently colorless.

Rayleigh scattering ($\omega << \omega_0$)

In the opposite limit that $\omega << \omega_0$ (as when visible light is directed at the same electrons-on-springs in molecules), the scattered intensity becomes

$$\lim_{(\omega << \omega_0)} I_{scatt} = \text{constant} \left(\frac{\omega}{\omega_0}\right)^4 \tag{1.45}$$

Equation 1.45 accounts for the blue color of the sky, because the highest visible frequencies (i.e., blue wavelengths) of sunlight are scattered much more intensely by atmospheric molecules than are the lower frequencies (red wavelengths) in directions away from a direct line from the observer to the sun. Sunsets are red for the same reason.

1.4.3 Zero-mass limit: dielectric or ultrasonic relaxation

A final limiting situation of the driven damped mass on a spring is the case for which $m \to 0$ (massless "weight" on a spring). The equation of motion now becomes either

$$f\frac{dx}{dt} + kx = F_0 \cos \omega t \qquad \text{(Steady-state experiment)} \tag{1.46}$$

or

$$f\frac{dx}{dt} + kx = 0 \qquad \text{(Transient experiment)} \tag{1.47}$$

The reader can verify (see Problems) that the following expressions are solutions to Equations 1.46 and 1.47.

$$x = x' \cos \omega t + x'' \sin \omega t \tag{1.48}$$

in which

$$x' = \frac{F_0}{k}\frac{1}{1+\omega^2\tau^2} \tag{1.49a}$$

(Steady-State)

$$x'' = \frac{F_0}{k}\frac{\omega}{1+\omega^2\tau^2} \tag{1.49b}$$

or

$$x = x_0 \exp(-t/\tau) \qquad \text{(Transient)} \tag{1.50}$$

in which

$$\frac{1}{\tau} = \frac{f}{k} \tag{1.51}$$

Eqs. 1.49 and 1.50 show that the massless spring behaves as if its "natural" frequency were zero—the transient response shows no oscillations, and the frequency-domain response is centered at zero frequency. There are two other important differences. First, although the form of Eqs. 1.49 is similar to that for the finite-mass oscillator (Eqs. 1.22), the functional forms of the absorption and dispersion are interchanged (see Figure 1.7). Second, the characteristic relaxation time, τ, has a different form for the massless case ($1/\tau = f/k$) than for the finite-mass case ($1/\tau = f/2m$).

The massless limit of the driven damped mass on a spring applies to dielectric relaxation (e.g., in-phase and 90°-out-of-phase a.c. capacitance) and ultrasonic relaxation (e.g., in-phase and 90°-out-of-phase response to oscillating pressure excitation) steady-state experiments, as well as to their transient analogs (E-jump, P-jump) and other transient experiments (temperature-jump; stopped flow).

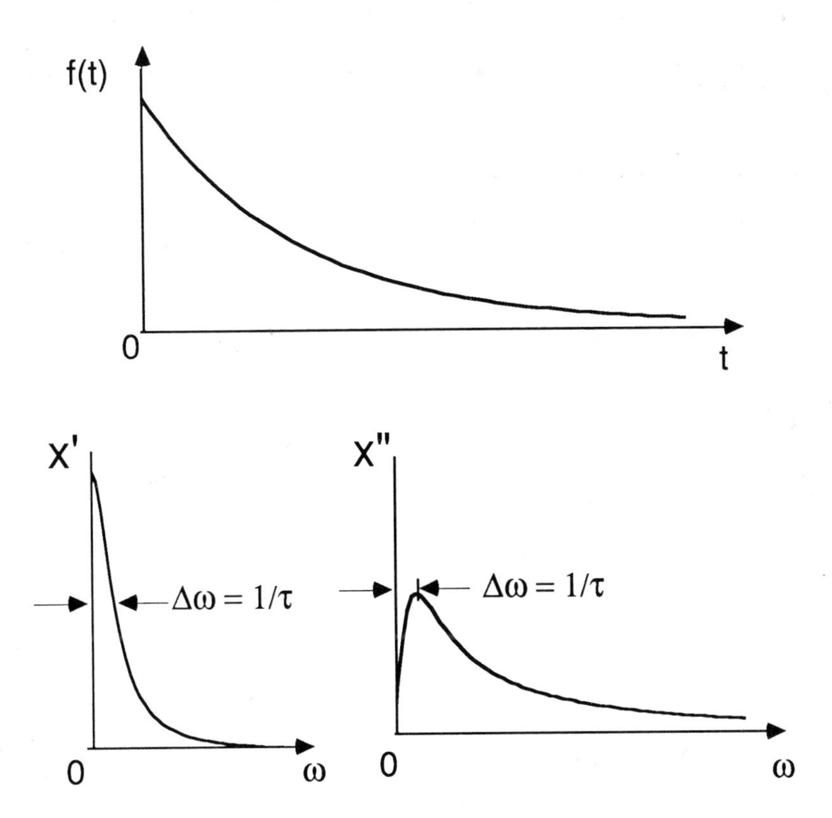

Figure 1.7 Behavior of a driven damped massless spring. Top: transient response to an impulse displacement. Bottom: amplitudes, x' and x'', of the in-phase and 90°-out-of-phase components of the steady-state displacement due to a sinusoidal driving force, $F_0 \cos \omega t$. Note the reversed shapes of the steady-state plots compared to the finite-mass case in Figure 1.5.

We close this section by pointing out that the driven damped mass on a spring applies directly to the analysis of electric circuits composed of a resistor, R, inductor, L, capacitor, C, and driving voltage source, $V_0 \cos \omega t$. The analogy between the equation of motion for the driven damped mass on a spring and the equation for motion of charge, q, through an electronic circuit is direct: f becomes R; m becomes L; k becomes $1/C$, and the driving mechanical force, F_0, becomes the driving voltage, V_0.

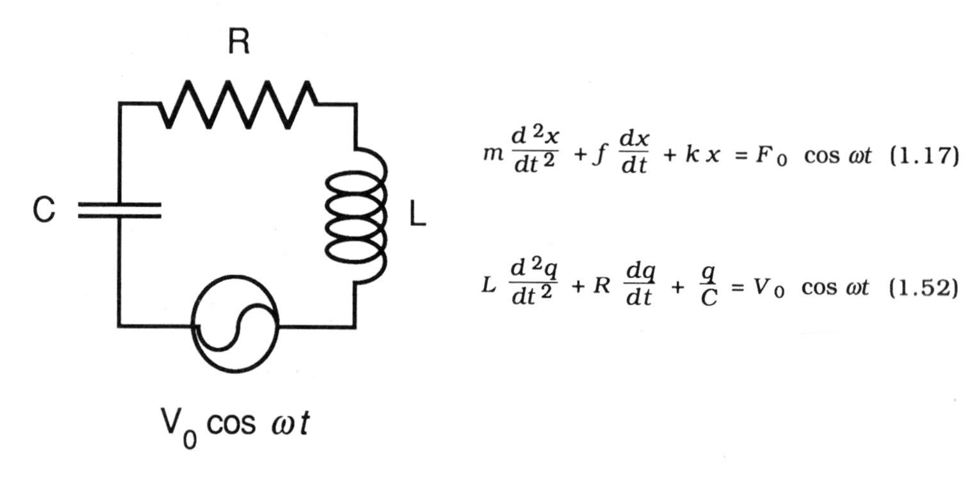

R
C
L
$V_0 \cos \omega t$
$m \frac{d^2x}{dt^2} + f \frac{dx}{dt} + kx = F_0 \cos \omega t$ (1.17)
$L \frac{d^2q}{dt^2} + R \frac{dq}{dt} + \frac{q}{C} = V_0 \cos \omega t$ (1.52)

In particular, the "natural" or "resonant" (angular) frequency of the illustrated circuit becomes

$$\omega_0 = \sqrt{\frac{1}{LC}} \tag{1.53}$$

with time constant,

$$\tau = \frac{2L}{R} \tag{1.54}$$

and the "massless" limit in the mechanical model corresponds to an "inductorless" electric circuit, with its familiar time constant,

$$\tau = RC \tag{1.55}$$

In this chapter, we have presented the physical and mathematical bases for spectroscopic line shapes for a *linear* system (i.e., response magnitude is proportional to excitation magnitude). The frequency-domain steady-state response ("spectrum") of a sinusoidally driven weight-on-a-spring approaches the Lorentz limit near resonance ($\omega \cong \omega_0$), the Thomson $\omega >> \omega_0$ or Rayleigh $\omega << \omega_0$ limits far from resonance, and the "relaxation" limit when the mass of the weight approaches zero. These models form the basis for quantitative description of virtually all spectroscopy and scattering experiments: in particular, the optical, magnetic resonance, and ion cyclotron resonance spectra which are the featured applications in this book.

We also showed that the Fourier transform provides the mathematical relationship between the time-domain response to an impulse excitation and the frequency-domain steady-state response to a continuous oscillating driving force. Therefore, since the frequency-domain is preferred for display, whereas the time-domain is preferred for experiments (see Chapter 4), the Fourier transform becomes necessary to convert time-domain data into the desired frequency-fdomain spectra. In the next two chapters, we will develop the continuous (analog) and discrete (digital) basics of Fourier transforms, in preparation for the general FT spectroscopy experiment described in Chapter 4. Then, following a brief treatment of the effects of noise in Chapter 5, and non-FT methods in Chapter 6, the reader will be prepared to address the applications of Fourier transforms to mass, optical, and NMR spectroscopy.

Further Reading

R. P. Feynman, R. B. Leighton, & M. Sands, *The Feynman Lectures on Physics*, Addison-Wesley, Reading, MA, 1964. (Weight-on-a-spring analysis)

A. G. Marshall, *Biophysical Chemistry: Principles, Techniques, and Applications*, Wiley, N.Y., 1978, Chapter 13. (Introduction to spectral line shapes)

A. G. Marshall, in *Fourier, Hadamard, and Hilbert Transforms in Chemistry*, Ed. A. G. Marshall, Plenum, N.Y., 1982, Chapter 1. (Introduction to FT)

A. G. Marshall, *Chemometrics & Intelligent Laboratory Systems* **1988**, *3*, 261-275. (Dispersion *vs.* Absorption plots for spectral phasing and line shape analysis)

Problems

1.1. The equation of motion for the displacement, x, of a mass, m, on an undriven, *undamped* spring (force constant, k) is given by Eq. 1.3:

$$m \frac{d^2 x}{dt^2} + k x = 0 \tag{1.3}$$

(a) Show that $x = x_0 \cos \omega_0 t$ is a (real) solution of Eq. 1.3, and determine the form of ω_0.

(b) Show that $\boldsymbol{x} = x_0 \exp(i \omega_0 t)$ is a (complex) solution of Eq. 1.3, and again determine ω_0. Note that the real part of the complex displacement is the same as the (real) solution of the original (real) Eq. 1.3.

1.2. The equation of motion for the displacement, x, of a weight of mass, m, on an undriven, *damped* (frictional constant = f) spring (force constant = k) is given by Eq. 1.12:

$$m \frac{d^2 x}{dt^2} + f \frac{dx}{dt} + k x = 0 \tag{1.12}$$

(a) Show that a (real) solution of Eq. 1.12 is, $x = x_0 \exp(-t/\tau) \cos W t$, and find the form of τ and W. Then find the limiting behavior of the solution when damping is weak ($1/\tau << \omega_0$).

(b) Show that a (complex) solution of Eq. 1.12 is $\boldsymbol{x} = x_0 \exp(i c t)$, and find the form of c. Show that the real part of this complex solution is the same as the (real) solution of the original (real) equation. This exercise is the solution to the transient experiment in spectroscopy.

1.3. Show that the sum of two sine waves of the same frequency (but different phase and different amplitude) gives a resultant sine wave of the same frequency. In other words, show that there exists a ψ, such that

$$A \sin \theta + B \sin(\theta + \varphi) = C \sin(\theta + \psi)$$

[A special case of this formula ($\varphi = \pi/2$) is the basis for breaking down an arbitrary sinusoid into sine and cosine components (Fig. 1.4), the process that forms the basis for understanding the origin of spectroscopic absorption and dispersion. The general case shows why superpositions of several light waves of the same wavelength will always give a resultant light wave of that wavelength—see Chapter 9.2.1.

1.4. Confirm the text expressions for x' and x'' ("dispersion" and "absorption") in the Lorentzian limit that the driving frequency is near the "natural" frequency of the weight on a spring. In other words, obtain Eqs. 1.20 from Eqs. 1.18. These two expressions describe the behavior of (for example) refractive index and power absorption as a function of incident radiation frequency in optical spectroscopy.

1.5. The normalized Lorentzian line shape may be written:

$$A(\omega) = \frac{\tau}{1 + (\omega_0 - \omega)^2 \tau^2} = \text{"Absorption" line shape}$$

$$D(\omega) = \frac{(\omega_0 - \omega)\, \tau^2}{1 + (\omega_0 - \omega)^2 \tau^2} = \text{"Dispersion" line shape}$$

(a) Plot $A(\omega)$ versus ω, and compute the peak height [maximum of $A(\omega)$] and the full width of the curve at half its maximum peak height. Then repeat the same calculation for the magnitude-mode and power spectra (use Eqs. 1.24 and 1.25). Compare the three results.

(b) Plot $D(\omega)$ versus ω, and compute the extrema [i.e., maximum and minimum values of $D(\omega)$]. Then compute the frequency separation between the two extrema. The above calculations comprise the principal properties of the Lorentzian spectral line shape.

(c) In certain spectroscopic experiments, it is convenient to detect the first derivative (with respect to driving frequency) of $A(\omega)$, as in electron spin resonance spectrosocopy. The resulting plot of $d A(\omega)/d\omega$ resembles the "dispersion" line shape of part (b). Calculate the locations, magnitudes, and frequency separation between the extrema of a plot of $d A(\omega)/d\omega$ *vs.* ω, and compare to the "dispersion" line shape of part (b).

1.6. Show that a plot of Lorentzian dispersion,

$$D(\omega) = \frac{(\omega_0 - \omega)\tau^2}{1 + (\omega_0 - \omega)^2 \tau^2}$$

versus Lorentzian absorption,

$$A(\omega) = \frac{\tau}{1 + (\omega_0 - \omega)^2 \tau^2}$$

gives a circle of diameter, τ, centered at $A(\omega) = \tau/2$ on the abscissa. In other words, show that

$$\left(A(\omega) - \frac{\tau}{2}\right)^2 + [D(\omega)]^2 = \left(\frac{\tau}{2}\right)^2$$

This "DISPA" plot is useful for spectral line shape analysis, and especially for phasing of Fourier transform spectra (see Further Reading).

1.7. The equation of motion for the displacement, x, of a mass, m, on a sinusoidally driven (driving force = $F_0 \cos \omega t$), damped (frictional coefficient = f), spring (force constant = k) is:

$$m \frac{d^2 x}{dt^2} + f \frac{dx}{dt} + kx = F_0 \cos \omega t \tag{1.17}$$

(a) Show that a (real) solution of Eq. 1.17 is $x = x' \cos \omega t + x'' \sin \omega t$, and determine the form of x' and x''.

(b) Begin again, this time from the complex form of Eq. 1.17:

$$m \frac{d^2 \mathbf{x}}{dt^2} + f \frac{d\mathbf{x}}{dt} + k\mathbf{x} = F_0 \exp(i\omega t) \tag{1.28}$$

and show that a (complex) solution of Eq. 1.28 is:

$$\mathbf{x} = \chi \exp(i\omega t)$$

in which

$$\chi = x' - i x''$$

and

$$\mathrm{Re}[\chi \exp(i\omega t)] = x' \cos \omega t + x'' \sin \omega t \quad \text{with } x' \text{ and } x'' \text{ as in part (a).}$$

In other words, show that the *real* part of the *complex* displacement is the same as the *real* solution of the original *real* equation. Note the simpler algebra involved in the mathematically "complex" manipulations. These calculations are the basis of the Lorentz line shape.

1.8. From Eqs. 1.36 and 1.38 compute the cosine and sine Fourier transforms, $A(\omega)$ and $D(\omega)$, of the (real) time-domain signal, $f(t)$. Hint: Use Eqs. A.75 and A.76.

$$f(t) = \exp(-t/\tau)\cos\omega_0 t \qquad \text{for } t \geq 0$$

$$f(t) = 0 \qquad \text{for } t < 0$$

Now solve the same problem with *complex* variables—i.e., apply Eq. 1.37 to obtain the real part of (complex) $\boldsymbol{F}(\omega)$ for the *complex* time-domain signal,

$$\boldsymbol{f}(t) = \exp(-t/\tau)\exp(i\omega_0 t) \qquad \text{for } t \geq 0$$

$$\boldsymbol{f}(t) = 0 \qquad \text{for } t < 0$$

Hint: $$\int_0^{\infty} \exp(-a x)\, dx = \frac{1}{a} \qquad \text{(A.69)}$$

Compare the final result and relative difficulty for the real and complex calculations.

1.9. The equation of motion for the displacement, x, of a damped (frictional constant = f), massless (m = 0) spring (force constant = k) is given by Eqs. 1.46 and 1.47

$$f\frac{dx}{dt} + kx = F_0 \cos\omega t \qquad \text{(Steady-state experiment)} \qquad (1.46)$$

$$f\frac{dx}{dt} + kx = 0 \qquad \text{(Transient experiment)} \qquad (1.47)$$

(a) Show that a (real) solution of Eq. 1.46 is $x = x'\cos\omega t + x''\sin\omega t$, and determine the form of x' and x''.

(b) Show that a (real) solution of Eq. 1.47 is: $x = x_0 \exp(-kt/f)$. .

(c) Show that a (complex) solution of the complex form of Eq. 1.46,

$$f\frac{d\boldsymbol{x}}{dt} + k\boldsymbol{x} = F_0 \exp(i\omega t)$$

is

$$\boldsymbol{x} = \chi \exp(i\omega t)$$

in which

$$\chi = x' - i x''$$

and x' and x'' are as in part (a). Again note the shorter algebra with mathematically complex notation. These calculations form the basis for dielectric and ultrasonic relaxation, as well as "RC" electronic circuits.

Solutions to Problems

Note: The answers to all but Problems 1.3 and 1.5 are outlined in the text.

1.3 $$A \sin\theta + B \sin(\theta + \phi) \stackrel{?}{=} C \sin(\theta + \psi) \tag{1.3.1}$$

Begin by dividing both sides by A; then let $D = B/A$ and $E = C/A$:

$$\sin\theta + D\sin(\theta + \phi) \stackrel{?}{=} E \sin(\theta + \psi) \tag{1.3.2}$$

Next, expand the second two terms with the trigonometric identity,

$$\sin(a + b) = \sin a \cos b + \cos a \sin b \tag{A.25}$$

to give:

$$\sin\theta\,(1 + D\cos\phi) + D\cos\theta\,\sin\phi \stackrel{?}{=} E\sin\theta\,\cos\psi + E\cos\theta\,\sin\psi \tag{1.3.3}$$

If Eq. 1.3.3 is true in general, then it is true in particular for $\theta = 0$ or $\pi/2$:

$$D\sin\phi = E\sin\psi \tag{1.3.4}$$

$$1 + D\cos\phi = E\cos\psi \tag{1.3.5}$$

Finally, divide Eq. 1.3.4 by Eq. 1.3.5 to obtain

$$\frac{D\sin\phi}{1 + D\cos\phi} = \tan\psi \tag{1.3.6}$$

Thus, the original Eq. 1.3.1 may be satisfied by choosing ψ such that

$$\psi = \arctan\left(\frac{D\sin\phi}{1 + D\cos\phi}\right) \tag{1.3.7}$$

C may then be found by solving Eq. 1.3.1 for a particular value of θ. **Q.E.D.**

1.5 To find the frequency at which $A(\omega)$, $D(\omega)$, or $dA(\omega)/d\omega$ is a maximum (or minimum), take the derivative of each function with respect to ω, and solve for ω. You should find that:

$A(\omega)$ has maximum value, τ, at $\omega = \omega_0$;

$D(\omega)$ has maximum and minimum values, $\pm\frac{\tau}{2}$, at $\omega = \omega_0 \pm \frac{\tau}{2}$;

$dA(\omega)/d\omega$ has maxima and minima, $\pm\frac{9\tau^2}{8\sqrt{3}}$, at $\omega = \omega_0 \pm \frac{\tau}{\sqrt{3}}$.

To obtain the absorption-mode peak width, $\Delta\omega$, at half-maximum peak height, simply set $A(\omega)$ equal to half its maximum value, namely, $(\tau/2)$, and solve for ω. You should find that $\omega = \omega_0 \pm (1/\tau)$, to give $\Delta\omega = (2/\tau)$. Finally, the same computation extended to magnitude-mode, $M(\omega)$, leads to the same maximum peak height as for $A(\omega)$ [since $D(\omega_0) = 0$], and a full line width at half-maximum peak height of $(2\sqrt{3}/\tau)$.

CHAPTER 2

Fourier transforms for analog (continuous) waveforms

2.1 Cosine, sine, and complex Fourier transforms: library of FT pairs

In the previous chapter, we introduced the Fourier transform as the recipe for converting a time-domain exponentially decaying sinusoid to frequency-domain Lorentzian absorption and dispersion spectra. In this chapter, we apply the same Fourier transform recipes (Eqs. 1.36 and 1.37) to several other simple time-domain functions. Along the way, we will note the effect of symmetry (even or odd) on the FT process. Our short library of FT pairs (see Appendix D) can then be expanded by use of the convolution theorem and phase shift theorem discussed later in this chapter.

2.1.1 Sinusoid/δ-function: even, odd, signum, and causal functions

It should seem obvious that a time-domain cosine or sine wave of angular frequency, ω_0, should be represented in the frequency-domain by a signal at angular frequency, ω_0. For an infinitely long time-domain sinusoid, the frequency-domain signal can be represented by a (Dirac) δ-function, $\delta(\omega_0 - \omega)$, which can be thought of as an infinitely narrow peak of unit area centered at $\omega = \omega_0$ (see Appendix B and Problems). (Strictly speaking, the δ-function is a linear *functional* rather than a *function*, but the distinction need not concern us here.) The algebra is shown below, with corresponding graphs in Figure 2.1. For the time-domain functions,

$$c(t) = \cos \omega_0 t = \cos 2\pi\nu_0 t, \qquad -\infty < t < +\infty \tag{2.1a}$$

or $$s(t) = \sin \omega_0 t = \sin 2\pi\nu_0 t, \qquad -\infty < t < +\infty \tag{2.1b}$$

or $$\mathbf{e}(t) = \exp(i\omega_0 t) = \exp(i 2\pi\nu_0 t), \qquad -\infty < t < +\infty \tag{2.1c}$$

the cosine, sine, and complex Fourier transforms (see Appendix B) are:

$$C(\nu) = \int_{-\infty}^{+\infty} \cos 2\pi\nu_0 t \ \cos 2\pi\nu t \ dt = \frac{1}{2}[\delta(\nu - \nu_0) + \delta(\nu + \nu_0)] \tag{2.2a}$$

and $$S(\nu) = \int_{-\infty}^{+\infty} \cos 2\pi\nu_0 t \ \sin 2\pi\nu t \ dt = 0 \tag{2.2b}$$

or $$C(\nu) = \int_{-\infty}^{+\infty} \sin 2\pi\nu_0 t \ \cos 2\pi\nu t \ dt = 0 \tag{2.3a}$$

and $$S(\nu) = \int_{-\infty}^{+\infty} \sin 2\pi\nu_0 t \ \sin 2\pi\nu t \ dt = \frac{1}{2}[\delta(\nu - \nu_0) - \delta(\nu + \nu_0)] \tag{2.3b}$$

$$\text{or} \quad \boldsymbol{F}(\nu) = \int_{-\infty}^{+\infty} \exp(i 2\pi \nu_0 t) \exp(-i 2\pi \nu t)\, dt = \delta(\nu - \nu_0) \tag{2.3c}$$

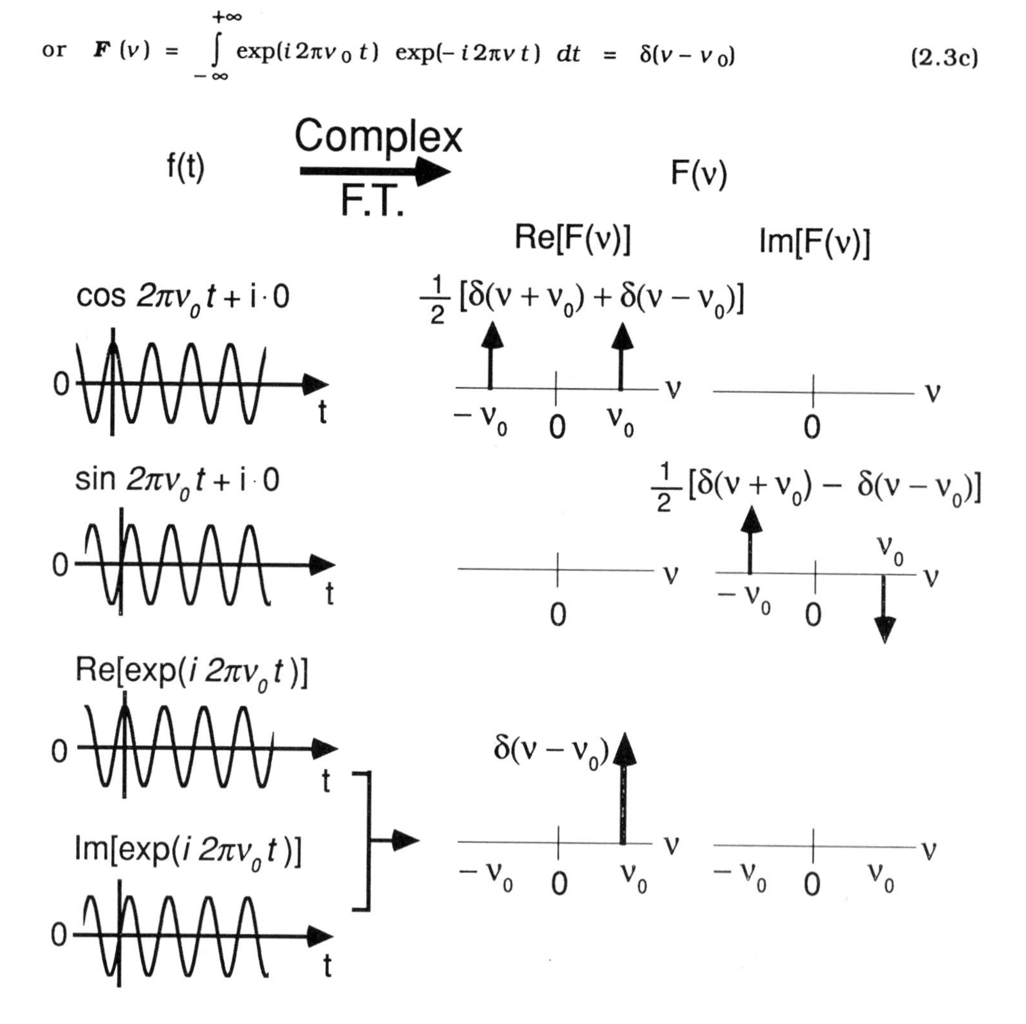

Figure 2.1 Time-domain sinusoids (left) and their complex Fourier transforms (right). Each solid vertical line in a frequency spectrum represents a Dirac δ-function, which can be thought of as an infinitely narrow peak of unit area. Note that the real part of the complex FT is the same as the (real) cosine FT, whereas the imaginary part of the complex FT is the negative of the (real) sine FT (*cf.* Eqs. 1.38)

At this stage, it is useful to digress briefly on the subject of *even* and *odd* functions. The reader will recall that most tables of integrals do not include every possible integral—the reader is expected to be able to change variables to compute, for example, $\int \exp(-ax)\, dx$ if $\int \exp(-x)\, dx$ is given. Similarly, tables of Fourier transform integrals are constructed with the assumption that the reader understands how to recognize even and odd functions. Even and odd functions are respectively symmetrical or antisymmetrical about the origin:

$$Even(-t) = Even(t), \text{ e.g., } t^2, t^4, \cos \omega t, \exp(-t^2), \text{ etc.} \tag{2.4a}$$

$$Odd(-t) = -Odd(t), \text{ e.g., } t, t^3, \sin \omega t, 1/t, \text{ etc.} \tag{2.4b}$$

For products of functions, the following simple rules apply.

$$Even \cdot Even = Even \tag{A.49}$$

$$Even \cdot Odd = Odd \tag{A.50}$$

$$Odd \cdot Odd = Even \tag{A.51}$$

Finally, it should be evident that

$$\int_{-a}^{+a} Even(t)\, dt = 2 \int_{0}^{a} Even(t)\, dt \tag{A.52}$$

$$\int_{-a}^{+a} Odd(t)\, dt = 0 \tag{A.53}$$

Therefore, it should now be clear that the integrals in Eqs. 2.2b and 2.3a vanish because each integrand is odd. Similarly, most tables of Fourier definite integrals will be defined only for $0 \le t \le \infty$ or $0 \le t \le T$, because the reader is supposed be able to apply Eqs. A.52 or A.53.

Most physical time-domain signals are *causal*, i.e.,

$$f(t) = 0 \text{ for } t < 0 \leftrightarrow f(t) \text{ is causal} \tag{2.5}$$

because ordinarily we have no knowledge of the system before the detector was turned on. (FT-IR is a special case, because an optical interferogram can be double-sided—see Chapter 9.2.) Any function (in particular, a causal function) may be analyzed into even and odd components:

$$\text{Causal(t)} = \text{Even(t)} + \text{Odd(t)} \tag{2.6}$$

For example, the reader can quickly confirm (see Problems) that the infinitely long causal time-domain cosine signal,

$$fcausal(t) = \cos \omega_0 t,\ t \ge 0; \quad fcausal(t) = 0,\ t < 0 \tag{2.7}$$

can be represented as the sum of the even and odd functions,

$$feven(t) = (1/2) \cos \omega_0 t,\ -\infty < t < +\infty \tag{2.8a}$$

and $fodd(t) = (1/2)\, \text{sgn}(t) \cos \omega_0 t,\ -\infty < t < +\infty$ (2.8b)

in which the "signum" function, sgn(t), is defined by sgn(t) = 1 for $t > 0$, sgn(t) = 0 for $t = 0$, and sgn(t) = −1 for $t < 0$. The effect of causality upon the real and imaginary parts of a complex Fourier transform is illustrated in Figure 2.2 for infinitely long-duration time-domain sinusoids. Again, because of the redundancy in the various displays of Figure 2.2, most pictorial libraries will show only some of the possible spectra.

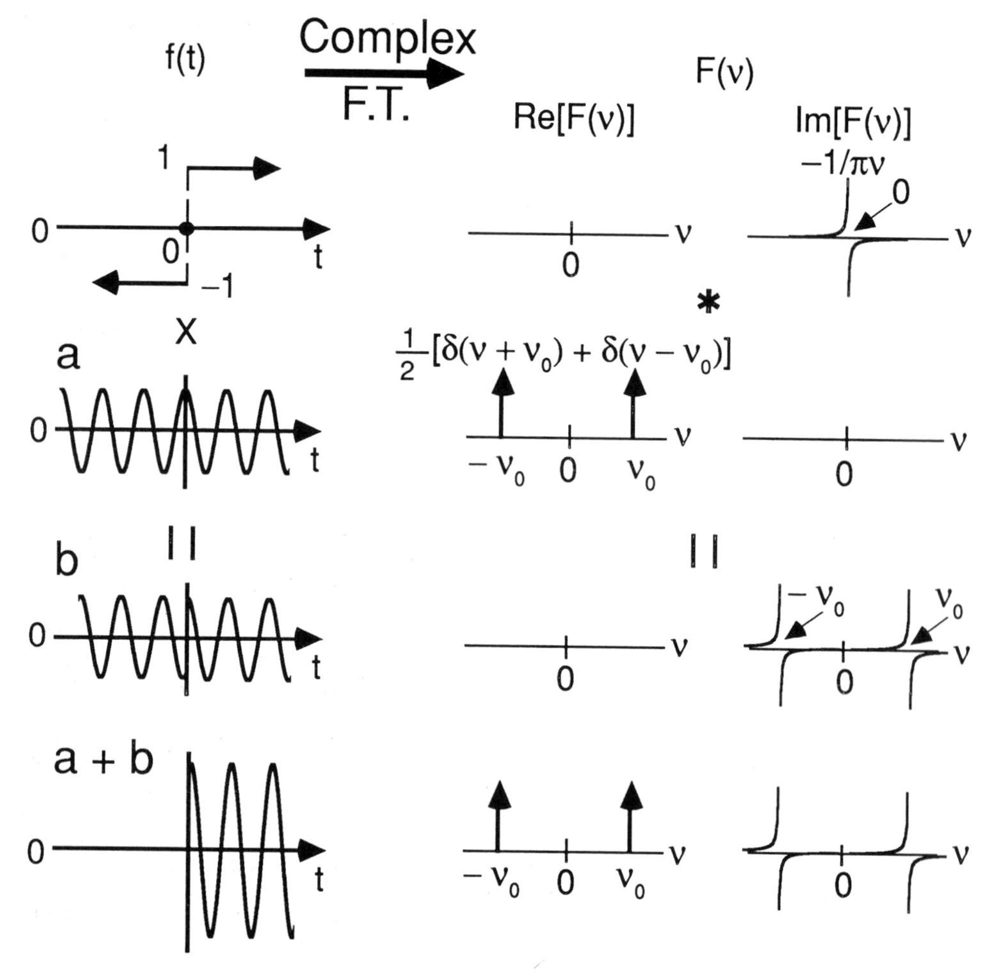

Figure 2.2 Real and imaginary parts of a complex Fourier transforms of the signum function (top row), and of the causal sine and cosine functions obtained by use of the signum function (Eqs. 2.6 to 2.8). The ✱ symbol denotes convolution (see Chapter 2.2.2.1).

2.1.2 Rectangle/sinc: effect of off-resonance excitation

A rectangular time-domain pulse of duration, T, is one of the most important waveforms for physical applications. Such a pulse can serve as a radiation source, as in FT/NMR. Alternatively, the response may sometimes be detected without loss in amplitude for an observation period, T, as in FT/ICR. In addition, this FT pair applies to diffraction from a rectangular slit. For the rectangular time-domain function,

$$f(t) = 1,\ 0 \le t \le T$$
$$f(t) = 0,\ t < 0 \text{ and } t > T \qquad (2.9)$$

the Fourier transform is readily computed from the definitions (Eqs. 1.36 to 1.38) and the integrals (see Appendix A),

$$\int \sin(a\,x)\,dx = -\frac{\cos(a\,x)}{a} + \text{constant} \qquad (A.58)$$

$$\int \cos(a\,x)\,dx = \frac{\sin(a\,x)}{a} + \text{constant} \qquad (A.59)$$

to give

$$A(\omega) = \frac{\sin(\omega T)}{\omega} \qquad (2.10a)$$

$$D(\omega) = \frac{1 - \cos(\omega T)}{\omega} \qquad (2.10b)$$

and $$\mathbf{F}(\omega) = \frac{\sin(\omega T)}{\omega} + i\,\frac{1 - \cos(\omega T)}{\omega} \qquad (2.10c)$$

The reader is left to compute (see Problems) the amplitudes and widths of the absorption, dispersion, and magnitude-mode spectra shown in Figure 2.3. $\sin(\pi x)/(\pi x)$ is important enough to have its own name—the "sinc" function.

At this stage, we should stop to point out a major conceptual difference between the spectra obtained directly by steady-state response to a continuous oscillating excitation and the spectra produced by FT of the time-domain response to a sudden rectangular pulse. In the steady-state experiment, *off-resonance excitation* (at $\omega \ne \omega_0$) produces a finite response at that excitation frequency (ω), because friction (relaxation) is "on" during the excitation. In contrast, we usually conduct FT experiments with an excitation period so short that relaxation during the excitation itself is negligible—e.g., FT/NMR excitation usually consists of a rectangular pulse whose duration, T, is a few μ s, compared to typical NMR relaxation times of ~1 s in liquids. Thus, there is essentially no "friction" during pulsed excitation, and the system therefore can respond only to power applied *at* its natural (angular) frequency, ω_0. However, in practice we observe that power applied at a frequency near ω_0 still excites a detectable signal.

Figure 2.3 resolves this paradox. Although the applied excitation may be precisely centered at a frequency, ω, which differs from ω_0, our act of turning the excitation power on at time zero and off at time, T, effectively broadens the spectral range of the excitation (to a bandwidth of ~$1/T$ Hz). The longer the time-domain signal duration, the narrower is its corresponding frequency-domain spectral width. Thus, if an off-resonance excitation at frequency, ω, is sufficiently brief that

$$(1/T) \ge |\omega - \omega_0| \qquad (2.11)$$

it can still excite a system whose natural angular frequency is ω_0. We shall return to this point in Chapters 7 and 8.

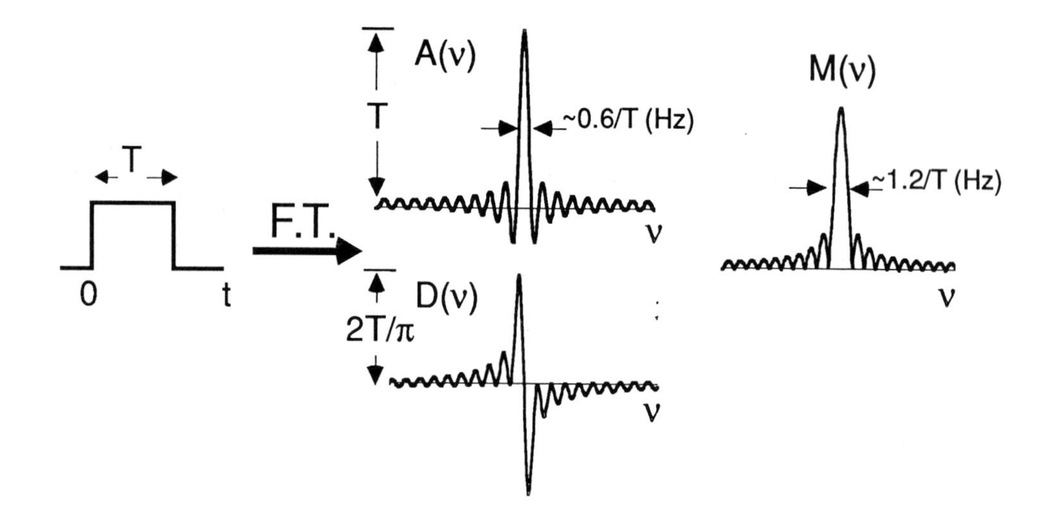

Figure 2.3 Absorption, Dispersion, and Magnitude spectra, generated by Fourier transformation of a rectangular time-domain signal. See text and Problems for the mathematical representations of these curves.

2.1.3 Exponential/Lorentz

In Chapter 1, we demonstrated the FT relationship between an exponentially decaying sinusoid oscillating at $\omega_0 = 2\pi\nu_0$ and a frequency-domain Lorentzian peak centered at $\omega = \omega_0$. From the definite integrals (see Appendix A),

$$\int_0^\infty \exp(-ax)\cos(bx)\,dx = \frac{a}{a^2+b^2} \tag{A.75}$$

$$\int_0^\infty \exp(-ax)\sin(bx)\,dx = \frac{b}{a^2+b^2} \tag{A.76}$$

the reader can readily show (see Problems) that a similar FT relationship holds between a time-domain causal non-oscillating decreasing exponential,

$$f(t) = \exp(-t/\tau), \qquad t \geq 0$$
$$\text{and } f(t) = 0, \qquad t < 0 \tag{2.12}$$

and a Lorentzian centered at $\omega_0 = 0$:

$$A(\omega) = \text{Re}[\mathbf{F}(\omega)] = \frac{\tau}{1+\omega^2\tau^2} \tag{2.13a}$$

$$D(\omega) = \text{Im}[\mathbf{F}(\omega)] = \frac{\omega\tau^2}{1+\omega^2\tau^2} \tag{2.13b}$$

As for the rectangle/sinc case, a longer-lasting time-domain signal (longer τ) leads to a narrower frequency-domain spectrum.

The similarity between Eqs. 1.40 and 2.13 (i.e., the spectra from time-domain oscillating and non-oscillating decreasing exponentials) illustrates further redundancy in the FT process (see Figure 2.4). Later in this chapter, we will use the convolution theorem to generate new FT functions from an existing library of simpler ones. For now, we will proceed to complete our preliminary library.

2.1.4 Gaussian/Gaussian

One of the most unusual Fourier transform pairs is the time-domain Gaussian function, whose Fourier transform yields a frequency-domain Gaussian function of the same shape. For the time-domain causal Gaussian function,

$$f(t) = \exp(-a^2t^2), \qquad t \geq 0$$
$$\text{and } f(t) = 0 \qquad t < 0 \tag{2.14}$$

the cosine Fourier transform spectrum can be generated from the definite integral (see Appendix A)

$$\int_0^{\infty} \exp(-a^2t^2)\cos(bx)\,dx = \frac{\sqrt{\pi}}{2a}\exp(-b^2/4a^2) \tag{A.72}$$

(The analytical expression for the sine FT of the causal half-Gaussian signal of Eq. 2.14 is much more complicated—see Appendix D.) The Gaussian time-domain signal can arise directly (as for FT/NMR in the presence of a linear magnetic field gradient) or can be generated intentionally as a weight function (see Chapter 2.3.3). Again, a broader time-domain signal (smaller value of a in Eq. A.72) corresponds to a narrower frequency-domain spectrum.

2.1.5 Comb/Comb

Although this chapter is concerned primarily with *continuous* functions (corresponding to analog signals), we will soon need to consider *discrete* functions (corresponding to digitized signals). The mathematical connection between these two types of functions is the "shah" function, $III(x)$, also known as a "comb" function from its appearance (see Figure 2.5). The "shah" function consists of an infinite number of δ-functions, equally spaced along the time- (or frequency-axis):

$$III(t) = \sum_{n=-\infty}^{n=+\infty} \delta(t - n\,\Delta t), \quad n \text{ is an integer} \tag{2.15a}$$

$$III(\nu) = \frac{1}{\Delta t}\sum_{n=-\infty}^{n=+\infty} \delta(\nu - n\,\Delta\nu), \quad n \text{ is an integer} \tag{2.15b}$$

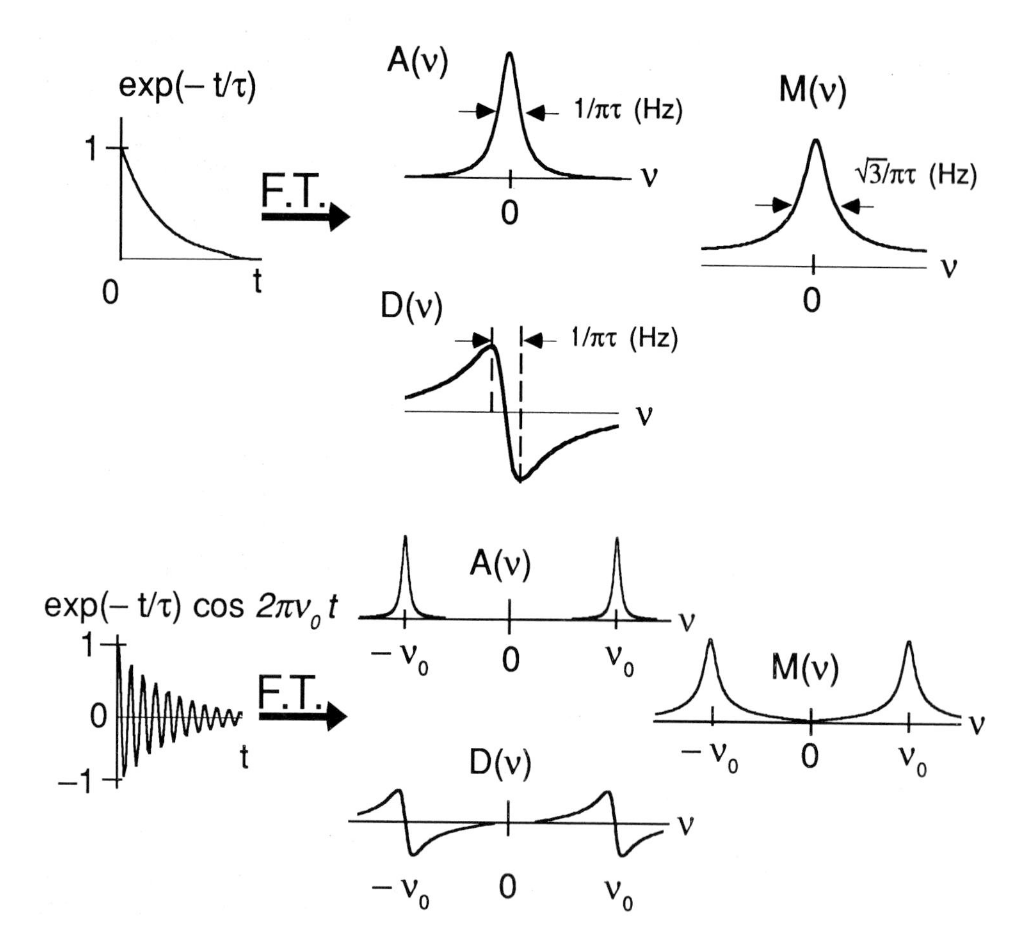

Figure 2.4 Left: Time-domain non-oscillatory (top) and sinusoidally oscillatory (bottom) decreasing exponentials. Right: Absorption, dispersion, and magnitude spectra obtained by Fourier transformation of those time-domain signals.

The shah function, $III(x)$, has the property that it is its own Fourier transform (to within a scale factor), as shown in Figure 2.5. Specifically, if $III(t)$ is given by Eq. 2.15a, then $III(\nu)$ is given by Eq. 2.15b (see Appendix D), in which

$$\Delta\nu = 1/\Delta t \tag{2.16}$$

As we might now expect from our prior examples of other FT pairs, Eq. 2.16 shows that a longer time-domain sampling interval (Δt) corresponds to a narrower frequency-domain sampling interval ($\Delta \nu$). This general effect can be thought of as a "classical uncertainty principle". In order to measure a frequency more accurately, it is necessary to "count" its oscillations for a longer time-domain period: e.g., one needs to observe for ~1 s to determine a frequency to within ~1 Hz (i.e., one oscillation), or observe for ~1 ms in order to determine a frequency to within ~1 kHz, etc. (See Chapter 5.1.3 for more about frequency precision.)

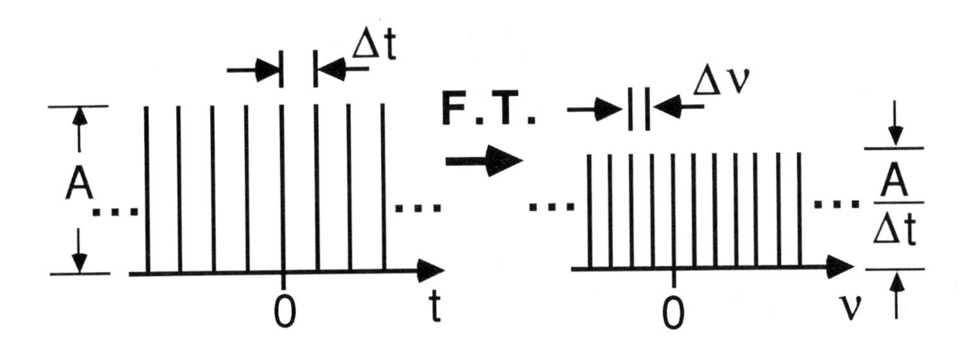

Figure 2.5 Time-domain (left) and frequency-domain (right) representations of the "shah" functions, $III(t)$ and $III(\nu)$.

2.1.6 Chirp/Frequency-Sweep

When it is necessary to excite a wide bandwidth (e.g., ≥ 1 MHz, as in FT/ICR), a convenient time-domain waveform is the "chirp" function, consisting of a constant-amplitude sinusoid whose instantaneous frequency increases linearly with time. As usual, the algebra is simpler for mathematically complex rather than real notation.

$$f(t) = \exp\left(i\left(\omega_A t + \frac{at^2}{2}\right)\right), \quad 0 \leq t \leq T \tag{2.17a}$$

$$\text{and } f(t) = 0, \qquad t < 0 \text{ or } t > T \tag{2.17b}$$

in which the frequency-sweep rate, a (in rad s^{-2}) is determined by the instantaneous angular frequencies, ω_A at $t = 0$ and ω_B at $t = T$: $\omega_B = \omega_A + aT$. From the "error function", erf(x) (see probability integral in Appendix A):

$$\text{erf}(x) = \frac{2}{\sqrt{\pi}} \int_0^x \exp(-t^2)\, dt \tag{A.77}$$

$$\int_{t_1}^{t_2} \exp[i(at^2 + bt + c)]\, dt = \frac{1}{2}\sqrt{\frac{\pi}{a}} \exp\left(i\left(-\frac{\pi}{4} + c - \frac{b^2}{4a}\right)\right) [\text{erf}(z') - \text{erf}(z)] \tag{2.18a}$$

in which

$$z = \left(t_1\sqrt{a} + \frac{b}{2\sqrt{a}}\right)\left(\frac{1}{\sqrt{2}} + \frac{i}{\sqrt{2}}\right) \text{ and } z' = \left(t_2\sqrt{a} + \frac{b}{2\sqrt{a}}\right)\left(\frac{1}{\sqrt{2}} + \frac{i}{\sqrt{2}}\right) \quad (2.18b)$$

(note that $a \neq 0$), the reader can confirm that the Fourier transform of the time-domain chirp function gives the frequency-sweep spectrum of Eq. 2.19.

$$\boldsymbol{F}(\omega) = \sqrt{\frac{\pi}{2a}} \exp\left(-i\left(\frac{\pi}{4} + \frac{(\omega_A - \omega)^2}{2a}\right)\right)$$
$$\times \left(\text{erf}\left(\frac{1}{\sqrt{a}}\frac{\omega_B - \omega}{2} + \frac{i}{\sqrt{a}}\frac{\omega_B - \omega}{2}\right) - \text{erf}\left(\frac{1}{\sqrt{a}}\frac{\omega_A - \omega}{2} + \frac{i}{\sqrt{a}}\frac{\omega_A - \omega}{2}\right)\right) \quad (2.19)$$

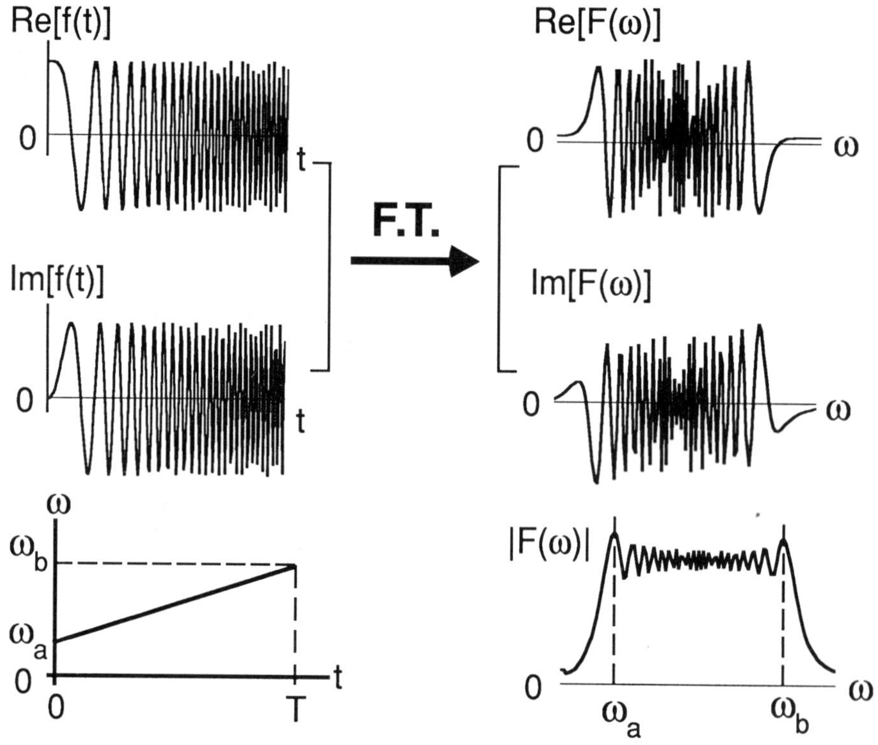

Figure 2.6 Time-domain (left) and frequency-domain (right) representations of a circularly-polarized (i.e., mathematically complex) linear frequency-sweep "chirp".

As might be expected, the frequency-sweep magnitude spectrum (see Figure 2.6) is relatively uniform over the scanned frequency range (i. e., from ω_A to ω_B). Because the chirp is limited to the duration, T, we find that the frequency-selectivity of the sweep (i.e., the width of the spectral edge near $\omega = \omega_A$ or $\omega = \omega_B$) is limited to ~$(1/T)$ Hz. Moreover, the amplitude of the "wiggles" in the frequency-sweep spectrum generally increases with increasing sweep rate (for a given sweep range).

2.1.7 Random Noise/"White" Spectrum

We will defer a detailed consideration of the frequency-domain spectrum of time-domain random and pseudo-random noise until Chapter 5. At this stage, we simply pause to note that a Fourier transform of a single time-domain noise signal of finite duration necessarily yields (on the average) noise of a specified bandwidth in the frequency-domain. In other words, if the frequency-domain spectrum has a *non*-random shape, then that spectrum can no longer be derived purely from time-domain noise!

Unfortunately, the mean value (averaged over many time-domain signals) of the noise at any given frequency value must be zero (otherwise the noise wouldn't be random); thus, the absorption and dispersion spectra obtained from the real and imaginary parts of a complex FT of the time-domain noise must *average* to zero. However, the mean *squared* noise does not average to zero at a given time or frequency, and can thus be used to characterize the noise spectrum. For example, the frequency-domain spectrum of time-domain random noise, averaged over many experiments, is shown in Figure 2.7. Such random noise is sometimes called "white" noise because its magnitude is relatively constant with frequency (particularly if one plots the noise *vs.* log ω—see Chapter 5).

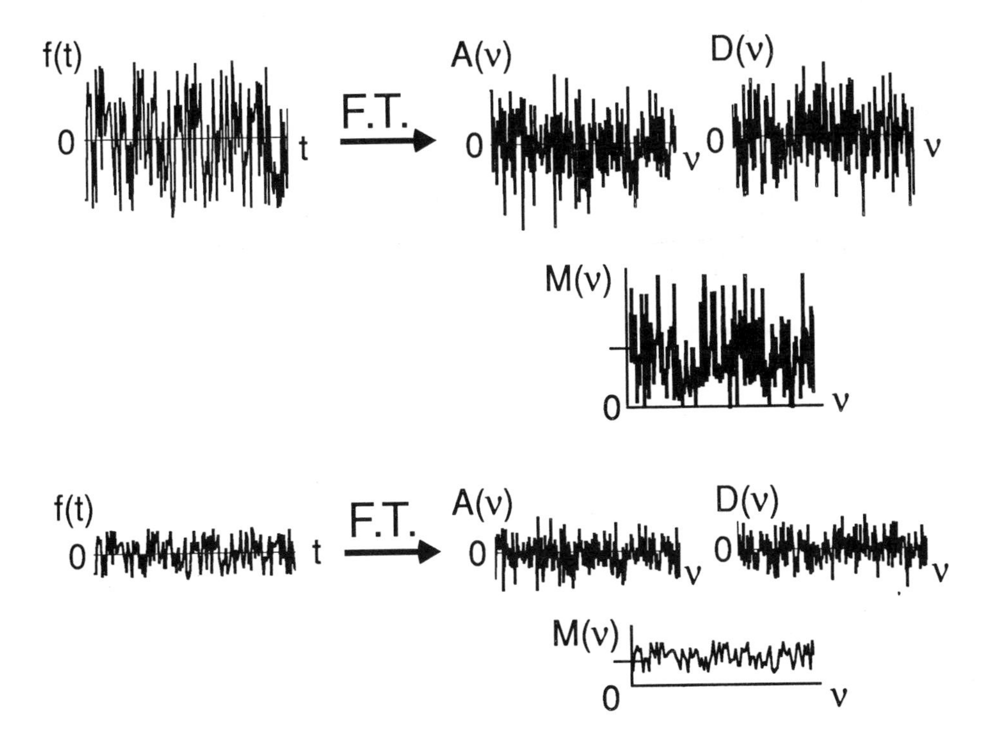

Figure 2.7 Time-domain random noise (left), and its frequency-domain absorption, dispersion, and magnitude spectra (right). Top: a single time-domain waveform. Bottom: sum (divided by N) of N time-domain noise waveforms.

2.2 Convolution

The concept of convolution is fundamental to the understanding of Fourier transform spectroscopy for three reasons. First, the observed time-domain response in an FT spectroscopy experiment can be regarded as the convolution of the (ideal) time-domain response to an impulse excitation and the actual time-domain excitation. Second, most tables of Fourier integrals or pictorial libraries of FT pairs are limited to a few simple functions. The reader is usually expected to be able to generate others by use of convolution, just as a user of ordinary integral tables is supposed to be able to integrate by parts, or evaluate definite integrals from indefinite integrals. Third, the "convolution theorem" connects the convolution process to the computationally faster Fourier transform, so that FT methods can be used to solve convolution problems. For example, if the actual time excitation waveform and observed time-domain response are known, the inverse process (deconvolution) can be conducted by (complex) division in the frequency-domain, so as to extract the true response spectrum, independent of the measuring device.

2.2.1 Ideal versus observed response

The basic idea of convolution can be understood from the operation of a conventional scanning spectrometer (Figure 2.8). In this device, a slit is scanned slowly across a dispersed spectrum. From measurement of the radiation transmitted through the sample, an absorption spectrum can be computed. [In the optical range, for example, the disperser might be a prism or grating.] For an infinitely narrow slit (neglecting diffraction effects), the observed absorption spectrum would represent the true absorption spectrum of the sample.

For a slit of *finite* width, however, the absorption peaks in the *detected* spectrum will be broader than those for the *true* spectrum. In order to obtain the detected spectrum from the true spectrum, one must sum (integrate) the intensity across the slit width for each slit position, as the slit is scanned across the spectral window. The detected response is an example of the *convolution* of the true response with the instrument function (in this case, the slit shape).

More generally, the observed response, $\boldsymbol{f}(t)$, of any linear detector can be described as the convolution of the true response, $\boldsymbol{h}(t)$, to an impulse excitation and the instrument excitation function, $\boldsymbol{e}(t)$, which actually produced the signal.

Mathematically, the convolution (denoted here by the symbol, $*$) of the two functions, $\boldsymbol{h}(t)$ and $\boldsymbol{e}(t)$, to give the observed response, $\boldsymbol{f}(t)$, is given by

$$\boldsymbol{f}(t) = \boldsymbol{e}(t) * \boldsymbol{h}(t) = \int_{-\infty}^{+\infty} \boldsymbol{e}(t')\, \boldsymbol{h}(t - t')\, dt' \qquad (2.20)$$

in which $\boldsymbol{e}(t)$ and $\boldsymbol{h}(t)$ may be complex.

Convolution can often be applied by means of simple graphical construction without resorting to the full algebra of Eq. 2.20. For example, Figure 2.9 shows a step-by-step graphical convolution of two rectangular (real) functions. One of the two functions is first reversed, left-to-right; the two functions are then multiplied point-by-point along the abscissa, and the resultant points added together. The process is repeated for all non-zero superpositions of one of the functions over the other.

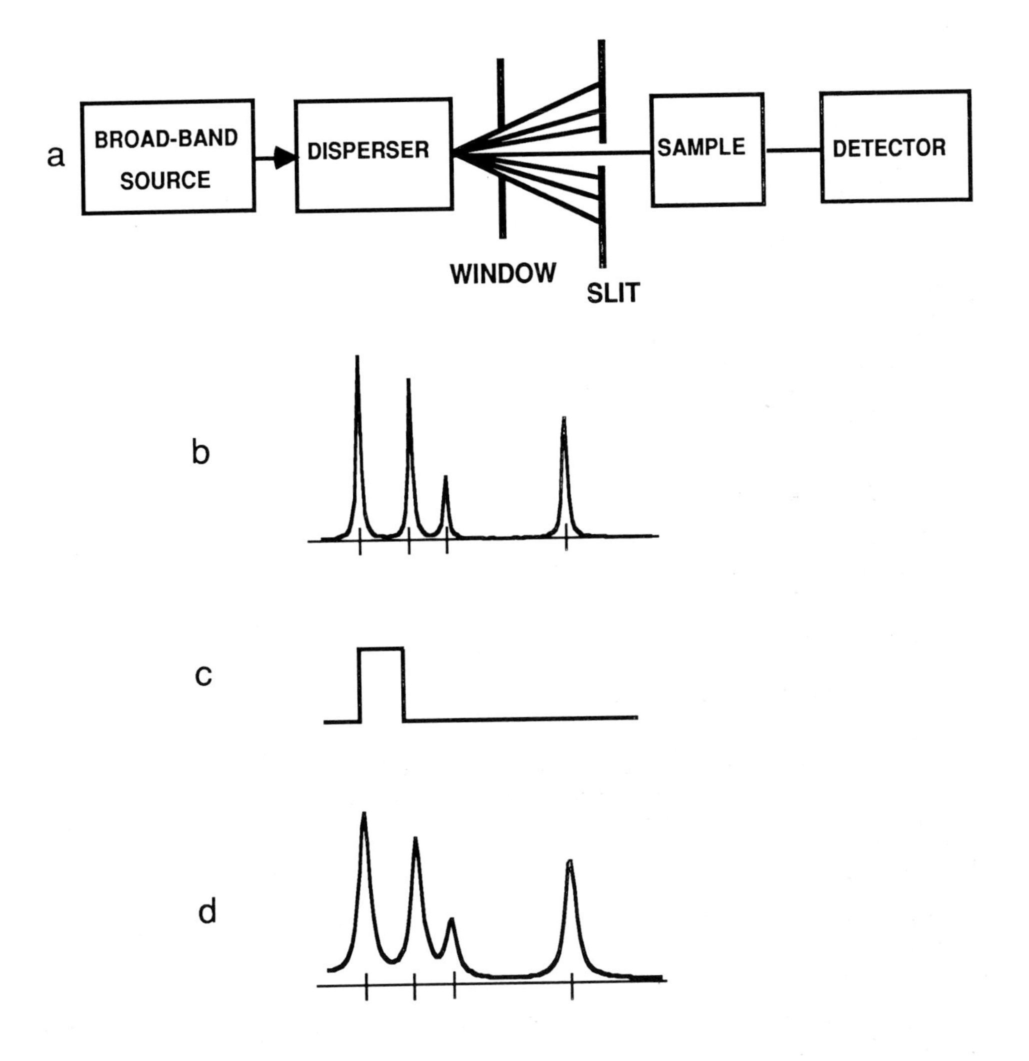

Figure 2.8 Convolution, illustrated by the detected absorption spectrum of a sample by a steady-state scanning spectrometer (a). The true spectrum (b), in passing through a slit of finite width (c), produces the detected spectrum (d). The detected spectrum (d) can be computed by multiplying (b) and (c) and integrating the resulting function for each slit position, as the slit is scanned across the spectrum (i.e., summing all of the light that passes through the slit at each slit position). Note the broadening effect of the "convolution" process [compare (d) to (b)].

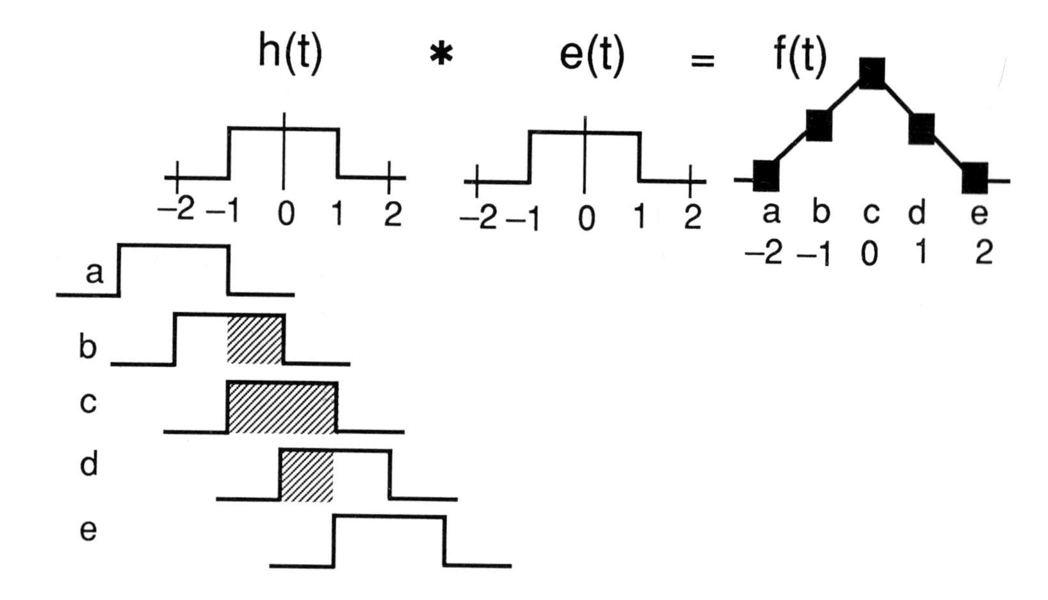

Figure 2.9 Graphical constructions to demonstrate the convolution of two simple rectangular (real) time-domain signals, $h(t)$ and $e(t)$, to yield a triangular function, $f(t)$. Holding the first rectangle fixed, the second rectangle (reversed left-to-right) is moved from left to right, and the two functions multiplied together as shown by the shaded areas of overlap. Each *point*, a to e, of the convolution thus represents the *area* of the product shown at the left (see text). $t = 0$ for $f(t)$ occurs when $t = 0$ for the reversed $e(t)$ exactly overlays $t = 0$ for $h(t)$ [i.e., zero offset between the horizontal axes for $h(t)$ and the reversed $e(t)$]. This example shows how the convolution of rectangular entrance and exit slits produces a triangular slit function in optical monochromators.

Several other graphical convolutions are given in the Problems. It may help to draw $e(t)$ on a piece of paper, then cut it out, flip it over left-to-right, and then overlay it on $h(t)$ and move the cut-out from left to right across $h(t)$, while mentally noting the relative area of overlap between the two functions. The reader is urged to work several examples until the convolution process becomes routine enough to carry out without intermediate pencil-and-paper steps (e.g., b to e in Figure 2.9).

In general, the convolution of two functions is *smoother* and *broader* than either of the two original functions, as seen for both the scanning slit spectrometer and the example of Figure 2.9.

2.2.2 Convolution and Fourier transforms: the convolution theorem

The relationships between convolution and the Fourier transform are shown in Figure 2.10. The mathematical theorem describing those relationships can be stated in several equivalent forms, of which Eq. 2.21 is perhaps the simplest.

If $\boldsymbol{f}(t) = \boldsymbol{h}(t) * \boldsymbol{e}(t)$ = convolution of $\boldsymbol{h}(t)$ with $\boldsymbol{e}(t)$,

and the (complex) Fourier transforms of $\boldsymbol{f}(t)$, $\boldsymbol{h}(t)$, and $\boldsymbol{e}(t)$ are $\boldsymbol{F}(\omega)$, $\boldsymbol{H}(\omega)$, and $\boldsymbol{E}(\omega)$,

then $\boldsymbol{F}(\omega) = \boldsymbol{H}(\omega) \cdot \boldsymbol{E}(\omega)$ (2.21)

In other words, the convolution operation (which is a form of integration) in one domain becomes the much simpler multiplication operation in the corresponding Fourier transform domain. Equation 2.21 is thus an extremely powerful tool. Just as logarithms effectively convert *multiplication* into *addition*,

$$\log a \cdot b = \log a + \log b \qquad \text{(A.3)}$$

convolution effectively converts *integration* into *multiplication*. In the following sections, we shall first use the convolution theorem to expand our library of FT pairs, and then apply the converse of the convolution theorem to "deconvolve" (i.e., remove the effect of) an imperfect excitation from an observed response.

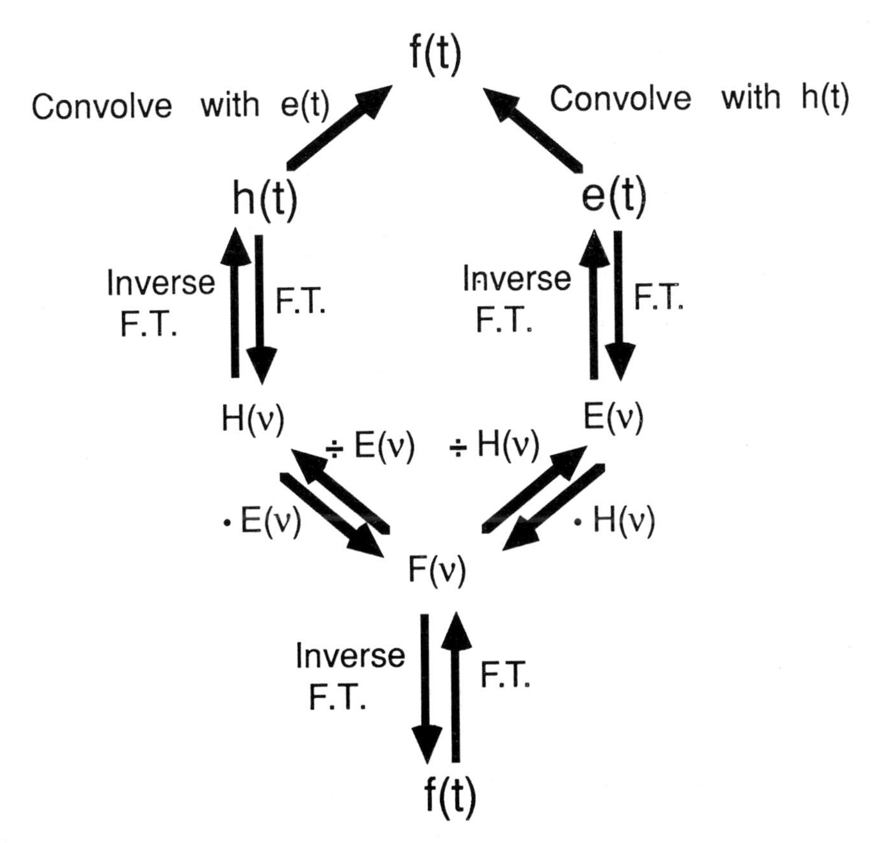

Figure 2.10 Interrelationships between convolution and Fourier transforms. Examples are shown in sections 2.2.2.1. and 2.2.2.2.

2.2.2.1 *Expansion of the library of FT pairs*

From the convolution theorem (Eq. 2.21), we can immediately expand our pictorial library of Fourier transform pairs to include any new time-domain function formed by multiplying together any two of our previous time-domain functions. For example (Figure 2.11), if we know the FT of a time-domain rectangle and of a time-domain sinusoid, then we immediately deduce the FT of their product (i.e., a rectangular time-domain sinusoid) from the convolution theorem. Even without going through the algebra of Eq. 2.21, we can in this case (as in many others) infer the approximate shape of the convolution by means of the graphical construction illustrated in Figure 2.9.

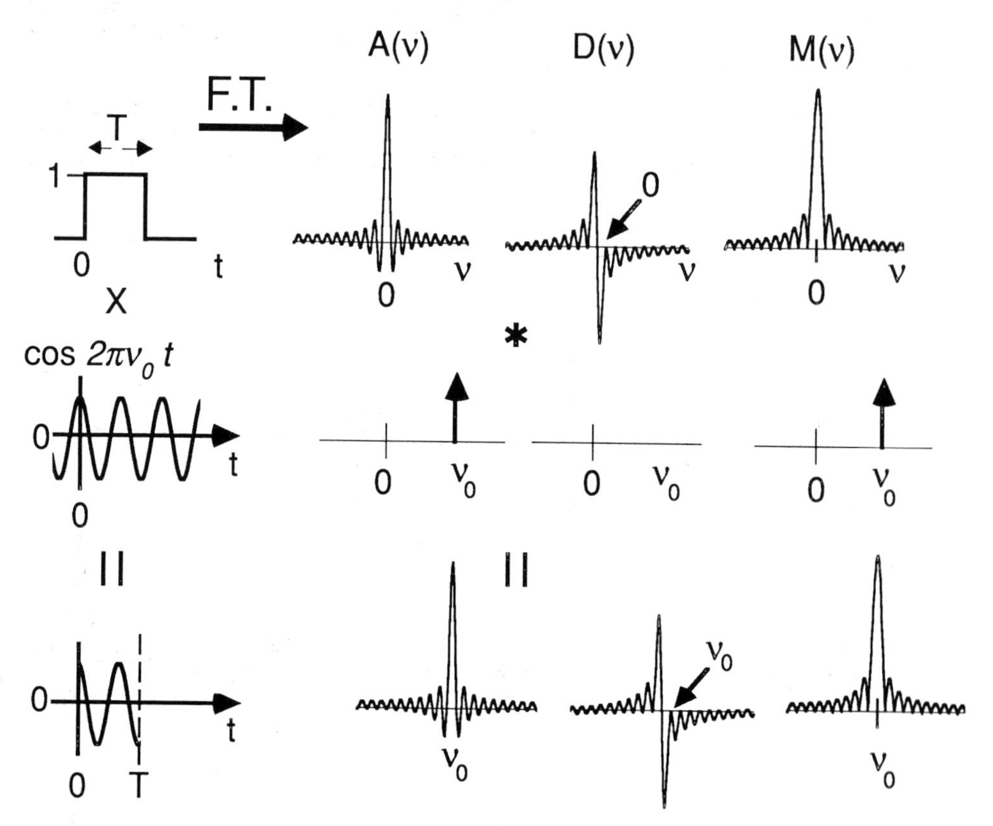

Figure 2.11 Construction of a new FT pair from two existing FT pairs. On the left-hand side of the Figure, a time-domain sinusoid having a rectangular amplitude envelope is represented as the product of a time-domain rectangle and a time-domain sinusoid (of frequency, ν_0) of infinite duration. According to the convolution theorem, the FT of the time-domain product function can be derived by convolution (right-hand side of figure) of the FT's of its two factors (i.e., a frequency-domain sinc function centered at zero frequency and a δ-function centered at ω_0), to give a frequency-domain sinc function centered at ν_0. In this case (as in many others), the convolution can be deduced graphically without recourse to algebra.

For actual spectroscopic time-domain signals, the time-domain observation period is neither infinite, leading to a Lorentz spectrum (Figure 2.12, top) nor so short that damping is negligible, to give a sinc spectrum (Figure 2.11, bottom). A physically more realistic time-domain signal might look like the bottom row of Figure 2.12, namely, a truncated damped sinusoid. Unfortunately (see Problems), the algebra required to generate the Fourier transform of such a function begins to become formidable.

However, we can gain an immediate qualitative picture of the desired spectral line shape by modeling the truncated signal as the product of an infinitely long damped sinusoid and a finite-period rectangle. From the convolution theorem, the spectral line shape for the product then becomes the convolution of the (known) FT's of the two time-domain product signals, to give the line shapes shown at bottom right in Figure 2.12.

Proceeding in this way (see Problems), the reader can readily use the convolution theorem to expand our library of FT pairs to any additional function which can be constructed as the product of two functions whose FT's are already known.

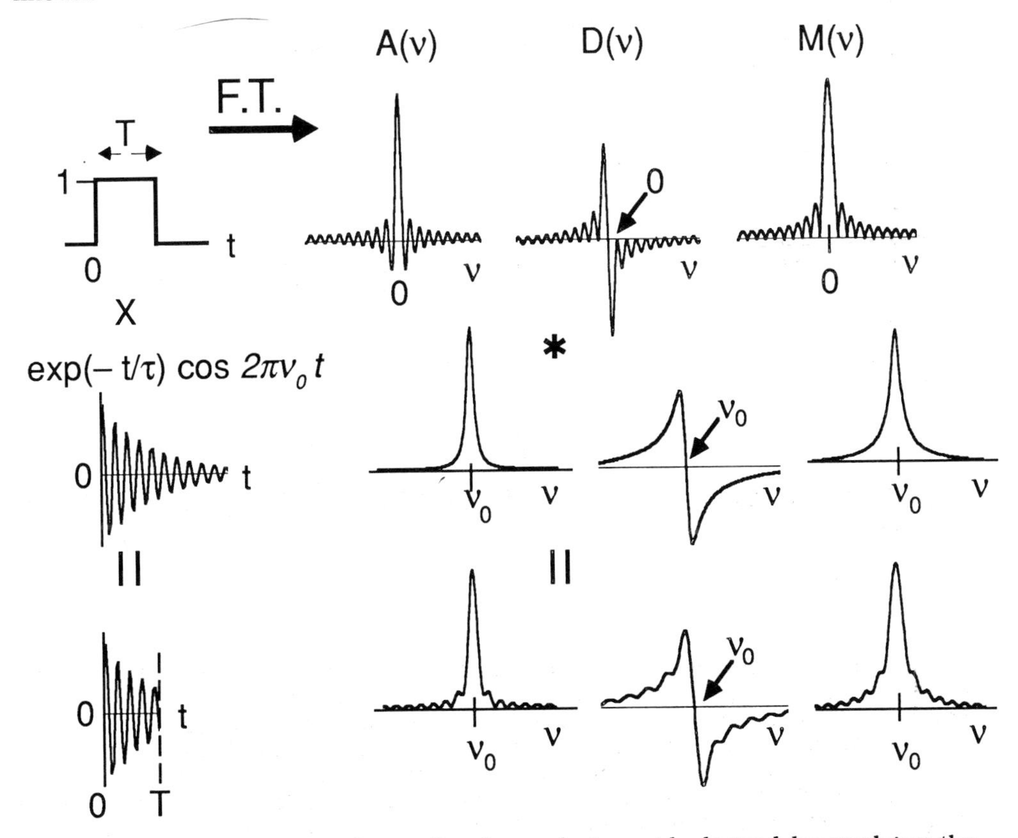

Figure 2.12 Fourier transform of a damped sinusoid, derived by applying the convolution theorem to the time-domain product functions from which the damped sinusoid can be generated (see text).

2.2.2.2 *Compensation for an imperfect experiment*

In the preceding examples, we were concerned with *convolution* of two frequency-domain functions. We will next consider applications for "*deconvolution*", to eliminate the spectral broadening and distortion introduced by the measuring device.

The "perfect" FT experiment would be one in which a time-domain signal, $h(t)$, is excited at time zero by a delta-function time-domain excitation, $\delta(t)$. $h(t)$ can be regarded as the "ideal" response—analogous to the signal from a scanned-slit spectrometer with a very narrow slit (Figure 2.8b). Provided that the system (i.e., sample and measuring device) is linear and noiseless, the experimentally observed time-domain response during an (imperfect) excitation, $e(t)$, of finite duration, is the *convolution*, $f(t)$, of $h(t)$ and $e(t)$. Thus, we would like to detect the "ideal" response, $h(t)$, but are forced to settle for experimentally observable response, $f(t)$.

Fortunately, if we know or can measure the actual excitation, $e(t)$, then we can compute the Fourier transforms, $\boldsymbol{F}(\omega)$ and $\boldsymbol{E}(\omega)$, we can then apply the convolution theorem (Eq. 2.21) in the (equivalent) form,

$$\boldsymbol{H}(\omega) = \frac{\boldsymbol{F}(\omega)}{\boldsymbol{E}(\omega)}, \quad \boldsymbol{E}(\omega) \neq 0 \tag{2.22a}$$

to recover the desired "ideal" spectrum, $\boldsymbol{H}(\omega)$, simply by (complex) division of $\boldsymbol{F}(\omega)$ by $\boldsymbol{E}(\omega)$. The extraordinary implication of this procedure is that we can in principle correct for any imperfections in our experiment, provided only that the system is linear (i.e., the magnitude of the response is proportional to the magnitude of the excitation), and that detection and excitation are simultaneous.

A good example of convolution is shown in the simulated time- and frequency-domain signals of Figure 2.13. For a time-domain frequency-sweep (chirp) excitation (Figure 2.13, upper left),

$$e(t) = \cos\left(\omega_A t + \frac{a t^2}{2}\right) \tag{2.23}$$

such as that used for FT/NMR "correlation" spectroscopy or for broadband FT/ICR excitation, the excitation frequency increases linearly with time. Therefore, the accumulated phase angle increases quadratically with time, leading to observed non-uniform excitation magnitude as a function of frequency, $|\boldsymbol{E}(\omega)|$ (Figure 2.13, upper right). [The frequency-domain spectral phase, $\phi(\omega)$,

$$\phi(\omega) = \arctan\left(\frac{\mathrm{Im}[\boldsymbol{E}(\omega)]}{\mathrm{Re}[\boldsymbol{E}(\omega)]}\right), \tag{2.24}$$

varies quadratically with frequency.

Figure 2.13 also shows the time-domain [$h(t)$ or $f(t)$] and frequency-domain magnitude-mode [$|\boldsymbol{H}(\omega)|$ or $|\boldsymbol{F}(\omega)|$] response of a system of four exponentially damped oscillators of equal amplitude and relaxation time (τ) either to an impulse excitation or to the time-domain frequency-sweep excitation, $e(t)$ of Eq. 2.23.

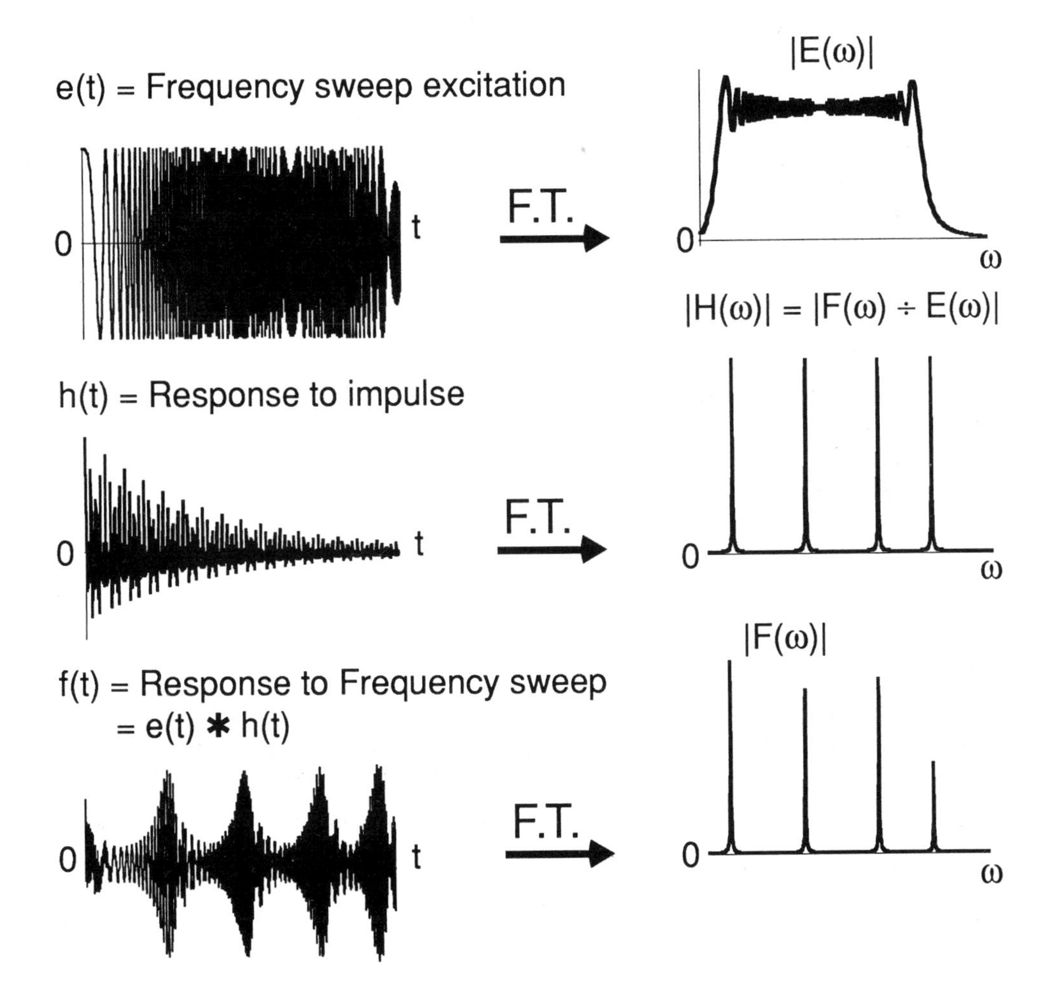

Figure 2.13 Deconvolution of an actual spectrum to yield the spectrum which would have been obtained with a perfect excitation having uniform magnitude and phase. Time-domain (left) and magnitude-mode frequency-domain (right) representations of: a frequency-sweep excitation, $e(t)$ and $\boldsymbol{E}(\omega)$; response, $h(t)$ and $\boldsymbol{H}(\omega)$, of a system of four equally abundant oscillators of equal relaxation time to an impulse excitation; and response, $f(t)$ and $\boldsymbol{F}(\omega)$, of the same system to the frequency-sweep excitation, $e(t)$. The spectral magnitude variation from the excitation itself (upper right) produces corresponding magnitude variations in the spectral response to that excitation (bottom right). However, because $f(t)$ is the convolution of $e(t)$ and $h(t)$, the convolution theorem allows us to recover the desired spectrum, $\boldsymbol{H}(\omega)$, by dividing the actual (complex) spectrum, $\boldsymbol{F}(\omega)$ by the (complex) spectrum of the excitation, $\boldsymbol{E}(\omega)$, to yield the magnitude-corrected spectrum shown at middle right.

The apparent relative peak heights in the magnitude-mode response to the frequency-sweep excitation are unequal because of the non-uniform magnitude of the excitation itself—see also section 4.2.2.

Complex division of the (complex) FT of the response, $\boldsymbol{F}(\omega)$ by the (complex) FT of the excitation, $\boldsymbol{E}(\omega)$ yields the "deconvolved" or "ideal" spectrum, $\boldsymbol{H}(\omega)$, whose magnitude spectrum is shown at the middle right of Figure 2.13. The "deconvolved" spectrum exhibits the correct relative peak magnitudes. Although not shown in the Figure, the deconvolution process also corrects for any spectral phase variation arising from the excitation process; thus, phasing (see Chapter 3.5) of the spectrum to yield an absorption-mode spectrum (with its narrower peak widths) is facilitated.

Important limitations apply to the application of deconvolution methods to experimental data. First, the time-domain excitation and response, $e(t)$ and $f(t)$, must be acquired *simultaneously* to avoid "phase-wrap" distortion (see Chapter 3.5.2). Second, deconvolution performs poorly when noise is present, because the noise in $\boldsymbol{F}(\omega)$ and $\boldsymbol{E}(\omega)$ is generally uncorrelated and therefore increases as a result of the complex division in Eq. 3.22a. Third, the frequency-domain division shown in Eq. 3.22a must be performed on the *complex* FT functions rather than the *magnitude-mode* functions as in Eq. 2.22b. However, Eq. 2.22b is reasonably accurate for spectra in which the peaks are well-separated.

$$|\boldsymbol{H}(\omega)| \neq \frac{|\boldsymbol{F}(\omega)|}{|\boldsymbol{E}(\omega)|} \tag{2.22b}$$

2.2.3 Cross-correlation and auto-correlation

The *cross*-correlation between two functions, $\mathbf{e}(t)$ and $\mathbf{f}(t)$, is defined by

$$\boldsymbol{g}(t) = \int_{-\infty}^{+\infty} \boldsymbol{e}(t')\boldsymbol{f}^*(t'-t)\,dt' = \int_{-\infty}^{+\infty} \boldsymbol{e}(t'+t)\,\boldsymbol{f}^*(t')\,dt' \tag{2.25a}$$

in which * denotes complex conjugate.

Cross-correlation differs from convolution (Eq. 2.20) by a sign change and complex conjugate (*) for the second function. When a function is correlated with itself, the resulting *auto*-correlation function, $\boldsymbol{g}(t)$, takes the form,

$$\boldsymbol{g}(t) = \int_{-\infty}^{+\infty} \boldsymbol{f}(t')\boldsymbol{f}^*(t'-t)\,dt' = \int_{-\infty}^{+\infty} \boldsymbol{f}(t'+t)\,\boldsymbol{f}^*(t')\,dt' \tag{2.25b}$$

Just as absorption and dispersion spectra, $A(\omega)$ and $D(\omega)$, are obtained directly from the real and imaginary parts of the complex Fourier transform of $f(t)$, it turns out (see Chapter 5) that the Fourier transform of the *autocorrelation function* yields the *power* spectrum, $P(\omega)$. Ordinarily, as shown in Figure 2.13, we prefer $A(\omega)$ to $M(\omega)$ or $P(\omega)$ for display of experimental spectra. Therefore, Eq. 2.25b is seldom applied to systems whose spectra have well-resolved peaks, except when it is experimentally more convenient to record $g(t)$ than $f(t)$.

However, as we saw in Figure 2.7, the usual absorption and dispersion representations fail to tell us anything about *noise* signals, whereas the magnitude or power spectra do inform us about frequency spectrum of the noise. Therefore, the autocorrelation function is a useful way to characterize a spectrum of random noise, and we shall defer its further discussion until our treatment of noise in Chapter 5 (see also Appendix D).

2.3 Apodization (windowing): tailoring of spectral peak shape

In a typical FT/spectroscopy experiment, time-domain (or interferogram) data is acquired for a finite (rather than infinite) period, T. The abrupt truncation of a time-domain signal results in corresponding frequency-domain spectra which exhibit auxiliary wiggles on either side of the main peak (see Figures 2.11 and 2.12). Apart from distorting the "true" spectral line shape (i.e., the line shape corresponding to an infinite data acquisition period), such wiggles make it difficult to detect a small peak next to a large peak.

In this section, we shall show that the troublesome wiggles may be reduced by any of several time-domain weight functions applied before Fourier transformation. Such weighting ("windowing") of the *time*-domain data is known as *apodization* (literally, removal of "feet" at the base of a spectral peak) in the *frequency* -domain.

2.3.1 Dirichlet (rectangle, boxcar)

The simplest apodization is that from the Dirichlet function (also known as a rectangle or boxcar function) shown in Figure 2.11. In that example, the time-domain signal before apodization is a sinusoid of infinite duration, corresponding to an infinitely narrow (δ-function) frequency-domain spectrum. The act of truncating the signal after acquisition period, T, leads to the frequency-domain sinc line shape with its characteristic wiggles (known as *Gibbs oscillations*). In general, the truncation of any time-domain signal (e.g., the damped sinusoid in Figure 2.12) generates auxiliary wiggles in the resulting frequency-domain spectrum.

2.3.2 Others: triangle, trapezoid, Hamming, Blackman-Harris, Bessel, etc.

Figure 2.14 shows that boxcar apodization of an infinite-duration time-domain sinusoid produces side lobes with a maximum amplitude of ~22% of the height of the frequency-domain central peak. Since the side lobes arise from the abrupt truncation of the time-domain signal, it is logical to try to reduce the frequency-domain wiggles by applying weight functions designed to reduce the time-domain signal amplitude smoothly from a maximum at $t = 0$ to zero at $t = T$. Figure 2.14 summarizes the effects of several weight functions in commercial or research use in various types of FT spectroscopy.

Frequency-domain (spectral) *resolution*, R, is usually defined as the ratio of the peak center frequency, ω_0, to the peak width, $\Delta\omega$, measured at some convenient fraction of the peak height (e.g., 50%, 10%, 1%, etc.).

$$R = \frac{\omega_0}{\Delta\omega} \tag{2.26}$$

Generally speaking (see Problems), apodization weight functions have two effects on spectral resolution: (a) reduction in the amplitude of the spectral side lobes, and (b) broadening of the spectral peak. By apodization alone, it is not possible to achieve one effect without the other. For example, the side lobe amplitude of the FT of a triangle weight function (see Figure 2.14) is reduced from ~22% to ~5% compared to a Dirichlet (boxcar) weight function, but at the cost of an increase in spectral peak width of ~47%. Moreover, apodization also affects *signal-to-noise ratio*, as discussed in Chapter 5.1.4.

In practice, the apodization functions most commonly applied to spectroscopy are the Hamming (e.g., FT/IR), exponential (e.g., FT/NMR), and Blackman-Harris 3-term (e.g., FT/ICR).

2.3.3 Line shape transformations (Fourier self-deconvolution)

If the spectral line shape is known, then one can design a weight function to eliminate the source of broadening of the peak to leave a peak whose width is determined only by the duration of the acquisition period, T. For example, consider a time-domain exponentially damped sinusoid, $f(t)$.

$$f(t) = \exp(-t/\tau) \cos \omega_0 t\,, \quad 0 \le t < \infty \tag{2.27}$$

If such a signal is acquired for an infinite period, then the corresponding frequency-domain spectrum will be a Lorentzian whose absorption-mode full width at half-maximum peak height is $\Delta\omega = 2/\tau$ rad s^{-1} (see Figure 2.15). Knowing that the time-domain signal decays exponentially, we might think to apply a countervailing weight function,

$$W(t) = \exp(+\,t/\tau) \tag{2.28}$$

in which τ is adjusted empirically until $W(t)$ exactly compensates for the decay of $f(t)$ to leave an undamped time-domain signal. In the absence of noise, such a procedure would generate a peak whose FT spectral width is as narrow as the acquisition period, T, permits.

However (see Chapter 5), time-domain *noise* persists with equal (root-mean-square) amplitude throughout the time-domain acquisition period, whereas the time-domain *signal* decreases with time. The weight function of Eq. 2.28 therefore has the undesirable effect of giving the strongest weight to the noisiest part (i.e., latest-time) portion of the time-domain signal. In order to counteract that effect, it is thus desirable to apodize again with a second weight function (e.g., the Gaussian shown in Figure 2.15) designed to give the greatest weight to the *earliest* portion of the time-domain signal.

The net effect of this procedure is to remove from the spectrum all effects of the original line-broadening mechanism, and to regenerate spectral peak(s) with an arbitrarily chosen line shape optimized for signal-to-noise ratio and resolution. From the convolution theorem (Section 2.2.2), we know that time-domain multiplication by a time-domain weight function is equivalent to frequency-domain convolution with the Fourier transform of the time-domain weight function. Thus, when the procedure of Figure 2.15 is conducted wholly in the frequency-domain by convolution, the process is known as "Fourier self-deconvolution". The method is also used to reduce spectral peak width with or without change in peak shape.

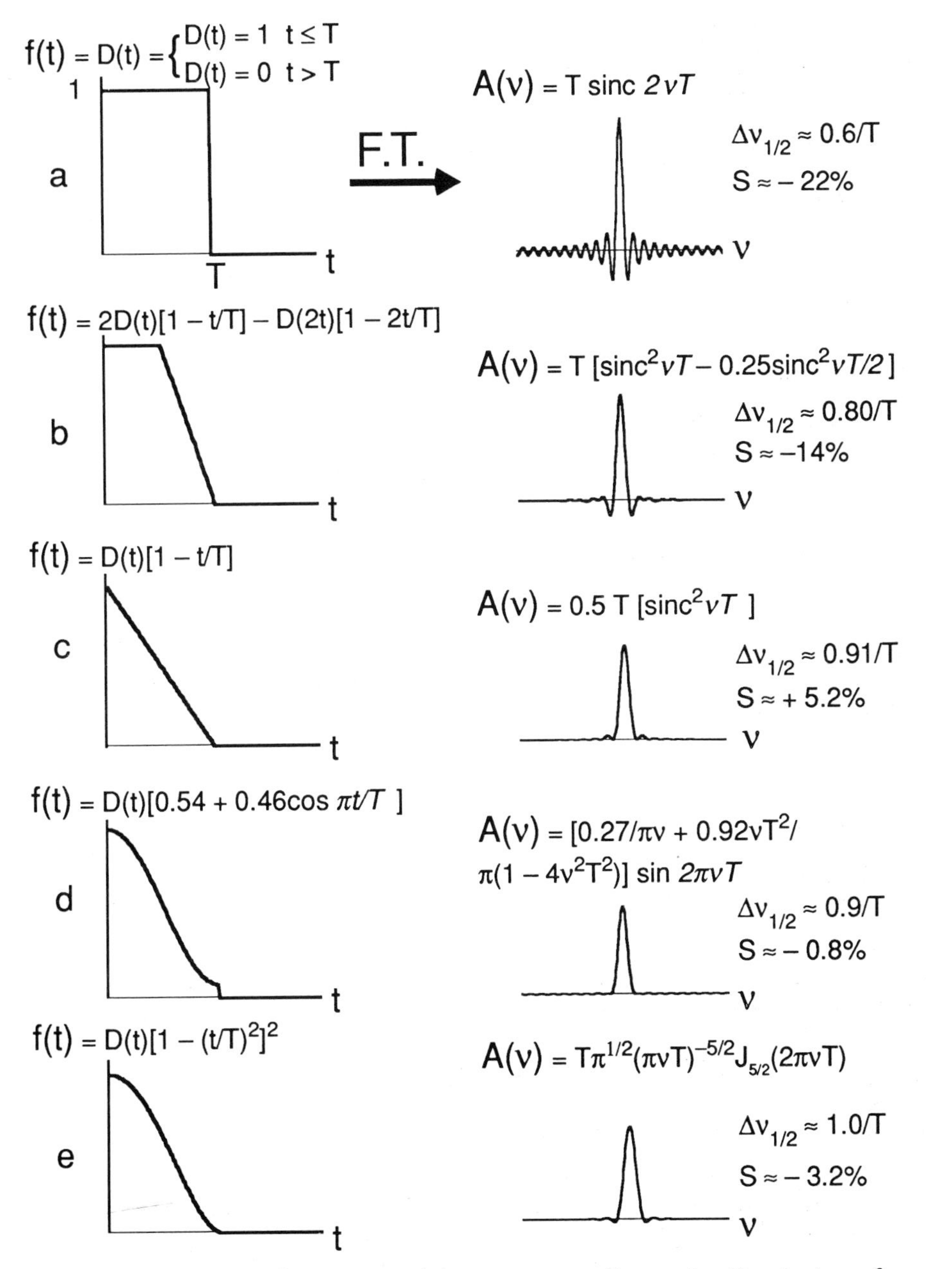

Figure 2.14 Apodization functions and their corresponding cosine Fourier transforms ("instrument line shape functions"), showing the frequency-domain full spectral width at half-maximum peak height, $\Delta\nu$ (Hz), and amplitude (relative to the maximum peak height) of the largest side lobe, S, (as a percentage of the maximum peak height).

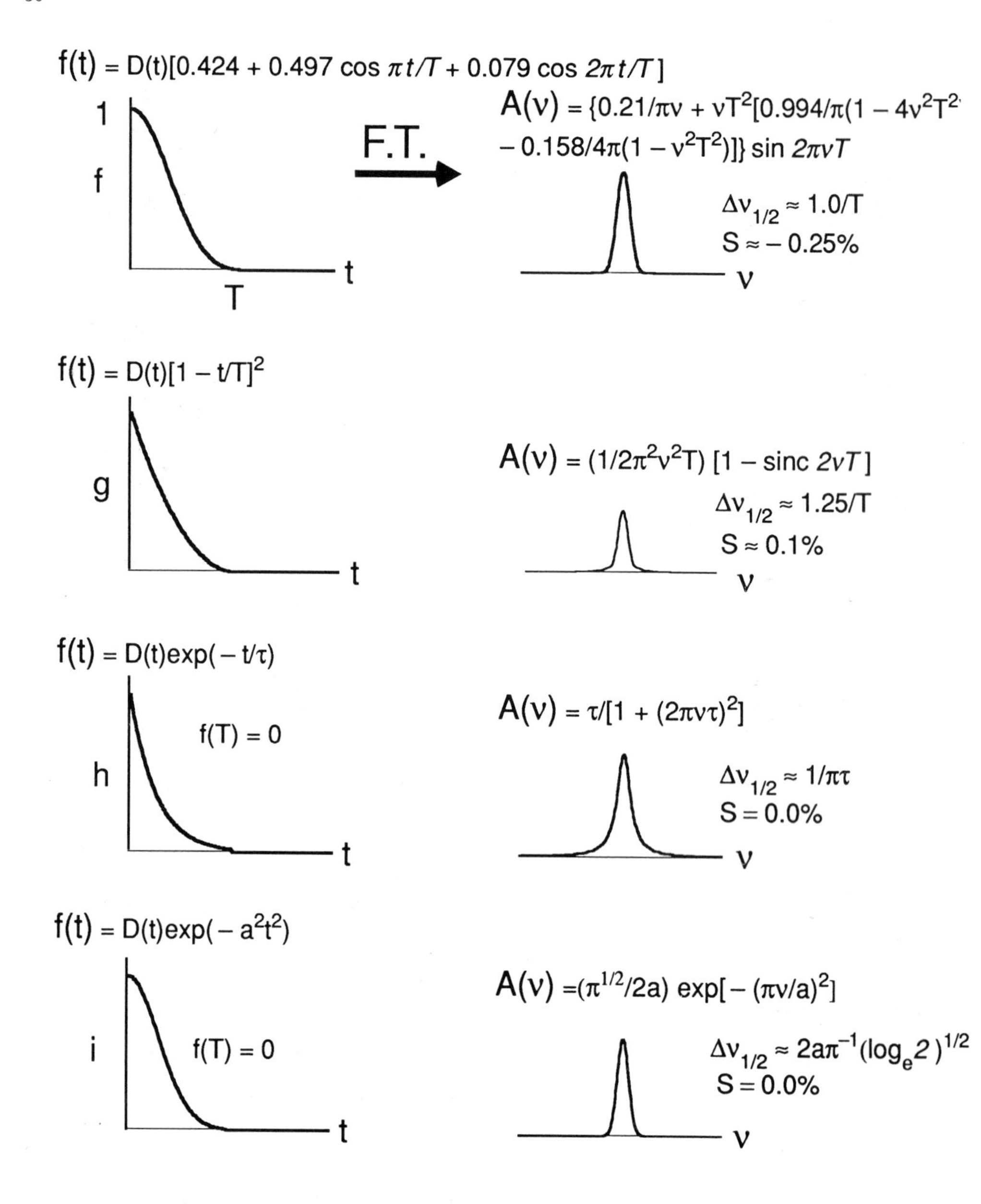

Figure 2.14, continued. (a) Dirichlet (boxcar); (b) trapezoid; (c) triangle; (d) Hamming (a = 0.54); (e) Bessel; (f) Blackman-Harris (3-term); (g) $(\text{triangle})^2$; (h) exponential; (i) Gaussian.

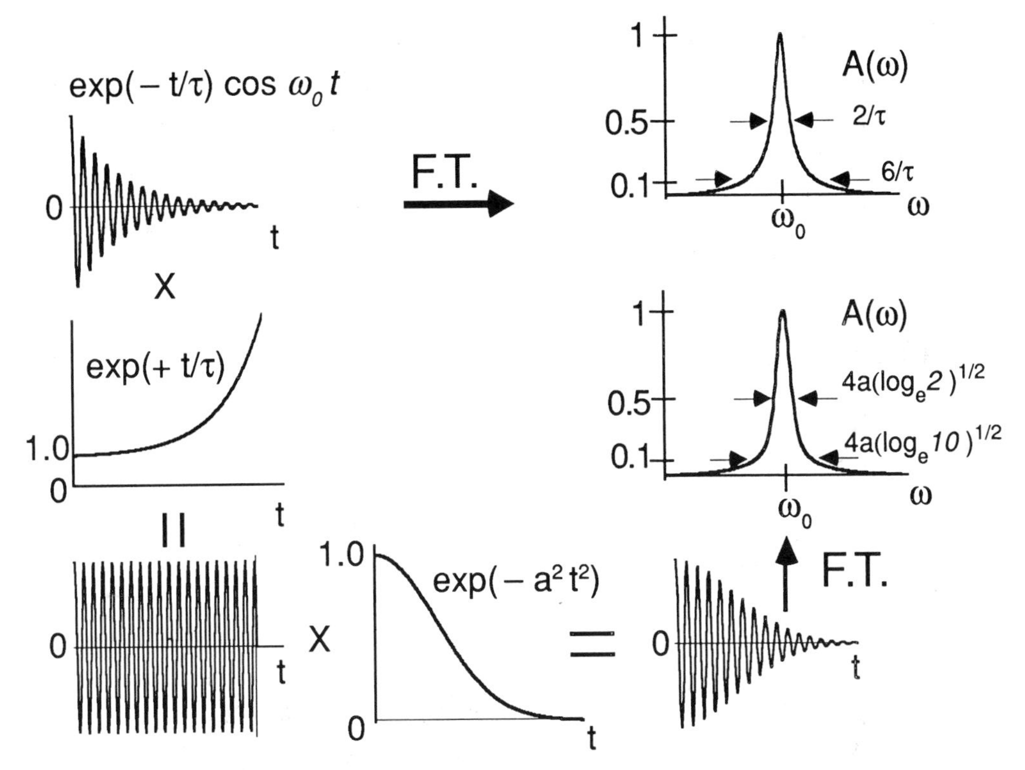

Figure 2.15 Step-by-step transformation from Lorentz to Gauss line shape (see text). This example of "Fourier self-deconvolution" is designed first to remove the inherent broadening due to exponential damping of the time-domain signal, and to regenerate a (Gaussian) line shape whose half-height width is similar to that for the original Lorentzian, but whose width at the base of the peak (e.g., at 10% of the peak height) is much narrowed. Fourier self-deconvolution is particularly useful for analyzing spectra of overlapped peaks.

2.4 Time delay and phase shift

2.4.1 Shift theorem: Relation between time delay and spectral phase

From the trigonometric formula (see Appendix A),

$$\cos(a + b) = \cos a \cos b - \sin a \sin b \tag{A.26}$$

we can generate a time-domain cosine wave, $c(t)$, as the special case of a sinusoid, $f(t)$, of arbitrary initial phase, ϕ_0,

$$f(t) = \cos(\omega_0 t - \phi_0), \qquad 0 \le t \le T \tag{2.29}$$

for which the initial phase, $\phi_0 = 0$. Similarly, a time-domain sine wave, $s(t)$, can be regarded as a sinusoid with $\phi_0 = \pi/2$:

$$c(t) = \cos \omega_0 t \,, \qquad 0 \le t \le T \tag{2.30a}$$

$$s(t) = \sin \omega_0 t \,, \qquad 0 \le t \le T \tag{2.30b}$$

We have previously shown that the absorption and dispersion spectra obtained from the real and imaginary parts of a complex Fourier transform of $c(t)$ and $s(t)$ yield frequency-domain sinc-shape absorption and dispersion spectra (Figure 2.16). For arbitrary initial phase, ϕ_0, (see Problems and Figure 2.16), the real and imaginary parts of the complex Fourier transform frequency-domain spectrum represent linear combinations of the absorption-mode and dispersion-mode sinc line shapes. For example, a phase shift, $\phi_0 = \omega_0 t_0$, could result from an experimental time delay, t_0, between the excitation event and the beginning of the detection period (see below).

or $$f(t) = \cos \omega_0 (t - t_0) = \cos(\omega_0 t - \phi_0) \tag{2.31}$$

The example shown in Figure 2.16 is a special case of the more general "shift theorem" of Fourier analysis, namely,

Shift Theorem: If $f(t)$ has the Fourier transform, $\boldsymbol{F}(\omega)$, then $f(t - t_0)$ the Fourier transform, $\boldsymbol{F}(\omega) \exp(-i \omega t_0)$ (2.32)

As we shall show in the next section, $A(\omega)$ and $D(\omega)$ can be recovered from $\mathrm{Re}[\boldsymbol{F}(\omega)]$ and $\mathrm{Im}[\boldsymbol{F}(\omega)]$ by the phase "correction" of Eq. 2.37.

Eq. 2.31 says that if the time-domain signal is shifted (delayed) by t_0, then its *magnitude-mode* frequency-domain spectrum will be unaffected ($|\exp(-i \omega t_0| = 1$ for all ω), but the real and imaginary spectra will be phase-shifted (see next section). In general, the magnitude-mode display is phase-independent.

The shift theorem is easy to prove:

If $$\int_{-\infty}^{+\infty} \boldsymbol{f}(t) \exp(-i \omega t) \, dt = \boldsymbol{F}(\omega) \tag{2.33}$$

then $$\int_{-\infty}^{+\infty} \boldsymbol{f}(t - t_0) \exp(-i \omega t) \, dt$$

$$= \int_{-\infty}^{+\infty} \boldsymbol{f}(t - t_0) \exp[-i \omega (t - t_0)] \exp(-i \omega t_0) \, d(t - t_0)$$

$$= \exp(-i \omega t_0) \int_{-\infty}^{+\infty} \boldsymbol{f}(t - t_0) \exp[-i \omega (t - t_0)] \, d(t - t_0)$$

$$= \exp(-i \omega t_0) \, \boldsymbol{F}(\omega) \qquad \textbf{Q.E.D.} \tag{2.34}$$

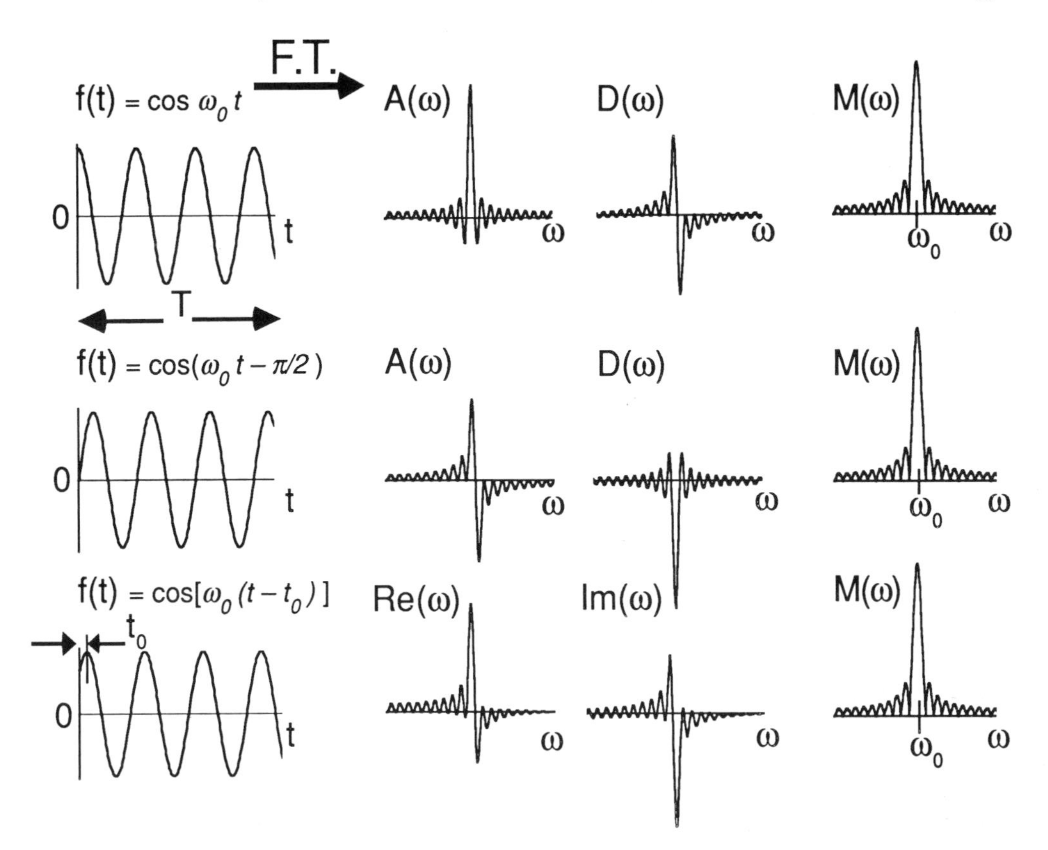

Figure 2.16 Time-domain sinusoids (left) and their corresponding frequency-domain spectra obtained from the real and imaginary parts (or magnitude) of the complex Fourier transform of each time-domain signal. Although the pure cosine or sine time-domain signals lead to pure absorption and dispersion spectra, $A(\omega)$, and $D(\omega)$, the FT of a time-delayed time-domain cosine gives real and imaginary spectra, $\mathrm{Re}[\boldsymbol{F}(\omega)]$ and $\mathrm{Im}[\boldsymbol{F}(\omega)]$, which consist of linear combinations (Eqs. 2.35) of $A(\omega)$ and $D(\omega)$, with a frequency-dependent phase,

$$\phi(\omega) = \omega t_0 = \arctan\left(\frac{\mathrm{Im}[\boldsymbol{F}(\omega)]}{\mathrm{Re}[\boldsymbol{F}(\omega)]}\right).$$

2.4.2 Frequency-domain phase correction

Following a δ-function excitation, oscillations at all frequencies are excited simultaneously, and thus each time-domain sinusoid begins with the same initial phase (say, zero, corresponding to a wave that starts as a cosine at $t = 0$). If there is a time-delay, t_0, between excitation and detection, then the corresponding phase shift, ωt_0, of Eq. 2.34 will be linearly proportional to t_0 (see Figure 2.16). In other words, a low-frequency signal will oscillate only a few times (and thus accumulate only a small phase angle) during t_0, whereas a high-frequency signal will oscillate many times (and accumulate a proportionately larger phase angle) during t_0.

It is left as an exercise (see Problems) for the reader to show that the real and imaginary parts, $\mathrm{Re}[\boldsymbol{F}(\omega)]$ and $\mathrm{Im}[\boldsymbol{F}(\omega)]$, of the complex FT of a time-delayed sinusoid are related to the absorption and dispersion spectra, $A(\omega)$ and $D(\omega)$, for the undelayed sinusoid according to Eqs. 2.35.

$$\mathrm{Re}[\boldsymbol{F}(\omega)] = \cos\phi\; A(\omega) + \sin\phi\; D(\omega) \tag{2.35a}$$

$$\mathrm{Im}[\boldsymbol{F}(\omega)] = -\sin\phi\; A(\omega) + \cos\phi\; D(\omega) \tag{2.35b}$$

in which $\phi = \omega t_0$ (2.36)

From Eqs. 2.35 (see Problems), the correctly "phased" absorption and dispersion spectra, $A(\omega)$ and $D(\omega)$, can be recovered from $\mathrm{Re}[\boldsymbol{F}(\omega)]$ and $\mathrm{Im}[\boldsymbol{F}(\omega)]$ according to the inverse transformation (reverse rotation in the complex frequency-domain plane) of Eqs. 2.37.

$$A(\omega) = \cos\phi\; \mathrm{Re}[\boldsymbol{F}(\omega)] - \sin\phi\; \mathrm{Im}[\boldsymbol{F}(\omega)] \tag{2.37a}$$

$$D(\omega) = \sin\phi\; \mathrm{Re}[\boldsymbol{F}(\omega)] + \cos\phi\; \mathrm{Im}[\boldsymbol{F}(\omega)] \tag{2.37b}$$

The reader may have observed that the transformation of Eqs. 2.35 has the same mathematical form as does the rotation of an ordinary vector in the x-y plane. In fact, the analogy is complete if we plot $\mathrm{Re}[\boldsymbol{F}(\omega)]$ and $\mathrm{Im}[\boldsymbol{F}(\omega)]$ in the complex plane, with ω as the third axis (Figure 2.17). The projection down the ω-axis of such a plot then yields the DISPA (dispersion-vs.-absorption) plot of Chapter 1.3.3. The phase transformation of Eqs. 2.35 is manifested as a rotation of the DISPA plot by angle, ϕ, about its origin (Figure 2.17).

The phase angle, $\phi(\omega)$, may thus be viewed in three equivalent ways:

Time-domain: $\phi(\omega) = \omega t_0$,

(t_0 = time-delay for a sinusoid of frequency, ω) (2.38a)

Frequency-domain: $\phi(\omega) = \arctan\left(\dfrac{\mathrm{Im}[\boldsymbol{F}(\omega)]}{\mathrm{Re}[\boldsymbol{F}(\omega)]}\right)$

[$\phi(\omega)$ = phase angle in complex plane] (2.38b)

DISPA-plot: $\phi(\omega)$ = rotation angle of a DISPA plot about its origin (provided that ϕ does not vary significantly from one side of the peak to the other [i.e., $(\omega - \omega_0)\, t_0 << 1$ over the ω-range near the peak] (2.38c)

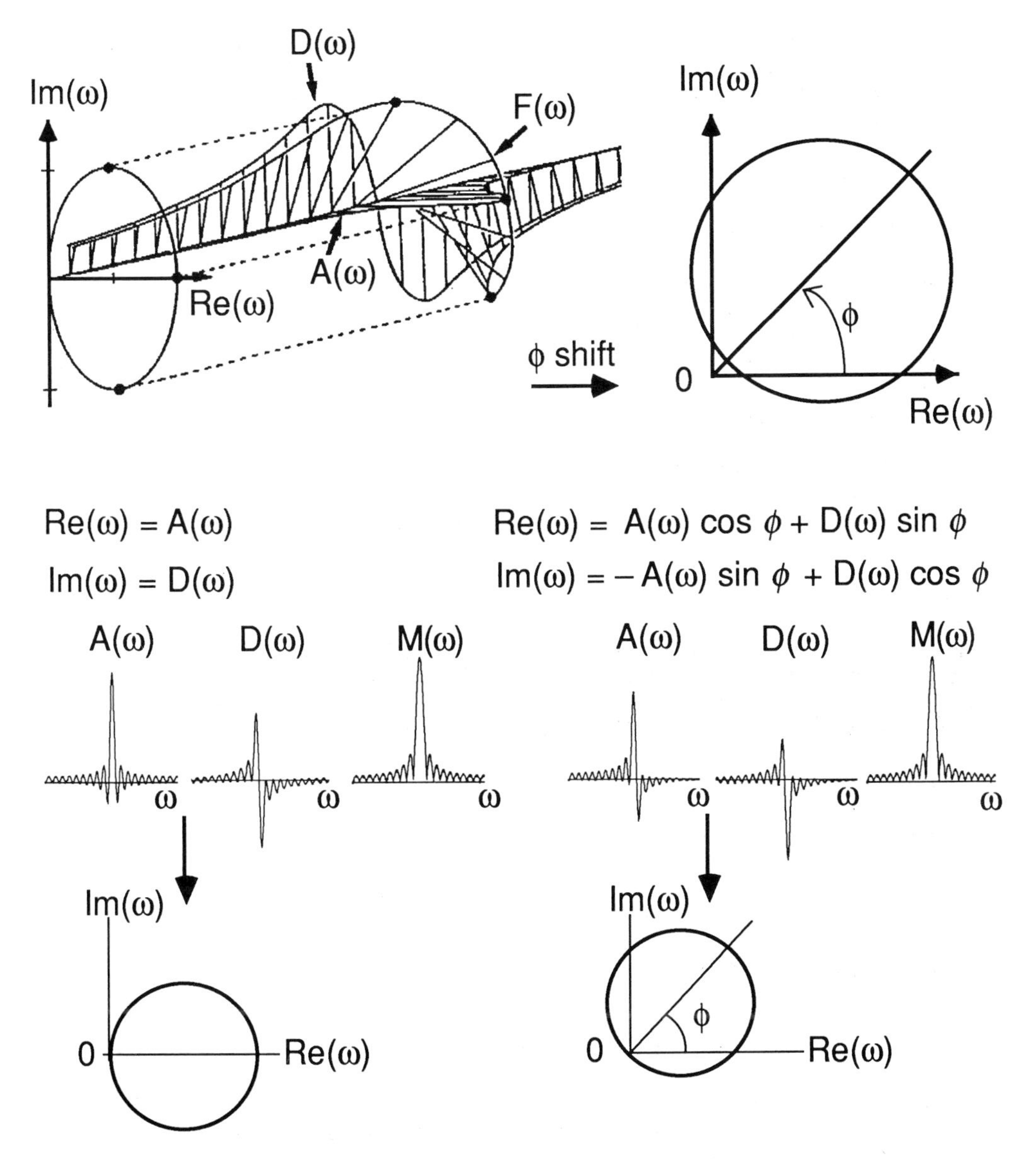

Figure 2.17. Complex Lorentzian spectrum, $\mathbf{F}(\omega) = A(\omega) + iD(\omega)$. The absorption and dispersion components, $A(\omega)$ and $D(\omega)$, are the projections of $\mathbf{F}(\omega)$ along the real and imaginary axes. A dispersion-versus-absorption (DISPA) plot of $D(\omega)$ *vs.* $A(\omega)$ is the projection obtained by looking directly down the ω-axis. A constant phase-shift across the spectrum corresponds to rotation of $\mathbf{F}(\omega)$ about the ω-axis (or, equivalently, a rotation of the DISPA plot about its origin).

In this section, we have considered the linear (as a function of ω) phase shift resulting from a fixed time-domain delay, t_0, between excitation and detection. The experimentally observed phase shift in FT spectroscopy may also include zero-order (i.e., a constant phase shift which is independent of ω) and higher-order phase shifts (e.g., quadratic phase shift from frequency-sweep excitation). Moreover, such phase shifts may prove difficult to distinguish from other sources of line shape distortion (e.g., baseline drift, inhomogeneously broadened line shape, etc.), as we will discuss further in sections 3.5 and 7.4.4. Finally, we will find that phase correction for the double-sided interferograms of FT/optical spectroscopy is somewhat different (Chapter 9.2.4) than for the causal (one-sided) time-domain signals of FT/NMR and FT/ICR.

Further Reading

M. Abramowitz & I. A. Stegun, Eds., *Handbook of Mathematical Fundctions with Formulas, Graphs, and Mathematical Tables*, National Bureau of Standards Applied Mathematics Series 55, U.S. Government Printing Office, Washington, D.C., 1964. (Comprehensive table of integrals)

R. Bracewell, *The Fourier Transform and Its Applications*, McGraw-Hill, N.Y., 1965. (Good discussion of FT and convolution; pictorial FT atlas)

D. C. Champeney, *Fourier Transforms and Their Physical Applications*, Academic Press, N.Y., 1973. (Good general FT text; pictorial FT atlas)

J. D. Gaskill, *Linear Systems, Fourier Transforms, and Optics*, J. Wiley & Sons, N.Y., 1978. (Good presentation of the mathematics of Fourier transforms)

I. S. Gradshteyn & I. M. Ryzhik, *Table of Integrals, Series, and Products*, Academic Press, Orlando, FL, 1980. (Comprehensive table of integrals)

F.J. Harris, Proc. IEEE **1978**, *66*, 51-83. (Review of 23 apodization functions)

J. P. Lee & M. B. Comisarow, *Appl. Spectrosc.* **1987**, *41*, 93-98; M. Aarstol and M. B. Comisarow, *Int. J. Mass Spectrom. Ion Proc.* **1987**, *76*, 287-297. (Apodization functions for magnitude-mode spectra: these papers show how the choice of optimal weight function varies with T/τ in Figure 2.12 and with dynamic range)

T.-C. L. Wang & A. G. Marshall, Anal. Chem. **1983**, *55*, 2348-2353. (Formulas for dispersion-mode Gaussian spectrum)

Problems

2.1. (a) Compute the cosine and sine Fourier transforms, $C(\omega)$ and $S(\omega)$, of the time-domain signal,

$$f(t) = \cos \omega_0 t \,, \qquad 0 \le t \le T$$

$$f(t) = 0, \qquad t < 0 \text{ and } t > T$$

(b) Now combine them to obtain the magnitude spectrum, $M(\omega)$.

(c) Compute the height (in units of T) for $C(\omega)$, $S(\omega)$, and $M(\omega)$.

(d) Next, compute (numerically, by iteration—it should take only a few steps) the width at half height for $C(\omega)$ and $M(\omega)$.

(e) Now sketch $C(\omega)$, $S(\omega)$, and $M(\omega)$ versus ω. It should then be clear why the absorption-mode display [$C(\omega)$ in this case] is preferred to the magnitude-mode.

This problem shows the origin and principal properties of the "sinc" function which characterizes such spectroscopic applications as pulse excitation, undamped detection, and diffraction from a slit.

2.2 (a) Compute and sketch the spectra of the cosine and sine Fourier transforms, $C(\omega)$ and $S(\omega)$, of the time-domain non-oscillatory exponential decay signal,

$$f(t) = \exp(-t/\tau), \qquad 0 \le t \le \infty$$

$$f(t) = 0, \qquad t < 0$$

(b) Now combine them to obtain and sketch the magnitude spectrum, $M(\omega)$.

2.3 Compute the sine and cosine Fourier transforms, $C(\omega)$ and $S(\omega)$, of the following three time-domain functions.

$$fcausal(t) = \cos \omega_0 t \quad \text{for } t \ge 0; \qquad fcausal(t) = 0 \quad \text{for } t < 0$$

$$feven(t) = (1/2) \cos \omega_0 t \,, \qquad -\infty < t < +\infty$$

and

$$fodd(t) = (1/2)\,\mathrm{sgn}(t) \cos \omega_0 t \,, \qquad -\infty < t < +\infty$$

in which $\mathrm{sgn}(t) = 1$ for $t \ge 0$; $\mathrm{sgn}(t) = 0$ for $t = 0$; and $\mathrm{sgn}(t) = -1$ for $t < 0$.

Now sketch $fcausal(t)$, $feven(t)$, and $fodd(t)$ *vs.* t and sketch $C(\omega)$ and $S(\omega)$ vs. ω for their corresponding Fourier transforms. This example illustrates the general result that the absorption-mode spectrum arises from the even part (and the dispersion-mode spectrum from the odd part) of a causal [$f(t) = 0$ for $t < 0$] time-domain (real) signal.

2.4 (a) Compute and sketch the spectra of the sine and cosine Fourier transforms, $C(\omega)$ and $S(\omega)$, of the truncated time-domain exponentially damped sinusoid,

$$f(t) = \exp(-t/\tau) \cos \omega_0 t , \qquad 0 \le t \le T$$

$$f(t) = 0, \qquad t < 0 \text{ and } t > T$$

(b) Now combine them to obtain and sketch the magnitude spectrum, $M(\omega)$.

(c) Compare this intermediate result to the two limiting cases of $\tau \to \infty$ ("sinc" line shape of Problem 2.1), and $T \to \infty$ (Lorentz line shape of Problem 2.2).

2.5 It should be apparent from Problem 2.2 that for a given time-domain damping constant, τ, the narrowest absorption-mode (or magnitude-mode) line width is obtained in the limit that the experimental data acquisition period, $T \to \infty$. However, it is desirable to know how long T must be in order that the line width is less than (say) 1% or 10% broader than the minimum line width, since in practice we can record only so much data before our patience or computer storage space is exhausted! Therefore, compute the full width at half-maximum height for $C(\omega)$ of Problem 2.2, for the following values of T :

(a) $T = \tau$
(b) $T = 2\tau$
(c) $T = 3\tau$
(d) $T \to \infty$

Then use these four values to plot a graph of absorption-mode line width *vs.* acquisition time, T. You should find that acquiring the time-domain signal for only 2-3 decay time constants suffices to yield an acceptably narrow peak.

2.6 Use Eq. 2.18 to compute the Fourier transform, Eq. 2.19, of the time-domain function of Eq. 2.17. This problem takes the reader through the algebra for the frequency-domain representation of a time-domain frequency-sweep ("chirp") signal. The reader with access to a microcomputer can plot the absorption, dispersion, and magnitude spectra shown in Figure 2.6 from Eq. 2.19.

2.7 Use graphical construction (as in Figure 2.9) to find the convolution for each pair of functions shown at the left. The Voigt function, (d), represents a Gaussian distribution of Lorentzian peaks of different position, and is commonly seen in inhomogeneously broadened spectra (e.g., NMR spectra of polymers or adsorbed species; optical spectra of gases).

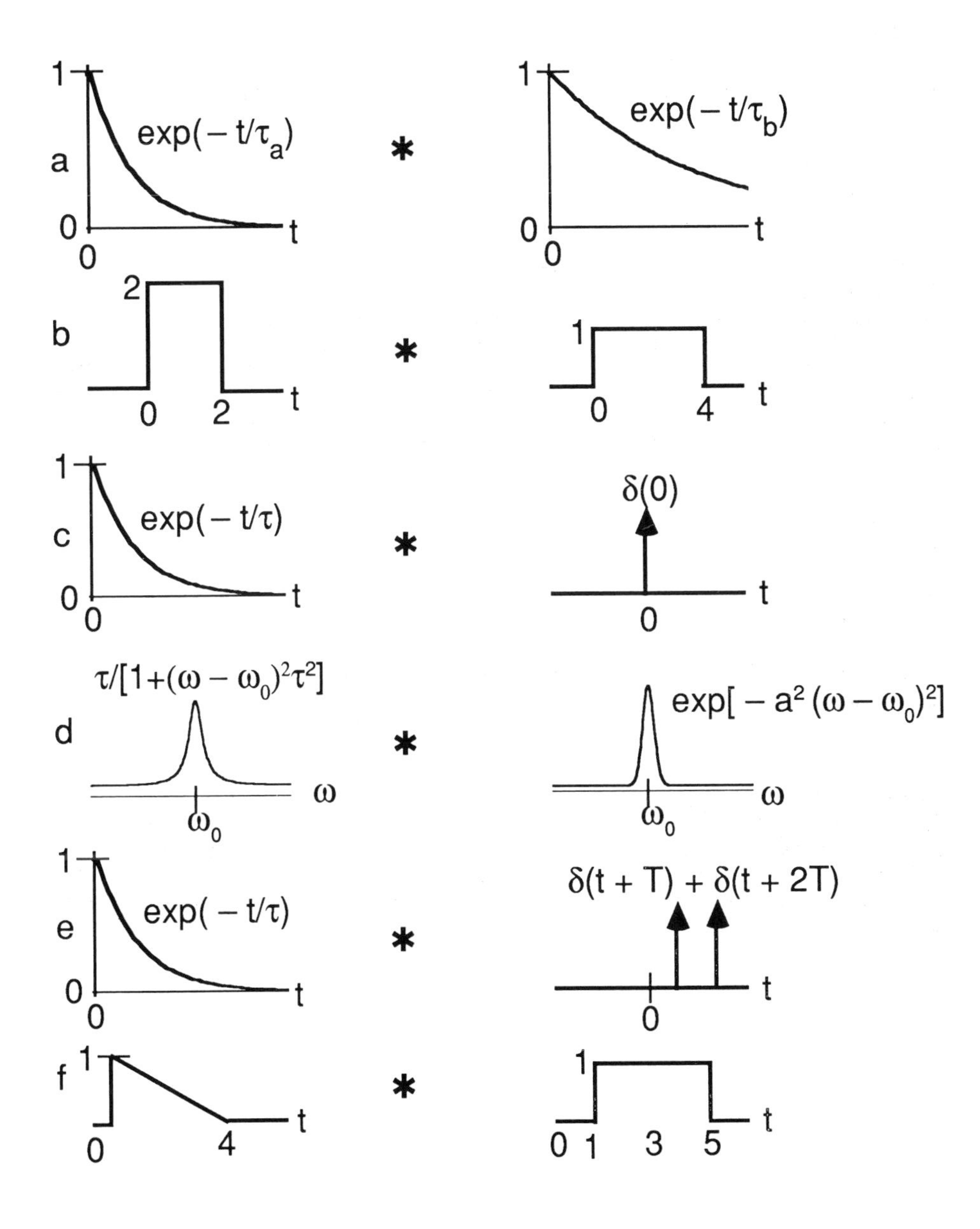

2.8 Use the convolution theorem (graphically, if you can) to sketch the Fourier transform spectra [Re[$F(\omega)$], Im[$F(\omega)$], and $M(\omega)$] for the following products of time-domain signals. These exercises show how to expand the library of FT pairs. [See Appendix D for FT's of signum and Heaviside functions, sgn(t) and H(t).] Example (a) shows how two competing relaxation processes combine to broaden a spectral peak. (b) to (d) show the relation between absorption and dispersion spectra and the even and odd parts of a time-domain signal. (e) is an example of sampling, to be discussed further in the next chapter. (f) shows how time-domain multiplication by a sinusoid shifts the frequency-domain spectrum along the frequency axis.

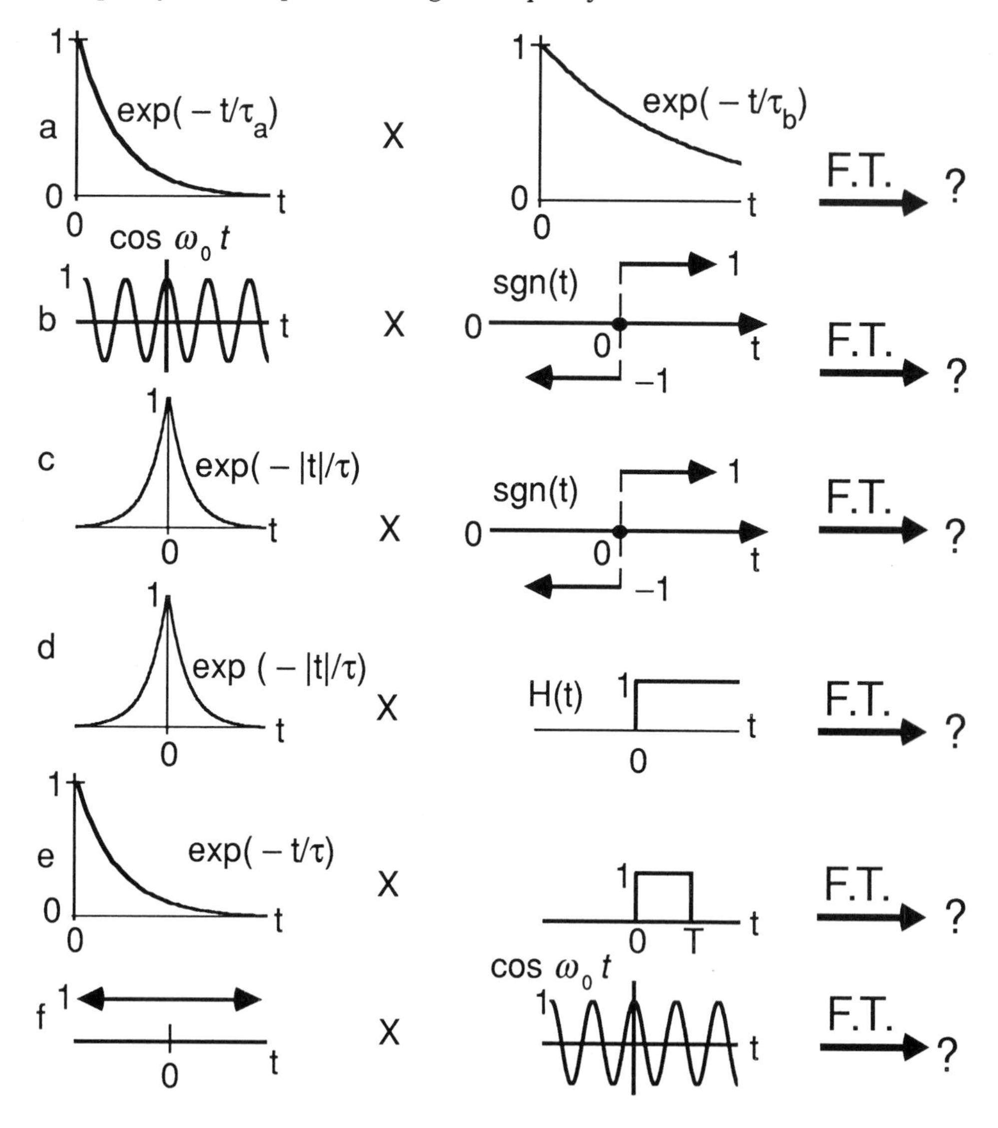

2.9 A good example of apodization used to change the shape of a spectral peak is the Noest-Kort function (A. J. Noest & C. W. F. Kort, *Computers & Chem.* **1982**, *6*, 115-119),

$$f(t) = \frac{1}{2} \exp\left(\frac{(t/\tau) - (t/\tau)^2}{4 \log_e(2)}\right) [\, 0.430 + 0.843 \cos(\pi t) - 0.239 \cos(2\pi t)$$

$$+ 0.043 \cos(5\pi t) - 0.602\, U\,],$$

$$\text{with } U = 1 - 4t \qquad \text{for } 0 < t < T/4$$

$$= 3 - 4t \qquad \text{for } 3T/4 < t < T$$

$$= 0 \qquad \text{elsewhere}$$

Write a short computer program to sketch $f(t)$ *vs.* t. Then solve analytically for $\mathrm{Re}[F(\omega)]$ and $\mathrm{Im}[F(\omega)]$, and write another short program to plot $\mathrm{Re}[F(\omega)]$, $\mathrm{Im}[F(\omega)]$,, and $M(\omega)$ *vs.* ω. You should find that the magnitude-mode spectrum is a broad and very flat peak. The Noest-Kort apodization is therefore useful for determining relative peak heights in discrete FT spectra (see next chapter), when there are relatively few spectral points per spectral peak width, so that the largest-magnitude (discrete) data point may not happen to fall exactly at the maximum of the (analog) peak shape.

2.10 The Hilbert transform relation

$$D(\omega) = \frac{1}{\pi} \int_{-\infty}^{+\infty} \frac{A(\omega')}{\omega - \omega'}\, d\omega \qquad (1.26)$$

between absorption and dispersion spectra can be thought of as the convolution between the two functions, $A(\omega)$ and $(1/\omega)$. Sketch the Fourier transforms of these two functions [the Fourier transform of $\mathrm{sgn}(t)$ is $-2i/\omega$]. Then apply the convolution theorem graphically to obtain $D(\omega)$ corresponding to a Lorentzian absorption-mode $A(\omega)$.

2.11 For each of the time-domain apodization functions in Figure 2.14, compute (see Appendix A) the corresponding frequency-domain spectra. For each function, determine the frequency-domain line width formula shown at the right in Figure 2.14.

2.12 For a time-domain sinusoidal signal,

$$f(t) = \exp(i\,\omega_0 t)\,, \qquad 0 \le t \le T,$$

we have already computed the cosine and sine Fourier transforms, $C(\omega)$ and $S(\omega)$, in Chapter 2.2.2.

(a) Compute (and sketch as a function of ω) the cosine and sine Fourier transforms of the time-delayed sinusoid,

$$f(t) = \exp[i\,\omega_0 (t + t_0)]$$

(b) Compare your result to that predicted by the shift theorem (Eq. 2.33).

(c) Express the result from (a) in terms of the original $A(\omega)$ and $D(\omega)$ of the undelayed time-domain signal—compare to Eqs. 2.35.

(d) Work backward from (c) to recover the desired "phased" spectra, $A(\omega)$ and $D(\omega)$, from the cosine and sine Fourier transforms computed in (a), and compare your result to Eqs. 2.37. This problem demonstrates the origin of (and correction for) linear phase shifts in FT spectroscopy.

2.13 Show that the effect of a constant phase shift, $\phi(\omega)$ = independent of ω, is to rotate a DISPA (dispersion-vs.-absorption) plot about its origin by ϕ radians (see Figure 2.17).

Note: The following problems are mathematical exercises which establish several important properties of Fourier transforms. The reader need not solve these problems in order to use Fourier transforms, but may find it useful to refer to the various results.

2.14 Inverse Fourier transform

From the following general definitions of the F.T., inverse F.T. and Appendix B, prove that the inverse F.T. of an F.T. produces the original function.

$$\mathbf{F}(y) = \int_{-\infty}^{+\infty} \boldsymbol{f}(x)\, \exp(-i2\pi y x)\; dx; \qquad \text{i.e., } \mathbf{F}(y) = \text{F.T. of } \boldsymbol{f}(x)$$

$$\boldsymbol{f}(x) = \int_{-\infty}^{+\infty} \mathbf{F}(y)\, \exp(-i2\pi x y)\; dy = \text{definition of inverse F.T.}$$

In other words, show that $(\text{F.T.})^{-1}$ of the F.T. of $\boldsymbol{f}(x)$ gives back $\boldsymbol{f}(x)$.

2.15 Linearity

If a and b are constants, show that

F.T. of $[a\boldsymbol{f}(x) + b\,\mathbf{h}(x)] = a\,[\text{F.T. of } \boldsymbol{f}(x)] + b\,[\text{F.T. of } \mathbf{h}(x)]$

2.16 Scaling

If $\boldsymbol{F}(y)$ is the F.T. of $\boldsymbol{f}(x)$, and a is a constant, show that

$$\boldsymbol{f}(x/a) \overset{\text{F.T.}}{\rightarrow} |a|\, \boldsymbol{F}(a\,y)$$

2.17 Shift/phase relation

If $\boldsymbol{F}(y)$ is the F.T. of $\boldsymbol{f}(x)$, and x_0 is a constant, show that the F.T. of $\mathbf{f}(x - x_0)$ is: $\exp(-i\,2\pi x_0\, y)\, \boldsymbol{F}(y)$. In other words, a scale (e.g., time) shift in one (e.g., time-) domain corresponds to a phase shift in the other (e.g., frequency-) domain.

2.18 Complex conjugate

If $\boldsymbol{F}(y)$ is the F.T. of $\boldsymbol{f}(x)$, show that

$$\boldsymbol{f}^*(x) \overset{\text{F.T.}}{\rightarrow} \boldsymbol{F}^*(-y).$$

2.19 Fourier transform of a convolution of two functions

Given that $\boldsymbol{f}(x)$ is the convolution of $\boldsymbol{e}(x)$ and $\boldsymbol{h}(x)$,

$$\boldsymbol{f}(x) = \boldsymbol{e}(x) * \boldsymbol{h}(x),$$

and that the F.T.'s of $\boldsymbol{f}(x)$, $\boldsymbol{h}(x)$, and $\boldsymbol{e}(x)$ are $\boldsymbol{F}(y)$, $\boldsymbol{H}(y)$, and $\boldsymbol{E}(y)$,

(a) show that

$$\boldsymbol{F}(y) = \boldsymbol{E}(y)\, \boldsymbol{H}(y)$$

(b) Now use this result to show that the F.T. of an isosceles triangle is a sinc^2 function. (Recall from Chapter 2 that an isosceles triangle can be generated by convolving two identical rectangles, and that the F.T. of a rectangle is a sinc function.)

2.20 Derivatives and integrals

Show that the F.T. of the k'th *derivative* of $\boldsymbol{f}(x)$ is $(i\,2\pi y)^k\, \boldsymbol{F}(y)$, in which $\boldsymbol{F}(y)$ is the F.T. of $\boldsymbol{f}(x)$. In other words, if

$$\boldsymbol{f}(x) \underset{\text{inverse F.T.}}{\overset{\text{F.T.}}{\rightleftarrows}} \boldsymbol{F}(y)\ ,$$

then

$$\frac{d}{dx^k}\, \boldsymbol{f}(x) \underset{\text{inverse F.T.}}{\overset{\text{F.T.}}{\rightleftarrows}} (i\,2\pi\, y)^k\, \boldsymbol{F}(y)$$

From the $k = -1$ case, it follows that the F.T. of the *integral* of $\boldsymbol{f}(x)$ is simply $[1/(i2\pi y)]\ \boldsymbol{F}(y) + [\boldsymbol{F}(0)/2]\ \delta(y)$. [See Appendix B for a discussion of the delta-functional, $\delta(y)$.]

2.21 Negative frequencies

Show that in order to recover a time-domain *real* function, $f(t)$, from its frequency-domain spectrum, $\boldsymbol{F}(\nu)$, only positive-frequency information is necessary.

Hint: Consider a complex time-domain function, $\mathrm{Re}[\boldsymbol{f}(t)] + i\ \mathrm{Im}[\boldsymbol{f}(t)]$, and its complex Fourier transform, $\mathrm{Re}[\boldsymbol{F}(\nu)] + i\ \mathrm{Im}[\boldsymbol{F}(\nu)]$.

2.22 Convolution with a Dirac δ-functional

From the definition of convolution (Eq. 2.20) and the properties of the Dirac δ-function (Appendix B), show that

$$f(x) \ast \delta(x) = f(x).$$

Solutions to Problems

2.1., 2.2, 2.3, 2.4, 2.6. See Appendix D.

2.5. See Figure 4.7.

2.7.

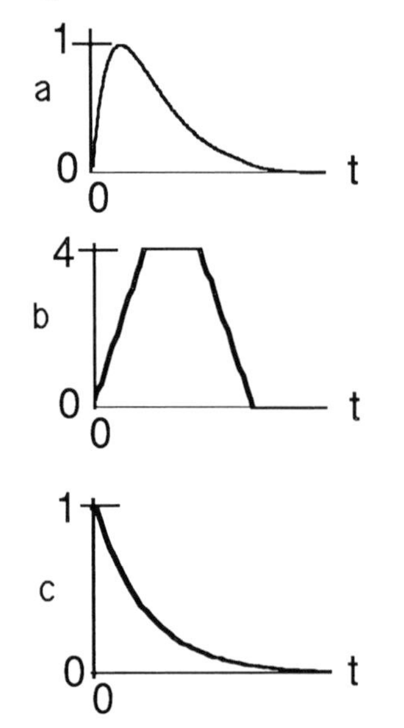

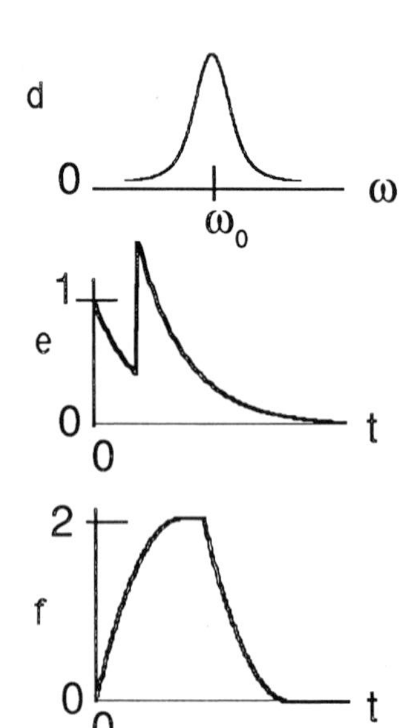

2.8.

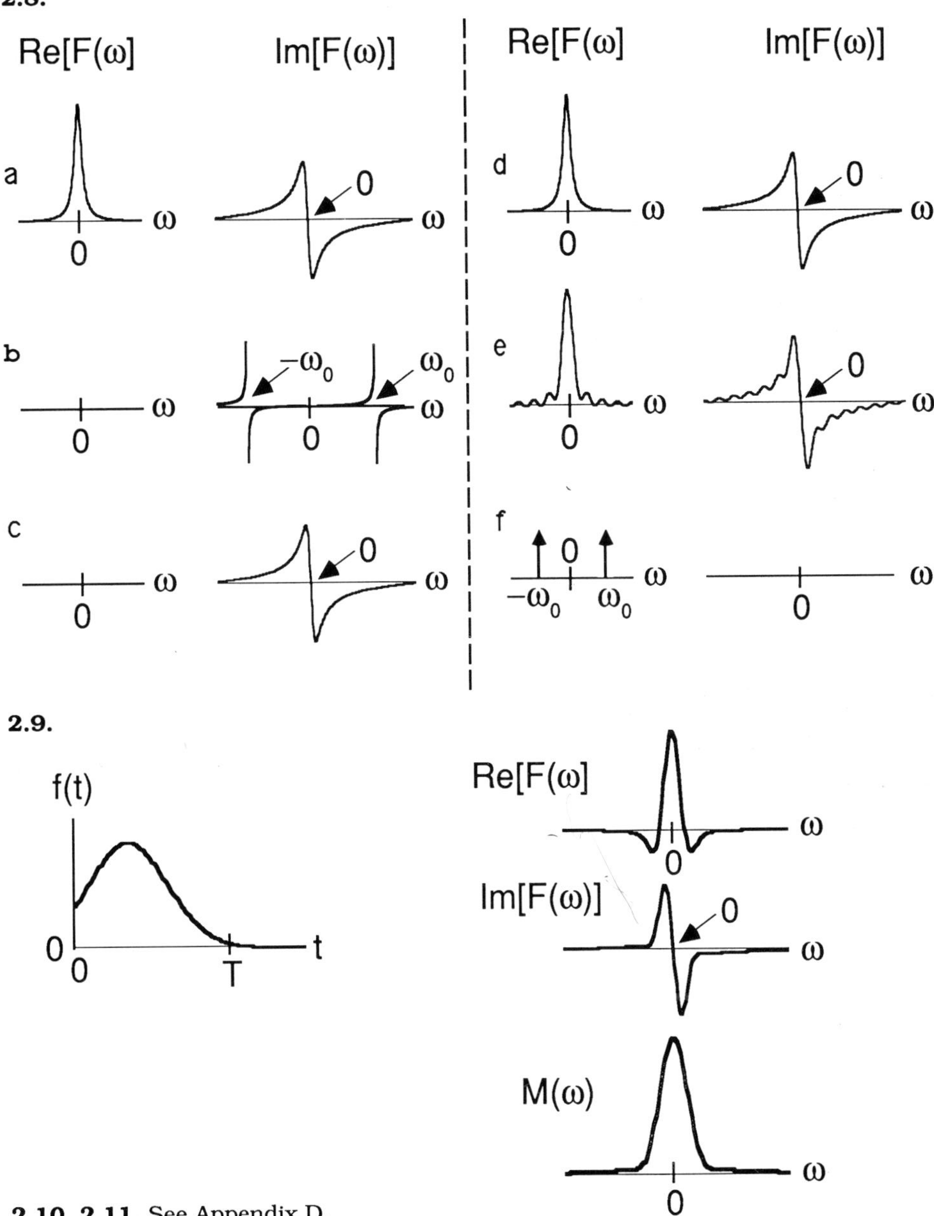

2.10, 2.11. See Appendix D.

2.12. See Figure 2.17, and Eqs. 2.35 and 2.37.

2.13. The plotted points should trace out the bottom right plot of Figure 2.17.

2.14 $$\mathrm{FT}^{-1}\,[\boldsymbol{F}(y)] = \int_{-\infty}^{+\infty}\int_{-\infty}^{+\infty} f(z)\,\exp(-i\,2\pi yz)\,dz\,\exp(i\,2\pi xy)\,dy$$

$$= \int_{-\infty}^{+\infty} f(z) \int_{-\infty}^{+\infty} \exp[-i\,2\pi y\,(z-x)]\,dy\,dz$$

But remember (see Appendix B) that

$$\int_{-\infty}^{+\infty} \exp[-i\,2\pi y\,(z-x)]\,dy = \delta(z-x)$$

Therefore,

$$\mathrm{FT}^{-1}\,[\boldsymbol{F}(y)] = \int_{-\infty}^{+\infty} f(z)\,\delta(z-x)\,dz = f(x) \qquad \textbf{Q.E.D.}$$

Note that if a factor, 2π, is used in the F.T. operator, no factor of $(2\pi)^{-1}$ appears in the inverse F.T.

2.15 $$\mathrm{FT}\,[a\boldsymbol{f}(x) + b\,\boldsymbol{h}(x)] = \int_{-\infty}^{+\infty} [a\,\boldsymbol{f}(x) + b\,\boldsymbol{h}(x)]\,\exp(-i\,2\pi xy)\,dx$$

$$= a\int_{-\infty}^{+\infty} \boldsymbol{f}(x)\,\exp(-i\,2\pi xy)\,dx + b\int_{-\infty}^{+\infty} \boldsymbol{h}(x)\,\exp(-i\,2\pi xy)\,dx$$

$$= a\,\boldsymbol{F}(y) + b\,\boldsymbol{H}(y) \qquad \textbf{Q.E.D.}$$

2.16 $$\mathrm{FT}\,[(\boldsymbol{f}(x/a)] = \int_{-\infty}^{+\infty} \boldsymbol{f}(x/a)\,\exp(-i\,2\pi xy)\,dx;$$ Let $z = x/a$. Then

$$= |a| \int_{-\infty}^{+\infty} \boldsymbol{f}(z)\,\exp[-i\,2\pi(a\,z)y]\,dz = |a| \int_{-\infty}^{+\infty} \boldsymbol{f}(z)\,\exp[-i\,2\pi\,z\,(ay)]\,dz$$

$$= |a|\;\boldsymbol{F}(a\,y) \qquad \textbf{Q.E.D.}$$

2.17 $$\mathrm{FT}\,[\boldsymbol{f}(x - x_0)] = \int_{-\infty}^{+\infty} \boldsymbol{f}(x - x_0)\, \exp[-i\,2\pi\, x\, y]\, dx\,; \quad \text{Let } z = (x - x_0). \text{ Then}$$

$$= \int_{-\infty}^{+\infty} \boldsymbol{f}(z)\, \exp[-i\,2\pi\,(z + x_0)\, y]\, dz$$

$$= \exp(-i\,2\pi\, x_0\, y) \int_{-\infty}^{+\infty} \boldsymbol{f}(z)\, \exp(-i2\,\pi\, z\, y)\, dz$$

$$= \exp(-i\,2\pi\, x_0\, y)\, \boldsymbol{F}(y)$$ **Q.E.D.**

Thus, the Fourier transform of (e.g.) a *time-shifted* time-domain signal function is just the spectrum of the *unshifted* time-domain signal multiplied by an exponential factor whose *phase* varies linearly with frequency.

2.18 $$\mathrm{FT}\,[\boldsymbol{f}^*(x)] = \int_{-\infty}^{+\infty} \boldsymbol{f}^*(x)\, \exp[-i\,2\pi\, x\, y]\, dx$$

The reader can quickly show that for any two complex numbers, a and b, $a^* \, b = (a\, b^*)^*$. Therefore, the original expression becomes

$$\mathrm{FT}\,[(\boldsymbol{f}^*(x)] = \left| \int_{-\infty}^{+\infty} \boldsymbol{f}(x)\, \exp(+i\,2\pi x\, y)\, dx \right|^*$$

$$= \left| \int_{-\infty}^{+\infty} \boldsymbol{f}(x)\, \exp[-i\,2\pi\, x\,(-y)]\, dx \right|^*$$

$$= \boldsymbol{F}^*(-y)$$ **Q.E.D.**

2.19 $$\boldsymbol{F}(y) = \mathrm{FT}\,[\boldsymbol{e}(x) \ast \boldsymbol{h}(x)] = \mathrm{FT}\left(\int_{-\infty}^{+\infty} \boldsymbol{e}(x')\, \boldsymbol{h}(x - x')\, dx' \right)$$

$$= \int_{-\infty}^{+\infty} \int_{-\infty}^{+\infty} \boldsymbol{e}(x')\, \boldsymbol{h}(x - x')\, \exp(-i\,2\pi\, y\, x)\, dx'\, dx$$

which, on application of the shift theorem, becomes

$$= \int_{-\infty}^{+\infty} \boldsymbol{e}(x')\, \boldsymbol{H}(y)\, \exp(-i\,2\pi\, y\, x')\, dx' = \boldsymbol{H}(y) \int_{-\infty}^{+\infty} \boldsymbol{e}(x')\, \exp(-i\,2\pi\, y\, x')\, dx'$$

$$= \boldsymbol{E}(y)\, \boldsymbol{H}(y)$$ **Q.E.D.**

2.20 Begin from the definition of the (inverse) Fourier transform of $\boldsymbol{F}(y)$, and differentiate k times on both sides of the equation:

$$\boldsymbol{f}(x) = \int_{-\infty}^{+\infty} \boldsymbol{F}(y) \exp(i 2\pi x y)\, dy \tag{2.20.1}$$

$$\frac{d^k}{dx^k}[\boldsymbol{f}(x)] = \frac{d^k}{dx^k}\left[\int_{-\infty}^{+\infty} \boldsymbol{F}(y) \exp(i 2\pi x y)\, dy\right]$$

$$\frac{d^k}{dx^k}[\boldsymbol{f}(x)] = \int_{-\infty}^{+\infty} (i 2\pi y)^k\, \boldsymbol{F}(y) \exp(i 2\pi x y)\, dy \tag{2.20.2}$$

Comparison of Eqs. 2.20.1 and 2.20.2 shows that $d^k[\boldsymbol{f}(x)]/dx^k$ and $(i 2\pi y)^k$ $\boldsymbol{F}(y)$ form a Fourier transform pair. Thus, the FT of Eq. 2.20.2 yields:

$$\mathrm{FT}\left(\frac{d^k}{dx^k}[\boldsymbol{f}(x)]\right) = (i 2\pi y)^k \boldsymbol{F}(y) \qquad \textbf{Q.E.D.}$$

2.21 This time, begin from the inverse FT representation of the complex time-domain function, $\boldsymbol{f}(t)$.

$$\boldsymbol{f}(t) = \int_{-\infty}^{+\infty} \boldsymbol{F}(\nu) \exp(i 2\pi \nu t)\, d\nu$$

$$= \int_{-\infty}^{+\infty} \Big(\mathrm{Re}[\boldsymbol{F}(\nu)] + i\, \mathrm{Im}[\boldsymbol{F}(\nu)]\Big)\, [\cos \nu t + i \sin \nu t]\, d\nu$$

Next, separate $\boldsymbol{f}(t)$ into its real and imaginary parts:

$$\mathrm{Re}[\boldsymbol{f}(t)] = \int_{-\infty}^{+\infty} \Big(\mathrm{Re}[\boldsymbol{F}(\nu)] \cos \nu t\Big) - \Big(\mathrm{Im}[\boldsymbol{F}(\nu)] \sin \nu t\Big)\, d\nu \tag{2.21.1}$$

$$\mathrm{Im}[\boldsymbol{f}(t)] = \int_{-\infty}^{+\infty} \Big(\mathrm{Re}[\boldsymbol{F}(\nu)] \sin \nu t\Big) + \Big(\mathrm{Im}[\boldsymbol{F}(\nu)] \cos \nu t\Big)\, d\nu \tag{2.21.2}$$

Now, if $\boldsymbol{f}(t)$ is real, then $\mathrm{Im}[\boldsymbol{f}(t)] = 0$. Since sine is an odd function and cosine is an even function, Eq. 2.21.2 is satisfied (see Eqs. 2.5 and 2.6) if $\mathrm{Re}[\boldsymbol{F}(\nu)$ is even and $\mathrm{Im}[\boldsymbol{F}(\nu)]$ is odd. Thus, the integrals in Eq. 2.21.1 become:

$$\mathrm{Re}[\boldsymbol{f}(t)] = 2 \int_{0}^{+\infty} \Big(\mathrm{Re}[\boldsymbol{F}(\nu)] \cos \nu t\Big) - \Big(\mathrm{Im}[\boldsymbol{F}(\nu)] \sin \nu t\Big)\, d\nu$$

In other words, only the positive-frequency spectrum is required in order to recover a (real) time-domain function, $f(t)$ from its (complex) spectrum, $\boldsymbol{F}(\nu)$, by means of inverse Fourier transformation.

CHAPTER 3

Fourier transforms of digital (discrete) waveforms

3.1 Numerical algorithms

3.1.1 Fourier series and fast Fourier transform

Experimental Fourier transform data reduction is performed with digital computers. Since computers are inherently designed to deal with *discrete* (rather than *continuous*) data, we must extend our prior discussion of Fourier transforms of *continuous* time-domain signals to the digitized (discrete) time-domain response to some sort of brief [compared to relaxation time(s)] excitation. Mathematically, we can represent such a discrete time-domain signal as the product of the analog (continuous) system response, $f(t)$, with a "shah" ("comb" or "sampling") function, $III(t)$, to give a sequence of time-domain numbers (see Figure 3.1). $III(t)$ is simply a series of equally spaced δ-functions extending to ± ∞ (see Appendix D).

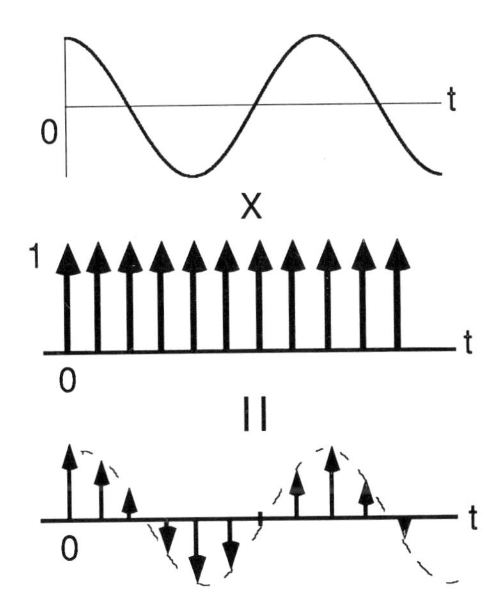

Figure 3.1 Discrete time-domain waveform (bottom), obtained by sampling (middle) a continuous time-domain response (top).

The key feature of such a measurement is that the time-domain sampling is conducted at *N equally spaced* "lag" times, t_n—we shall return to this point in Chapter 4. T is the total time-domain sampling period.

$$t_n = \frac{nT}{N}, \qquad n = 0, 1, 2, \cdots, N-1 \tag{3.1}$$

The sampling interval, Δt,

$$\Delta t = \frac{T}{N} \tag{3.2}$$

is also known as the "dwell time" or "dwell period", and it defines (see below) the highest frequency which can be correctly represented by Fourier analysis.

The time-domain sampling thus yields the series of numbers,

$$f(t_0), f(t_1), f(t_2), \cdots, f(t_{N-1}) \tag{3.3}$$

It turns out that a Fourier relation between time- and frequency-domain representations still holds, but as a (Fourier) *series* of numbers rather than the *integrals* of Eqs. 1.36, 1.37 and 1.39. The recipe for generating the discrete frequency-domain spectrum, $F(\nu_m)$,

$$F(\nu_m) = \frac{m}{T} \qquad m = -N/2, \cdots, N/2 \tag{3.4}$$

is given to within a scale factor (see Chapter 3.6) by

$$F(\nu_m) = \sum_{n=0}^{N-1} F_{nm}\, f(t_n)\,, \qquad m = -N/2, \cdots, N/2 \tag{3.5a}$$

$$= F_{0m}\, f(t_0) + F_{1m}\, f(t_1) + \cdots + F_{N-1,m}\, f(t_{N-1})$$

By combining Eqs. 3.1 and 3.4 (remember that $\omega_m = 2\pi\nu_m$), we obtain

$$F_{nm} = \exp(-i\omega_m t_n) = \exp(-i\,2\pi nm/N) \tag{3.5b}$$

At this stage, we digress briefly to note that the discrete Fourier transformation of Eqs. 3.5 converts a series of N (complex) time-domain data into another series of N (complex) frequency-domain data, ranging in frequency from $-\nu_m = -m/T$ to $\nu_m = m/T$. [For the special (and important) case that the N time-domain data points are real, the discrete FT still generates N complex frequency-domain data points.] However, half of those data points are redundant. Therefore, it is usual to display only *half* of the complex frequency-domain data obtained by (complex) FT of real data, as discussed in section 4.3.2 (quadrature detection), and in Appendix C.

For example, a 4-point time-domain sampling would lead to the following Fourier series.

$$\begin{aligned}
F(\nu_0) &= 1 \cdot f(t_0) + 1 \cdot f(t_1) + 1 \cdot f(t_2) + 1 \cdot f(t_3) \\
F(\nu_1) &= 1 \cdot f(t_0) + (-i) \cdot f(t_1) + (-1) \cdot f(t_2) + i \cdot f(t_3) \\
F(\nu_2) &= 1 \cdot f(t_0) + (-1) \cdot f(t_1) + 1 \cdot f(t_2) + (-1) \cdot f(t_3) \\
F(\nu_3) &= 1 \cdot f(t_0) + i \cdot f(t_1) + (-1) \cdot f(t_2) + (-i) \cdot f(t_3)
\end{aligned} \tag{3.6}$$

Eq. 3.6 can be expressed more compactly in matrix notation (see Appendix A) as

$$\begin{pmatrix} F(\nu_0) \\ F(\nu_1) \\ F(\nu_2) \\ F(\nu_3) \end{pmatrix} = \begin{pmatrix} 1 & 1 & 1 & 1 \\ 1 & -i & -1 & i \\ 1 & -1 & 1 & -1 \\ 1 & i & -1 & -i \end{pmatrix} \begin{pmatrix} f(t_0) \\ f(t_1) \\ f(t_2) \\ f(t_3) \end{pmatrix} \tag{3.7}$$

in which the coefficient matrix can be thought of as a Fourier "code" for converting time-domain data into frequency-domain data.

The reader can quickly confirm (see Problems) that the *inverse* Fourier code matrix for the 4-point example is given by

$$\begin{pmatrix} f(t_0) \\ f(t_1) \\ f(t_2) \\ f(t_3) \end{pmatrix} = \begin{pmatrix} 1 & 1 & 1 & 1 \\ 1 & i & -1 & -i \\ 1 & -1 & 1 & -1 \\ 1 & -i & -1 & i \end{pmatrix} \begin{pmatrix} F(\nu_0) \\ F(\nu_1) \\ F(\nu_2) \\ F(\nu_3) \end{pmatrix} \tag{3.8}$$

In general, the inverse discrete Fourier transform is given by

$$f(t_n) = \sum_{m=0}^{N-1} F_{nm}^{-1} \, F(\nu_m) \ , \tag{3.9a}$$

in which

$$F_{nm}^{-1} = (1/N) \exp(i\,2\pi nm/N) \tag{3.9b}$$

In Equations 3.5 and 3.9, there is no restriction on the integer, N. For example, the reader can easily determine (see Problems) the forward and inverse Fourier "code" matrices for $N = 3$ to $N = 8$. When computational speed is not important (as when $N < 1{,}000$ or so), modern microcomputers can produce Fourier transforms from Eqs. 3.5 and 3.9 in a few minutes. However, for spectroscopic data sets which may range to $N = 500{,}000$ or more, the computation time for Eqs. 3.5 or 3.9 can become prohibitively long, because Eq. 3.5 (for example) requires N multiplications and $N-1$ additions to compute each of the N frequency-domain spectral data points, $F(\nu_m)$.

Fortunately (see Appendix C), there is considerable redundancy in the Fourier code, as evident even in the simple 4-point example. For example, from Eqs. 3.5b and 3.9b, we see that the forward (or inverse) Fourier code matrix is symmetric about its diagonal:

$$F_{nm} = F_{mn} \ ; \qquad F_{n,m}^{-1} = F_{m,n}^{-1} \tag{3.10}$$

For larger N, even more redundancy becomes apparent (see Problems for $N = 8$). By exploiting such redundancy, Cooley and Tukey were able to reduce the number of multiplications from N^2 to $N \log_2 N$ for the special case that the number of time-domain data points is a power of two:

$$N = 2^p, \quad p = 2, 3, 4, \cdots \tag{3.11}$$

The rapid development of FT/NMR and FT/IR spectroscopies dating from about 1966 is a direct consequence of the 1965 Cooley-Tukey fast Fourier transform (FFT) algorithm, which has since become the most highly cited paper in all of mathematics!

The FFT algorithm is not unique, and a number of variants is now available. The best-choice algorithm depends on (for example) the relative speed of addition *vs.* multiplication for a particular computer's hardware. Representative FFT Basic and Fortran algorithms are listed in Appendix C. Recent improvements have further increased the speed of the FFT (particularly for the special case that the input data is mathematically real—see Hartley transform in Appendix C), and have extended the computational speed advantage to values of N which can be constructed from the product of prime factors (i.e., other than 2). The interested reader can pursue the details of the FFT algorithm in Appendix C and in the publications cited at the end of this chapter. Hardwired FFT processors can now perform a 1024-point FFT in ~1 ms, and optical-electronic combinations can perform even faster.

3.1.2 Discrete convolution and fast algorithm

Just as we proceeded from the analog to the discrete Fourier transform, we can express the discrete convolution, $f(t)$,

$$f(t) = [f(t_0), f(t_1), f(t_2), \ldots, f(t_{2N-1})] \tag{3.12}$$

of two discrete functions, $h(t)$ and $e(t)$, each of which is defined at a series of equally spaced times, $t_{i+1} = t_i + \Delta t$, $\quad 0 \leq i \leq (N-2)$:

$$e(t) = [e(t_0), e(t_1), e(t_2), \cdots, e(t_{N-1})] \tag{3.13}$$

and
$$h(t) = [h(t_0), h(t_1), h(t_2), \cdots, h(t_{N-1})] \tag{3.14}$$

according to

$$f(t_i) = \sum_{j=0}^{N-1} h(t_j) \cdot e(t_{i-j}), \text{ in which } 0 \leq (i-j) \leq N-1 \tag{3.15a}$$

e.g.,
$$\begin{aligned} f(t_0) &= h(t_0) \cdot e(t_0) \\ f(t_1) &= h(t_0) \cdot e(t_1) + h(t_1) \cdot e(t_0) \\ f(t_2) &= h(t_0) \cdot e(t_2) + h(t_1) \cdot e(t_1) + h(t_2) \cdot e(t_0) \\ &\vdots \\ f(t_{2N-1}) &= h(t_{N-1}) \cdot e(t_{N-1}) \end{aligned} \tag{3.15b}$$

In practice, the convolution described by Eq. 3.15 is easy to use. One begins by writing the first sequence, $h(t_i)$ on a piece of paper. The second sequence is then reversed left-to-right, and written on a second (movable) piece of paper. The second paper strip is then displaced one position at a time with respect to the first piece of paper, while the sum of the point-by-point products of the sequences at each position is computed. This process is illustrated in Figure 3.2 for two short (three-point) sequences.

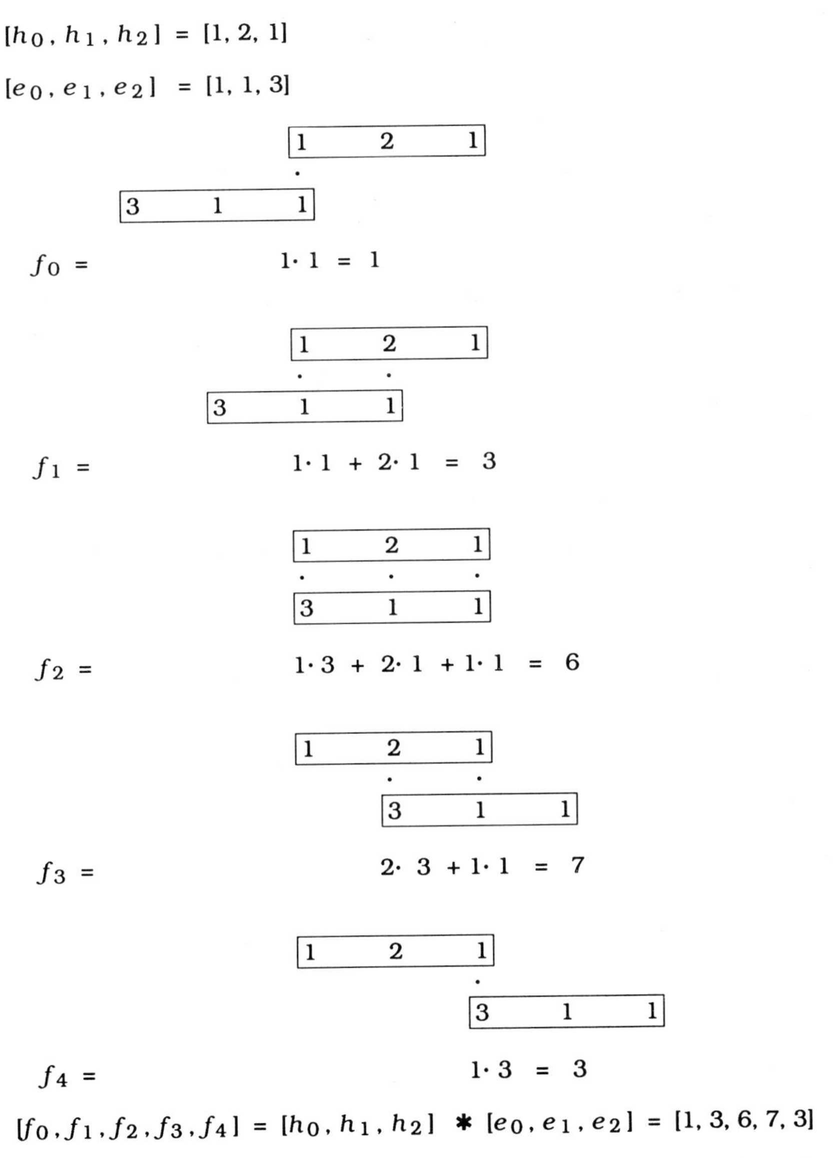

$[h_0, h_1, h_2] = [1, 2, 1]$
$[e_0, e_1, e_2] = [1, 1, 3]$
1 2 1
3 1 1
$f_0 = \quad 1 \cdot 1 = 1$
$f_1 = \quad 1 \cdot 1 + 2 \cdot 1 = 3$
$f_2 = \quad 1 \cdot 3 + 2 \cdot 1 + 1 \cdot 1 = 6$
$f_3 = \quad 2 \cdot 3 + 1 \cdot 1 = 7$
$f_4 = \quad 1 \cdot 3 = 3$
$[f_0, f_1, f_2, f_3, f_4] = [h_0, h_1, h_2] * [e_0, e_1, e_2] = [1, 3, 6, 7, 3]$

Figure 3.2 Convolution of two discrete $N = 3$-point sequences, h and e, to give the resultant $(2N-1) = 5$-point sequence, f. See text for details.

In most examples of interest in FT spectroscopy, both of the convolved data sets will have the *same* number of points (namely, N), to give a convolution sequence of $(2N-1)$ points, as in Figure 3.2. Practice examples are provided in the Problems. More generally, the convolution of two sequences of N_a and N_b points will yield a resultant sequence of $(N_a + N_b - 1)$ points.

Just as for the direct discrete Fourier transform (Eq. 3.5a), the direct computation of the convolution of two large (say, >1,024-point) data sets can be time-consuming, even for a microcomputer. Therefore, it is faster to use the convolution theorem to perform discrete fast Fourier transforms on each of the two convolved functions, followed by N complex point-by-point multiplications, and then an inverse discrete FFT to produce the desired convolution.

3.2 Nyquist criterion

In dealing with discrete data, it is important to realize that, even in the absence of noise, we no longer "know" what a time-domain waveform is, except at the instants of measurement (sampling). Fortunately, it can be shown that a discrete Fourier transform still yields the correct spectral frequency and amplitude at that frequency, provided that we have "enough" time-domain data points per unit time. The Nyquist criterion tells us how many points are "enough".

Specifically, the Nyquist criterion simply requires that it is necessary to sample a sinusoid at least *twice per cycle* in order to know its true frequency. In other words, we need to know that the signal went up and down once during one of its cycles. Figure 3.3 shows what happens when the sampling rate is smaller than the Nyquist limit. Because our knowledge of the time-domain signal is limited to its sampled values, we (and our discrete Fourier transform algorithm) must conclude that the signal frequency is *smaller* than its true value.

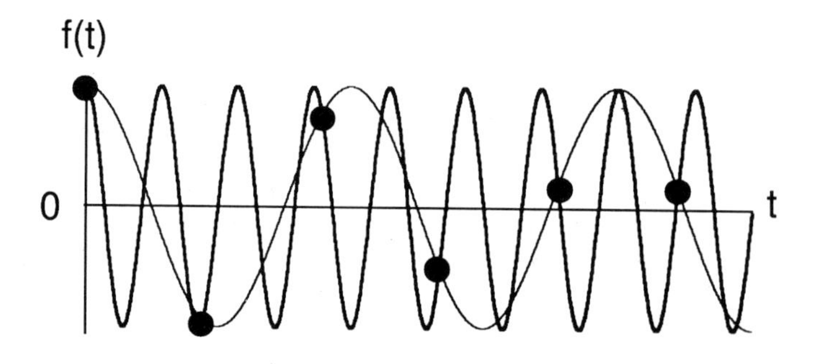

Figure 3.3 Effect of sampling rate on the apparent frequency of a time-domain sinusoid. "Under"–sampling (●) at a rate less than the Nyquist limit of two points per cycle yields a sinusoid whose *apparent* frequency (thin line, inferred from the sampled values) is *lower* than its *true* (analog) frequency (thick line).

3.2.1 Foldover (aliasing)

The effect shown in Figure 3.3 is variously known as "aliasing" (since the apparent frequency is false) or "foldover" (see Figure 3.4). As a familiar example, consider the apparent movement of spoked stagecoach wheels in a motion picture. The movie camera samples at a fixed number of frames per second. As the stagecoach accelerates from a standing start, each wheel first appears to move faster and faster up to the Nyquist limit (i.e., two film frames during the time for a given spoke to rotate by one spoke position). The wheel then appears to slow down as the stagecoach continues to accelerate. When the wheel moves so fast that a given spoke rotates by exactly one spoke position during the interval between successive film frames (i.e., twice the Nyquist limit), then the wheel appears to stop (i.e., zero frequency). A stroboscope produces the same effect.

Figure 3.4 shows how to predict the apparent signal frequency, $\nu_{apparent}$, if its true frequency, ν_{true}, and the sampling frequency, $\nu_{sampling}$, are known. For a given sampling rate, the highest signal frequency that can be represented by discrete Fourier transform of a sampled time-domain signal is the Nyquist frequency, $\nu_{Nyquist}$:

$$\nu_{Nyquist} = \frac{\nu_{sampling}}{2} \qquad (3.16)$$

Signals whose frequencies are smaller than $\nu_{Nyquist}$ will show up at their correct positions in the FT spectrum. Signals whose frequencies are larger than the Nyquist frequency will be "folded"-back to lie between 0 and $\nu_{Nyquist}$, as shown in Figure 3.4. Specifically,

$$\nu_{apparent} = |\nu_{true} - n\,\nu_{sampling}|, \qquad (3.17)$$

in which n is $(\nu_{true}/\nu_{Nyquist})$, rounded off to the next lowest integer. In particular, for $n = 0$ (i.e., $\nu_{true} < \nu_{Nyquist}$),

$$\nu_{apparent} = \nu_{true} \quad \text{for} \quad \nu_{true} \leq \nu_{Nyquist} \qquad (3.18)$$

The next obvious question is how to distinguish "true" from "folded-over" peaks in an experimental spectrum. From Equations 3.17 and 3.18, it should be clear that the position of a folded-over peak will vary with $\nu_{Nyquist}$, which is in turn determined by the sampling rate according to Eq. 3.16. Thus, one need simply compare spectra produced by Fourier transformation of time-domain data sets acquired at two different sampling frequencies—folded-over peaks will shift to different apparent frequency, but true-frequency peaks will stay at the same apparent frequency.

Of course, if the true frequencies of the spectral signals are known, then it may be desirable to *undersample* deliberately, in order to increase the number of data points per peak width. For further examples of undersampling, see Chapter 7.4.1 and Chapter 9.3.2.1.

Although sampling at $2\nu_{Nyquist}$ Nyquist criterion suffices for accurate positioning of the FT spectral signal *frequencies*, visually accurate display of time-domain signal *magnitude* may require *oversampling* at a rate of ~$20\nu_{Nyquist}$. Oversampling (see Chapter 5.15) may also be useful for reducing "quantization" noise.

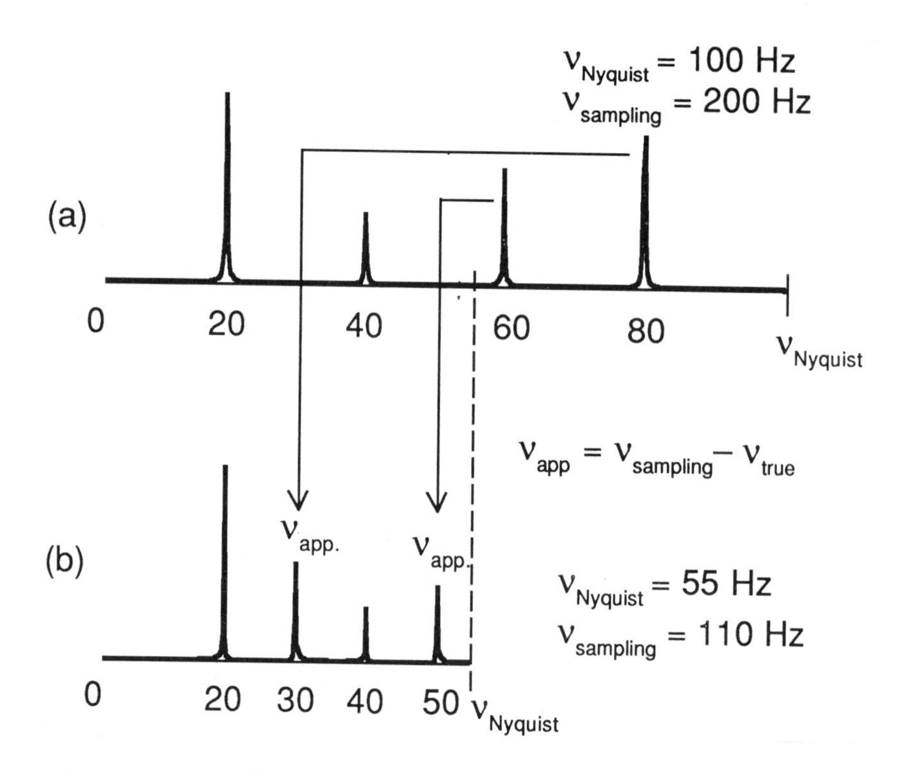

Figure 3.4 The foldover ("aliasing") effect of undersampling. (a) Hypothetical spectrum, with peaks located at their true frequencies. (b) Discrete FT of the time-domain signal corresponding to (a), with sampling and Nyquist frequencies as shown. The peaks in (b) have the correct relative magnitudes, but appear to be folded-back to smaller apparent frequencies. (See Eq. 3.17.)

3.2.2 Digital spectral resolution

The Nyquist criterion sets the *minimum* time-domain sampling (digitizing) rate, $\nu_{sampling}$, required to yield a frequency-domain spectrum with peaks at the correct frequencies. However, the *maximum* number of data points, N, that can be acquired is in turn limited by the size of the random access memory (RAM) in the host computer or its associated buffer memory. Therefore, the maximum data acquisition period, T, is constrained by Eq. 3.19:

$$\nu_{sampling}\, T = N \tag{3.19}$$

According to Eq. 3.1, the frequency-domain points in a discrete spectrum are spaced every $(1/T)$ Hz. Therefore, Eq. 3.19 limits the digital resolution in such a spectrum to a frequency-domain spacing of $(\nu_{sampling}/N)$ Hz/point. Finally, for a given spectral bandwidth (i.e., highest minus lowest spectral frequency), Eq. 3.16 leads to the final result:

$$\text{Freq-domain spacing (in Hz/point)} = 2\cdot(\text{spectral bandwidth in Hz})/N \tag{3.20}$$

Equation 3.20 poses an immediate dilemma. The *analog* frequency-domain peak line width (e.g., Lorentz or sinc line shape) varies inversely with data acquisition period, T. Thus, we would like to make T longer in order to produce narrower peaks and thus higher analog resolution. A longer T also gives higher *digital* resolution (i.e., more closely-spaced frequency-domain points). However, Eq. 3.20 limits T according to the desired bandwidth and available computer memory. Finally, the precision with which we can determine frequency-domain peak parameters (position, width, height) in the presence of noise depends on the number of data points per analog line width (see Chapter 5.1.3). For now (see Problems), we simply note that spectral peak width is determined by *digital* resolution for *slightly damped* time-domain signals, whereas peak width is determined by *analog* resolution for *highly damped* time-domain signals. Finally, both digital and analog resolution are enhanced by extending the data acquisition period as long as possible.

3.3 Vertical dynamic range

In the preceding discussion, we considered only *horizontal* (i.e., frequency) spectral resolution. In this section, we shall consider resolution in the *vertical* (amplitude) axis of a spectrum.

3.3.1 Fixed-point versus floating-point arithmetic

We must first realize that FFT data processors typically employ *fixed-point* arithmetic, in which each data point is represented by a binary "word" ranging in length from 16-32 bits. [An n-bit word is an n-digit number in binary notation—e.g., 12 in base (modulo) 10 becomes the four-bit number, 1100, in base (modulo) 2.] Moreover, the analog-to-digital converter (ADC) which samples the time-domain signal typically has an even more limited range of 8-16 bits per word.

In the absence of noise, time-domain *dynamic range* can be defined as the ratio of the largest signal to the smallest signal from the sample. For example, if the smallest observable signal corresponds to 1 bit, then the largest signal from an 8-bit/word ADC will be 2^8 = 256. Thus, the ADC immediately limits the available dynamic range for a single time-domain acquisition (to 256:1 for an 8-bit ADC).

One might think to add many time-domain data sets together, since a typical FT computer has a random-access-memory (RAM) word length of 16-32 bits: e.g., adding 2^8 scans together will produce a signal that is 2^8 times larger (8 bits longer) than in a single ADC data set. However, a little thought reveals that no matter how many time-domain data sets are added together, the *ratio* of the largest to smallest signal in the time-domain (and thus in the frequency-domain) will be the same (in the absence of noise). However, as we shall see in Chapter 5.1.2, signal-averaging (individually or in blocks) can increase dynamic range, provided that sufficient noise is present in the time-domain signal.

Even in the absence of noise, the FFT algorithm itself can distort the spectrum because of round-off errors during fixed-point computations, particularly when the spectrum exhibits very large dynamic range ($>10^4$:1). In such cases, *floating-point* arithmetic, in which each data point is represented in exponential notation (e.g., 3.57×10^3, expressed in binary notation, with say, an 8-bit argument and 8-bit exponent), can offer a major improvement by reducing round-off errors. Figure 3.5 shows a simulated example of reduction in round-off errors by use of floating-point in place of fixed-point arithmetic. Floating-point arithmetic becomes especially important for processors of limited word length (<20 bits/word) and/or for signals with very large dynamic range.

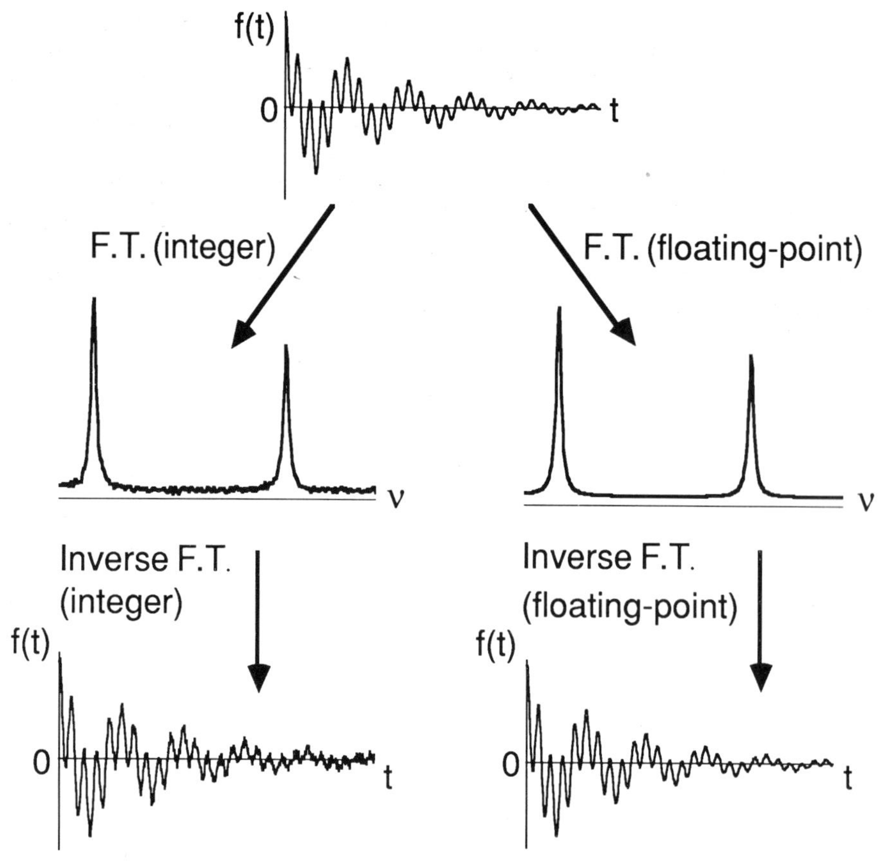

Figure 3.5 Comparison of FFT processing of the same simulated time-domain data, based on fixed-point (top) and floating-point (bottom) arithmetic. Note the reduction of artifacts in the floating-point case. (See Verdun & Marshall and Liu & Keneko references in Further Reading for quantitative treatment.)

3.3.2 Clipped data sets

An immediate problem with digital representation of data is the need to store a potentially vast number of points for a single experiment (e.g., $\geq 10^6$ points for a single GC/FT/ICR or two-dimensional FT/NMR experiment). Once the FFT computation has been performed, one could choose to store only certain segments of the spectrum (or peak heights and positions), but then the rest of the spectrum would be lost for future reference.

Fortunately, a little thought reveals that we do not need to digitize a sinusoidal time-domain signal with high (vertical) accuracy in order to determine its *frequency*. All we really need to know is whether the signal was positive or negative at a given instant, in order to count the time-domain oscillations and thus locate the frequency-domain peak position. Thus, we can "clip" each time-domain data point all the way down to 1 bit per time-domain word, FFT that data set, and still hope to locate the peaks in the spectrum.

For example, Figure 3.6 shows unclipped and clipped time-domain data sets and their corresponding frequency-domain spectra. In the absence of noise, such clipping produces severe harmonic and intermodulation distortion, much as from overdriving the speakers in an audio hi-fi system—i.e., additional peaks at multiples and combinations of the frequencies of the true peaks (see Chapter 4.4.). For clipped data, a Nyquist-like criterion requires that the number of time-domain zero-crossings be twice as large as the highest frequency in the spectrum. Thus, the presence of some noise actually reduces harmonic and intermodulation distortion, by increasing the number of zero-crossings in the clipped data set! Clipped data may find application for library storage of time-domain data sets, which can then be searched by comparison to a experimental clipped time-domain data set, without ever having to FFT the data to obtain a spectrum.

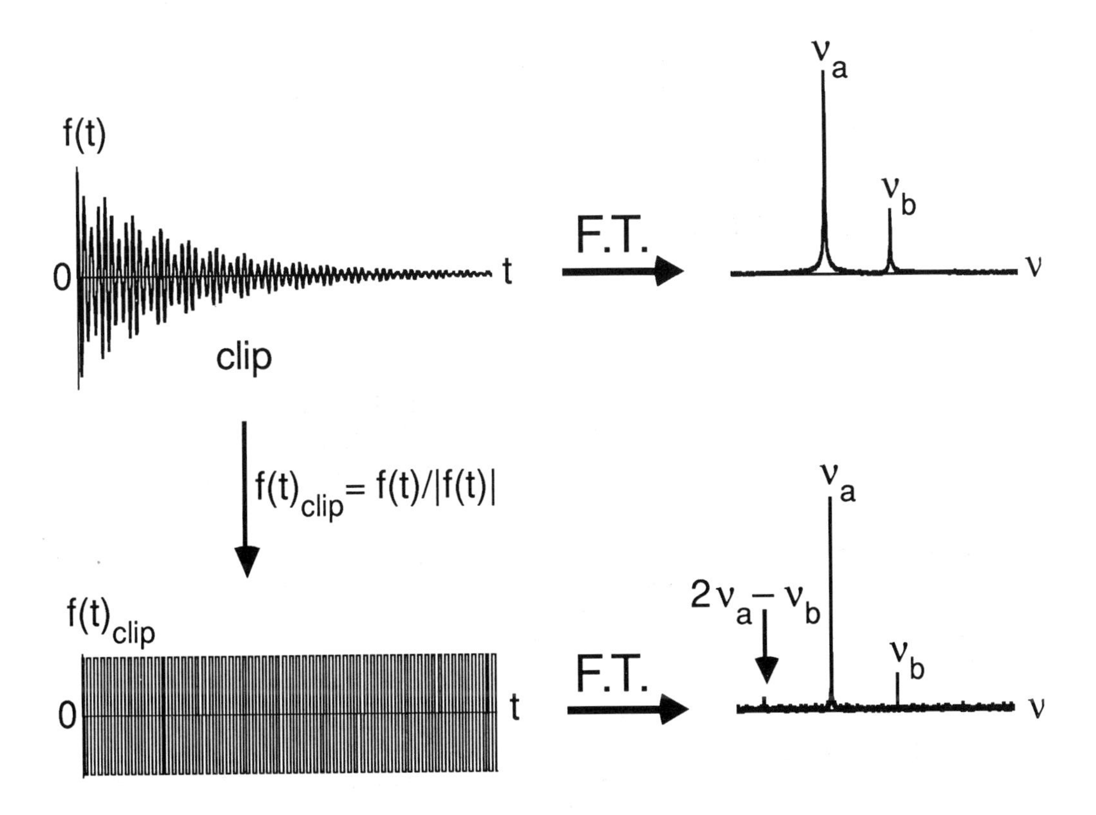

Figure 3.6 Unclipped (top left) and clipped (bottom left) simulated time-domain damped noisy sinusoids and their spectra (right) obtained by discrete FFT. Although 95% of the information (and required data storage space) has been removed by clipping (1-bit vs. 20-bit time-domain words), the two spectra are qualitatively highly similar, particularly for the larger peaks. A small intermodulation peak at $(2\nu_a - \nu_b)$ is barely visible.

3.4 Uses for zero

There are two kinds of uses for zero in Fourier analysis. On the one hand, zeroes may be added ("padded", "zero-filled") to a time-domain data set, in order to increase the digital resolution of the discrete frequency spectrum resulting from subsequent Fourier transformation and (at least from the first zero-fill) add information to the absorption-mode spectrum. Alternatively, we may filter out high-frequency noise (or low-frequency baseline curvature) by *substituting* (rather than *adding*) zeroes for some of the data in either the time- or frequency-domain.

3.4.1 Zero-filling

In Fourier transform spectrometry, a time-domain signal is sampled at N equally spaced intervals over a data acquisition period, T.

$$f(t) = f(0), f(T/N), f(2T/N), \cdots, f[(N-1)T/N] \tag{3.21}$$

Discrete Fourier transformation of a causal time-domain discrete signal (Eqs. 3.5) produces $N/2$ unique complex frequency-domain data. There are only $N/2$ frequencies, because the discrete frequencies are spaced $(1/T)$ Hz apart and the highest frequency that can be represented according to the Nyquist criterion is half the sampling frequency (which is N/T—see Eq. 3.19).

$$\boldsymbol{F}(\nu) = F(0), F(1/T), F(2/T), \cdots, F[(N-1)/2T] \tag{3.22}$$

At first glance, it might appear that information has been lost in the Fourier transform process, since we recover only $N/2$ (complex) frequency-domain data from an initial time-domain (real, for now) data set of N data. However, the frequency-domain data are *complex* numbers, $\boldsymbol{F}(\nu_i)$, each of which may be represented either as the product of a *magnitude*, $|\boldsymbol{F}(\nu_i)|$, and *phase*, $\phi(\nu_i)$,

$$\boldsymbol{F}(\nu_i) = |\boldsymbol{F}(\nu_i)| \exp[i\,\phi(\nu_i)] \tag{3.23}$$

or as the sum of *real* and *imaginary* components,

$$\boldsymbol{F}(\nu_i) = \mathrm{Re}[\boldsymbol{F}(\nu_i)] + i\,\mathrm{Im}[\boldsymbol{F}(\nu_i)]$$

$$= |\boldsymbol{F}(\nu_i)| \cos\phi(\nu_i) + i\,|\boldsymbol{F}(\nu_i)| \sin\phi(\nu_i)\,;\;\; \phi(\nu_i) = \arctan\left(\frac{\mathrm{Im}[\boldsymbol{F}(\nu_i)]}{\mathrm{Re}[\boldsymbol{F}(\nu_i)]}\right) \tag{3.24}$$

Thus, each complex frequency-domain data point actually contains two independent kinds of information.

Although we could represent the spectrum resulting from a discrete Fourier transform as a three-dimensional plot (i.e., real and imaginary amplitudes *vs.* frequency as in Fig. 2.17—see also Chapter 6.2), it is more convenient to limit the display to a two-dimensional plot (e.g., real amplitude *vs.* frequency or imaginary amplitude *vs.* frequency). For a properly phased spectrum, the real amplitude *vs.* frequency is the absorption-mode spectrum and the imaginary amplitude *vs.* frequency is the dispersion spectrum. If we now limit our attention to the absorption-mode spectrum only, then information has indeed been lost, because only $N/2$ independent values are specified.

Consider next what happens if the original time-domain data set is extended by adding N zeroes after the last experimental data point. Fourier transformation of the new $2N$-point time-domain data set now gives an N-point absorption (and N-point dispersion) spectrum. Although the spectral frequency range is unchanged (since the time-domain sampling rate, and hence the Nyquist frequency is the same), the new frequency-domain data are now spaced at discrete intervals of $(1/2T)$ rather than $(1/T)$ Hz, because the time-domain acquisition period has effectively been extended from T to $2T$ seconds (as far as frequency-domain digital resolution is concerned).

Although it might appear that all we have done is to interpolate between the original frequency-doman spectral data points, it can be shown that the first zero-fill effectively recovers the other half of the information initially located in the $N/2$ dispersion-mode spectrum (see Chapter 5.1.3). However, subsequent zero-fills (to $4N$, $8N$,$\cdots$ time-domain points) do not add any information to the absorption-mode spectrum, and can actually degrade the spectral information because of roundoff errors in the FFT computation. The effect of zero-filling may be understood in terms of convolution (see Problems).

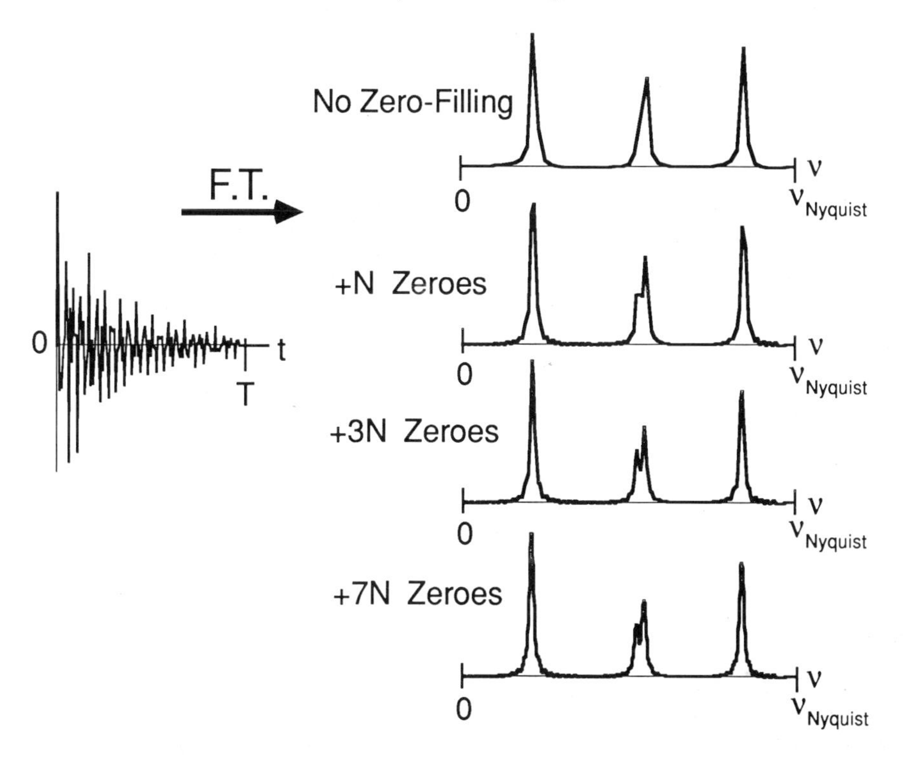

Figure 3.7 Fourier transform spectra (right) of a simulated time-domain N-point discrete signal consisting of a sum of exponentially damped sinusoids of different frequencies (left), to which 0, N, $3N$, or $7N$ zeroes were added before discrete Fourier transformation. The first zero-fill adds information to the spectrum; subsequent zero-fills improve digital resolution (thus making it easier to resolve the peak positions, heights, and widths visually), but do not increase the spectral information content (as evaluated by a computer).

Even though computer analysis of a multiply zero-filled absorption spectrum does not add any inherent information beyond that available from a once zero-filled spectrum, multiple zero-filling can render that information much more *visually* obvious, as seen from Figure 3.7.

3.4.2 Digital filters

The usual purpose of a filter is to remove (or at least diminish) the amplitudes of the low-frequency (or high-frequency, or both) components of a spectrum. *Analog* filters are often used to suppress signals from frequencies outside the bandwidth limited by the Nyquist frequency; such signals would otherwise fold into the detected bandwidth, as described in section 3.2.1. However, analog filters cannot be perfectly selective, and always introduce distortion in amplitude and phase across the spectrum. Moreover, because an analog filter acts on the signal *before* detection, the effect of the filter is not easily removed afterward.

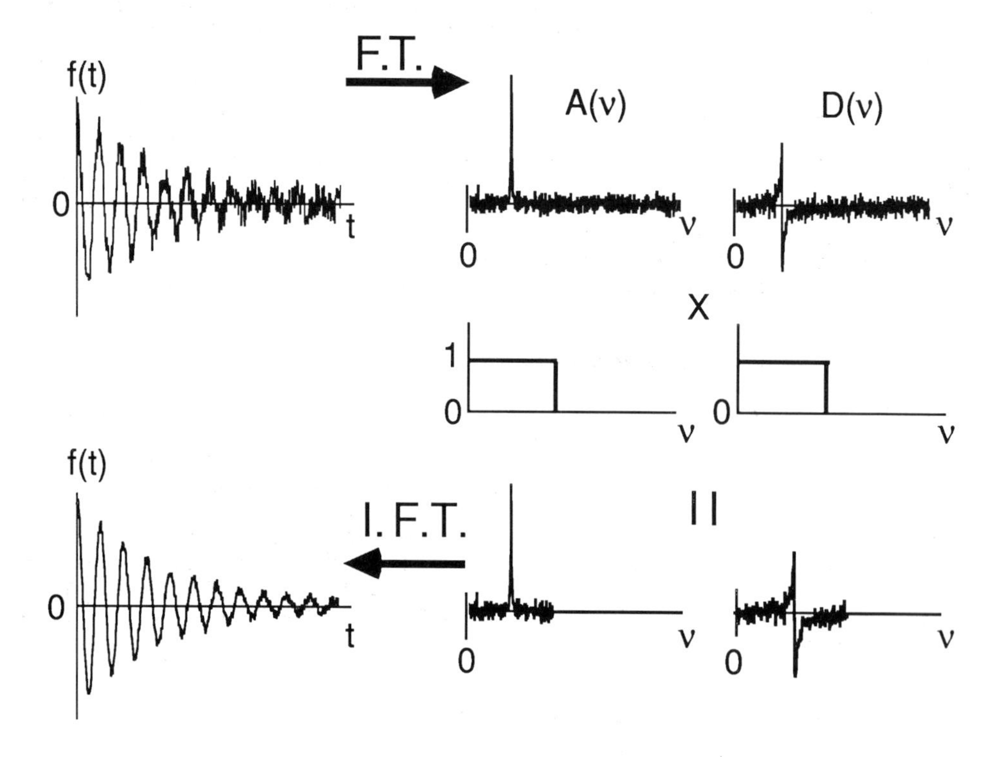

Figure 3.8 Digital filtering, to remove high-frequency noise from a more slowly-varying time-domain signal. In this case, the digital filter is the frequency-domain weight function which zeroes the high-frequency components of the spectrum obtained by FT of the digitized original time-domain signal. Inverse Fourier transformation then yields a smoothed time-domain signal from which the high-frequency noise has been removed.

Digital filtering, on the other hand, may be applied *after* the data has been acquired and stored; thus, we can try out various kinds of filtering without destroying the original data. Also, the shape of a digital filter function can be arbitrarily specified, in contrast to the analog filter profile which is limited by its electronic circuitry.

Digital filtering can be thought of as frequency-domain apodization. For example, Figure 3.8 shows how digital filtering can remove high-frequency noise from a low-frequency time-domain signal.

Digital filtering as illustrated in Figure 3.8 is a software "batch" manipulation of previously acquired data. In electrical engineering, a hardware "real-time" digital filter operates continuously on an analog signal by ADC, followed by a programmed manipulation of the data, followed by digital-to-analog conversion to yield a time-delayed ("digitally filtered") version of the original time-domain analog signal.

Digital filtering need not be confined to Fourier transform spectroscopy. Continuous-wave (e.g., Raman) spectra typically exhibit a baseline with pronounced curvature. For such a baseline, it is difficult to determine (by either computer or manual methods) the peak positions and amplitudes. For example, one could try to represent the baseline by a best-fit polynomial curve, but it is not easy to decide where baseline leaves off and peaks begin. In Raman spectra, peaks are usually relatively sharp and the baseline curvature is usually gradual as a function of frequency. Thus, if we think of the continuous-wave spectrum as a "time-domain" data set, then it makes sense to Fourier transform the continuous-wave discrete spectrum, then digitally filter out the high-"frequency" components in the FT domain, and then inverse Fourier transform to yield a spectrum of the baseline only. The baseline can then be subtracted from the original spectrum to yield a baseline-flattened spectrum, as shown by the simulated example in Figure 3.9.

3.5 Phase correction and its artifacts

As noted in Chapter 2.4, the phase of a spectral peak (e.g., pure absorption-mode, pure dispersion-mode, or something in between) can vary with frequency. Ideally, we desire a spectrum with pure absorption-mode phase at all frequencies, for optimum resolution and peak height-to-noise ratio. However, a fixed time-delay between excitation and detection leads to a linear ("first-order") variation of phase with frequency in the final frequency-domain spectrum. Other experimental effects can produce either a "zero-order" phase shift which is constant (but not zero) across the spectrum, or a higher-order phase shift which varies non-linearly with frequency (e.g., quadratic phase variation resulting from a frequency-sweep excitation).

3.5.1 Zero- and first-order phase correction

Real-time (i.e., computed as fast as parameters can be entered from a keyboard or knob) zero- and first-order phase correction can be performed manually or by any of several automated algorithms operating on two well-separated peaks, A and B, in a discrete Fourier transform spectrum (see Figure 3.10 for a typical FT/NMR example). In the manual method, the trial absorption-mode spectrum, $A(\omega)$, obtained from the raw real and imaginary data, $\mathrm{Re}[\boldsymbol{F}(\omega)]$ and $\mathrm{Im}[\boldsymbol{F}(\omega)]$, of Eqs. 2.37 is displayed continuously on a CRT screen, while ϕ_0 in Eqs. 3.25

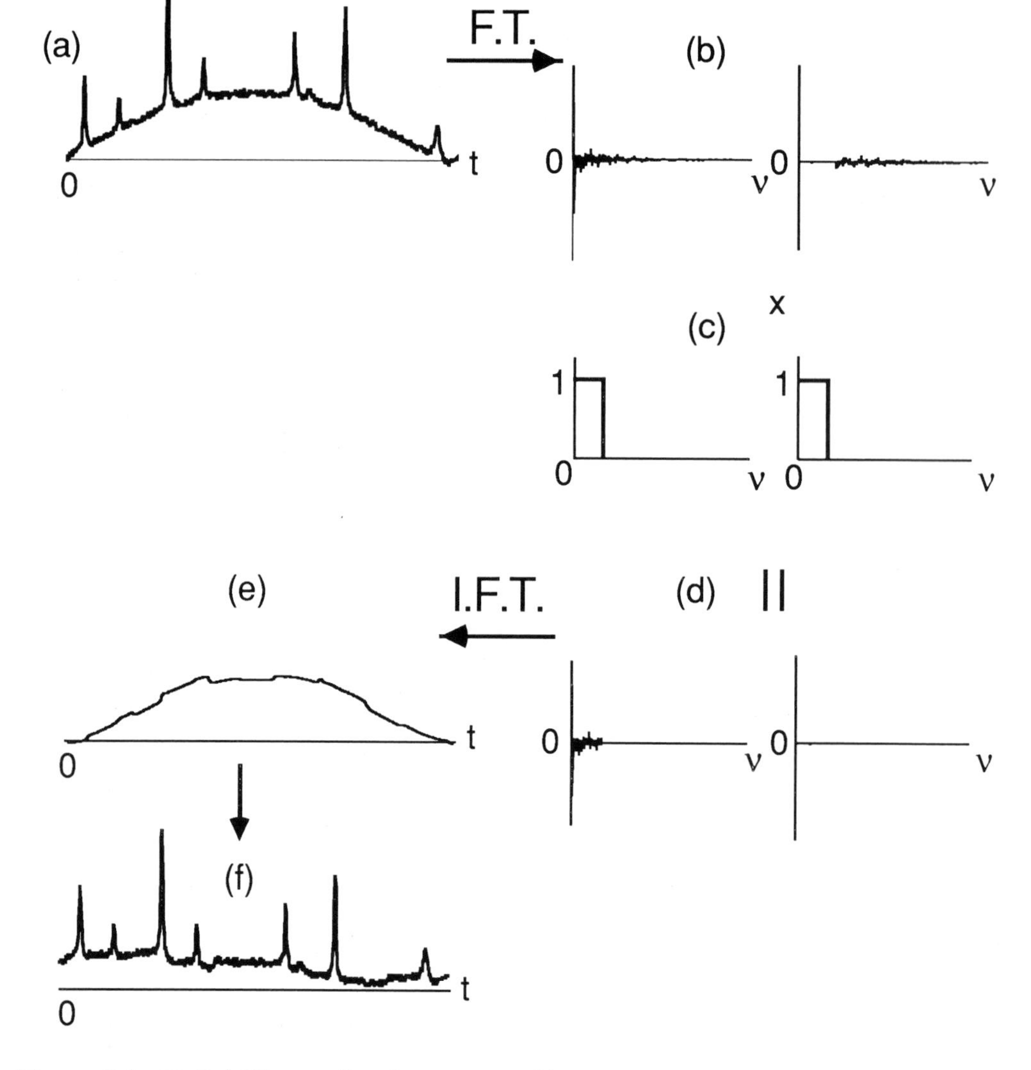

Figure 3.9 Digital filtering for elimination of baseline curvature from a simulated continuous-wave (e.g., Raman) spectrum.
(a) Digitized continuous-wave 1,024-point spectrum with curved baseline.
(b) Discrete Fourier transformation of (a).
(c) Rectangular weight function to remove high-"frequency" components.
(d) Multiplication of (b) by (c) to give a digitally filtered data set in the FT domain, containing only the low-"frequency" baseline drift.
(e) Inverse Fourier transform of (d) to yield a profile of the low-"frequency" spectral baseline alone.
(f) Baseline-flattened spectrum obtained by subtracting (e) from (a).

$$A(\omega) = \cos\phi_0 \quad \mathrm{Re}[\boldsymbol{F}(\omega)] - \sin\phi_0 \quad \mathrm{Im}[\boldsymbol{F}(\omega)] \tag{3.25a}$$

$$D(\omega) = \sin\phi_0 \quad \mathrm{Re}[\boldsymbol{F}(\omega)] + \cos\phi_0 \quad \mathrm{Im}[\boldsymbol{F}(\omega)] \tag{3.25b}$$

is varied by keyboard or knob entry, until one of the two selected peaks (centered at ω_A) becomes pure absorption-mode. A second knob or keystroke then adjusts the rate of change, k, of ϕ with frequency,

$$\phi(\omega) = \phi_0 + k(\omega - \omega_A) \tag{3.26}$$

until the second peak (centered at ω_B) also exhibits pure absorption-mode phase. In this method, the "best" phase adjustment is taken to be that which yields the most symmetrical absorption-mode spectral peak.

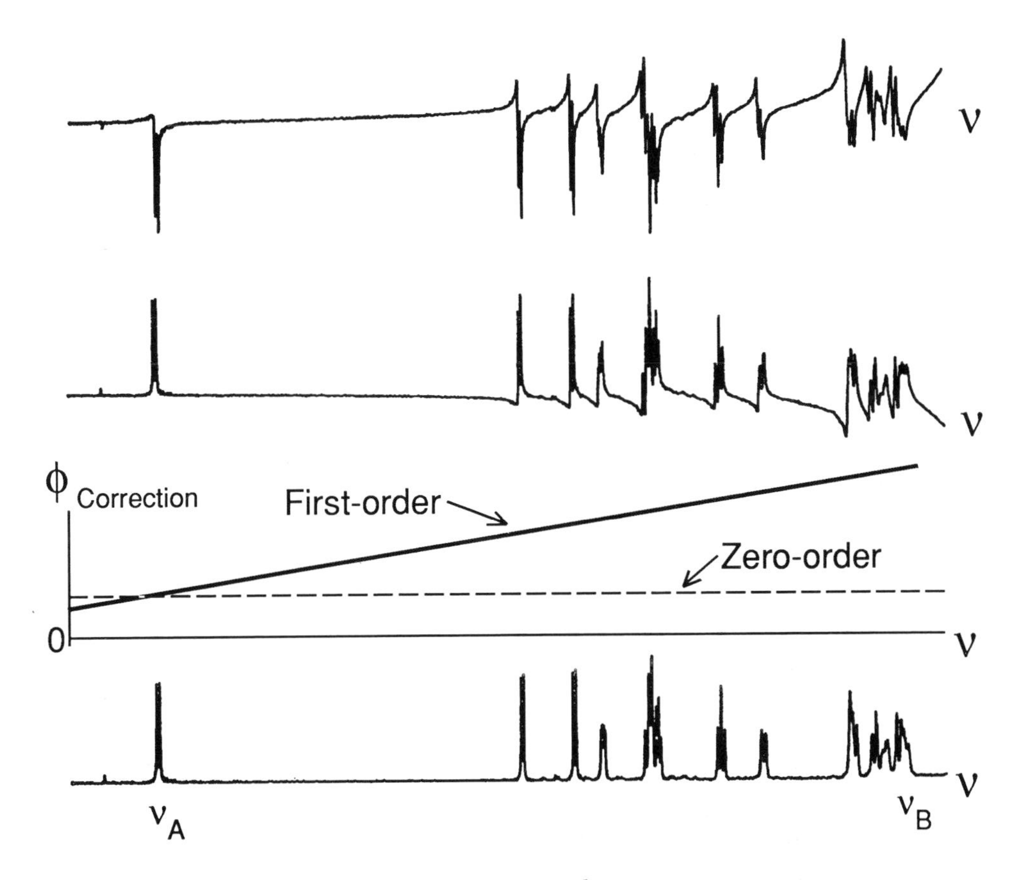

Figure 3.10 Phase correction of a discrete ^{1}H 500 MHz FT/NMR spectrum (provided by C. E. Cottrell) of acetylated uridine-diphosphoglucose (provided by E. J. Behrman). Top: uncorrected real spectrum. Middle: zero-order phase correction, adjusted to give pure absorption-mode phase at ν_A. Bottom: phase is varied linearly about ν_A to give pure absorption-mode phase at ν_B, thereby phase-correcting the whole spectrum. See text for details.

Automated phase adjustment resembles the above procedure, but with a less subjective criterion for "best" phase: e.g:, (a) absorption-mode spectrum with equal area on either side of the peak maximum; (b) dispersion-mode spectrum with minimum peak area; or (c) the phase adjustment which produces zero rotation angle in a dispersion-vs.-absorption (DISPA) plot (see Figure 2.17).

3.5.2 Artifacts

All phasing methods can be defeated when the spectral baseline has non-zero vertical offset and/or non-zero slope. Therefore, it is important to zero and flatten the baseline of any spectrum before attempted to phase-correct it. Also, peak overlap can lead to spurious phase correction. Finally, even after perfect phase correction, some absorption-mode discrete Fourier transform spectral peaks can still appear asymmetrical, as discussed in Further Reading and in Chapter 4.2.1.

3.6 Scaling of discrete FT energy and power spectra

Up to now, we have been able to neglect the scale factor in Eqs. 3.5, because (see below) we have been concerned only with discrete absorption, dispersion, and magnitude spectra. In this section, we shall justify that neglect, and then show that it is necessary to include the scale factor in Eqs. 3.5 when the *discrete power spectrum* is considered.

We begin by representing the *discrete* time-domain signal, $f_s(t)$, as a series (shown as a sum) of samples of the *continuous* causal real time-domain signal, $f(t)$, defined over the acquisition period, $0 \leq t < T$.

$$f_s(t) = \sum_{n=0}^{\infty} f(t_n) \tag{3.27}$$

Alternatively, $f_s(t)$ may be represented as the product of the continuous time-domain signal, $f(t)$, and a sampling function consisting of a time-domain "shah" function (Eq. 2.15a) scaled by the factor, Δt, where $\Delta t = T/N$ is the time-domain sampling interval (see Eq. 3.2):

$$f_s(t) = f(t)\, \Delta t\; III(t)$$

$$= f(t)\, \Delta t \sum_{n=-\infty}^{\infty} \delta(t - n\,\Delta t) \tag{3.28}$$

We may now compute the (continuous) FT of $f_s(t)$ in the usual way (Eq. 2.3c), to obtain the sampled (complex) frequency-domain function, $\boldsymbol{F}_s(\nu)$:

$$\boldsymbol{F}_s(\nu) = \int_{-\infty}^{\infty} f_s(t)\, \exp(-i\,2\pi\nu t)\; dt$$

$$= \Delta t \int_{-\infty}^{\infty} f(t)\, III(t)\, \exp(-i\,2\pi\nu t)\; dt$$

$$= \Delta t \bullet \text{FT of } [f(t)\, III(t)] \tag{3.29}$$

We can now use the convolution theorem (Figure 2.10) to evaluate the FT of the product, $f(t)\,III(t)$ as the convolution of the Fourier transforms of the two component functions, to yield

$$\boldsymbol{F}_s(\nu) = \Delta t \;\; \boldsymbol{F}(\nu) * III(\nu) \tag{3.30}$$

in which

$$III(\nu) = \frac{1}{\Delta t} \sum_{n=-\infty}^{\infty} \delta(\nu - n\,\Delta\nu) \tag{2.15b}$$

$$\text{and } \boldsymbol{F}(\nu) = \int_{-\infty}^{\infty} f(t)\; \exp(-i\,2\pi\nu t)\;\; dt \tag{2.3c}$$

From the definition of a δ-function (Appendix B), Eq. 3.30 then becomes

$$\boldsymbol{F}_s(\nu) = \Delta t \left(\frac{1}{\Delta t}\right) \sum_{n=-\infty}^{\infty} \boldsymbol{F}(\nu_n) = \sum_{n=-\infty}^{\infty} \boldsymbol{F}(\nu_n) \tag{3.31}$$

In other words, the scale factor, Δt, disappears for discrete FT calculations involving the usual spectral representations: Re[$\boldsymbol{F}(\nu)$], Im[$\boldsymbol{F}(\nu)$], and $|\boldsymbol{F}(\nu)|$. Many authors therefore omit (as we did) the scale factor from their definition of a discrete FT (Eqs. 3.5).

The importance of the above discussion is for scaling of the discrete *power* spectrum, as we shall now show. Since $f(t)$ is defined only for $0 \le t \le (N-1)\,\Delta t$, we can rewrite Eq. 3.28 as,

$$f_s(t) = \Delta t \sum_{n=0}^{N-1} f(t)\;\; \delta(t - n\,\Delta t\,) \tag{3.32}$$

The square of the magnitude spectrum then becomes,

$$|\boldsymbol{F}(\nu)|^2 = (\Delta t)^2 \left(\text{Discrete FT of } \sum_{n=0}^{N-1} f(t)\, \delta(t - n\,\Delta t\,)\right)^2 \tag{3.33}$$

In physical problems, the square of the frequency-domain magnitude represents *energy*. Since *power* is energy per unit time, we must divide Eq. 3.33 by Δt in order to obtain the correctly scaled discrete power spectrum, $P(\nu)$:

$$P(\nu) = \Delta t \left(\text{Discrete FT of } \sum_{n=0}^{N-1} f(t)\, \delta(t - n\,\Delta t\,)\right)^2 \tag{3.34}$$

Although most FT applications in spectroscopy do not require the power spectrum, we shall nevertheless need to consider the power spectrum when we consider *non-FT* methods for estimating spectral parameters (see Eq. 6.40b and accompanying discussion).

Further Reading

J. W. Adams, *Trans. Circuits & Systems*, **1987**, CAS-34, 768-770. (Describes use of zero-filling as an FT interpolation technique)

E. Bartholdi & R. R. Ernst, *J. Magn. Reson.* **1973** *11*, 9-19. (Zero-filling and its relation to absorption and dispersion spectra)

S. E. Bialkowski, *Anal. Chem.* **1988**, *60*, 355A-361A. (Excellent short discussion of digital filters)

R. Bracewell, *The Hartley Transform*, Oxford Univ. Press, N.Y., 1986. [Alternative to FT, based upon the real function, cas(θ) = cos(θ) + sin(θ), rather than the complex exp($i\theta$) = cos(θ) + isin(θ); the corresponding fast algorithm runs in half the time of the complex FFT]

E. O. Brigham, *The Fast Fourier Transform*, Prentice-Hall, Englewood Cliffs, NJ, 1974. (Excellent explanation of the FFT algorithm)

M. Comisarow & J. D. Melka, *Anal. Chem.* **1979** *51*, 2198-2203. (Effect of zero-filling on peak height and position precision)

E. Garcia-Torano, *Comp. Phys. Commun.* **1983** *30*, 397-402. (FFT with data sets factorable as a product of prime numbers other than 2)

P. R. Griffiths & J. A. de Haseth, *Fourier Transform Infrared Spectrometry*, Wiley, N.Y., 1986. (Good discussion of apodization functions, and brief explanation of FFT algorithm)

A. T. Hsu & A. G. Marshall, *Analyt. Chim. Acta* **1985** *178*, 27-41. (Clipped data sets)

B. Liu & T. Kaneko, *Proc. IEEE*, **1975**, *63*, 991-992. (FFT roundoff errors)

S. L. Marple, Jr., *Digital Spectral Analysis with Applications*, Prentice-Hall, Englewood Cliffs, NJ, 1987. (Good connections between continuous and discrete FT)

H. J. Nussbaumer, *Fast Fourier Transform and Convolution Algorithms*, 2nd ed., Springer-Verlag, Berlin, 1982.

A. V. Oppenheim & R. W. Schafer, *Discrete Signal Processing*, Prentice-Hall, Englewood Cliffs, NJ, 1989. (Very clear presentation of FT methods)

R. W. Page & A. S. Foster, *Hewlett-Packard J.*, **1988** (2), 26-31. (Good examples of oversampling to improve visual amplitude accuracy of time-domain waveforms)

R. Skarjune, *Comput. & Chem.* **1986** *10*, 241-251. (Efficiency of FFT algorithms)

F. R. Verdun & A. G. Marshall, *Appl. Spectrosc.* **1988** *42*, 199-203. (Beating the Nyquist limit by interleaved alternate delay sampling)

F. R. Verdun, C. Giancaspro, & A. G. Marshall, *Appl. Spectrosc.* **1988**, *42*, 715-721. (Effect of noise and FFT roundoff errors on FT spectra)

M. Vetterli, *I.E.E.E. Trans. Acoust., Speech, Signal Processing*, **1989** AASP-37(1), 57-64. (Generalization of the FFT algorithm for length, p^m)

Problems

3.1 Generate the forward and inverse Fourier transform "code" matrices (see Eqs. 3.7 and 3.8 for the $N = 4$ case),

$$F_{nm} = \exp(-i\ 2\pi nm/N) \tag{3.5b}$$

$$F_{nm}^{-1} = (1/N)\exp(i\ 2\pi nm/N) \tag{3.9b}$$

for $N = 3, 4, 5, 6, 7$, and 8. Then confirm that F^{-1} is indeed the inverse of F. In other words (see Eq. A.110), show that

$$\sum_{j=0}^{N-1} F_{nj} F_{jm}^{-1} = 1 \text{ for } n = m; \qquad = 0 \text{ for } n \neq m$$

3.2 In order to see why the FFT algorithm is so useful, consider the simple time-domain function,

$$f(t) = \cos 2\pi\nu t\ \exp(-t/\tau), \qquad 0 \leq t < T, \text{ with } T = 5\tau,$$

in which the period, T, is sampled at 1024 discrete points beginning at $t = 0$.

(a) Write a (short) computer program to compute the "longhand" (discrete) Fourier transform from Eq. 3.5.

(b) Next, compute the Fourier transform from one of the "fast" algorithms listed in Appendix C. Compare the two computation periods, to see if you realized the full theoretical time-saving advantage of the FFT algorithm.

3.3 Compute the (discrete) convolution for each of the following pairs of time-domain sequences, using Eq. 3.15.

(a) $h(t) = (1, 1, 1, 1)$; $e(t) = (1, 1, 1, 1)$

(b) $h(t) = (2, 4, 6, 8)$; $e(t) = (i, 0, -i, 0)$

(c) $h(t) = (1, 1, 0, 0, 1, 0, 1)$; $e(t) = (1, 1, 0, 0, 1, 0, 1)$

(d) $h(t) = (1, 1, 0, 0, 1, 0, 1)$; $e(t) = (1, 0, 1, 0, 0, 1, 1)$

3.4 Consider an undamped sinusoidal time-domain signal, which is acquired for a period, T

$$f(t) = \cos \omega_0 t, \quad 0 \leq t < T$$

For FT/NMR experiments with spectral bandwidths of 100,000 Hz, 10,000 Hz, or 1,000 Hz, and a computer equipped with 32,768 random access memory (RAM),

(a) Determine the analog line width for the frequency-domain absorption-mode spectrum of $f(t)$ for each bandwidth.

(b) Determine the digital resolution (i.e., points/Hz) for each of those spectra.

Next, suppose that the time-domain signal is damped according to

$$f(t) = \cos \omega_0 t \quad \exp(-t/\tau) , \qquad 0 \leq t < T$$

with $\tau = 1$ s

and repeat the above calculations for the same RAM and sampling rates. This exercise shows that digital resolution determines spectral peak width when $T << \tau$ whereas analog resolution determines spectral peak width when $T >> \tau$.

3.5 After Fourier transform of a time-domain data set which has been zero-filled one or more times, additional data points are interpolated at frequencies spaced equally between the original discrete frequencies. In this problem, we will show how to determine where those new points lie.

(a) For the undamped sinusoidal time-domain signal,

$$f(t) = \cos 2\pi \nu_0 t \quad , \ 0 \leq t < T$$

Assume that $f(t)$ is sampled at a rate, $4\nu_0$ points/second during the acquisition period, T, to give N data points, and plot the discrete spectrum obtained by Fourier transformation of that data set, over the (eight-point) frequency range,

$$\nu_0 - (4/T) \leq \nu \leq \nu_0 + (4/T).$$

(b) Next suppose that $f(t)$ lasts for twice as long (i.e., $2T$), and repeat the calculation and plot.

(c) Now suppose that $f(t)$ is sampled as in part (a) for T seconds, and that N zeroes are added to give a time-domain data set lasting for $2T$ seconds. This zero-filled data set is equivalent to one obtained by multiplying $f(t)$ in part (b) by the weight function (see diagram on the next page),

$$W(t) = 1 \ \text{ for } \ 0 \leq t \leq (N-1)T/N$$

$$W(t) = 0 \text{ for } t > T$$

Therefore, since multiplication in the time-domain becomes convolution in the frequency-domain, compute the FT of the zero-filled data set as the convolution of the FT of $W(t)$ with the FT of $f(t)$ from part (b).

(d) See if you can repeat this exercise for a second zero-fill (i.e., adding $3N$ zeroes to the original data set). You should begin to see why multiple time-domain zero-filling adds wiggles to the final spectrum.

Diagram for Problem 3.5:

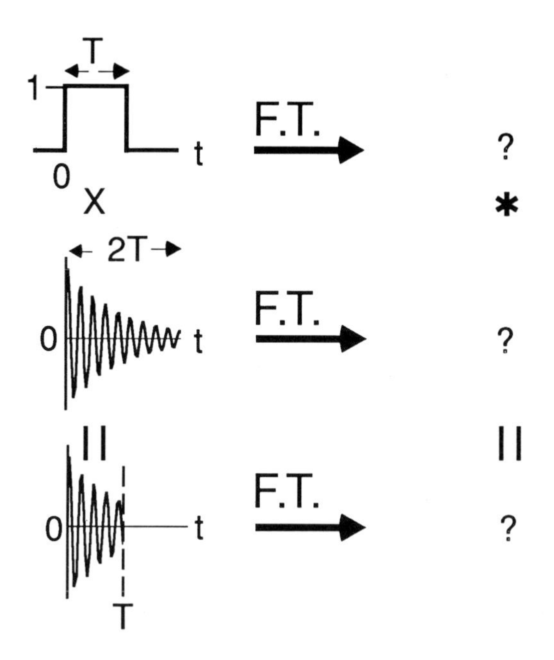

Solutions to Problems

3.1 For the $N = 3$ case,

$$F_{nm} = \exp(-i\omega_m t_n) = \exp(-i2\pi nm/N) = \exp(-i\ 2\pi nm/3)$$

Thus,

$$F_{00} = \exp(-i\ 2\pi\cdot 0\cdot 0/3) = \exp(0) = 1$$
$$F_{01} = \exp(-i\ 2\pi\cdot 0\cdot 1/3) = \exp(0) = 1$$
$$F_{02} = \exp(-i\ 2\pi\cdot 0\cdot 2/3) = \exp(0) = 1$$

$$F_{10} = \exp(-i\ 2\pi\cdot 1\cdot 0/3) = \exp(0) = 1$$
$$F_{11} = \exp(-i\ 2\pi\cdot 1\cdot 1/3) = \exp(-i\ 2\pi/3) = \cos(2\pi/3) - i\sin(2\pi/3) = -(1/2) - i(\sqrt{3}/2)$$
$$F_{12} = \exp(-i\ 2\pi\cdot 1\cdot 2/3) = \exp(-i\ 4\pi/3) = -(1/2) + i(\sqrt{3}/2)$$

$$F_{20} = \exp(-i\ 2\pi\cdot 2\cdot 0/3) = \exp(0) = 1$$
$$F_{21} = \exp(-i\ 2\pi\cdot 2\cdot 1/3) = \exp(-i\ 4\pi/3) = -(1/2) + i\sqrt{3}/2)$$
$$F_{22} = \exp(-i\ 2\pi\cdot 2\cdot 2/3) = \exp(-i\ 8\pi/3) = -(1/2) - i(\sqrt{3}/2)$$

$$\begin{pmatrix} F(\nu_0) \\ F(\nu_1) \\ F(\nu_2) \end{pmatrix} = \begin{pmatrix} 1 & 1 & 1 \\ 1 & -(1/2) - i(\sqrt{3}/2) & -(1/2) + i(\sqrt{3}/2) \\ 1 & -(1/2) + i(\sqrt{3}/2) & -(1/2) - i(\sqrt{3}/2) \end{pmatrix} \begin{pmatrix} f(t_0) \\ f(t_1) \\ f(t_2) \end{pmatrix}$$

For the inverse Fourier code,

$$F_{nm}^{-1} = (1/N)\,[\exp(i\,2\pi nm/N)] = (1/N)\exp(i\,2\pi nm/3) = (1/N)\,F_{nm}^{*}$$

The inverse matrix elements are obtained as $(1/N)\,F_{nm}^{*}$. For example,

$$F_{12}^{-1} = F_{12}^{*} = (-1/2) - i\,(\sqrt{3}/2)$$

$$\begin{pmatrix} f(t_0) \\ f(t_1) \\ f(t_2) \end{pmatrix} = \begin{pmatrix} 1 & 1 & 1 \\ 1 & -(1/2) + i(\sqrt{3}/2) & -(1/2) - i(\sqrt{3}/2) \\ 1 & -(1/2) - i(\sqrt{3}/2) & -(1/2) + i(\sqrt{3}/2) \end{pmatrix} \begin{pmatrix} F(\nu_0) \\ F(\nu_1) \\ F(\nu_2) \end{pmatrix}$$

The code elements for the $N = 4$ case are given as Eqs. 3.7 and 3.8, and the codes for other values of N are obtained similarly.

3.3 (a) (1, 2, 3, 4, 3, 2, 1)

(b) ($2i$, $4i$, $4i$, $4i$, $-6i$, $-8i$, 0)

(c) (1, 2, 1, 0, 2, 2, 2, 2, 1, 0, 2, 0, 1)

(d) (1, 1, 1, 1, 1, 1, 4, 1, 1, 1, 1, 1, 1)

3.4 Acquisition period, $T = \nu_{\text{sampling}}/N$. For an undamped signal, $\Delta\nu \cong 0.61/T$ Hz. For a damped signal of infinite duration, $\Delta\nu = (1/\pi\tau) = 0.32$ Hz. For the line width of a damped signal of finite duration, see Figure 4.7.

ν_{sampling}	T	$\Delta\nu_{\text{undamped}}$	$\Delta\nu_{\text{damped}}$
200,000 Hz	0.164 s	3.7 Hz	~3.7 Hz
20,000	1.64	0.37	~0.5
2,000	16.4	0.037	~0.32

Note that the line width for a damped signal is limited by acquisition period, T, when $T \ll \tau$, but is limited by the damping constant, τ, when $T \gg \tau$.

3.5

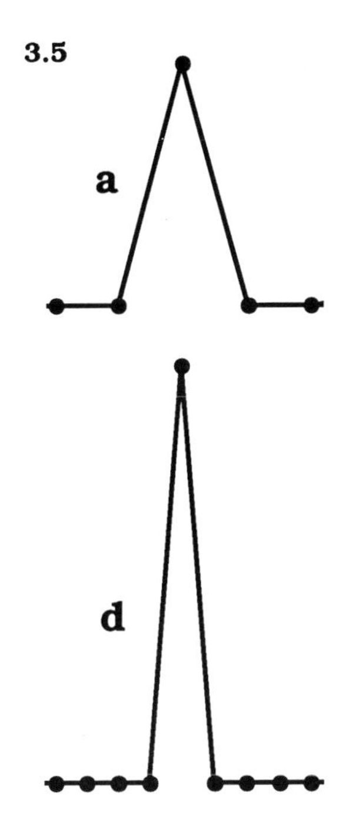

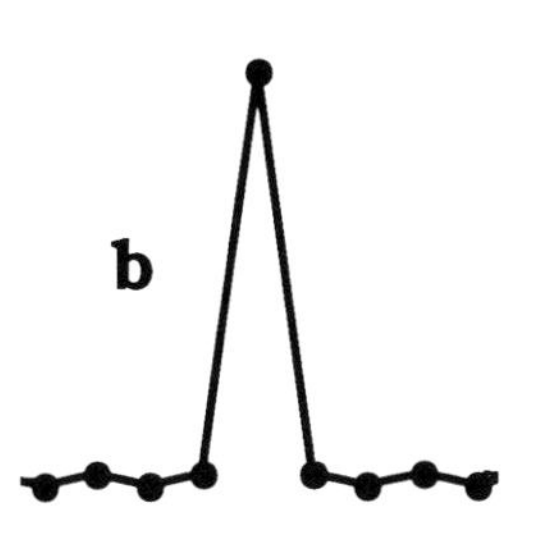

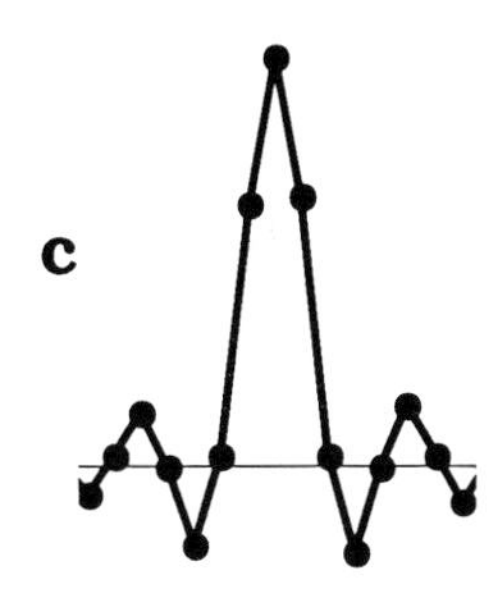

a: Acquisition period = T ; No zero-filling
b: Acquisition period = T ; Add N zeroes
c: Acquisition period = T ; Add $3N$ zeroes
d: Acquisition period = $2T$; No zero-filling

CHAPTER 4

Fourier Transform Spectrometry: Common Features

4.1 Multichannel spectrometry

In this chapter, we present those features of Fourier transform spectroscopy which are shared by all of its forms (in this book, FT/IR, FT/NMR, and FT/ICR). First, the principal practical advantage of Fourier transform spectrometry is the simultaneous detection of the whole spectrum at once—the so-called Fellgett or multichannel or multiplex advantage. However, in order to *detect* the whole spectrum at once, one must first devise a broadband source to *excite* the whole spectrum at once. We shall therefore begin by analyzing the operation of a schematic single-channel ("scanning") spectrometer with broadband or narrowband excitation, and extend to multidetector and single-detector "coded" ("multiplex") instruments. We shall find that the choice of multiplex code (i.e., Hadamard or Fourier) depends on whether the excitation source emits coherent or incoherent radiation.

Second, we shall show how the quality (e.g., resolution) of an FT spectrum depends on both the analog (Chapter 2) and discrete (Chapter 3) line shape. Third, we will discover that the design and interpretation of an FT experiment depends on whether the signal is linearly polarized (e.g., FT/IR) or circularly polarized (e.g., FT/NMR; FT/ICR). Finally, we will explain how non-linear responses can produce harmonic signals (e.g., peaks at multiple or combination frequencies), forming a basis for laser doubling, double-quantum NMR, and heterodyne detection to achieve vast improvement in spectral resolution.

4.1.1 Single-channel spectrometry

In a single-channel spectrometer (see top two diagrams of Figure 4.1), each "channel" detects the signal from only a narrow band of frequencies. The schematic "slit" in Figure 4.1b could represent an actual aperture (as in optical spectroscopy) or a mixer-filter detector whose bandwidth corresponds to the slit width (as in NMR or ICR spectroscopy—see Chapter 4.4.1). A discrete spectrum of N magnitudes corresponding to N slit positions may then be acquired by scanning sequentially from one slit position (channel) to the next. Although the single-channel spectrometer yields the point-by-point spectrum directly (i.e., without further data reduction), its obvious disadvantage is that a given measurement detects only a small fraction of the desired spectrum.

4.1.2 Multidetectors

An obvious improvement would be to open the slit width to the full spectral "window" of the desired spectral range, and replace the single broadband detector of Figure 4.1b with an array of N narrowband detectors, each positioned (or tuned) to detect radiation from just one channel (Figure 4.1c).

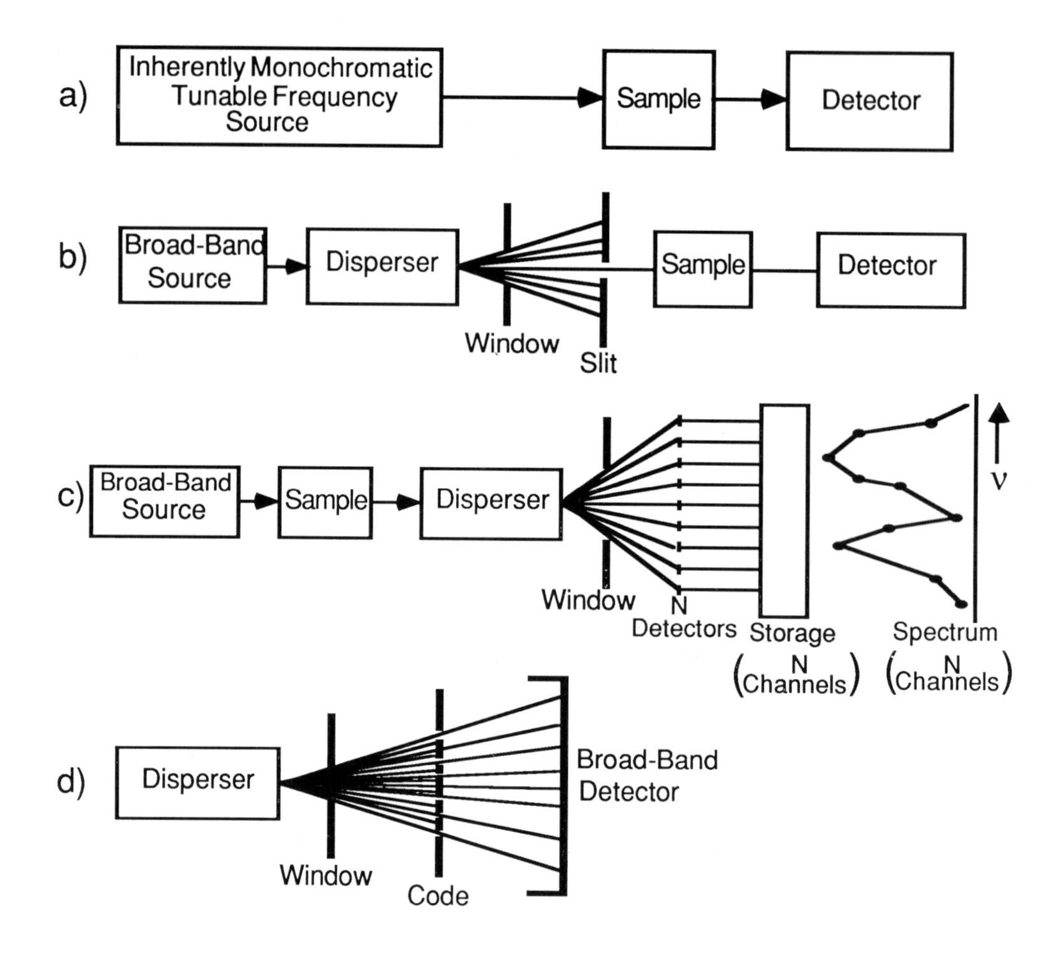

Figure 4.1 Schematic diagrams of (a), (b) single-channel, (c) multichannel, and (d) multiplex spectrometers. (a) Source provides inherently monochromatic radiation (e.g., laser). (b) Source provides inherently broad-band radiation (e.g., lamp filament), which is rendered monochromatic by passage through a disperser/window/slit combination as shown. Although the single detector may be either broadband (detects radiation anywhere within the window) or narrowband (detects radiation only from a single channel), it is usual to associate a broadband detector with a broadband source (b) and a narrowband detector with a narrowband source (a). (c) The multidetector spectrometer is similar to that of (b) with the slit removed, so that the full N- channel spectrum impinges on N individual detectors, each positioned (or tuned) to detect the radiation from just one spectral channel. The spectral signal accumulates in a storage unit, and is ultimately displayed in the point-by-point magnitude spectrum shown at the far right. (d) Schematic encoding of the signals of an N- channel spectrum, by means of insertion of a coding mask between the spectral window and the (single, broadband) detector. In the code shown here, the detected signal from any one channel is coded by a zero (slit closed for that channel) or one (slit open for that channel). Specific codes are discussed in section 4.1.3.

Because the multidetector spectrometer of Figure 4.1c is conceptually simple, we will next consider its feasibility. Any multichannel spectrometer must be able to resolve spectral features as narrow as the width of a typical spectral peak. Thus, the minimum number of channels is simply the width of the entire spectral range of interest divided by the width of the narrowest expected peak (see Table 4.1).

Table 4.1 Minimum number of channels required for various multidetector spectrometers. The minimum number of channels is the spectral range divided by the width of one spectral peak. All frequencies are listed in Hz.

Type of spectrum	Largest usual frequency	Typical spectral frequency range	Typical peak width	Approximate Minimum number of channels
Mossbauer[a]	6×10^{18}	10^{8}	10^{7}	10
X-ray photoelectron	3.5×10^{17}	10^{17}	10^{14}	1,000
UV Photoelectron	5×10^{15}	3×10^{15}	10^{12}	3,000
Electronic	1.5×10^{15}	1.2×10^{15}	10^{9}	1,250,000
Vibrational	2×10^{14}	1.5×10^{14}	3×10^{9}	50,000
Rotational	4×10^{10}	3×10^{10}	10^{5}	300,000
ENDOR[b]	5×10^{7}	5×10^{7}	5×10^{4}	1,000
^{13}C NMR[c]	8×10^{7}	2×10^{4}	0.2	100,000
^{1}H NMR[d]	5×10^{8}	5×10^{3}	0.3	15,000
ICR[e]	6×10^{6}	6×10^{6}	1	6,000,000

a ^{119m}Sn
b electron-nuclear double resonance for ^{1}H
c 7.5 tesla (e.g., 300 MHz for ^{1}H)
d 11.75 tesla (e.g., 500 MHz for ^{1}H)
e 7 tesla, with minimum ionic mass-to-charge ratio of 15

From Table 4.1, multidetector electronic (ultraviolet/visible) spectroscopy would appear to be one of the least feasible. However, a fine-grain photographic plate can in fact resolve the required huge number of required channels, since the spectrum may be dispersed over the necessary distance (a few meters) without undue effort. When the resolution need not be so high (as for liquid chromatography ultraviolet/visible detectors) any of several types of electronic multidetector arrays (e.g., vidicon, self-scanning photodiode arrays, charge-coupled devices, etc.) are available. For example, in the vidicon (television camera) device, photons of different frequency can be spatially dispersed to strike different pixels of an array. Although the electronic optical multidetector has relatively few channels compared to a photoplate, the electronic device can furnish a spectrum immediately without the delays of photographic processing and densitometric reading of the developed photoplate.

In photoelectron spectroscopy—also known as ESCA (Electron Spectroscopy for Chemical Analysis), electrons are dislodged from atoms or molecules by X-ray or ultraviolet radiation, and the released electrons possess a translational energy that depends on the energy of the bound state occupied by that electron in the original atom or molecule. Those electrons may be spatially dispersed (by an applied electric field) according to their velocity, to strike different pixels of a multidetector array, just as in Figure 4.1c. Because of the relatively small number of required channels (Table 4.1), multidetector methods are feasible for the XPS and UPS cases.

For the other forms of spectroscopy listed in Table 4.1, direct multidetection is simply not feasible. For rotational (microwave) spectroscopy, for example, no broadband radiation source is available—a black-body source such as that used for higher frequencies (e.g., xenon or hydrogen discharge for ultraviolet, hot tungsten wire for visible, globar for near- and mid-IR, mercury vapor for far-IR) would have to be operated at an unreasonably high temperature in order to obtain sufficient radiation flux. The cost of a huge array of individual narrowband microwave transmitters (at ~$1,000 each) as the "broadband" source would also be unreasonable. For infrared spectroscopy, on the other hand, the required broadband source is available, but one would have to disperse the spectrum over ~50 meters in order to resolve the desired spectral detail with existing (thermopile) individual detectors of ~1 mm aperture (at ~$200 each!). (Photographic detection does not extend above ~12,000 Å, and is thus unavailable for IR.) Finally, for NMR and ICR spectroscopy, broadband sources are again available, but the cost of an array of tens of thousands of individual narrowband mixer-filter detectors (as also for ENDOR) is again unreasonably high.

In conclusion, the multidetector approach is feasible for low-resolution particle and optical spectroscopy, but fails on both geometric and economic grounds for microwave and radiofrequency spectroscopy. Fortunately (see next section), it is possible to employ a single (inexpensive) broadband detector if one can somehow "code" the incoming spectrum and then "decode" it later.

4.1.3 Multiplex methods (Fellgett advantage)

The design and advantages of multiplex spectroscopy can be understood from the much simpler weighing experiment shown schematically in Figure 4.2. The object of the experiment is to determine the masses of three unknown objects by use of a double-pan balance. Let the masses of the unknown objects be x_1, x_2, and x_3. Most of us would probably place each unknown one at a time in (say) the left pan, and determine the appropriate weight of each from an appropriate combination of knowns, y_1, y_2, and y_3 in the right pan in each case. The obvious advantage of the one-at-a-time weighing procedure is that no data reduction is required: the result of each measurement (y_1, y_2, and y_3) yields each unknown mass directly:

$$y_1 = x_1 \quad (4.1a)$$

$$y_2 = x_2 \quad (4.1b)$$

$$y_3 = x_3 \quad (4.1c)$$

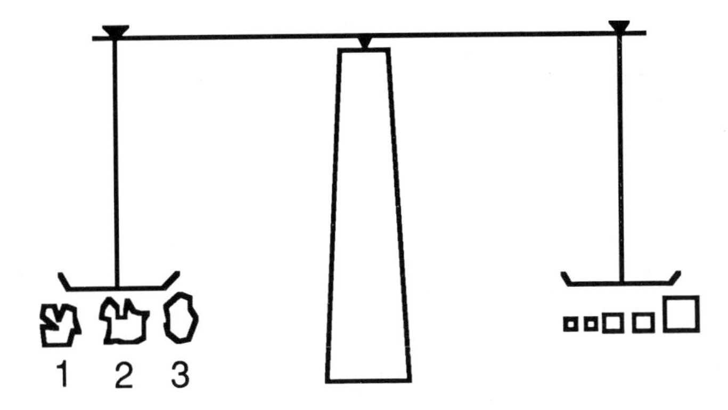

Figure 4.2 Schematic diagram of a double-pan balance, with three unknown weights on the left, and a set of standard (known) weights on the right. The unknowns may be weighed one at a time, or in "multiplex" combination (see text).

The disadvantage of the one-at-a-time weighing experiment becomes obvious from Table 4.2. Although the experiment gives the right answer, only *one* unknown is present in each measurement. The one-at-a-time weighing experiment is analogous to scanning a spectrum by one channel (e.g., slit width) at a time, as in Figure 4.1a,b. The multidetector spectrometer would be analogous to weighing each object simultaneously on three balances.

Table 4.2 Analysis of one-at-a-time weighing experiment. Each unknown is observed with a weight factor of 1 (if placed on the left pan) or 0 (if not). Thus, although *three* experiments are performed, each unknown object is weighed only *once.*

Observed weighing result	Unknown mass x_1	x_2	x_3
y_1	1	0	0
y_2	0	1	0
y_3	0	0	1
Number of times each unknown is weighed:	1	1	1

4.1.3.1 *Hadamard code*

The *Hadamard* multiplex method consists of placing approximately half of the unknown objects on left side of balance in each weighing, as listed below.

$$y_1 = x_1 + x_2 \qquad (4.2a)$$

$$y_2 = x_1 \qquad + x_3 \qquad (4.2b)$$

$$y_3 = \qquad x_2 + x_3 \qquad (4.2c)$$

The reader should verify that the known masses (x_1, x_2, and x_3) can be recovered from the observed weighing results (y_1, y_2, and y_3) by solving the three equations in three unknowns of Eq. 4.2:

$$2x_1 = y_1 + y_2 - y_3 = 1 \cdot y_1 + 1 \cdot y_2 - 1 \cdot y_3 \qquad (4.3a)$$

$$2x_2 = y_1 - y_2 + y_3 = 1 \cdot y_1 - 1 \cdot y_2 + 1 \cdot y_3 \qquad (4.3b)$$

$$2x_3 = -y_1 + y_2 + y_3 = -1 \cdot y_1 + 1 \cdot y_2 + 1 \cdot y_3 \qquad (4.3c)$$

Although additional computations (Eqs. 4.3) are required, the advantage of the Hadamard scheme (see Table 4.3) is that each unknown mass ("signal") is measured *twice*. Moreover, imprecision (noise) is expected to be independent of signal in each weighings. Since three weighings are required to measure each unknown mass, the signal thus increases by a factor of 2 and the "detector-limited" noise increases by a factor of $\sqrt{3}$ (see section 5.1.1.), for a net improvement in signal-to-noise ratio of a factor of $2/\sqrt{3}$. More generally, for N unknowns, one would place $(N+1)/2$ unknowns on the balance in each weighing, to achieve a net improvement in signal-to-noise ratio by a factor of $(N+1)/(2\sqrt{N})$.

Table 4.3 Hadamard multiplex weighing experiment, in which two unknowns are placed together on the balance for each weighing (see text).

Observed weighing result	Unknown mass x_1	x_2	x_3
y_1	1	1	0
y_2	0	1	1
y_3	1	0	1
Number of times each unknown is weighed:	2	2	2

The Hadamard scheme can be applied directly to spectroscopy as shown schematically in Figure 4.3 for a 7-channel case. Although the Hadamard method was originally implemented by physically translating a series of masks across the spectral window, W. G. Fateley (Chapter 9.3.3.2) recently introduced liquid-crystal electro-optical encoding (much like the display on a wrist watch) to create an "open" (transparent) or "closed" (black) slit at each desired slit position.

There is more to the Hadamard scheme than the idea of using N linearly independent combinations of open and shut slit combinations. In fact, each row of the matrix of Hadamard "code" coefficients (i.e., each row in Table 4.3) is generated from the preceding row by *cyclic permutation* (see Figure 4.3 and Problems). The magical property of such a code is that its *inverse* code (Eqs. 4.3) is obtained from the original code (Eqs. 4.2) simply by changing 0 to −1. Thus, one does not need to invert a matrix in order to solve the series of N equations in N unknowns (e.g., Eqs. 4.2), and the computation can be performed by a rapid algorithm on a microcomputer. The 7-channel Hadamard code is left as an exercise (see Problems). [Strictly speaking, a true Hadamard code matrix contains only +1 and −1 elements. However, its physical implementation would require both reflective and transmissive slits, with separate detectors. Therefore, our treatment here is limited to Hadamard "S-matrix" codes (see Problems).]

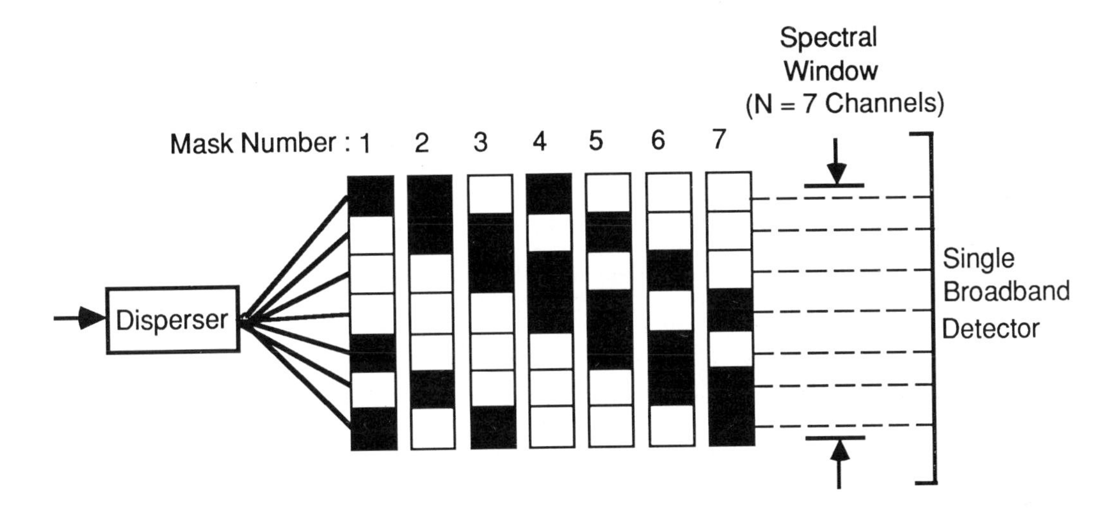

Figure 4.3 7-channel Hadamard spectrometric encoding scheme. The first 7-channel mask is interposed between the spectral disperser and a single broad-band detector. The total transmitted signal, y_1, is recorded, and the procedure is repeated for each of the 7 masks. Each mask pattern is related to that of the previous mask by cyclic permutation. Proceeding in this way, one obtains $N = 7$ total transmitted signals, consisting of 7 linearly independent combinations of the (desired) unknown individual spectral signals from each of the 7 channels (slit positions). The desired 7-point spectrum may then be recovered from the 7 measurements, without matrix inversion, simply by changing 0 to −1 in each row of the 7-channel Hadamard code (see Problems).

4.1.3.2 *Fourier code*

The multichannel advantage of discrete Fourier transform analysis can now be understood as a "code", much as the one-at-a-time weighing or Hadamard codes. In FT spectrometry, *each* sampled time-domain data point, $f(t_n)$, represents a linear combination of *all* of the frequency-domain amplitudes, $F(\nu_m)$,

$$f(t_n) = \frac{1}{N} \sum_{n=0}^{N-1} \exp(i\,2\pi nm/N)\ F(\nu_m) \tag{3.8}$$

Moreover, since *each* frequency-domain channel in the Fourier experiment is weighted in Eq. 3.8 by a factor whose absolute value is unity,

$$|\exp(i\,2\pi nm/N)| = 1 \tag{4.4}$$

it is as if *each* channel in the Fourier "code" or "mask" is fully open during each of the N time-domain measurements. Thus, compared to a single-channel experiment, the Fourier measurement will acquire up to N times as much signal (and $\sqrt{N}$ times as much noise), for a net multiplex (Fellgett) advantage of a factor of $N/\sqrt{N} = \sqrt{N}$. It is important to recognize that the Fellgett advantage depends markedly on the origin of the noise—in fact, for noise which is proportional to the signal strength, Fourier transform spectrometry can even yield a Fellgett *disadvantage* of a factor up to $\sqrt{N}$, as we shall explain in Chapter 5.

A major advantage of the Hadamard experiment over the Fourier experiment is that Hadamard multiplexing does not require *coherent* excitation. However, the consequent disadvantage of the Hadamard experiment is that the experiment yields only spectral *magnitudes*. Therefore, the only way to distinguish one channel from another is to fully open or fully close the slit for that channel, and the Hadamard code (with approximately half of the slits open) turns out to be the optimal choice. If, for example, we were to open only one slit in each mask, then we would be back to the single-channel experiment, with no Fellgett advantage. In the other extreme that all but one slit is open during each observation, we would obtain almost the full Fellgett advantage in each observation, but each observation would also yield almost the same value, thereby reducing the precision with which we could determine the signal strength from any one channel.

Because the Fourier experiment inherently yields frequency-domain magnitude *and* phase for each channel, the Fourier experiment effectively opens each channel all of the time (i.e., each channel has full magnitude) by *phase-encoding* the signal from each discrete spectral frequency channel (Eq. 3.9). Therefore, *coherent* excitation is needed in order to "set" the phase for each frequency channel for later decoding by the discrete Fourier transform (Eq. 3.5).

Although Fourier transform interferometry is performed with an *incoherent* radiation source, the interferometer effectively compares the signal to its time-delayed self, thereby effectively converting an incoherent radiation source into a coherent detector. Because the interferometer adds an additional complication to the Fourier transform experiment, we defer its discussion until last (Chapter 9). Figure 4.4 summarizes the distinctions between several common types of single-channel, multi-channel, and multiplex spectrometry.

4.2 Absorption, dispersion, magnitude, and power spectra

As noted in Chapter 3, discrete Fourier transformation of a sampled N-point time-domain signal yields absorption and dispersion discrete spectra, $A(\omega)$ and $D(\omega)$, each of which contain $N/2$ data points and thus half of the available spectral information. Since the absorption-mode display is usually preferred over the dispersion because of narrower and more symmetrical absorption-mode peak shape, it is usual to discard or ignore the dispersion data. Alternatively, one may recover the dispersion information while retaining absorption-mode peak shape by padding the original time-domain data set with another N zeroes before Fourier transformation (see Figure 3.8 for an example). The new N- point absorption-mode spectrum then contains all of the spectral information. The effect of one zero-fill is to interpolate (with correct amplitude) spectral data points at frequencies midway between originally adjacent discrete frequencies. Additional zero-fills do not increase the spectral information content, but may enhance the visual appearance of the spectrum (see Chapter 3.4.1).

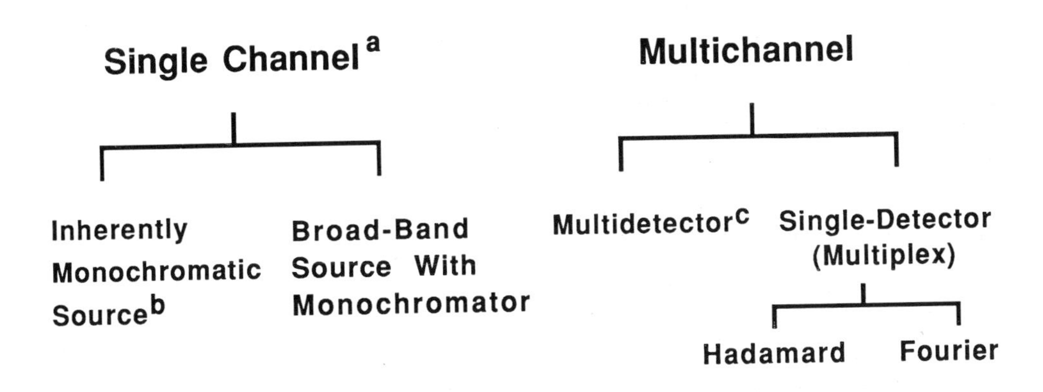

Figure 4.4 Relations between single-channel, multi-channel, and multiplex spectrometers. (a) Although single-channel spectrometers in this diagram are classified according to whether the *source* is inherently monochromatic or broadband, single-channel *detectors* can also be inherently monochromatic or broadband. (b) In these cases, the source and/or detector usually operate with coherent radiation. (c) In these cases, the source and detector usually operate with incoherent radiation which is spatially dispersed according to frequency. The Michelson interferometer is a special case (see text and Chapter 9).

The *discrete* nature of FT spectra cannot be overemphasized. For example, Figure 4.5 shows that even a perfectly phased absorption-mode spectrum can appear asymmetrical, if (as in ^{13}C FT/NMR) there are only a few data points per line width. Although the analog peak shape is indeed perfectly symmetrical in this case, the discrete frequencies at which we are able to sample the spectrum may not be symmetrically spaced with respect to the analog peak maximum. If one were to attempt to "phase" such a spectrum so as to give a symmetrical peak shape, then the spectrum would not be correctly phased. Figure 4.5 is a good example of why one should *not* judge the quality of a spectrum by *visual* criteria alone. In the worst case (see Problems), the FT spectral magnitude error can be 100% (i.e., even a large-magnitude peak can disappear entirely from the FT spectrum)!

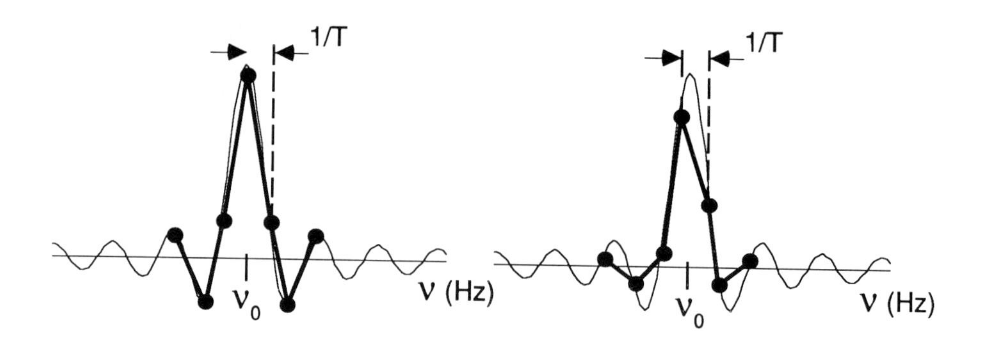

Figure 4.5 Absorption-mode discrete Fourier transform spectra (data points connected by solid lines) and their corresponding analog spectra (dotted curves), for a $T\,\mathrm{sinc}(2\nu_0 T) = \sin(2\pi\nu_0 T)/(2\pi\nu_0)$ line shape. The sampled discrete frequencies may distribute symmetrically (left) or asymmetrically (right) with respect to the analog peak maximum. Thus, a perfectly phased absorption-mode peak may appear asymmetrical (or even disappear entirely—see Problems), because we can observe only the sampled data points rather than the full underlying smooth curve. (See M. B. Comisarow, *J. Magn. Reson.* **1984** *58*, 209-218.)

The difficulty with absorption-mode display is that it requires accurate phase correction over the full spectral range. Although phase correction is relatively straightforward for well-resolved spectra spanning relatively narrow frequency range (e.g., high-resolution ^1H or ^{13}C FT/NMR), phase correction of a broadband spectrum and/or a spectrum containing broad peaks can be much more difficult, because it becomes hard to distinguish true spectral peaks from artifacts (e.g., FT/ICR or broadline NMR). Phasing is also problematic when the delay period between excitation and detection exceeds a period of the Nyquist frequency. If phase correction is not feasible, magnitude-mode display, $M(\omega)$, is preferred:

$$M(\omega) = \sqrt{([A(\omega)]^2 + [D(\omega)]^2)} \tag{1.24}$$

$$P(\omega) = [A(\omega)]^2 + [D(\omega)]^2 = [M(\omega)]^2 \tag{1.25}$$

For a well-resolved peak, the magnitude-mode peak area is directly proportional to the number of molecules (FT/IR), ions (FT/ICR), or nuclei (FT/NMR). Since the power spectrum is the square of the magnitude spectrum, the power spectrum, $P(\omega)$, is generally unsuitable for FT spectroscopy, because the relative peak areas in the power spectrum are no longer proportional to the relative numbers of species having those frequencies.

Two disadvantages of magnitude-mode display emerge when spectral peaks begin to overlap, as shown in Figure 4.6. First, because the magnitude-mode includes dispersion-mode data, a magnitude-mode spectral peak is broader than its absorption-mode component spectrum (by a factor ranging from $\sqrt{3}$ to 2 for peak shape ranging from Lorentz to sinc). Thus, two peaks which are just resolved in absorption-mode will not be resolved in magnitude-mode.

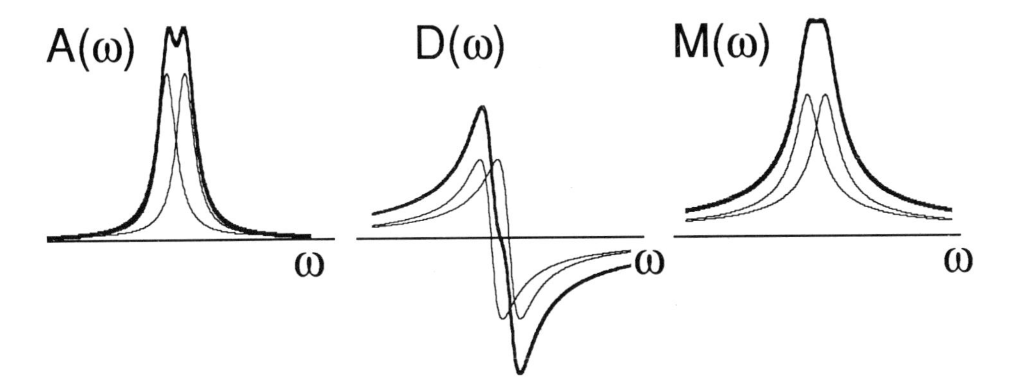

Figure 4.6 Effect of peak overlap on Lorentzian absorption, dispersion, and magnitude spectra. Although the absorption-mode spectrum is a simple sum of the absorption-mode spectra of its components, the dispersion-mode spectra of two overlapping peaks partly cancel and distort the magnitude-mode spectrum of the mixture. The magnitude-mode spectrum is also broader than the absorption-mode spectrum, making it more difficult to resolve overlapping peaks.

Second, because a correctly phased absorption-mode Lorentzian spectrum is everywhere positive-valued, the absorption-mode spectrum of a mixture of several species is the sum of the absorption-mode spectra of all of its components. However, because the dispersion-mode spectra of two nearby peaks partly cancel each other, the magnitude-mode spectrum of a mixture of two or more species can be distorted at frequencies where the dispersion spectral components overlap (see Further Reading). Thus, magnitude-mode spectra are unsuitable for direct analysis of the relative amounts of components of a mixture, unless one is prepared to simulate and add together the component absorption and dispersion spectra for each component before computing Eq. 1.24 (see Problems).

4.2.1 Digital versus analog resolution

Frequency-domain spectral resolution is limited by: (a) analog peak width (i.e., the width of the peak at half-maximum height if the data points were spaced so closely as to approach a continuous curve) and (b) the frequency spacing between spectral points. We will consider the two aspects separately, and then show that the ultimate precision with which spectral peak position (or width or height) can be determined depends on the number of data points per analog peak width.

Analog spectral resolution is defined as $\omega_0/\Delta\omega = \nu_0/\Delta\nu$, in which ω_0 is the angular frequency in rad s^{-1} (and ν_0 is the oscillation frequency in Hz) at which the absorption-mode peak is a maximum, and $\Delta\omega$ (or $\Delta\nu$) is a specified measure of peak width (e.g., full width at half-maximum absorption-mode peak height). Unfortunately, many different conventions for $\Delta\omega$ are in use (e.g., full width at 1%, 10%, or 50% of absorption-mode or magnitude-mode peak maximum height; height of the valley between two peaks of equal width and height; etc.). The reader may therefore need to convert from one peak width definition to another in order to compare resolutions based on different criteria (see Problems).

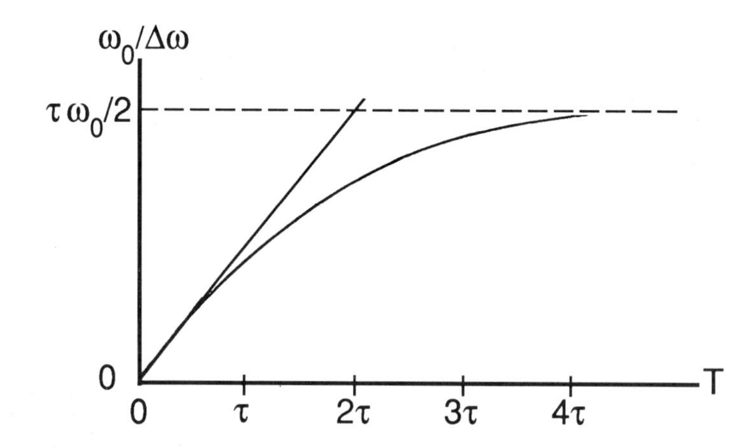

Figure 4.7 Fourier transform analog spectral resolution, $\omega_0/\Delta\omega$, as a function of time-domain acquisition period, T, for an exponentially damped time-domain signal, $\cos \omega_0 t \ \exp(-t/\tau)$. T is measured in multiples of the exponential damping constant, τ. $\Delta\omega$ is defined as the full width at half-maximum height of the absorption-mode spectral peak.

Analog frequency-domain FT spectral resolution increases (linearly at first, then directly but decreasingly) with increased time-domain acquisition period, T (see Figure 4.7). Thus, *analog* peak width sets the upper limit to FT spectral resolution, and *discrete* sampling further limits our knowledge only to the spectral values at the sampled frequencies.

4.2.2 Time-domain and frequency-domain excitation magnitude

In Chapter 1, we discovered that a suddenly displaced mass-on-a-spring oscillates only at its "natural" frequency, with an amplitude that decays exponentially with time due to frictional damping. A Fourier transform of the time-domain displacement yields a frequency-domain spectrum whose width is determined by the time-domain acquisition period, T, and time-domain exponential damping constant, τ.

However, we also showed that the same mass-on-a-spring can be driven into steady-state motion at any driving frequency, with a frequency-domain spectrum which is the same as that of the pulsed-excitation experiment in the limit of very long acquisition period, $T >> \tau$. The question then becomes, when does a system respond to excitation at its natural frequency (the pulse/FT experiment) and when does the same system respond at all frequencies? The answer is simple. Provided that the excited signal does not have time to decay *during* the excitation period, and that the excitation magnitude is sufficiently long duration (see Chapter 4.4), the system can respond *only* to excitation at its natural frequency.

Thus, the relative peak magnitudes in an FT spectrum depend directly on how much excitation magnitude is present at each frequency. For example, Figure 2.13 shows the FT magnitude spectrum of a time-domain frequency sweep excitation and the corresponding magnitude spectrum of the response of four equally abundant oscillators. Clearly, the apparent relative height of each peak is proportional to the excitation magnitude at that peak frequency.

The excitation magnitude spectrum is readily computed from the time-domain excitation waveform. At this stage, it is useful to introduce the FT property that the zero-frequency spectral magnitude, $\boldsymbol{E}(0)$, is simply the area under the time-domain excitation, $\boldsymbol{e}(t)$—see "central ordinate theorem" in Appendix D.

$$\boldsymbol{E}(\omega) = \int_{-\infty}^{+\infty} \boldsymbol{e}(t) \exp(-i\,\omega\, t)\; dt \qquad \text{[Definition of FT of } \boldsymbol{e}(t)\text{]}$$

$$\boldsymbol{E}(0) = \int_{-\infty}^{+\infty} \boldsymbol{e}(t) \exp(-i \cdot 0 \cdot t)\; dt = \int_{-\infty}^{+\infty} \boldsymbol{e}(t)\; dt \qquad (4.5)$$

For example (section 2.2.2.), the frequency-domain magnitude spectrum of a time-domain Dirichlet (rectangular, d.c.) pulse of unit height and duration, T, is a sinc-type function, $\sin(\omega T)/\omega$, of height $= 1 \cdot T = T$ at $\omega = 0$ (see Eq. A.13 in the limit that $\omega \rightarrow 0$), as shown in Figure 2.3. Similarly, a time-domain rectangular sinu-soidal pulse of duration, T, and unit time-domain magnitude produces maximum frequency-domain excitation magnitude, $1 \cdot T = T$ at $\omega = \omega_0$.

4.2.3 Inverse FT: tailored (stored waveform) excitation

Up to now, we have used the *forward* Fourier transform to compute the *frequency* domain spectrum of a previously specified *time* domain waveform. However, in designing an optimal time-domain excitation waveform, it would be better to proceed in reverse order: i.e., begin by specifying the optimal frequency-domain magnitude discrete spectrum, and then perform a discrete *inverse* Fourier transform to generate the corresponding time-domain excitation (discrete) waveform. That stored discrete time-domain data set can then be passed through a digital-to-analog converter to yield an analog time-domain excitation waveform for use as an excitation source. The procedure is illustrated in Figure 4.8 for the particular case of frequency-domain excitation which is uniform in magnitude at all frequencies except for an arbitary-width "window" within which the excitation magnitude is zero. Only the real part of the complex time-domain excitation waveform is needed to produce *linearly*-polarized radiation; both the real and imaginary ("quadrature") components are needed to produce *circularly*-polarized radiation (see section 4.3.2).

The great advantage of tailored (stored waveform) excitation is that it can provide an excitation magnitude spectrum of nearly arbitrary shape, for applications to be discussed in subsequent chapters. However, if the excitation bandwidth is wide, the dynamic range of the computed time-domain waveform can be unacceptably large (Figure 4.9). When the phase of each of the excitation signal frequency components is zero, as in Figure 4.8, for example, the corresponding time-domain signal consists of a superposition of cosines of different frequency. Since the cosine function is maximum at time zero, the time-domain signal is initially very large, and quickly decreases in amplitude as the various cosinusoids begin to interfere destructively with each other.

The key to solving the dynamic range problem is to somehow "scramble" the initial frequency-domain phases, so that the corresponding time-domain sinusoids interfere destructively throughout the time-domain period. One might first think to destroy the time-domain phase coherence by *linearly* varying the initially specified frequency-domain excitation phase before the inverse FT step (Figure 4.9, middle). However, consideration of the "shift theorem" of Fourier analysis (Problem 2.16) reveals that a linear *phase* shift in the *frequency* domain simply shifts the corresponding *time*-domain origin—the effect is the converse of the linear frequency-domain phase shift caused by a time-delay in first-order spectral phasing (Chapter 2.4.). Therefore, a non-linear (e.g., quadratic) phase modulation is required, as illustrated in Figure 4.9, bottom (see Problems). We shall return to the dynamic range issue in the applications Chapters 7–9.

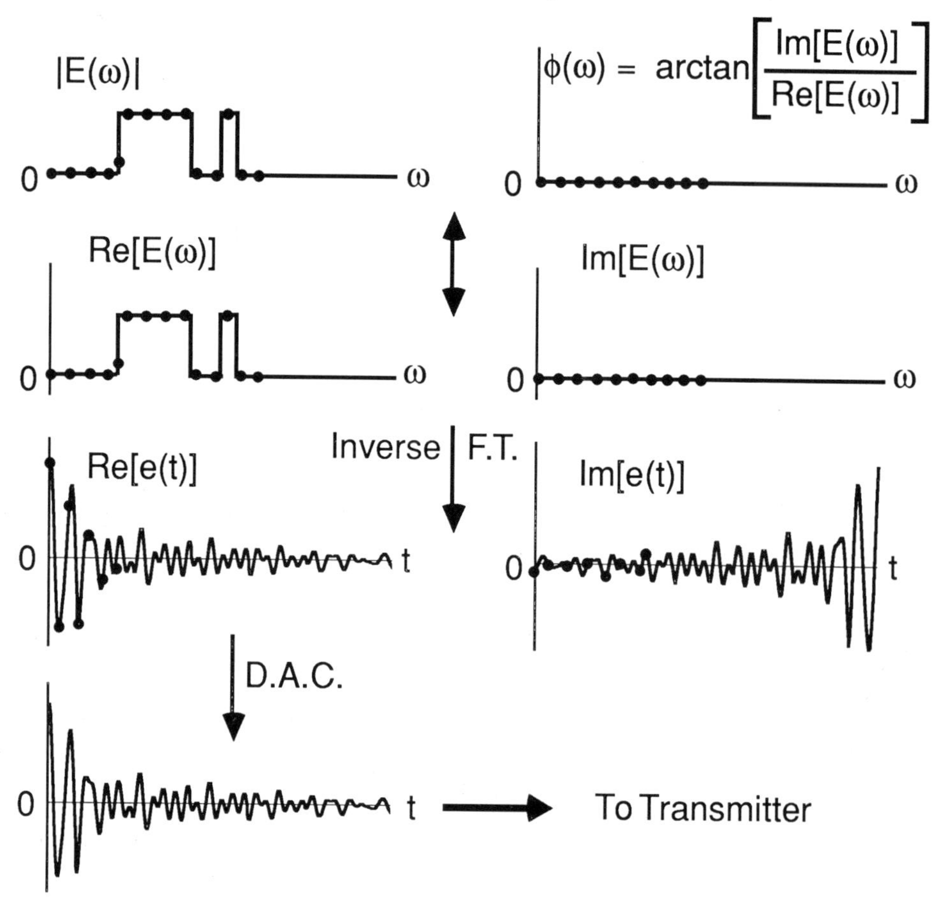

Figure 4.8 Tailored (stored waveform) excitation, for generating a time-domain excitation waveform whose frequency-domain magnitude spectrum is specified in advance. Proceeding from top to bottom in the Figure, the desired frequency-domain excitation is first specified as a series of magnitudes (left) and phases (right) at equally spaced frequency increments. Each magnitude/phase data pair is then converted to its equivalent real/imaginary pair. An inverse discrete FT then generates the corresponding discrete time-domain data set, which is digital-to-analog converted to yield an analog time-domain excitation waveform whose effect will be to produce the excitation spectrum specified initially at the top of the Figure. The point spacing has been exaggerated for purposes of illustration.

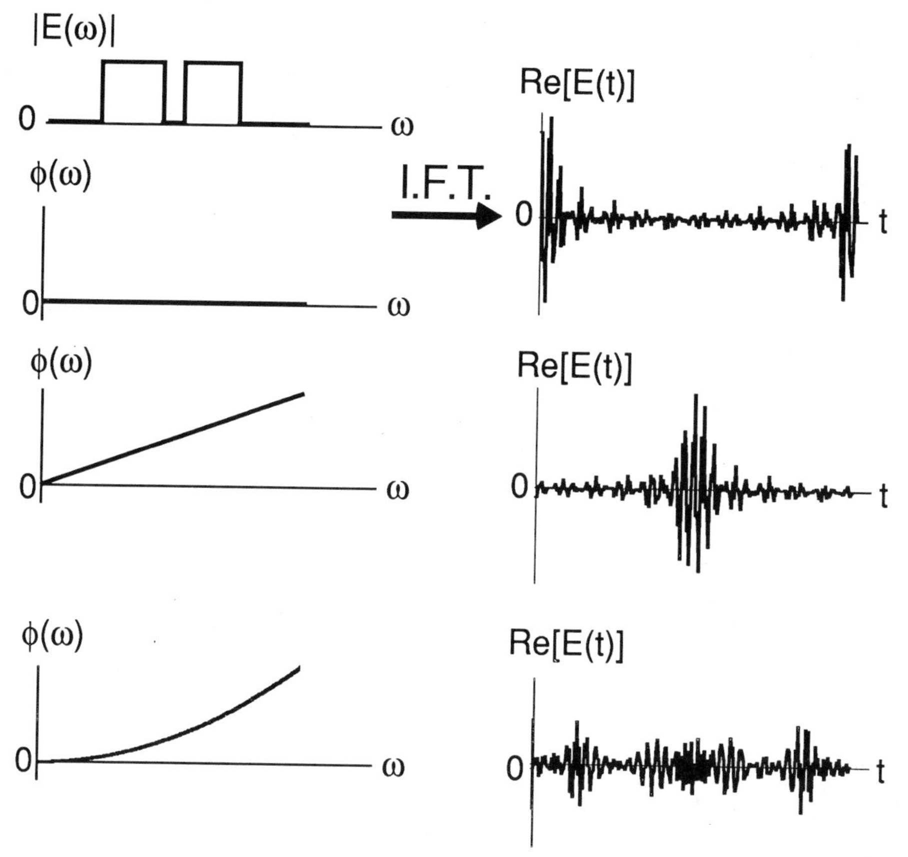

Figure 4.9 Phase-modulated stored waveform excitation. Top: The time-domain data set (upper right) generated by inverse discrete FT of a constant-phase frequency-domain discrete spectrum (upper left) exhibits impractically high dynamic range, because of the common initial phase of all of its component time-domain cosinusoids. Middle: Linear phase variation in the frequency-domain before inverse FT simply shifts the time-domain signal without changing its shape. Bottom: However, if the desired excitation magnitude spectrum is first phase-modulated (in this case, by quadratic variation of phase with frequency), then the resulting time-domain data set (lower right) has optimally reduced dynamic range.

4.3 Linear versus circular polarization

Natural motions probed by Fourier transform spectroscopy can be either inherently *linear* (e.g., molecular vibrations in FT/IR) or inherently *circular* (e.g., Larmor precession in FT/NMR; ion cyclotron motion in FT/ICR). Linear motion (or one of the two linear components of circular motion) is represented by a mathematically real (or the real part of a mathematically complex) function, and the x- and y- components of circular motion about the z- axis are represented by the real and imaginary parts of a complex rotating vector (also known as a "phasor"), as shown in Figure 4.10.

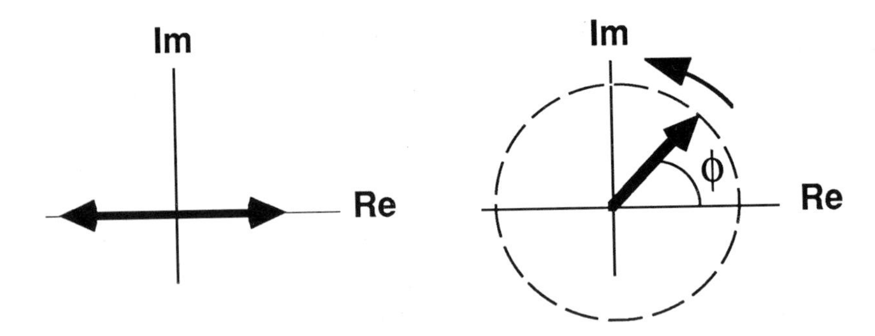

Figure 4.10 Linear motion (left) and circular motion (right), represented by real and complex vectors. Even when motion is inherently linear, its instantaneous displacement is conveniently represented by the real part of a rotating vector in the complex plane. For linear or circular motion, the accumulated phase (angle), $\phi = \omega t$, is clearly visualized (to within a multiple of 2π radians) in the circular representation.

Let us represent linear motion (or the x-axis component of an oscillating electric or magnetic field) as the real function,

$$f(t) = A_0 \cos(\omega_0 t - \phi_0) \tag{4.6}$$

and circular motion (or the x- and y-components of an electric or magnetic field vector rotating about the z-axis) by a complex function,

$$\boldsymbol{f}(t) = A_0 \exp(i\omega_0 t - \phi_0) \tag{4.7}$$

in which ω_0 is the natural angular frequency, A_0 is the magnitude, and ϕ_0 is the time-domain initial phase (angle) of the motion. We will now examine the frequency-domain spectra of such functions.

4.3.1 Negative-frequency components

The mathematical Fourier transform is defined to include negative times and negative frequencies. Actual physical observables are, of course, defined only for positive times, since we begin our measurements by definition at time zero. In order to complete the correspondence between physical reality and its mathematical representation, we need to decide how to treat negative time and negative frequency.

All physical time-domain excitation or response signals are "causal": i.e. defined only for times, $t \geq 0$. Mathematically, such a signals is simply defined as zero for $t < 0$. The effect of causality on Fourier transforms has been discussed in section 2.1.1 (see also Appendix D).

Similarly, the Fourier transform of a (real or complex) time-domain signal yields a frequency-domain spectrum extending over the frequency range, $-\infty < \omega < +\infty$. When the FT frequencies correspond to the true spectral frequencies (in so-called "direct-mode" experiments—see section 4.4.2.), then the negative-frequency part of the spectrum is ignored. However, in "heterodyne-mode" experiments, the frequency-domain spectrum is shifted down in frequency, so that mathematically negative frequencies can correspond to physically "real" motions. Direct-mode and heterodyne-mode spectra will be discussed in Chapter 4.4.2.

Figure 4.11 shows the positive- and negative-frequency spectra of right- or left-circularly-polarized signals (represented as vectors rotating counterclockwise or clockwise in the complex plane, as in Figure 4.10) and a linearly-polarized signal (represented as a one-dimensional oscillation along a mathematically real axis). Fourier transformation of a right- (or left-) *circularly* polarized time-domain signal (Figure 4.11, top two diagrams) yields a complex spectrum with only *positive* (or only *negative*) components. In contrast, FT of a *linearly* polarized signal produces a complex spectrum whose positive- and negative-frequency components are mirror images (Figure 4.11, bottom two diagrams). Since a *linearly polarized* time-domain signal, $\cos \omega_0 t \exp(-t/\tau)$ in this case, can be represented mathematically as the sum of equal-magnitude right- and left-circularly polarized components (see section 4.3.2.), e.g.,

$$\cos \omega_0 t = \frac{1}{2} [\exp(i \omega_0 t) + \exp(-i \omega_0 t)] \tag{4.8}$$

its FT complex spectrum thus contains real and imaginary components at both *positive and negative* frequencies, as noted in Chapter 1.3.1 and Appendix D.

From Figure 4.11, we see that the negative-frequency spectrum contains either zero (right-circularly-polarized signal) or redundant (linearly-polarized signal) information. Nevertheless, the negative-frequency components of a linearly-polarized signal can affect the physically relevant positive-frequency spectrum in two ways. First, comparison of the spectra of Figure 4.11 shows that if we wish to excite or detect (right-) *circularly* polarized motion, then half of the frequency-domain signal from a *linearly* polarized time-domain waveform is "wasted" (i.e., unobservable) by appearing at negative spectral frequency. The presence of the negative-frequency spectral component thus reduces our observable spectral signal strength by a factor of two when *linearly* polarized excitation or detection is applied to *circularly* polarized motion (as in NMR or ICR spectroscopy). In the next section, we shall discuss the "quadrature" method for recovering that factor of two (in the absence of noise).

The second effect of negative-frequency peaks is also shown in Figure 4.11. If the natural angular frequency, ω_0, approaches to within a few multiples of $(1/\tau)$ radians of zero frequency (in "direct-mode", or near the carrier frequency in "heterodyne"-mode), then the negative-frequency signal spills over and adds to (or subtracts from) the positive-frequency component. Thus, another advantage of circularly-polarized detection of circularly-polarized motion is that the positive-frequency spectrum is undistorted by the presence of negative-frequency components, even when the "natural" frequency, ω_0, approaches the spectral line width ($2/\tau$ for a Lorentzian peak).

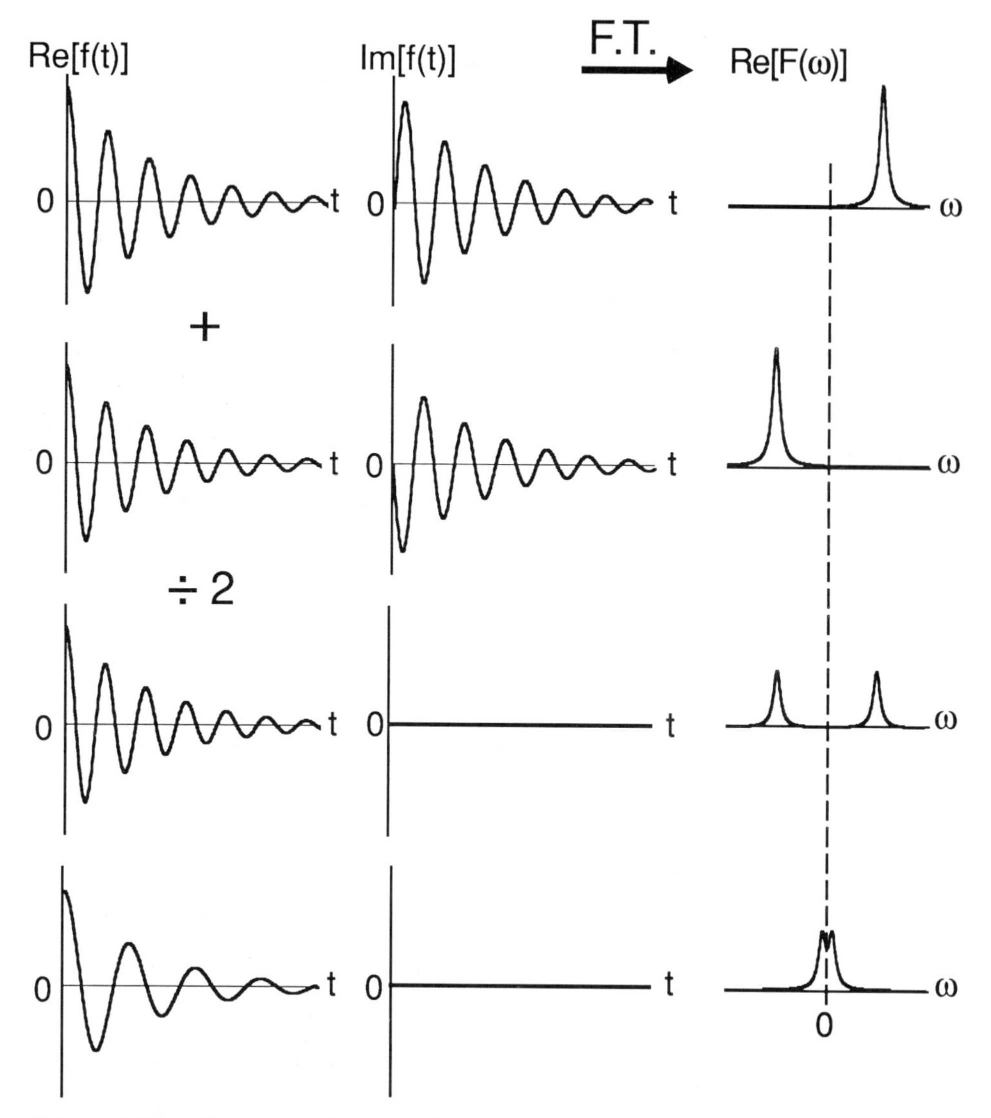

Figure 4.11 Frequency-domain real (absorption, in this case) and imaginary (dispersion, in this case) spectra (right) obtained by Fourier transformation of time-domain signals (left). Top row: The x- and y-components of a right-circularly polarized time-domain signal are represented by the real and imaginary parts of a complex function of time. Second row: left circularly-polarized time-domain signal. Third row: linearly-polarized time-domain signal, obtained as half the sum of the signals in the top two rows. Bottom row: as for the preceding case, except that the natural frequency of the motion approaches to within about one spectral absorption-mode peak width of zero frequency—note the spillover of the (physically meaningless) negative-frequency signal into the (physically observable) positive-frequency spectrum.

4.3.2 Quadrature excitation and detection

Figure 4.11 shows that when a time-domain (excitation or detection) signal, $f(t)$, is *circularly* polarized,

$$f(t) = A_0 \exp(i\omega_0 t) = A_0 \cos \omega_0 t + i A_0 \sin \omega_0 t \tag{4.9}$$

then the Fourier transform of just one of its two *linearly* polarized components (say, $A_0 \cos \omega_0 t$) effectively distributes half of the signal to negative frequency where it is unavailable experimentally (see Eq. 4.8).

For example, although NMR and ICR natural motions are inherently *circular*, we can nevertheless excite an NMR (or ICR) signal by irradiation with an oscillating *linearly* polarized magnetic (or electric) field (see Chapters 7 and 8). The mechanism (see Eq. 4.8 and Figure 4.12) is that such a linearly oscillating field can be analyzed into two field vectors rotating in opposite senses. Only the component rotating in the same sense (and at the same frequency) as the resonant nuclear magnetic moment (or cyclotron resonant ion) will excite (drive) the NMR or ICR system. The component rotating in the wrong sense corresponds to the negative-frequency signal in Figure 4.11 and has (almost) no direct effect on an NMR or ICR system (see, e.g., Chapter 7.5.1).

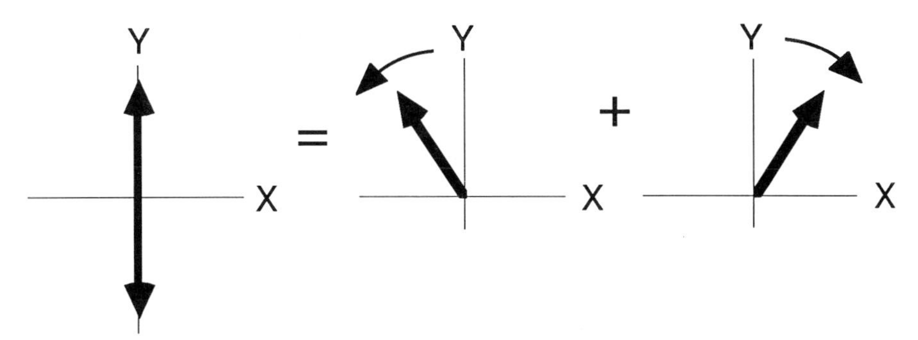

Figure 4.12 Analysis of a linearly-polarized oscillating magnetic (or electric) field into two counter-rotating components. In general, only the component rotating in the same sense as the NMR (or ICR) signal is experimentally significant.

Figures 4.11 and 4.12 should make clear that *linearly* polarized excitation or detection of inherently *circularly* polarized motion effectively throws away half of the available signal. Thus, it would be preferable to excite and detect with *circularly* polarized radiation. Experimentally, circularly polarized signals may be excited by separately generating their two linearly polarized components (Eq. 4.9): e.g., $A_0 \cos \omega_0 t$ along the x- direction and $A_0 \sin \omega_0 t$ along the y- direction to produce a signal of magnitude, A_0, rotating at angular frequency, ω_0, in the x-y plane, as shown in Figure 4.13. The construction (or analysis) of a circularly polarized signal from (or into) its two linearly polarized components is called *quadrature*.

Quadrature *excitation* reduces the required excitation magnitude by a factor of 2—often a critical advantage in FT/NMR or FT/ICR. Quadrature *detection* increases the signal strength by the same factor of two, but also increases the root-mean-square noise level by a factor of $\sqrt{2}$ (see Chapter 5), for a net gain in signal-to-noise ratio of a factor of $2/\sqrt{2} = \sqrt{2}$.

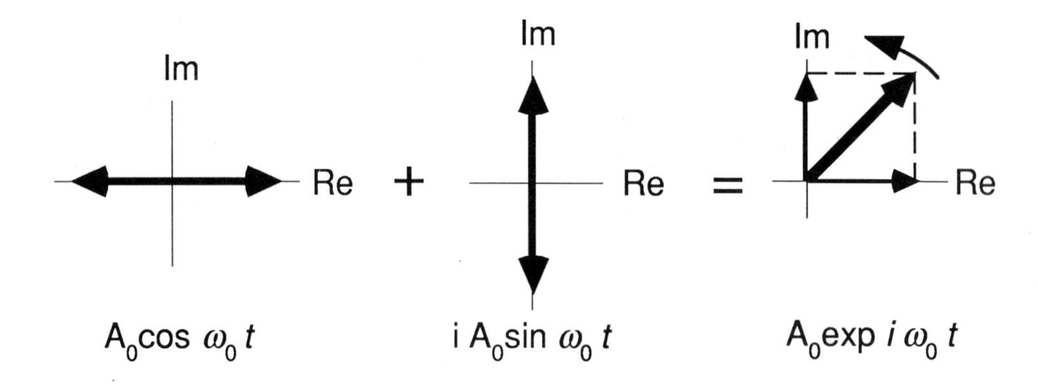

Figure 4.13 Quadrature construction of a circularly polarized signal rotating in the x-y plane from two components linearly polarized along the x- and y- axes, respectively.

We shall next consider three major experimental artifacts associated with quadrature operation. Ideally, we would like to excite or detect simultaneously, with equal signal strength on both quadrature "channels" (i.e., the x- and y-components of Figure 4.13. Failure to meet these constraints leads to three distinct artifacts (sections 4.3.2.1-3). Fortunately, we can show how to compensate for quadrature artifacts by co-adding successive time-domain signals of different phase (section 4.3.2.4). Finally, for heterodyne-mode operation, we will show how quadrature operation can reduce the frequency-domain bandwidth by a factor of 2 (section 4.4.2, and Chapter 7.4.6).

4.3.2.1 *Baseline offset between quadrature channels: the center glitch*

A common artifact in quadrature detection is the presence of a prominent (false) peak at the center of the spectrum (i.e., at zero frequency). (See Chapter 4.4.2 for further discussion of negative-frequency signals.) The artifact arises from a difference in d.c. offset between the two time-domain quadrature signals, as shown in Figure 4.14.

4.3.2.2 *Gain difference between quadrature channels: image peaks*

If the two quadrature components have different magnitude (see Figure 4.15), as from different gain in the two quadrature-detected channels, then their negative-frequency components will not quite cancel, and an "image" peak will appear at that negative frequency. The image peak can be identified by shifting the center (carrier) frequency: a "true" peak will have the same apparent frequency (after correcting for heterodyne shift in heterodyne mode), but an "image" peak will shift in apparent frequency (see Chapter 4.4.2). Both the baseline offset and gain difference artifacts can be eliminated by appropriate phase-cycling (Chapter 4.3.2.4).

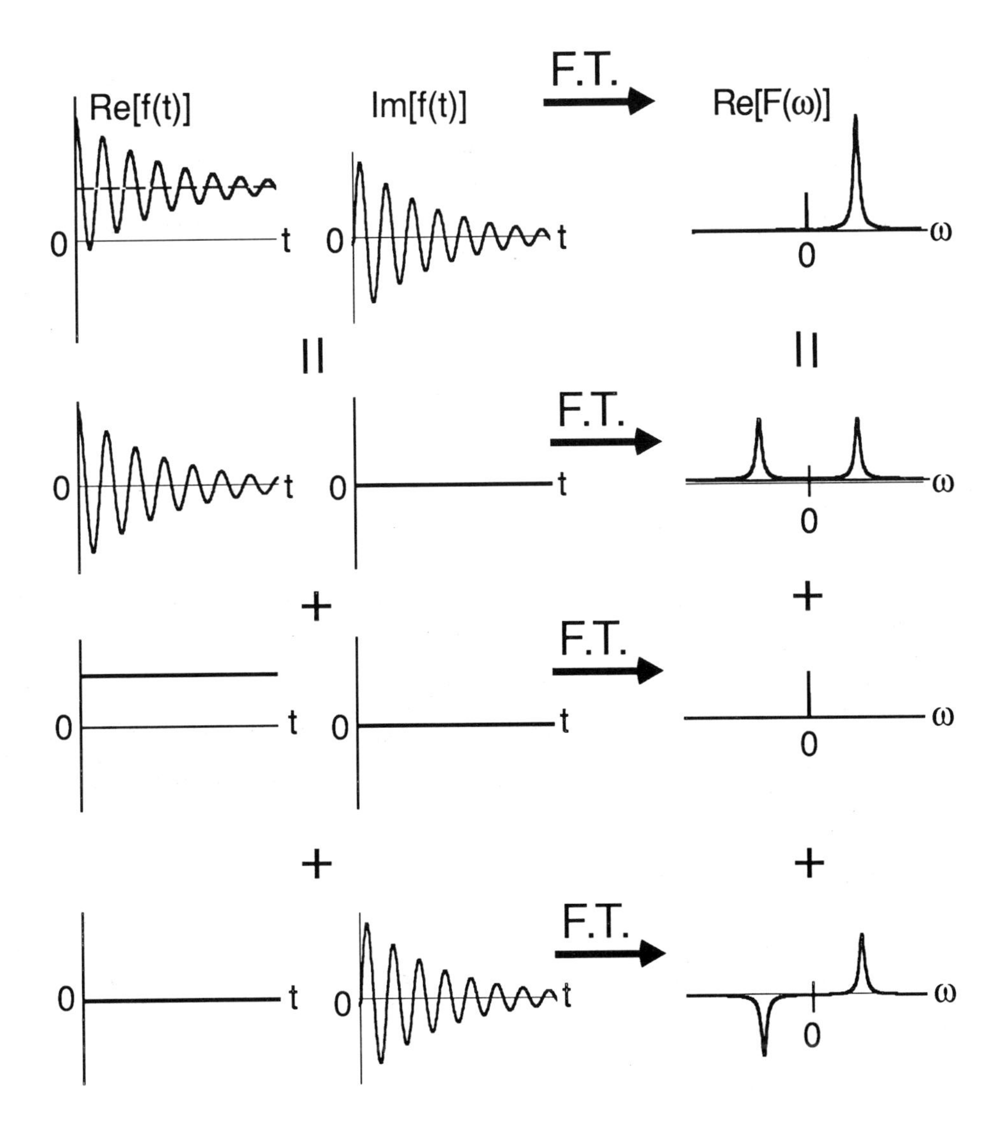

Figure 4.14 Origin of the "glitch" at zero frequency in the FT real spectrum resulting from quadrature detection. Proceeding counterclockwise from top left in the Figure, a difference in d.c. offset between the two quadrature channels effectively adds an excess d.c. signal to one channel. The Fourier transform of that d.c. (zero-frequency) signal then produces a peak at the center of the quadrature spectrum.

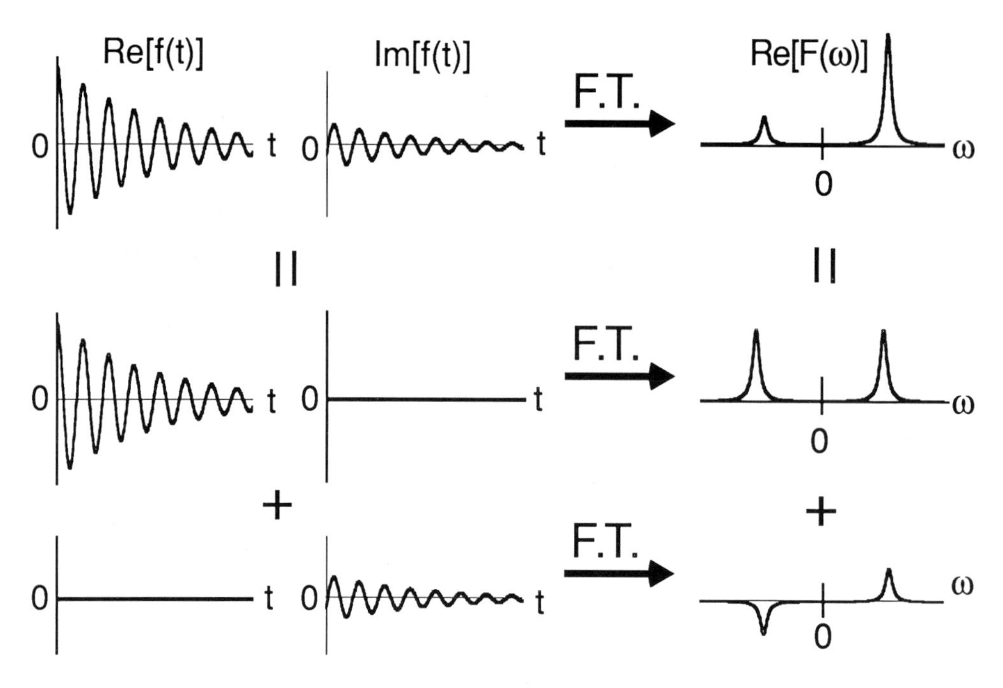

Figure 4.15 Origin of "image" peaks in quadrature FT real spectra. A difference in magnitude between the two quadrature time-domain components leads to incomplete cancellation of the negative-frequency spectral peak, which thus appears as an "image" (with reduced magnitude of the true peak.

4.3.2.3 *Time delay between quadrature channels: alternate* vs. *simultaneous sampling*

As described up to this stage, quadrature is understood to denote *simultaneous* transmission or detection of two signals differing in phase by 90° (either by mechanical or electronic orthogonality—see Chapter 7.4.6). Experimentally, one could achieve such an arrangement by duplicating the transmitter or receiver channels as shown for a schematic receiver in Figure 4.16 (top diagram).

However, in practice it can prove difficult (and/or expensive) to make some or all of the components of the two quadrature channels identical with respect to d.c. offset, gain, phase, and time-delay between the raw signal and the analog-to-digital converter (ADC) output. Therefore, quadrature detection is often accomplished by time-sharing a single analog-to-digital converter (and/or other channel components), as illustrated for two possible schemes in Figure 4.16 (middle and bottom). In time-shared detection, the digitized quadrature signals are sampled-and-held either simultaneously or alternately, and the ADC output is routed alternately to the "real" and "imaginary" data sets stored in the computer. (For the simultaneous time-shared detection case, one of the two sampled-and-held signals is delayed by one ADC sampling period with respect to the other.) In both methods shown in Figure 4.16 (middle and bottom), the ADC must sample at twice the sampling rate of the ADC in Figure 4.16 (top), because only half of the ADC output data points end up in either of the quadrature time-domain discrete data sets.

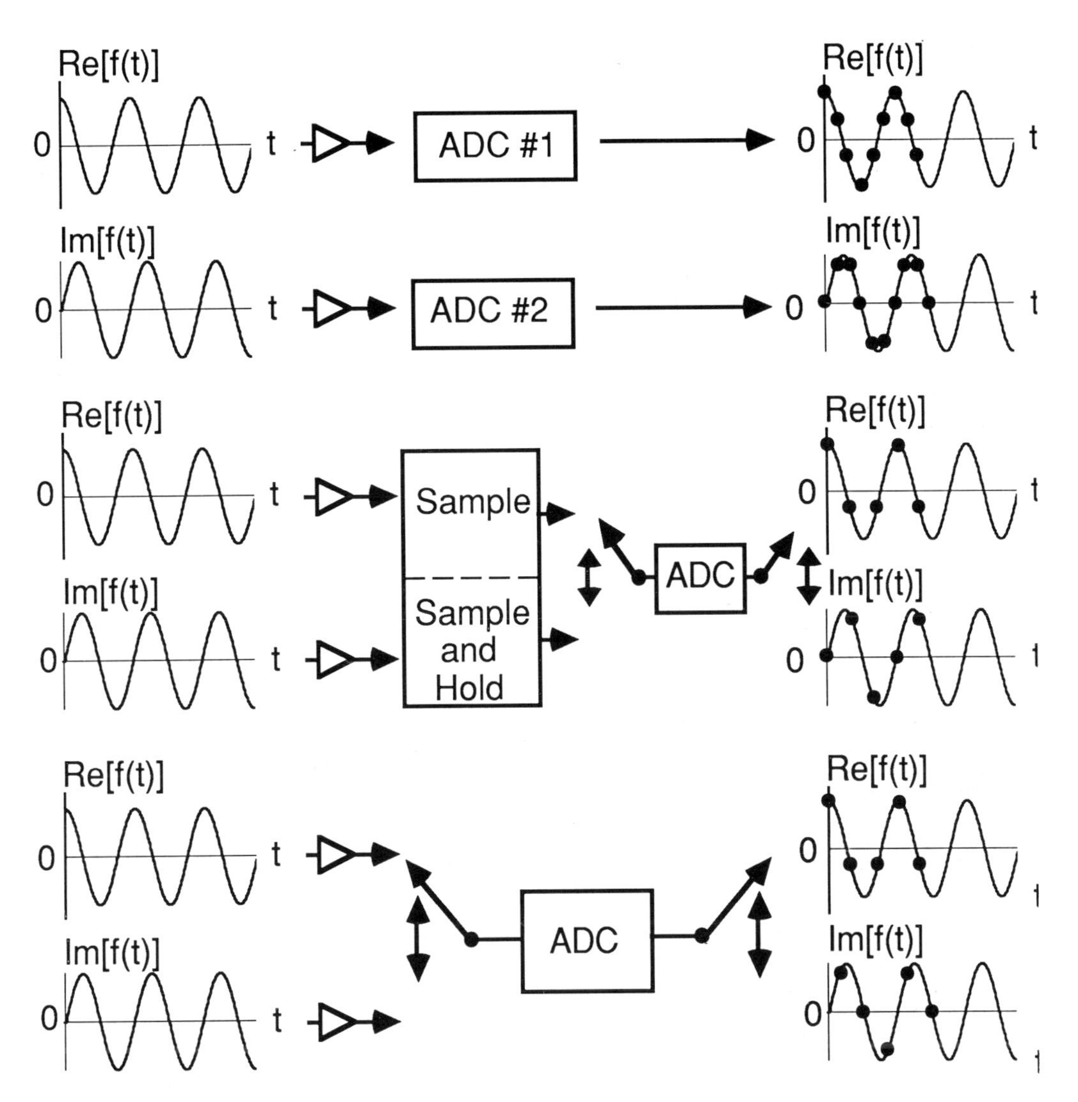

Figure 4.16 Schematic quadrature detection modes. Top: two independent analog-to-digital converters. Middle and Bottom: one analog-to-digital converter, with simultaneous (middle) or alternating (bottom) sample-and-hold—in each case, the odd-numbered data points are taken as the "real" and the even-numbered points as the "odd" parts of a complex time-domain data set. (See text).

If the two quadrature channels in any of the schemes of Figure 4.16 do not differ by exactly 90° in phase at each spectral frequency, then the "real" and "imaginary" data sets will no longer be perfectly independent, and complex Fourier transformation of those data sets will produce a spectrum with image peaks, as in Figure 4.15. Accurate quadrature operation requires that the two channels be separated (geometrically or electronically) by precisely 90° in phase.

The immediate problem posed by alternate sampling is that every other data point (2nd, 4th, 6th, etc., forming the "imaginary" time-domain data set) is acquired after a delay of one dwell time with respect to its quadrature partner (e.g., 1st, 3rd, 5th, etc. point, forming the "real" time-domain data). As noted in section 2.4.1, a constant time-domain delay leads to a linear frequency-domain phase shift for the spectrum of the even-numbered time-domain data points. Therefore, one might think to FT the odd- and even-numbered time-domain data sets separately, then correct for the linear phase-shift in the spectrum of the even-numbered data only, and then add the spectra from the odd- and even-numbered time-domain data. Such a procedure does in fact eliminate image peaks due to the time-shift between the alternately sampled channels, but at the cost of reducing the spectral bandwidth by a factor of two if foldover is to be avoided (see Problems). A simpler and better solution is to multiply successively acquired time-domain samples by the sequence: $1, 1, -1, -1, \cdots$. Conventional complex FT then yields a spectrum free of image peaks (see Problems).

The most important practical difference between simultaneous and alternated quadrature detection is the location of folded-over (aliased) peaks, whose true frequencies exceed the Nyquist frequency (see Chapter 3.2.1), as shown in Figure 4.17. For simultaneous acquisition, the folded-over peaks are "wrapped around" the spectral bandwidth, to appear at the opposite end of the spectrum from their true position. For alternated acquisition, on the other hand, the aliased peaks are "folded" about the nearest end of the spectral bandwidth.

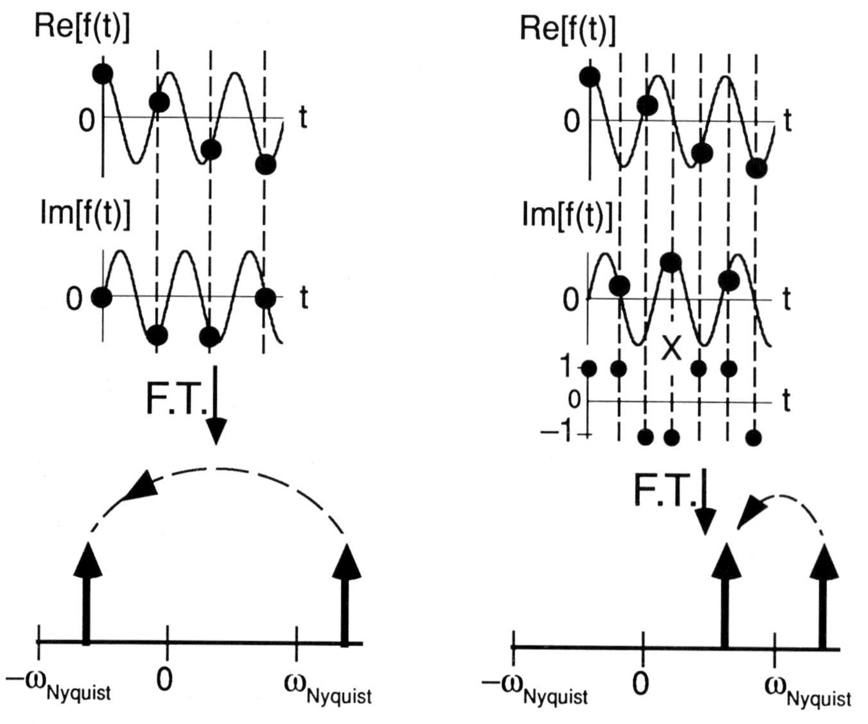

Figure 4.17 Simultaneous (left) and alternated (right) quadrature data acquisition. The Fourier transform spectrum of a signal whose frequency exceeds the Nyquist limit will "fold over" differently in the two cases (see text).

4.3.2.4 *Phase cycling*

If time-domain data acquisition can be repeated identically many times (e.g., as in FT/NMR), then the various quadrature detection anomalies (e.g., differential d.c. level, gain, phase, and time-delay between the two channels) may largely be corrected by phase cycling sequences. The basic principle of phase cycling is that the phase (including sign) and magnitude of the desired *signal* is determined directly by the phase and magnitude of the *excitation*, whereas the phase and magnitude of undesired artifact signals are unrelated to the excitation. Also, the signal must be identical (except for phase) from scan to scan in order for phase-cycling methods to succeed.

For example (see Figure 4.18), suppose that the d.c. offset level differs between the two quadrature channels, leading to the zero-frequency "glitch" of section 4.3.2.a. We need simply shift the transmitter phase by 180° between each two successive acquisitions, and then subtract the second acquisition from the first. Since a 180° phase shift corresponds to changing the sign of the time-domain excitation signal, the detected signal will also change sign on every other acquisition. However, the d.c. level will remain the same from one acquisition to the next, and will therefore cancel out when we subtract the second acquisition from the first.

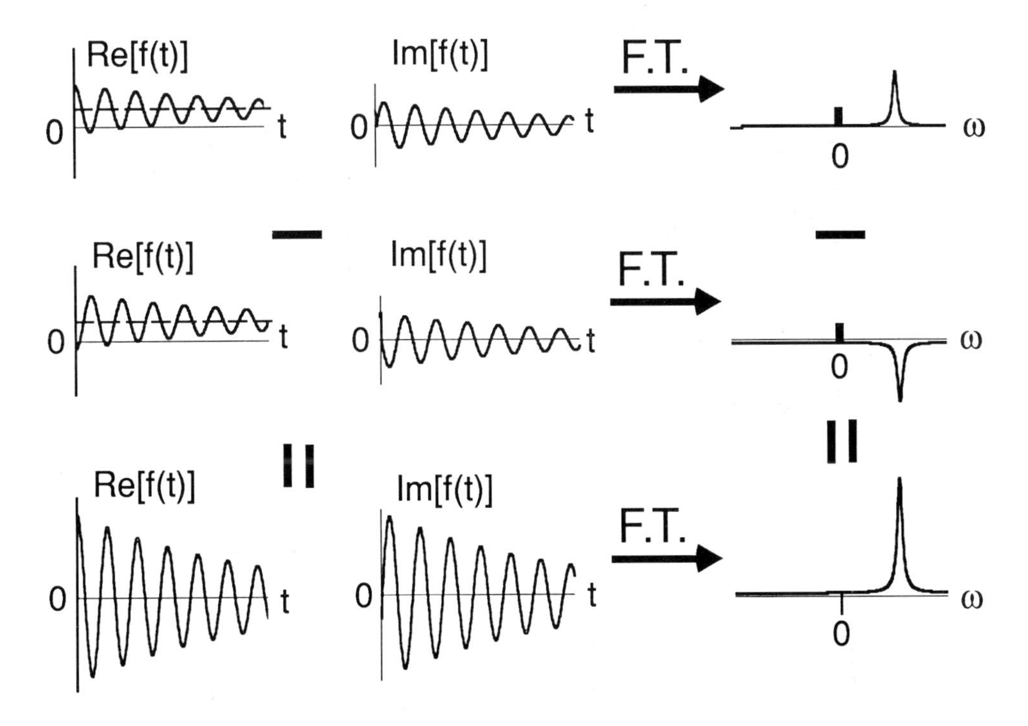

Figure 4.18 0° - 180° phase alternation for removal of the frequency-domain zero-frequency "glitch" arising from different d.c. offset in the two channels of a quadrature FT experiment. (See text.)

Similarly, a slightly more elaborate phase-cycling scheme involving 0°, 90°, 180°, and 270° excitation phase shifts can be used to compensate for unequal gain in the two quadrature channels (see Problems). Finally, a general scheme is available for unscrambling several types of quadrature detection artifacts (see Further Reading). For example, as previously noted, image peaks produced by alternate sampling can be eliminated by changing the sign of every 3rd and 4th time-domain point (1, 1, −1, −1; 1, 1, −1, −1; etc.). Quadrature signal analysis is an excellent test of the reader's command of Fourier transforms, because it involves so many aspects: real and imaginary data, even *vs.* odd waveforms, foldover, positive *vs.* negative frequencies (see Chapter 4.4.2), phase, etc.

4.4 Non-linear effects

Fourier analysis is inherently suited to *linear* systems: i.e., a system for which the output magnitude is proportional to the input magnitude at each frequency (or, equivalently, a system that can be modeled by a differential equation with constant coefficients—see Chapter 6.2). In this section, we discuss non-linear systems for three reasons. First, all experimental systems are somewhat non-linear. We therefore need to be able to recognize the artifacts of non-linear response: namely, signals at multiple and difference frequencies of the true spectrum (Chapter 4.4.1). Second, some of the most useful spectroscopic applications are based upon components designed expressly for their non-linearity: mixers for heterodyning; doublers as sources of new laser frequencies; and double- and other multiple- quantum techniques for simplifying NMR, ICR, and optical spectra (Chapter 4.4.2). Finally, the familiar amplitude-modulation (AM) and frequency-modulation (FM) methods of radiofrequency transmission and reception depend on non-linearity, and are rapidly spreading from their long-known applications in Raman and NMR spectroscopy to other fields (Chapter 4.4.3).

4.4.1 Origin of harmonic and intermodulation frequencies

Figure 4.19 shows the response output magnitude versus signal input magnitude for a typical non-linear system. Suppose that the input signal consists of a sum of two cosinusoids of different frequency:

$$In(t) = A_0 \cos \omega_A t + B_0 \cos \omega_B t \tag{4.10}$$

If the time-domain output signal, *Out* (*t*), is proportional to the input, *In* (*t*), their frequency-domain spectra will be identical except for a vertical scale factor.

For a non-linear system, however (e.g., Figure 4.24), we may in general express *Out* (*t*) as a (non-linear) power series in *In* (*t*):

$$Out(t) = c_1 \, In(t) + c_2 \, [In(t)]^2 + c_3 \, [In(t)]^3 + \cdots \tag{4.11}$$

To a first approximation, consider only the first two terms on the right-hand side of Eq. 4.11, for the particular input function given by Eq. 4.10:

$$\begin{aligned} Out(t) &= c_1 A_0 \cos \omega_A t + c_1 B_0 \cos \omega_B t \\ &+ c_2 A_0^2 \cos^2 \omega_A t + c_2 B_0^2 \cos \omega_B t \\ &+ 2 c_2 A_0 B_0 \cos \omega_A t \cos \omega_B t \end{aligned} \tag{4.12}$$

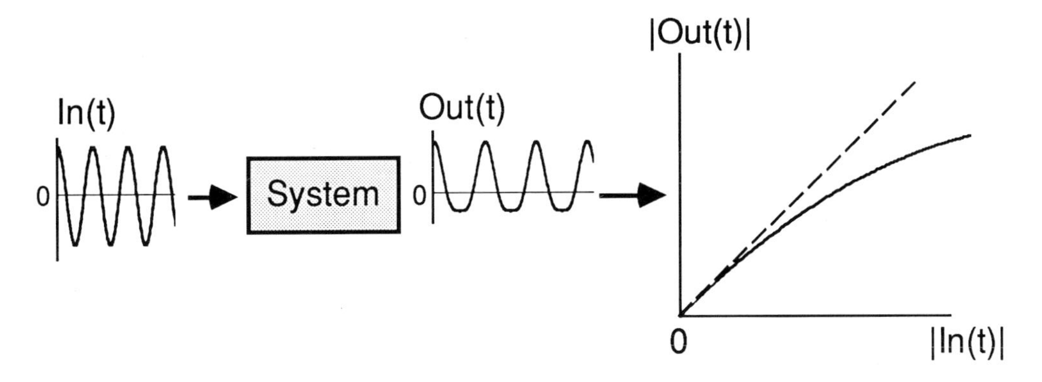

Figure 4.19 Plot of output (or response) magnitude versus input (or excitation) magnitude for a non-linear device (or system). Note that for sufficiently small input magnitude, the output magnitude is proportional to the input magnitude (i.e., the system is linear). The non-linearity produces harmonic and intermodulation distortion (see text).

From the by now familiar trigonometric identities (see Appendix A),

$$\cos^2 a = \frac{1}{2}(1 + \cos 2a) \tag{A.28a}$$

$$\cos a \cos b = \frac{1}{2}[\cos(a + b) + \cos(a - b)] \tag{A.26b}$$

it should be clear that Eq. 4.12 can be re-expressed in the form,

$$\begin{aligned} Out(t) &= A \cos \omega_A t + B \cos \omega_B t \\ &+ C \cos 2\omega_A t + D \cos 2\omega_B t \\ &+ E \cos(\omega_A + \omega_B)t + E \cos(\omega_A - \omega_B)t + F \end{aligned} \tag{4.13}$$

in which $F = \frac{c^2}{2}(A_0{}^2 + B_0{}^2)$.

Figure 4.20 shows the frequency-domain spectra corresponding to the time-domain *linear* output (Eq. 4.10) and the linear and quadratic terms of the *non-linear* output (Eq. 4.13). It is obvious that the "true" (or "fundamental") signals at ω_A and ω_B are still present in the spectrum of the non-linear system. However, two new kinds of frequencies are also present. First, there are "*harmonic*" signals at twice the fundamental frequencies: i.e., $2\omega_A$ and $2\omega_B$. Second, there are "*intermodulation*" signals at frequencies formed from the sum and difference of the "true" frequencies: i.e., $(\omega_A - \omega_B)$ and $(\omega_A + \omega_B)$. Perhaps the most familiar example of harmonic and intermodulation distortion is in the *non-linear* conversion of an audiofrequency voltage signal into sound by a loudspeaker.

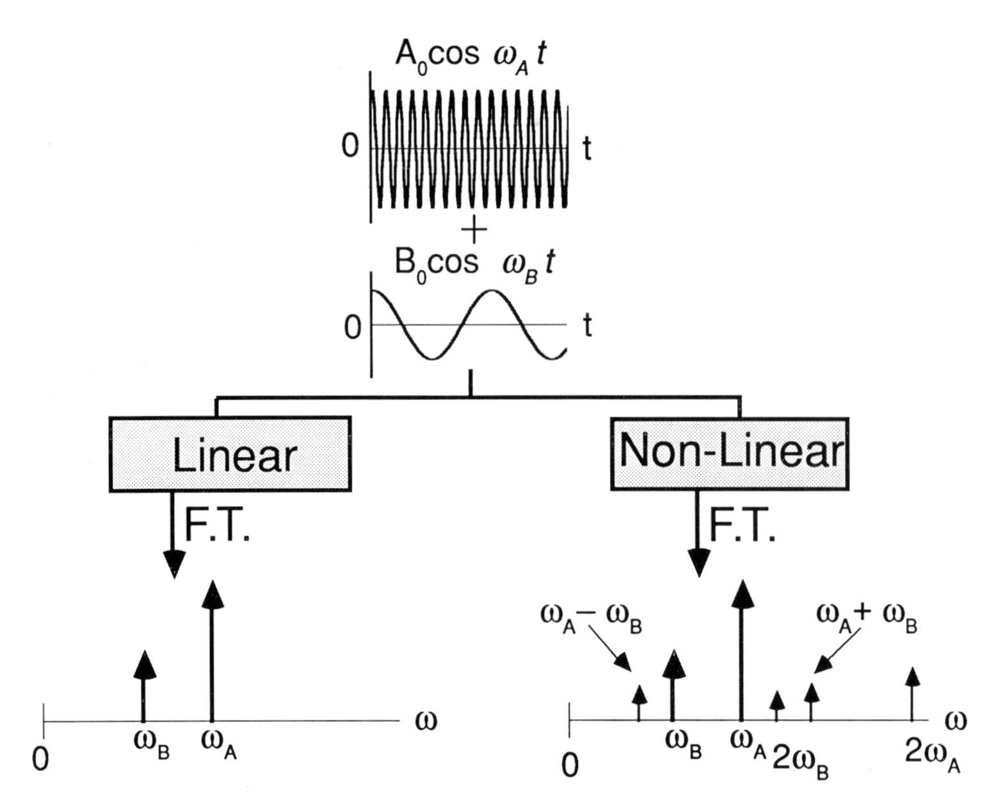

Figure 4.20 Frequency-domain spectra resulting from passage through a linear (Eq. 4.10) or (quadratically) non-linear (Eq. 4.13) device of a time-domain signal consisting of a sum of two cosinusoids of frequencies, ω_A and ω_B. Note that the non-linear system introduces new signals at harmonic ($2\omega_A$, $2\omega_B$) and inter-modulation [($\omega_A + \omega_B$) and ($\omega_A - \omega_B$)] frequencies.

Higher-order harmonics are generated from the higher-order terms in the expansion of Eq. 4.11 {e.g., $\omega_A{}^3$ from c_3 $[In(t)]^3$}, as are more exotic intermodulation frequencies (e.g., combinations of three or more frequencies). We will next consider practical applications for the new frequency signals made available by non-linear devices.

4.4.2 Mixers and doublers: heterodyne detection

A frequency "doubler" results from the special case that $\omega_A = \omega_B$ in Eqs. 4.10 and 4.11. In other words, passage of a sufficiently strong *single-frequency* input signal, $A_0 \cos \omega_A t$, through a non-linear device will generate an output signal with a component oscillating at $2\omega_A$:

$$\cos^2 \omega_A t = \frac{1}{2}[1 + \cos 2\omega_A t] \qquad (4.14)$$

For example, if an intense Nd/YAG laser beam (1064 nm wavelength) passes through a special crystal, such as potassium dihydrogen phosphate (KH_2PO_4 or "KDP") or "beta-barium borate" (β-BaB_2O_4), the non-linear optical response of the crystal is such that up to 30-45% of the light leaving the crystal is at the second harmonic frequency (i.e., 532 nm wavelength)! In the absence of phase-matching problems (see texts on non-linear spectroscopy), a smaller fraction of the crystal output intensity may appear at tripled or quadrupled frequency (355 or 266 nm).

It is interesting to examine the mechanism for laser frequency doubling. A sufficient condition for a crystal to act as a frequency doubler is that the electrons of its component molecules be more easily forced (by the electric field of the incident light) to move in one direction than in the opposite direction. For example, consider an incident light wave whose electric field is plane-polarized as shown in Figure 4.21. On striking a crystal of aromatic molecules oriented with electron-donor and electron-acceptor groups *para* to each other, the electric field of the light wave will displace the aromatic π-electrons farther toward the acceptor than toward the donor end of the molecule. Thus, when those π-electrons are set into motion, the electromagnetic wave that they broadcast ("scatter") will propagate with an asymmetric electric field oscillation. That asymmetry distorts the wave from its original sinusoidal shape, and results in the generation of harmonics at multiples of the fundamental frequency). Unfortunately, even a highly optically anisotropic molecule may exhibit negligible doubling properties as a crystal, if the molecule happens to crystallize in a centrosymmetric crystal lattice. Thus, the "magical" property that makes KDP and "beta-barium borate" crystals unique is that they are composed of highly optically anisotropic molecules whose crystals are not centrosymmetric.

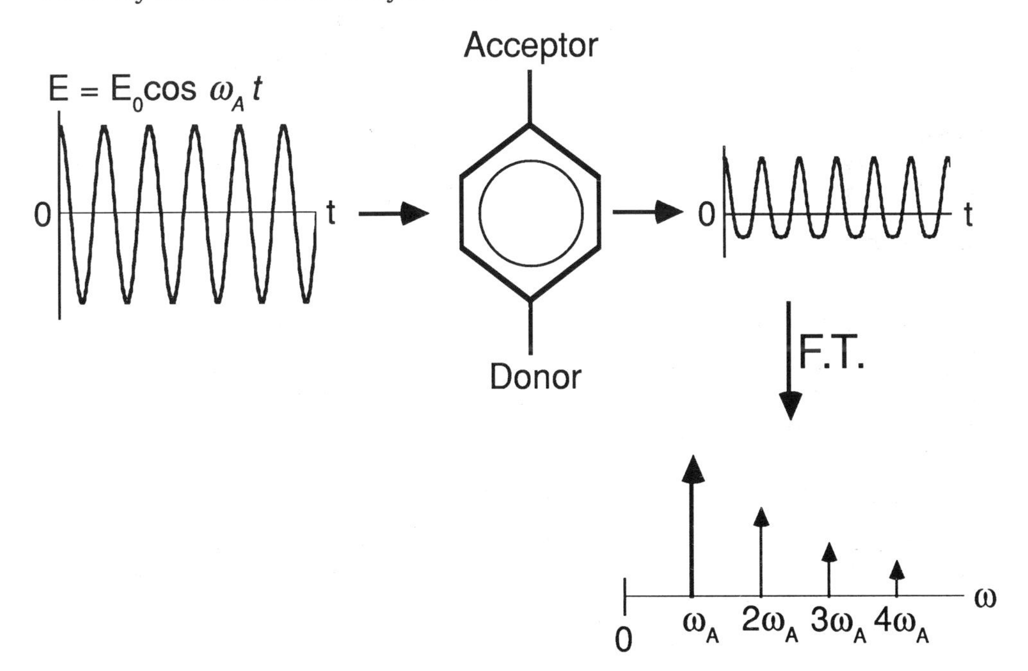

Figure 4.21 Schematic mechanism for doubling of laser light frequency by an ordered array of optically anisotropic molecules (see text).

Let us now return to the more general case that $\omega_A \neq \omega_B$. Specifically, suppose that one frequency component represents the frequency, ω_0, of the spectral peak of interest, and that the second component at frequency, ω_{ref}, is generated by a reference oscillator, ω_{ref}, is chosen sufficiently close to ω_0 that

$$|\omega_{ref} - \omega_0| << |\omega_{ref} + \omega_0| \tag{4.15}$$

Next, let the two signals be added to form the input to a suitable non-linear device. The device output (as we have previously noted) will contain signals at frequencies, ω_0, ω_{ref}, $2\omega_0$, $2\omega_{ref}$, $(\omega_0 + \omega_{ref})$, and $(\omega_0 - \omega_{ref})$, as shown in Figure 4.22. Finally, let the output signal be subjected to a low-pass filter (i.e., a device which will pass signals only up to an upper frequency limit). The low-pass filter removes all of the fundamental, harmonic, and sum-frequency components, and leaves only the much lower-frequency difference-frequency components.

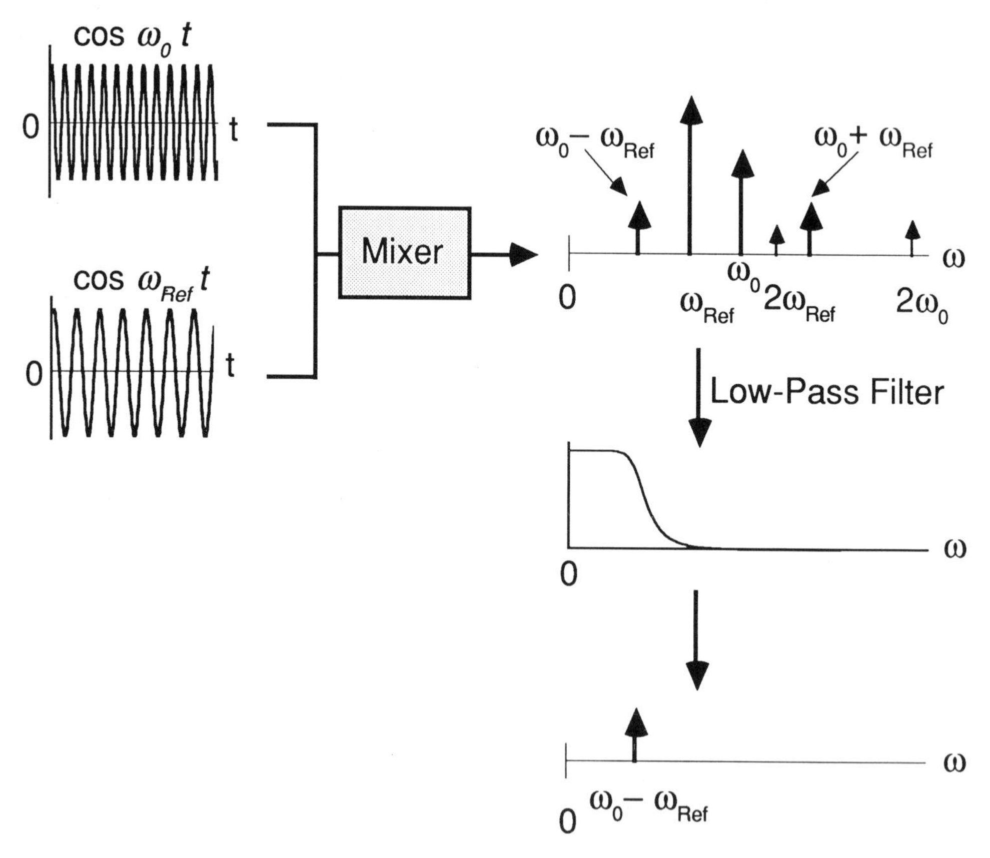

Figure 4.22 Schematic heterodyne detector, consisting of a reference oscillator, mixer, and low-pass filter. The mixer is a non-linear device which has an effect equivalent to multiplying the two input signals together. The mixer output has frequency components at multiple (harmonic) and combination (intermodulation) frequencies. The low-pass filter removes all but the difference frequencies, leaving a narrow-band frequency-domain spectrum of potentially high digital resolution (i.e., all of the spectral data points packed into a narrow bandwidth).

In physical terms, the effect of the mixer/filter combination is to serve as a spectral "slit", whose position is centered at the frequency of the reference oscillator, and whose width is equal to the bandwidth of the low-pass filter. Because the mixer has an effect similar to multiplying ("beating") the two input signals together, the experiment is called "heterodyning".

Apart from any FT applications, the heterodyne experiment has several major advantages for spectroscopy. First, the "slit" width is adjusted electronically, and may therefore be made arbitrarily wide or narrow without any mechanical adjustment to the spectrometer. Second, the reference oscillator frequency may be specified with high accuracy (e.g., a laser for the case of optical mixing, or an rf oscillator for NMR or ICR mixers). Since the frequency scale is also fixed electronically, without internal calibration, it can therefore be extremely accurate.

4.4.3 Amplitude modulation

Following pulsed excitation, a single *hypothetical* oscillator will move sinusoidally with a well-defined magnitude and frequency. However, the instantaneous magnitude (amplitude), frequency, and/or phase of the detected response to an *experimental* excitation may themselves vary sinusoidally with time during the data acquisition period, as shown in Figure 4.23.

Amplitude (or magnitude) modulation can be described mathematically by Eq. 4.16:

$$f(t) = A(t) \cos \omega_0 t \tag{4.16a}$$

in which

$$A(t) = A_0 (1 + m \cos \omega_m t) \tag{4.16b}$$

A_0 is the amplitude of the fundamental oscillation at frequency ω_0; ω_m is the modulation frequency; and m is the modulation "index" (see Figure 4.23, left).

$$0 \leq m \leq 1 \tag{4.17}$$

From the trigonometric identity,

$$\cos a \cos b = \frac{1}{2} [\cos(a + b) + \cos(a - b)] \tag{A.26b}$$

Eq. 4.16 can be rewritten in the form,

$$f(t) = A_0 \cos \omega_0 t + \frac{m A_0}{2} \cos(\omega_0 + \omega_m)t + \frac{m A_0}{2} \cos(\omega_0 - \omega_m)t \tag{4.18}$$

Thus, the frequency-domain spectrum obtained from the Fourier transform of $f(t)$ is simply a peak at the fundamental ("carrier") frequency, ω_0, and two new "sideband" peaks of equal magnitude, $(m A_0/2)$ (relative to the central peak), shifted away from the fundamental frequency by $\pm \omega_m$ rad s^{-1}, as shown in Figure 4.24 (see also Appendix D).

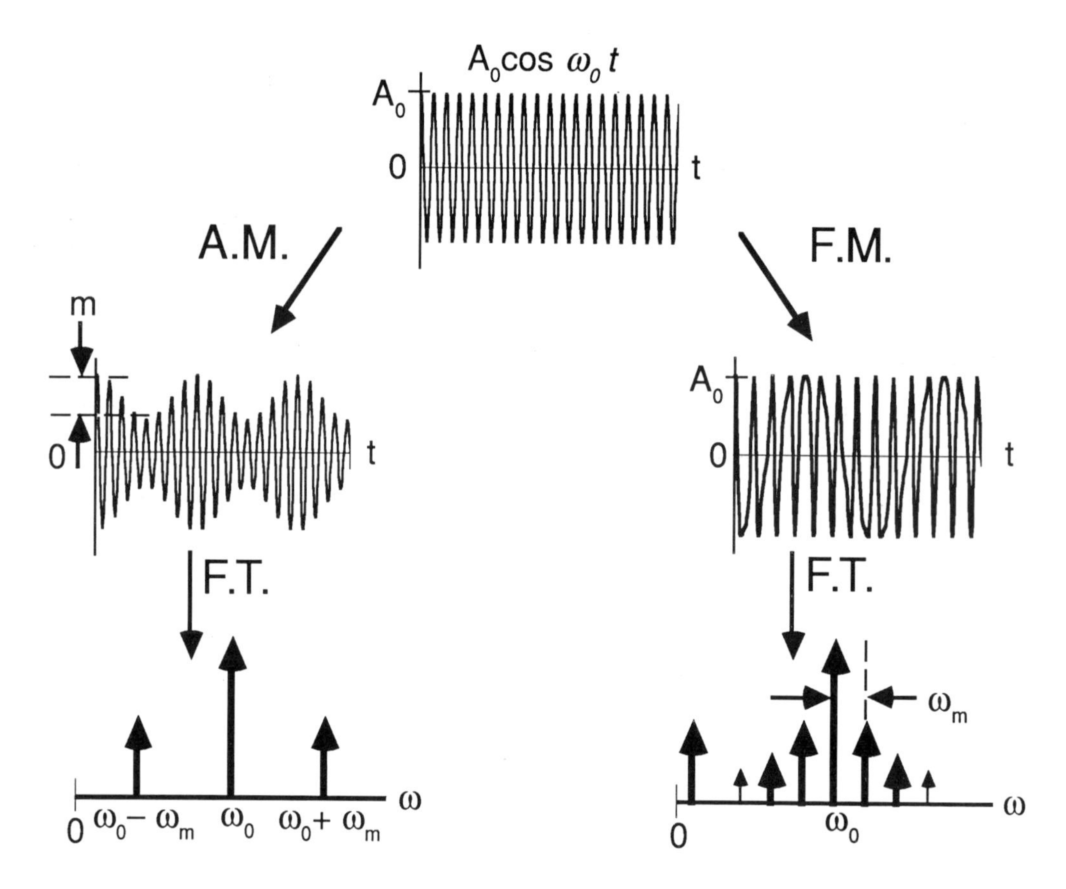

Figure 4.23 Amplitude modulation (left) and frequency modulation (right) of a sinusoidal time-domain signal (top), and corresponding frequency-domain spectra (bottom). In both cases, the spectrum shows a peak at the fundamental frequency of the unmodulated signal, as well as sideband peaks at frequencies spaced by (for AM) or at integral multiples of (for FM) the modulation frequency above and below the fundamental frequency. The relative magnitudes of the sidebands are determined by the modulation "index", m, which is defined differently for the two cases (see text).

Amplitude modulation is a common source of sidebands in spectroscopy. For example, electronic motion in a molecule can be modeled as an electron bound by a spring to a rigid molecular frame, vibrating at a natural frequency in the uv-visible frequency range. However, the nucleus to which a given electron is bound can itself vibrate at a much lower (infrared) frequency. The nuclear motion therefore modulates the amplitude of the electron displacement. In the spectrum of light radiated (scattered) by that electron, the peak at the fundamental frequency, ω_0, is called the Rayleigh line, and the two sidebands are called the Stokes ($\omega_0 - \omega_m$) and anti-Stokes ($\omega_0 + \omega_m$) Raman peaks (see Chapter 9.1.2). As another example, NMR signals can be amplitude-modulated by sinusoidal variation in the "Q" of the receiver circuit during sample spinning.

4.4.4 Frequency and phase modulation

For a time-domain signal of fundamental frequency, ω_0, the instantaneous frequency, $\omega(t)$, can in certain situations vary sinusoidally with time:

$$\omega(t) = \omega_0 + A_m \cos \omega_m t \tag{4.19}$$

It would be tempting (but wrong) to express the corresponding time-domain signal as

$$f(t) = A_0 \cos[\omega(t) \cdot t]$$

The above equation is unnumbered, because it is incorrect, just as it is incorrect to say that the linear distance, $s(t)$, traveled by a moving car is equal to the product of instantaneous velocity, $v(t)$ and time, t, if the velocity is not constant.

$$s(t) \neq v(t) \cdot t, \text{ if } v(t) \text{ varies with time}$$

Just as the correct expression for *linear* distance traveled is obtained from Eq. 4.20,

$$s(t) = \int_0^t v(t')\, dt' \tag{4.20}$$

the correct phase angle for *circular* motion (or linear motion represented as one of the two linear components of circular motion) should be written as:

$$\phi(t) = \int_0^t \omega(t')\, dt'$$

$$= \int_0^t [\omega_0 + A_m \cos \omega_m t']\, dt'$$

$$= \omega_0 t + \frac{A_m}{\omega_m} \sin \omega_m t' \tag{4.21}$$

If we further define a frequency modulation "index", m, as

$$m = \frac{A_m}{\omega_m} \tag{4.22}$$

then Eq. 4.21 can be rewritten as

$$\phi(t) = \omega_0 t + m \sin \omega_m t \tag{4.23}$$

and the frequency-modulated time-domain signal can be expressed as

$$f(t) = A_0 \cos[\phi(t)] \tag{4.24a}$$

$$= A_0 \cos(\omega_0 t + m \sin \omega_m t) \tag{4.24b}$$

From the above reasoning, it is now clear that cosinusoidal modulation of the instantaneous *frequency* (i.e., frequency modulation, Eq. 4.19) is equivalent to sinusoidal modulation of the instantaneous *phase* (phase modulation, Eq. 4.24b). Our interest is in the frequency-domain spectrum, $F(\omega)$, obtained by Fourier transformation of $f(t)$ in Eqs. 4.24. First, however, we must convert Eq. 4.24b into a function which corresponds to one of the Fourier integrals in Appendix A.

The first step is to apply the trigonometric identity,

$$\cos(a + b) = \cos a \cos b - \sin a \sin b \tag{A.26}$$

to Eq. 4.24b, with $a = \omega_0 t$ and $b = m \sin \omega_m t$:

$$f(t) = A_0 \, [\cos \omega_0 t \cos(m \sin \omega_m t) - \sin \omega_0 t \sin(m \sin \omega_m t)] \tag{4.25}$$

Since $\sin \omega_m t$ varies sinusoidally at frequency ω_m, it is logical to think of representing $\cos(m \sin \omega_m t)$ as an infinite (Fourier) series (i.e., a discrete Fourier transform with an infinite number of terms) of cosines whose frequencies are integral multiples of ω_m. [$\cos \phi(t)$ is an even function; thus the corresponding sine terms will vanish.]

$$\cos(m \sin \omega_m t) = \sum_{n=0}^{\infty} a_n(m) \cos n \omega_m t \tag{4.26}$$

In other words, if we can find an expression for $a_n(m)$ which satisfies Eq. 4.26, then substitution of Eq. 4.26 into Eq. 4.25 will yield terms like $\cos \omega_0 t \cos n \omega_m t$, which (see Eq. A.26b or Chapter 4.4.3.) will produce a spectrum of sidebands located at frequencies, $\omega_0 \pm n \omega_m$, with amplitudes proportional to $a_n(m)$. All that remains is to compute the relative amplitudes of those sidebands.

The magnitude, $a_n(m)$, of each term in the series is found by multiplying both sides of Eq. 4.26 by $\cos n' \omega_m t$ and integrating over one period, $T = 2\pi/\omega_m$. From the properties of the Dirac δ-function (see Appendix B), the right-hand side of Eq. 4.26 then reduces to

$$\sum_{n=0}^{\infty} \frac{1}{T} \int_0^t a_n(m) \cos n' \omega_m t \cos n \omega_m t \; dt = a_{n'}(m) \tag{4.27}$$

Thus, $a_{n'}(m)$ must be given by the integral arising from the left-hand side of Eq. 4.26:

$$a_{n'}(m) = \frac{1}{T} \int_0^T \cos n' \omega_m t \cos(m \sin \omega_m t) \; dt \tag{4.28}$$

The integrand of Eq. 4.28 can be expressed as a power series in $(m \sin \omega_m t)$, which, upon integration, becomes a power series in m. The final result of the above manipulations is the generation of Bessel functions of integer order:

$$\cos(m \sin \omega_m t) = J_0(m) + 2 \sum_{k=1}^{\infty} J_{2k}(m) \cos 2k\omega_m t \tag{A.93}$$

A similar argument leads to an expansion of $\sin(m \sin \omega_m t)$ in Eq. 4.25 in a sine series, resulting in a related series of Bessel functions:

$$\sin(m \sin \omega_m t) = 2 \sum_{k=0}^{\infty} J_{2k+1}(m) \sin(2k+1)\omega_m t \tag{A.94}$$

Substitution of Eqs. A.93 and A.94 into Eq. 4.25, simplified by use of Eq. A.26, produces the final result:

$$f(t) = A_0 \, [J_0(m) \cos \omega_0 t + \sum_{n=-\infty}^{-1} (-1)^n J_{|n|} \cos(\omega_0 - n\,\omega_m)t + \sum_{n=1}^{\infty} J_n \cos(\omega_0 + n\,\omega_m)t \,] \tag{4.29}$$

The frequency-domain spectrum of the time-domain function, $f(t)$, resulting from frequency (or phase) modulation is now obvious from inspection of Eq. 4.29. There is a spectral signal of relative magnitude, $J_0(m)$, at the fundamental frequency, ω_0, and signals of relative magnitude, $J_{|n|}(m)$, at an infinite number of sideband frequencies, $\omega_0 \pm n\,\omega_m$, $n = 1,2,3,\cdots$, spaced at integer multiples of the modulation frequency away from the fundamental frequency.

The magnitudes of the frequency-modulation sidebands vary with modulation index, m, as seen from a plot of the first few Bessel functions of the first kind in Figure 4.24. In the limit of very small modulation amplitude ($m \ll 1$), $J_0(m) \to 1$, and the spectral magnitude is all concentrated at the carrier (centerband, fundamental) frequency. As the modulation index increases, the centerband magnitude decreases and the first sidebands appear with relative magnitude, $J_{|1|}(m)$. On further increase in modulation index, the higher-order sidebands appear. Moreover, because the Bessel functions oscillate in magnitude as a function of m, the relative magnitudes of the frequency-modulated sidebands also oscillate as a function of modulation index, m, as shown in Figure 4.25 (see also Appendix D).

Frequency modulation pervades spectroscopy and electronics. For example, 60-Hz sidebands are commonly encountered as a result of modulation of an electronic signal by the a.c. line voltage. As another example, the magnetic field strength in an NMR experiment varies slightly from one part of the sample to another, leading to an inhomogeneously broad spectral peak. Rapid spinning of the sample has the effect of narrowing the spectral peak, but also produces "spinning sidebands" spaced at multiples of the spinning frequency away from the true signal, because the NMR frequency oscillates with time as nuclei rotate into and out of regions of different magnetic field strength.

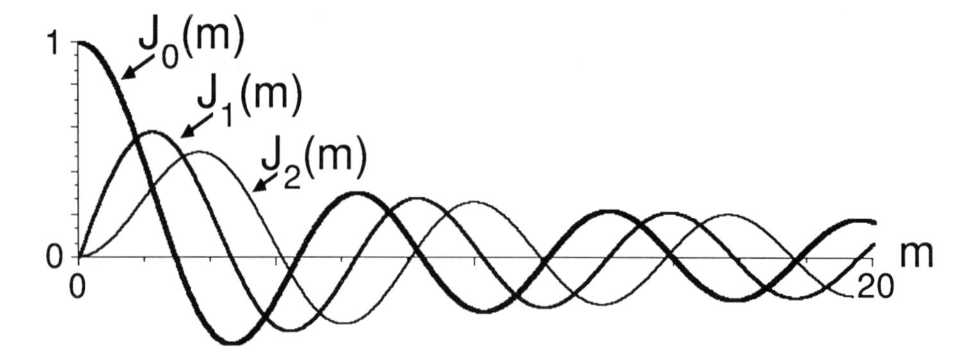

Figure 4.24 Bessel functions of the first kind, $J_n(m)$, for n = 0, 1, and 2. The relevance of these curves is that Bessel functions describe the relative magnitudes of sidebands resulting from frequency (or phase) modulation (see text).

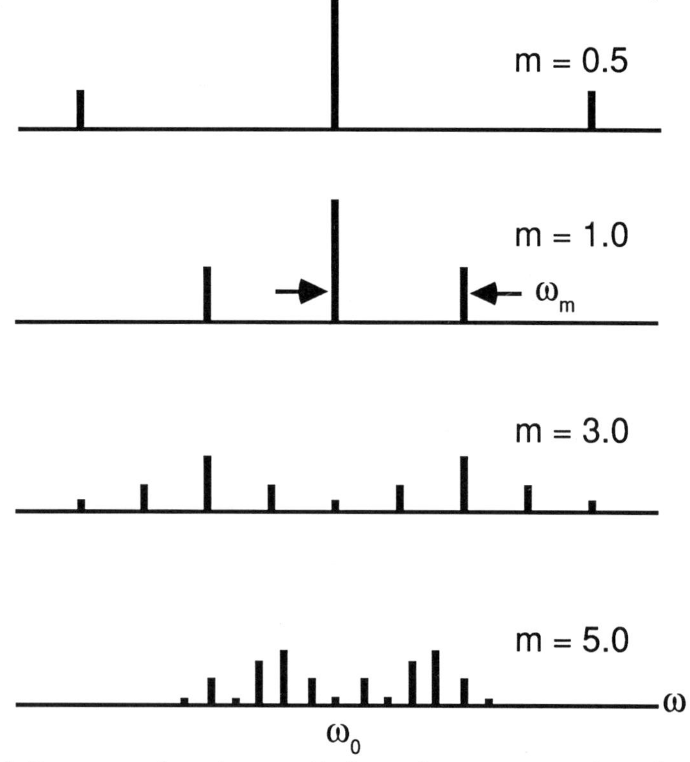

Figure 4.25 Frequency-domain magnitude-mode spectra resulting from frequency (or phase) modulation of a time-domain cosinusoid (see Eqs. 4.19, 4.24, and 4.29). Note that the sideband signal frequencies are spaced at integral multiples of the modulation frequency, ω_m, away from the carrier (fundamental) frequency, ω_0. The relative magnitudes of the carrier and sideband signals depend on the ratio, m, of modulation amplitude to modulation frequency.

Further Reading

K. M. Chow & M. B. Comisarow, *Int. J. Mass Spectrom. Ion Proc.* **1989**, *89*, 187-203. (Overlap effects for magnitude-mode spectral peaks)

R. M. Hammaker, J. A. Graham, D. C. Tilotta, & W. G. Fateley, *Vibrational Spectra and Structure*, vol. 15, Ed. J. R. Durig (Elsevier, Amsterdam, 1986). (Good discussion of Hadamard spectroscopy)

M. Harwit & N. J. A. Sloane, *Hadamard Transform Optics* (Academic Press, NY, 1979). (Thorough treatment of Hadamard transform infrared spectrometry based on mechanical mask translation. Includes tables of Hadamard codes, and fast Hadamard transform algorithm)

S. I. Parks & R. B. Johannesen, *J. Magn. Reson.* **1976**, *22*, 265-267. (Algorithm for automatic correction of quadrature detection artifacts—see Problems)

D. Tilotta, R. D. Freeman, & W. G. Fateley, *Appl. Spectrosc.* **1987**, *41*, 1280-1287; D. Tilotta and W. G. Fateley, *Spectroscopy* **1988**, *3*, 14-25. (Describes Hadamard transform visible and Raman spectrometry based on electronically encoded mask)

C. J. Turner & H. D. W. Hill, *J. Magn. Reson.* **1986**, *66*, 410-421. (Excellent review of quadrature detection artifacts)

Problems

4.1 For the 7-channel Hadamard code shown below, 4 channels (slits) are open in each of 7 measurements, y_1 to y_7, of a spectral window spanning 7 individual slit positions.

$$\begin{pmatrix} y_1 \\ y_2 \\ y_3 \\ y_4 \\ y_5 \\ y_6 \\ y_7 \end{pmatrix} = \begin{pmatrix} 1&1&1&0&1&0&0 \\ 1&1&0&1&0&0&1 \\ 1&0&1&0&0&1&1 \\ 0&1&0&0&1&1&1 \\ 1&0&0&1&1&1&0 \\ 0&0&1&1&1&0&1 \\ 0&1&1&1&0&1&0 \end{pmatrix} \cdot \begin{pmatrix} x_1 \\ x_2 \\ x_3 \\ x_4 \\ x_5 \\ x_6 \\ x_7 \end{pmatrix}$$

Each measurement, y_i, represents a linear combination of the spectral magnitudes, x_i from each of the individual slit positions, e.g.,

$$y_1 = x_1 + x_2 + x_3 + x_5$$

Show that the inverse Hadamard code can be generated by replacing 0 by −1 in the original code. In other words, show that

$$\begin{pmatrix} x_1 \\ x_2 \\ x_3 \\ x_4 \\ x_5 \\ x_6 \\ x_7 \end{pmatrix} = (1/4) \begin{pmatrix} 1 & 1 & 1 & -1 & 1 & -1 & -1 \\ 1 & 1 & -1 & 1 & -1 & -1 & 1 \\ 1 & -1 & 1 & -1 & -1 & 1 & 1 \\ -1 & 1 & -1 & -1 & 1 & 1 & 1 \\ 1 & -1 & -1 & 1 & 1 & 1 & -1 \\ -1 & -1 & 1 & 1 & 1 & -1 & 1 \\ -1 & 1 & 1 & 1 & -1 & 1 & -1 \end{pmatrix} \cdot \begin{pmatrix} y_1 \\ y_2 \\ y_3 \\ y_4 \\ y_5 \\ y_6 \\ y_7 \end{pmatrix}$$

You should confirm that the spectral magnitude from each slit position, x_i, can be recovered from the 7 measured magnitudes transmitted by each mask, e.g.,

$$x_1 = \frac{1}{4}\,(y_1 + y_2 + y_3 - y_4 + y_5 - y_6 - y_7)$$

4.2 This problem clarifies the differences between Hadamard H- and S-matrices. The elements of Hadamard H-matrices are exclusively 1 and −1. H-matrices have the following properties:

$$H_nH_n^T = H_n^TH_n = nI_n$$

in which I_n is the unit matrix of rank, n (i.e., diagonal elements are all unity and all other matrix elements are zero).

Hadamard matrices of any rank may readily be generated from the definitions,

$$H_1 = [1] \text{ and } H_2 = \begin{bmatrix} 1 & 1 \\ 1 & -1 \end{bmatrix}, \text{ and the generating rule, } H_{2n} = \begin{bmatrix} H_n & H_n \\ H_n & -H_n \end{bmatrix}$$

(a) Construct H_4 and confirm that H_4 is its own inverse [if multiplied by (1/4)]. The Hadamard H-matrix code defines the appropriate linear combinations of weights for the two-pan weighing problem of Figure 4.2. (An H-matrix element of 1 or −1 corresponds to placing that object on the left or right side of the balance.)

(b) For Hadamard optical spectroscopy, the code elements must be +1 and 0 (corresponding to placing unknowns only in one of the two pans in the analogous weighing experiment) rather than +1 and −1. In that case, the appropriate Hadamard code is the Hadamard S-matrix, which is generated by deleting the first row and column of a Hadamard H-matrix, and then changing all of the 1's to 0's and all of the −1's to 1's.

Construct the H_8 H-matrix, and use the above method to generate S_7. Then generate the inverse matrix, $(S_7)^{-1}$, by changing each 0 to −1 in S_7. Your result should resemble that of the previous problem, although the two matrices are not identical.

4.3 An interesting consequence of the discrete nature of FT spectra is that relative apparent peak heights can be significantly distorted from their true values. As a simple example, consider a noiseless undamped sinusoidal time-domain signal,

$$f(t) = A_0 \cos 2\pi\nu_0 t \quad , \quad 0 \le t \le T$$

for which the *analog* absorption-mode spectral peak maximum height is T. Determine the highest *discrete* spectral magnitude if $\nu_0 = (n + \delta)/T$ Hz, in which n is an integer, for

(a) $\delta = 0$ (i.e., ν_0 is an exact integral multiple of $(1/T)$;

(b) $\delta = 0.5$ (i.e., ν_0 falls exactly midway between two discrete FT spectral frequencies;

(c) $\delta = 0.25$

4.4 Overlap of signals of two closely spaced frequencies is particularly problematic for magnitude-mode spectra. For example, consider the time-domain signal,

$$f(t) = \left(\cos 2\pi\nu_1 t + \cos 2\pi\nu_2 t\right) \exp(-t/\tau), \quad 0 \le t \le T \ , \quad T/\tau = 1.0$$

From the analytical magnitude-mode FT spectrum of $f(t)$ (see Problem 2.4), compute (iteratively, preferably with a short computer program) the peak maxima for the following "true" peak separations:

$$\nu_2 - \nu_1 = 1/T, \ 1.5/T, \text{ and } 2.0/T$$

You should find that the apparent magnitude-mode peak separation can be *smaller* or *larger* than the true difference in frequency between the two component signals. Perhaps even more surprising, it turns out that the apparent magnitude-mode peak positions also depend on the relative *phase* of the two component signals, even though we normally think of magnitude-mode display as phase-independent (for an isolated peak). Apodization can further shift the apparent magnitude-mode peak positions. All of these effects are discussed in detail in the Chow and Comisarow reference in Further Reading. The related effect of peak overlap on magnitude-mode spectral peak heights is discussed in J. P. Lee and M. B. Comisarow, *Anal. Chem.* **1988**, *60*, 2212-2218.

4.5 As noted in Chapter 4.2.3., stored-waveform inverse FT (SWIFT) excitation can produce a signal of almost any frequency-domain magnitude profile, but may require very high time-domain dynamic range. Show that for a rectangular frequency-domain magnitude-mode spectrum,

$$|\boldsymbol{F}(\nu)| = 1.0, \quad N/8T \le \nu \le 3N/8T$$

$$|\boldsymbol{F}(\nu)| = 0, \quad 0 \le \nu \le N/2T = \nu_{\text{Nyquist}}$$

the frequency-domain phase-encoding (see Figure 4.9) which produces a time-domain signal with optimally flat magnitude is a quadratic phase-variation across the original frequency-domain spectrum.

Hint: Divide the spectrum from $N/8T \leq \nu \leq 3N/8T$ into n equal segments. For any one of those segments, the (inverse) FT is a time-domain sinc function centered at a t- value determined by the "shift" theorem (Problem 2.17). Moreover, a linear phase shift across a given original spectral segment will result in a fixed time-shift of the corresponding time-domain sinc function. Therefore, you can produce a series of equally spaced time-domain sinc functions, by applying linear phase shifts of increasing slope across each successive original spectral segment. You can then determine the appropriate phase *vs.* frequency slopes which produce a series time-domain sinc functions which are separated by one sinc peak width (i.e., which just coalesce) to produce the desired optimally flat-magnitude time-domain signal. You should find that the variation of phase *vs.* frequency of the spectral segments is a quadratic curve. A similar method can be used to produce optimally flat-magnitude time-domain SWIFT waveforms from a frequency-domain excitation of nearly arbitrary shape, by dividing the frequency-domain spectrum into segments of equal area, and then proceeding as above.

4.6 In attempting to devise a measure of spectral resolution, spectroscopists have unfortunately come to rely on several different criteria. Most generally, spectral resolution is reported either as width [$\Delta\omega$ (rad s^{-1} or $\Delta\nu$ (Hz)] or as the ratio of peak center frequency (ω_0 or ν_0) to peak width. Unfortunately, there is no general agreement on how to measure peak width:

(a) peak full width at 50% (or 10% or 1%) of peak maximum height;

(b) difference between the actual frequencies (ω_A and ω_B) of two peaks of equal height, shape, and width, separated such that the minimum height of the valley between them is 100% (or 50%, or 10%, or 1%) of the height of either individual peak (see diagram).

(c) same as (b), except that the valley is 100% (or 50%, or 10%, or 1%) of the maximum height of the either peak in the actual spectrum when both peaks are present—see diagram.

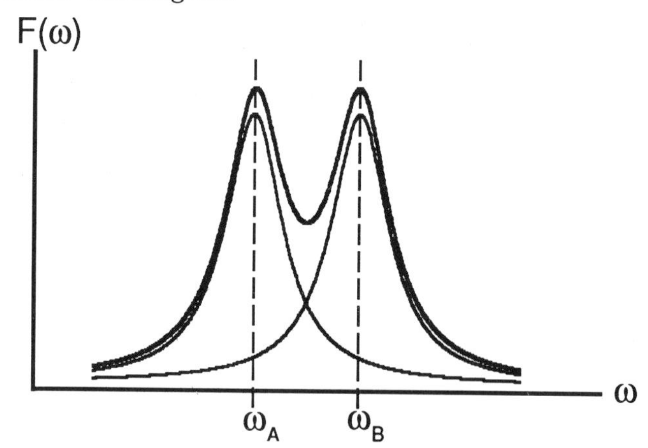

Moreover, several different line shapes apply (or are fitted) to spectral peaks: e.g.,

$$F(\omega) = \frac{\tau}{1 + (\omega_0 - \omega)^2 \tau^2} \qquad \text{(Lorentz absorption-mode)}$$

$$F(\omega) = A_0 \exp\{-a[(\omega - \omega_0)/\Delta\omega]^2\} \qquad \text{(Gaussian)}$$

$$F(\omega) = \frac{\sin(\omega_0 - \omega)T}{(\omega_0 - \omega)} \qquad \text{(sinc-type absorption-mode)}$$

For each of the three listed line shapes, obtain expressions for spectral resolution based on criteria (a), (b), and (c). In particular, note the peak separation required to satisfy the 50% (or 10% or 1%) valley definitions for the various line shapes. You should find that resolution can vary by a factor of more than 100, depending on peak shape and the choice of resolution criterion.

Finally, resolution in optical spectroscopy was originally defined from the sinc^2-type line shape,

$$F(\omega) = \frac{1 - \cos(\omega_0 - \omega)T}{[(\omega_0 - \omega)T]^2}$$

according to:

(d) Rayleigh criterion: two adjacent peaks of equal height and width, separated such that the center of one peak is at the same frequency as the first zero of the other; or

(e) Dawes criterion (used in astronomy): twice the Rayleigh peak separation

Compute the peak separation (as a function of T) for the Rayleigh and Dawes criteria, and compare to the other resolution criteria.

This problem offers good practice in manipulating the common spectral line shape expressions. The exercise also demonstrates the danger of comparing resolution defined by different criteria on different instruments or samples.

4.7 In section 4.3.2.4, we showed that a difference in d.c. offset between the two quadrature receiver channels could be eliminated by alternating the phase of the transmitter by 180° on every other acquisition, and then alternately adding and subtracting successive acquisitions in memory:

Scan #	Transmitter phase	Output of quadrature channels A and B sent to Real memory	Imaginary memory
1	0°	A	B
2	180°	$-A$	$-B$
3	0°	A	B
4	180°	$-A$	$-B$
.	.	.	.
.	.	.	.
.	.	.	.

Now suppose that the two quadrature channels have unequal gain (see Figure 4.15). Show (with diagrams, as in Figure 4.18) that the following "CYCLOPS" phase-cycling sequence will compensate for the unequal gain in the two channels, provided that the experiment is exactly reproducible from scan to scan, and that we can repeat the experiment $4n$ times, $n = 1, 2, 3, \cdots$. The reader is referred to the Parks & Johannesen reference for a more general algorithm for correction of quadrature detection artifacts.

Scan #	Transmitter phase	Output of quadrature channels A and B sent to Real memory	Imaginary memory
1	0°	A	B
2	90°	$-B$	A
3	180°	$-A$	$-B$
4	270°	B	$-A$
5	0°	A	B
6	90°	$-B$	A
7	180°	$-A$	$-B$
8	270°	B	$-A$
.	.	.	.
.	.	.	.
.	.	.	.

4.8 In 1975, Redfield and Kunz presented a simple procedure for obtaining an FT of quadrature time-domain data acquired alternately rather than simultaneously. They multiplied, element-by-element, the alternately-acquired experimental time-domain data,

$$(x_0, y_0, x_1, y_1, \ldots x_{N-1}, y_{N-1})$$

by the repeated sequence, $[+1, -1, -1, +1, \cdots]$. Treat the resultant data set as a mathematically real time-domain discrete signal and show that the complex FT of that data generates the same discrete spectrum that would have been obtained by complex FT of simultaneously acquired quadrature data treated as pairs of complex numbers.

Hint: Begin by representing the j 'th alternately acquired quadrature data set $f(t_j)$ in pairs: (x_{2n}, y_{2n+1}) as:

$$x_{2n} = \cos[2\pi \nu_0 n/N] \quad \text{for } j = 2n \quad (\text{even } j) \qquad (4.8.1a)$$

$$y_{2n+1} = \sin[2\pi \nu_0 (n + 1/2)/N] \quad \text{for } j = 2n + 1 \ (\text{odd } j) \qquad (4.8.1b)$$

$$n = 0, 1, 2, 3, \cdots, N - 1 \qquad (4.8.1c)$$

Then write out the discrete FT of $(x_0, y_0, x_1, y_1, \ldots x_{N-1}, y_{N-1})$; apply the weight function, $(-1)^n = \exp(-i n \pi)$; and compare to the discrete FT of the simultaneously acquired quadrature data,

$$x_j = \cos[2\pi \nu_0 n/N] \qquad (4.8.1a)$$

$$y_j = \sin[2\pi \nu_0 n/N] \qquad (4.8.1b)$$

Solutions to Problems

4.1 For example,

$$x_1 = \frac{1}{4}(y_1 + y_2 + y_3 - y_4 + y_5 - y_6 - y_7)$$

$$\begin{aligned} = \frac{1}{4}(&x_1 + x_2 + x_3 + 0 + x_5 + 0 + 0 \\ &+ x_1 + x_2 + 0 + x_4 + 0 + 0 + x_7 \\ &+ x_1 + 0 + x_3 + 0 + 0 + x_6 + x_7 \\ &+ 0 - x_2 + 0 + 0 - x_5 - x_6 - x_7 \\ &+ x_1 + 0 + 0 + x_4 + x_5 + x_6 + 0 \\ &+ 0 - 0 - x_3 - x_4 - x_5 - 0 - x_7 \\ &+ 0 - x_2 - x_3 - x_4 + 0 - x_6 + 0) \end{aligned}$$

$$= \frac{1}{4}(4x_1) = x_1$$

The computation for x_2 through x_7 is similar.

4.2 $H_2 = \begin{bmatrix} 1 & 1 \\ 1 & -1 \end{bmatrix}$, and $H_n = \begin{bmatrix} H_n & H_n \\ H_n & -H_n \end{bmatrix}$. Thus, $H_4 = \begin{bmatrix} 1 & 1 & 1 & 1 \\ 1 & -1 & 1 & -1 \\ 1 & 1 & -1 & -1 \\ 1 & -1 & -1 & 1 \end{bmatrix}$; and

$$H_8 = \begin{bmatrix} 1 & 1 & 1 & 1 & 1 & 1 & 1 & 1 \\ 1 & -1 & 1 & -1 & 1 & -1 & 1 & -1 \\ 1 & 1 & -1 & -1 & 1 & 1 & -1 & -1 \\ 1 & -1 & -1 & 1 & 1 & -1 & -1 & 1 \\ 1 & 1 & 1 & 1 & -1 & -1 & -1 & -1 \\ 1 & -1 & 1 & -1 & -1 & 1 & -1 & 1 \\ 1 & 1 & -1 & -1 & -1 & -1 & 1 & 1 \\ 1 & -1 & -1 & 1 & -1 & 1 & 1 & -1 \end{bmatrix}, \text{ from which}$$

$$S_7 = \begin{bmatrix} 1 & 0 & 1 & 0 & 1 & 0 & 1 \\ 0 & 1 & 1 & 0 & 0 & 1 & 1 \\ 1 & 1 & 0 & 0 & 1 & 1 & 0 \\ 0 & 0 & 0 & 1 & 1 & 1 & 1 \\ 1 & 0 & 1 & 1 & 0 & 1 & 0 \\ 0 & 1 & 1 & 1 & 1 & 0 & 0 \\ 1 & 1 & 0 & 1 & 0 & 0 & 1 \end{bmatrix} \text{ and } (S_7)^{-1} = (1/4)\begin{bmatrix} 1 & -1 & 1 & -1 & 1 & -1 & 1 \\ -1 & 1 & 1 & -1 & -1 & 1 & 1 \\ 1 & 1 & -1 & -1 & 1 & 1 & -1 \\ -1 & -1 & -1 & 1 & 1 & 1 & 1 \\ 1 & -1 & 1 & 1 & -1 & 1 & -1 \\ -1 & 1 & 1 & 1 & 1 & -1 & -1 \\ 1 & 1 & -1 & 1 & -1 & -1 & 1 \end{bmatrix}$$

4.3 (a) Maximum discrete absorption-mode value = T, i.e., identical to the *analog* absorption-mode peak height.

(b) All of the discrete absorption-mode *discrete* spectral points fall at the zeroes of the sinc function. Therefore, the *discrete* absorption-mode spectrum is identically zero!

(c) Maximum discrete absorption-mode value $\cong 0.64T$. For the effect of zero-filling on this computation, see M. B. Comisarow and J. D. Melka, *Anal. Chem.* **1979**, *51*, 2198-2203.

4.4 Refer to the cited publications.

4.5 Following the "hint", the full solution may be found in: S. Guan, *J. Chem. Phys.* **1989**, *91*, 775-777.

4.6 See S. L. Mullen and A. G. Marshall, *Anal. Chim. Acta* **1985**, *178*, 17-26.

4.7 The solution is essentially a combination of Figures 4.15 and 4.18.

4.8 The discrete Fourier transform of the time-domain signal, $f(t_j)$, is given by

$$F_k = \sum_{j=0}^{2N-1} f(t_j)\, e^{-i 2\pi jk/2N}, \quad k = 0, 1, 2, \cdots, N-1 \tag{4.8.2}$$

or
$$F_k = \sum_{n=0}^{N-1} \left(x_{2n} + y_{2n+1}\, e^{-i\pi k/N}\right) e^{-i 2\pi jk/2N}$$

$$= \frac{1}{2} \sum_{n=0}^{N-1} \{(1 - i\, e^{-i\pi(k-\nu_0)/N})\, e^{-i 2\pi n (k-\nu_0)/N}\}$$

$$+ \{(1+i\, e^{-i\pi(k+\nu_0)/N})e^{-i2\pi n(k+\nu_0)/N}\}$$

If we replace i by its equivalent, $e^{i\pi/2}$, the above expression becomes:

$$F_k = \frac{1}{2} \sum_{n=0}^{N-1} \left([1 - e^{-i\pi(k-\nu_0-N/2)/N}]\, e^{-i 2\pi n (k-\nu_0)/N}\right)$$

$$+ \frac{1}{2} \sum_{n=0}^{N-1} \left([1 + e^{-i\pi(k+\nu_0-N/2)/N}]\, e^{-i 2\pi n (k+\nu_0)/N}\right)$$

Since the two exponential terms do not go to zero for the same argument, image peaks appear in the spectrum. At this stage, Redfield and Kunz introduced the idea of multiplying $f(t_j)$ by $(-1)^n$ before the discrete FT process. Since $(-1) = e^{-in\pi}$, $F_k' = -1^n\ F_k$ becomes

$$F_k' = \frac{1}{2}\sum_{n=0}^{N-1}\left([1 - e^{-i\pi(k - \nu_0 - N/2)/N}]\ e^{-i2\pi n\,(k - \nu_0 - N/2)/N}\right)$$

$$+\ \frac{1}{2}\sum_{n=0}^{N-1}\left([1 + e^{-i\pi(k + \nu_0 - N/2)/N}]\ e^{-i2\pi n\,(k + \nu_0 - N/2)/N}\right)$$

Note: Multiplication of $f(t_j)$ by $(-1)^n$ is equivalent to the multiplication of the alternately acquired data set by the sequence [+1, −1, −1, +1, . . .]. Check the indices j and n : the cosine terms will remain positive; the sine terms will become negative.

From the definition of the δ-function (see Appendix B) ,

$$\sum_{n=0}^{N-1} e^{-i2\pi n\,(k - k')/N} = N\,\delta(k - k')$$

F_k' becomes:

$$F_k' = \frac{N}{2}\left([1 - e^{-i\pi(k - \nu_0 - N/2)/N}]\ \delta(k - \nu_0 - N/2) + [1 + e^{-i\pi(k + \nu_0 - N/2)/N}]\ \delta(k + \nu_0 - N/2)\right)$$

$= N\,\delta(k + \nu_0 - N/2\ )$, which produces positive-frequency peaks $(0 \leq \nu_0 < N/2)$ at the left side of the spectum $(0 \leq k < N/2)$, without any image peaks (i.e., as from a conventional F.T.).

This procedure can be extended to the more general case in which the delay between the channels in quadrature is not necessarily equal to a single sampling period. If the delay between the readings is Δ, then F_k' becomes:

$$F_k' = \frac{N}{2}\left([1 - e^{-i\pi(k - 2\nu_0\Delta - N/2)/N}]\right)\ \delta(k - \nu_0 - N/2)$$

$$+\ \frac{N}{2}\left((1 + e^{-i\pi(k + 2\nu_0\Delta - N/2)/N})\ \delta(k + \nu_0 - N/2)\right)$$

Taking advantage of a property of the δ-function, we find that the only term which will not be equal to zero is the second term when:

$$k + \nu_0 - N/2 = 0$$

so:

$$\nu_0 = N/2 - k$$

and $k - 2\nu_0\,\Delta - N/2 = (k - N/2)(1\ - 2\Delta)$

$F_k{}'$ can be re-written as:

$$F_k{}' = \frac{N}{2}\left([1 - e^{-i\pi(k - N/2)(1 - 2\Delta)/N}]\ \delta(k - \nu_0 - N/2)\right)$$
$$+ \frac{N}{2}\left([1 + e^{-i\pi(k - N/2)(1 - 2\Delta)/N}]\ \delta(k + \nu_0 - N/2)\right)$$

Finally, in order to cancel the image peaks, one must form the complex conjugate of $F'_{N-k} = H_k$ (i.e., multiply all of the *dispersion* data by −1) and apply the following readily generated formula:

$$F_k = \frac{1}{2}\left((F_k{}' + H_k{}^*) + (F_k{}' - H_k{}^*)\ e^{i\pi(k - N/2)(1 - 2\Delta/N)}\right)$$

For more details, see:

A. G. Redfield & S. D. Kunz, *J. Mag. Res.* **1975**, *19*, 250;
A. Keller & U. H. Haeberlen, *Hewlett-Packard Journal*, **1988** (Dec.), 74.

An experimental FT/ICR example is given in:

A. G. Marshall, F. R. Verdun, S. L. Mullen, & T. L. Ricca , *Adv. Mass Spectrom.* **1989**, *11*, 670-671.

Chapter 5

Noise

5.1 Noise in the detected spectrum

In the previous chapter, we considered those practical aspects of Fourier transforms which are common to the *signal* in all forms of Fourier transform spectroscopy. However, the quality of any spectral measurement depends on the ratio of signal to *noise*. Therefore, in this chapter, we will examine the effects of various types of noise on Fourier transform spectra.

As we shall show, the degree of improvement in signal-to-noise ratio provided by multichannel or multiplex detection methods is determined by: (a) the relation (if any) between root-mean-square average noise level and signal strength, and (b) the way in which the signal and noise are distributed across the spectral bandwidth (e.g., one large peak or many small ones, etc.). Therefore, this chapter begins with a discussion of types of noise (Chapter 5.1.1).

In Chapter 5.1.2, we proceed to show how signal-to-noise ratio may in general be improved by co-adding repeated measurements. However, in Fourier transform spectroscopy, the maximum dynamic range (i.e., ratio of largest to smallest spectral peak magnitude) in any single measurement (or in a repeated measurement of a noiseless signal) is limited by the dynamic range (e.g., 8-16 bits) of the analog-to-digital converter. Surprisingly, we will find that the dynamic range of a repeated measurement can actually be increased to well above the ADC word length if random noise is present along with (or deliberately added to) the signal in each measurement! In Chapter 5.1.3, we show how the act of digitizing a signal can introduce "quantization" noise, even for a noiseless signal.

Apodization can increase signal-to-noise or resolution (but not both simultaneously) in a Fourier transform spectrum computed from a given time-domain data set. Although such trade-offs appear visually obvious (Chapter 5.1.5), we shall show (Chapter 5.1.4) that the true precision in determination of spectral peak height, area, width, and center frequency depends only on the peak shape, signal-to-noise ratio, and number of data points per peak width. In other words, when noise is present, apodization can alter the *appearance* of a spectrum, but not its inherent *information* content (except to reduce the available information).

Noise (in the form of randomly fluctuating electric and magnetic fields) is present in the environment surrounding a given ion (ICR), molecular oscillator (IR) or magnetic nucleus (NMR). Although such noise has a *mean* value of zero, the *mean square* noise is non-zero, and we may use Fourier transform methods to determine its (power) spectrum. We can then think of such noise as a spectral "source" (Chapter 5.2.1), and then show how it generates a spectral response. We can even generate noise intentionally, by use of "pseudorandom" number sequences (see also Hadamard matrices in Chapter 4) which behave like random noise but are reproducible from one measurement to the next (Chapter 5.2.2).

It is natural to wonder whether or not the Fourier transform provides the "best" spectrum available from a given time-domain data set. The answer to that question depends upon the signal-to-noise ratio, the number of time-domain data points, and the definition of "best". For example, "best" might mean a spectrum computed in such a way that the magnitudes of all of the peaks are independent of each other. Or, we might choose to take advantage of the assumed damped sinusoidal form of the time-domain signal components. In Chapter 6, we describe some of the alternative methods (maximum-entropy, autoregression, linear prediction) to Fourier transformation, based on new criteria for "best" spectrum.

5.1.1 Source-limited versus detector-limited noise

The two principal advantages of a multichannel (i.e., multiplex or multi-detector) spectrometer over a single-channel ("scanning") spectrometer are enhanced speed and/or signal-to-noise ratio. However, both advantages depend critically upon the nature of the random noise which accompanies the spectral signal.

In this section, we consider frequency-domain random noise, $n(\omega)$, which fluctuates randomly about zero from one data acquisition to the next. At any given frequency, ω, the *average* (over many data acquisitions) noise value, $< n(\omega) >$, is therefore zero. However the *root-mean-square* noise, $N(\omega)$,

$$N(\omega) = \sqrt{< [n(\omega)]^2 >} \tag{5.1}$$

is non-zero. From here on, "noise" will denote $N(\omega)$ defined by Eq. 5.1.

Frequency-domain noise is conveniently classified according to its relation to the frequency-domain *signal* strength, $S(\omega)$. "Detector-limited" noise, $N_D(\omega)$, is independent of signal strength:

$$N_D(\omega) = \text{independent of } S(\omega) = \text{"detector-limited" noise} \tag{5.2a}$$

Detector-limited noise is typically dominant when the energy of a detected photon is less than kT, in which k is the Boltzmann constant and T is absolute temperature, as in NMR and ESR experiments or IR detection at room temperature.

When the energy of a detected photon exceeds kT (as for uv/visible photons detected at room temperature, or infrared photons detected at very low temperature), then it becomes possible to detect (count) individual photons. The imprecision in counting randomly arriving photons is determined by the same Poisson statistics (see Problems) which govern radioactive decay count rates. The root-mean-square imprecision (i.e., noise) in number of counts is proportional to the square root of the number of counts (i.e., square root of the signal strength in the spectroscopic experiment). Such "source-limited" noise, $N_S(\omega)$, is related to the (square root of the) signal strength,

$$N_S(\omega) \propto \sqrt{S(\omega)} = \text{"source-limited" (e.g., "shot") noise} \tag{5.2b}$$

In detection of low-level and/or low-frequency signals, there is a third type of noise, $N_F(\omega)$, which is directly proportional to the signal strength:

$$N_F(\omega) \propto S(\omega) \tag{5.2c}$$

= "fluctuation", "modulation", or "scintillation" noise

$N_F(\omega)$ is variously known as "fluctuation", "modulation", or "scintillation" noise, according to its origin. The corresponding type of noise in electronic circuits is called $(1/f)$ noise, because its magnitude varies inversely with frequency. When fluctuation noise is the dominant type of noise, the spectral signal-to-noise ratio cannot be improved by repeating the measurement (i.e., time-averaging), because the noise accumulates at a rate proportional to the rate of accumulation of the signal.

Figure 5.1 shows the effects of the three types of noise on single-channel and multiplex (e.g., Fourier or Hadamard) spectra, when only a few peaks are present.

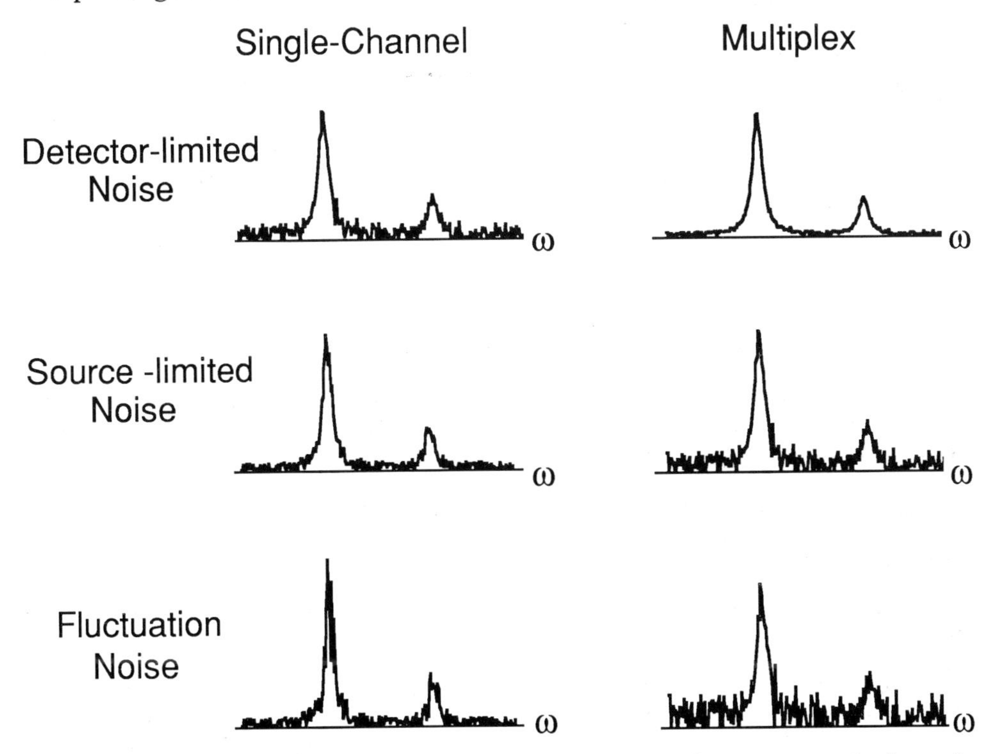

Figure 5.1 Simulated spectra for the same frequency-domain magnitude-mode signals in the presence of each of three types of noise (Eqs. 5.2a-c). In single-channel operation (left), the noise in a given spectral channel is either independent of (top) or determined by (middle, bottom) the signal *in that channel*. In multiplex (Fourier or Hadamard) operation (right), the total noise is independent of (top) or determined by (middle, bottom) the sum of the signals from *all* spectral channels, so that the frequency-domain noise is distributed evenly (on the average) *throughout* the spectrum. Thus, when noise is directly related to signal strength (middle, bottom), multiplex spectrometers in general enhance the signal-to-noise ratio for large peaks but can actually decrease (multiplex disadvantage) the signal-to-noise ratio for small peaks, compared to single-channel detection. (See text and Table 5.1).

Table 1 summarizes the multidetector and multiplex advantages for both signal-to-noise ratio and speed, for each of three types of noise of Eqs. 5.2. The basic principle underlying Table 1 is that the multiplex experiment effectively distributes the time-domain noise evenly *throughout* the frequency-domain spectral channels, whereas the noise in a given frequency-domain channel in a single-channel (or multidetector) experiment is determined only by the signal and noise *at that frequency*.

Thus, for source-limited or fluctuation noise, the noise level in a given frequency-domain channel in a multiplex experiment is directly related to the sum of *all* of the frequency-domain signal components. The signal-to-noise ratio in any given frequency-domain channel therefore depends both upon the signal in that channel and the number and magnitudes of signals at *other* frequencies in the same spectrum, as illustrated in Figure 5.1 for a spectrum containing strong and weak signals of different frequency.

5.1.2 Noise, data word length, signal-averaging, and dynamic range

Normally, we think of noise as an undesirable parameter to be minimized in any spectroscopic experiment. However, when a spectrum exhibiting large dynamic range is represented by discrete numbers (as in Fourier transform spectroscopy), it turns out that the spectral dynamic range can actually be improved by *adding noise* to the detected signal! Another surprise is that time-averaging with a *less* precise ADC (i.e., shorter word length) can actually produce a *more* precise (i.e., higher signal-to-noise ratio) final spectrum! In this section, we explore and explain these anomalies, for detector-limited noise (i.e., noise that is independent of signal magnitude).

The typical situation in Fourier transform spectroscopy is that the maximum dynamic range (i.e., vertical resolution or word length) for a *single* time-domain transient is limited by the resolution (word length) of the analog-to-digital converter (e.g., 8-16 bits). Fortunately, the word length of the random-access memory in which the *accumulated* time-domain transient is stored is typically much larger (16-32 bits or more). Thus, by co-adding many successive time-domain transients in random-access memory, we can increase the signal-to-noise ratio. Since the signal-to-noise ratio, SNR_{final}, of a signal accumulated from n acquisitions is $\sqrt{n}$ times larger than the signal-to-noise ratio, SNR_0, for a single acquisition,

$$SNR_{final} = \sqrt{n}\, SNR_0 \qquad (5.3)$$

it is clearly desirable to acquire the maximum possible number of transients.

Suppose that the ADC word length (for a single acquisition) is d, and the random-access memory word length (for the accumulated signal) is w. In the absence of noise, the maximum number of acquisitions would be

$$n_{max} = 2^{(w-d)} \qquad \text{(absence of noise)} \qquad (5.4)$$

since the signal magnitude increases by one bit for each factor of 2 increase in n. For example, if the ADC word length is 12 bits and the random-access memory word length is 16 bits, then the random-access memory word will be filled when $2^{(16-12)} = 2^4 = 16$ acquisitions are co-added in memory.

Table 5.1 Signal-to-noise enhancement (for the same total time-domain observation period) and time required to obtain a spectrum (of the same signal-to-noise ratio) by use of a multidetector or multiplex spectrometer compared to a single-channel scanning spectrometer, as a function of the type of noise and the distribution of signal across the spectrum. The total noise, $N(\omega) = \sqrt{[N_D(\omega)]^2 + [N_S(\omega)]^2 + [N_F(\omega)]^2}$. $S(\omega)$ is the signal magnitude at frequency, ω.

Dominant type of noise	Signal-to-noise (*SNR*) enhancement factor			Time factor		
	n-channel multidetector	*n*-channel multiplexer Fourier	*n*-channel multiplexer Hadamard	*n*-channel multidetector	*n*-channel multiplexer Fourier	*n*-channel multiplexer Hadamard
$\mathbf{N_D(\omega)}$ **= independent of signal** ("detector-limited" noise)	$\sqrt{n}$	$\sqrt{n}$ [a]	$\frac{\sqrt{n}}{2}$	$\frac{1}{n}$	$\frac{1}{n}$ [a]	$\frac{4}{n}$
$\mathbf{N_S(\omega) \propto \sqrt{S(\omega)}}$ ("source-limited" or "shot" noise)	$\sqrt{n}$	$\sqrt{n}$ to 1 [a,b]	$\sqrt{\frac{n}{2}}$ to $\sqrt{\frac{1}{2}}$ [b]	$\frac{1}{n}$	$\frac{1}{n}$ to 1 [a,b]	$\frac{2}{n}$ to 2 [b]
$\mathbf{N_F(\omega) \propto S(\omega)}$ ("fluctuation", "modulation", or "scintillation" noise)	1	1 to $\left(\frac{1}{\sqrt{n}}\right)$ [a,b]	1 to $\left(\sqrt{\frac{2}{n}}\right)$ [b]	c	1 to n [a,b,d]	1 to $\frac{n}{2}$ [b,d]

a The table entries apply to Fourier transform NMR and ICR spectroscopy. For optical (uv/visible or infrared) interferometry, half of the signal is lost at the beam splitter, and the Fourier advantages are reduced by a factor of $\sqrt{2}$ (signal-to-noise) and 2 (speed).

b In each case, the first value corresponds to a spectrum with signal only in one channel (*SNR* computed for that channel); the second value corresponds to a spectrum with approximately equal signal magnitude in each channel.

c In this case, time-averaging cannot increase the signal-to-noise ratio, because the noise accumulates at a rate proportional to the rate of accumulation of signal.

d In these cases, time-averaging again has no effect on signal-to-noise ratio.

However, when detector-limited noise dominates, then the accumulated signal-to-noise ratio is proportional to $\sqrt{n}$ rather than n, and many more transients can be co-added before the random-access memory is full. Eq. 5.4 then takes the form (see Cooper reference in Further Reading),

$$n_{max} = \left[\frac{-1 + \left[1 + 4\, SNR_0\, (SNR_0 + 1)\, 2^{(w-d)}\right]^{1/2}}{[2\, SNR_0]} \right]^2 \tag{5.5}$$

Eq. 5.5 has several interesting consequences. First, if no noise is present, and if the signal is smaller than half of one ADC bit, then the sampling process will record a zero for each time-domain data point, and signal-averaging has no effect (see Figure 5.2).

Second, if noise is added to the time-domain transient signal at a magnitude greater than half of one ADC bit, then even a signal smaller than one ADC bit will be detectable, because the noise will occasionally "boost" that signal to a level larger than one ADC bit (see Figure 5.2). Of course, the more noise we add to each transient, the more transients we need to accumulate in order to increase the signal-to-noise ratio back to a value bigger than the ADC word length (see Table 5.2). Similarly, a large signal (or additional amplification of a small signal) can render a small signal observable, because the small signal will produce oscillations above and below each bit-level threshold.

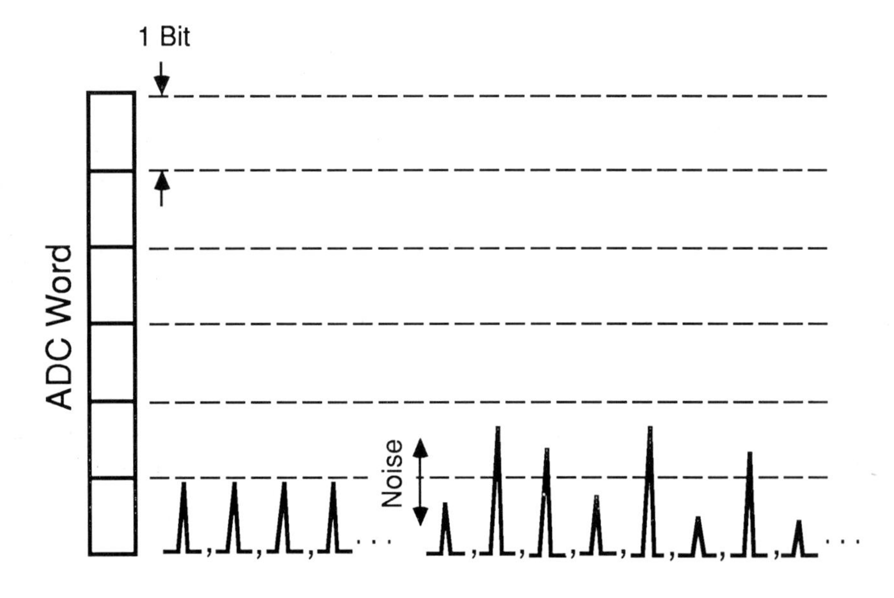

Figure 5.2 Effect of noise on analog-to-digital conversion for low-magnitude signals. In the absence of noise or other signals (left) a signal smaller than one ADC bit can never be detected. However, when ≥ 0.5 bit of noise (or other signal) is present (right), then a signal component smaller than one bit can trigger an ADC response. The signal-to-noise ratio can then be increased by signal-averaging (see text).

Third, since signal-to-noise ratio increases as $\sqrt{n}$ (Eq. 5.3), and since we can accumulate more transients by using a smaller ADC word length in each transient, it is clear (see Eq. 5.5 and Table 5.2) that the signal-to-noise ratio of the final accumulated data set actually *increases* when we reduce the ADC word length (i.e., precision per transient). In fact, the first Fourier transform microwave pure rotational spectra (see Flygare reference) were obtained with a *one-bit* ADC, in order to provide for extended time-averaging of millions of transients!

In practice, the ADC word length is usually limited by the minimum acceptable ADC sampling rate: faster ADC sampling rate requires higher cost and/or smaller ADC word length. For a given ADC word length, Table 5.2 illustrates the trade-offs between signal-to-noise ratio, dynamic range, and number of acquisitions.

Table 5.2 Maximum number, n_{max}, and accumulated signal-to-noise ratio, SNR_{final}, of time-domain acquisitions [each of signal-to-noise ratio, SNR_0, and analog-to-digital converter word length, d] whose accumulated dynamic range will just fill a random-access memory word of length, w. (Adapted from J. W. Cooper, *Anal. Chem.* **1978** *50*, 801A-812A.)

Initial SNR_0 =	1		0.1		0.01	
$(w - d)$	n_{max}	SNR_{final}	n_{max}	SNR_{final}	n_{max}	SNR_{final}
2	5	2.2	10	0.31	15	0.038
4	26	5.1	84	0.91	200	0.14
6	117	10.8	484	2.20	1,996	0.44
8	489	22.1	2,332	4.83	14,016	1.18
10	2,003	44.7	10,251	10.1	75,878	2.75
12	8,101	90.0	42,982	20.7	354,182	5.95
14	32,587	180.5	176,028	42.0	1,531,040	12.4
16	130,710	361.5	712,455	84.4	6,366,810	25.2
18	523,564	723.5	2,866,650	169.3	25,966,900	51.0

5.1.3 Quantization noise and oversampling

It is important to recognize that the sampling process itself produces time-domain magnitude errors which appear as frequency-domain "quantization" noise, even for a *noiseless* time-domain signal, as shown in Figures 5.3 and 5.4. An ADC must necessarily assign each analog time-domain value to one of a finite number (namely, 2^n, in which n is the number of ADC bits/word) of (discrete) values. The ADC effectively "rounds off" the true analog signal to the nearest available discrete ADC value. The differences between discrete and analog values define a time-domain error known as "quantization" noise. The quantization errors (see Problems) have some usual qualities of random noise: signal-independent frequency-independent (up to the Nyquist bandwidth) root-mean-square frequency-domain magnitude (i.e., detector-limited white noise). In FT spectroscopy, quantization noise is generally quite small (e.g., $\leq$ 1 part in 8,192 for a 12-bit ADC).

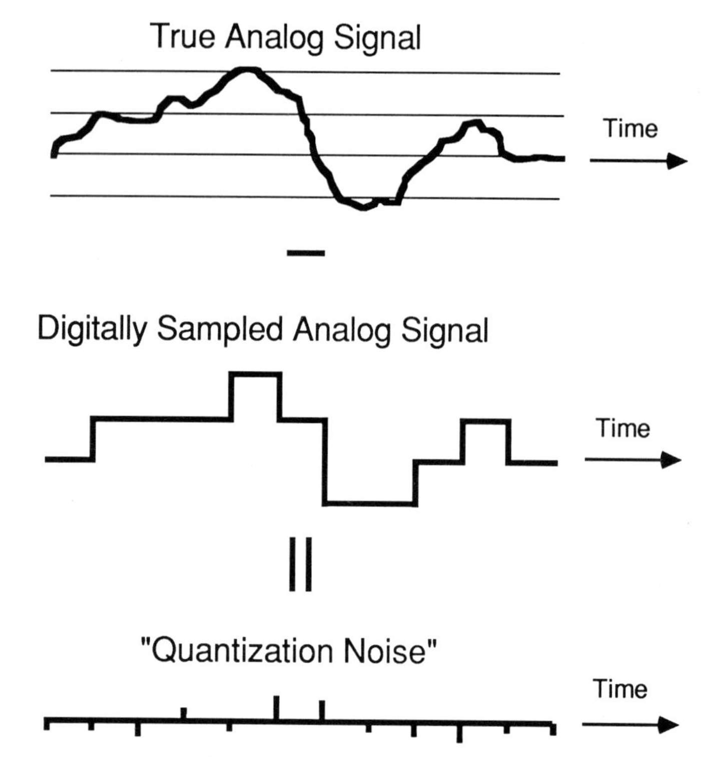

Figure 5.3 Origin of "quantization" noise (bottom), as the difference between the time-domain analog ("true") signal (top) and its digitized representation (middle). Because the digitization process rounds off each time-domain value to the nearest available discrete ADC bit, the digitization process itself introduces magnitude errors (known as "quantization" noise) in the discrete time-domain signal. (Figure provided by G. M. Alber)

Quantization noise is evident in the frequency-domain spectra shown in Figure 5.4 for *noiseless* time-domain signals. It is clear that quantization noise may be reduced by increasing the number of ADC bits/word (i.e., by subdividing the *vertical* discrete scale in Figure 5.3), as shown in Figure 5.4. Each additional ADC bit reduces quantization noise by $\sqrt{2}$, provided that the time-domain signal fills the ADC word fully in both cases (i.e., that the largest time-domain signal magnitude approximately matches the largest ADC discrete value).

Interestingly, quantization noise may also be reduced by *oversampling* (i.e., by subdividing the *horizontal* discrete scale in Figure 5.3, as shown in Figure 5.4. The quantization root-mean-square noise is reduced by a factor of $\sqrt{2}$ for every doubling of sampling rate. The process is designated as "over"-sampling because it requires sampling at a rate higher than the usual Nyquist criterion. Finally, a practical advantage of oversampling (exploited in audio compact disk players) is that noise above the Nyquist frequency is no longer "folded back" into the discrete frequency spectrum; thus, the analog low-pass filter used to remove high-frequency noise need no longer have a sharp cut-off (associated with spectral magnitude and phase distortion) at the Nyquist frequency.

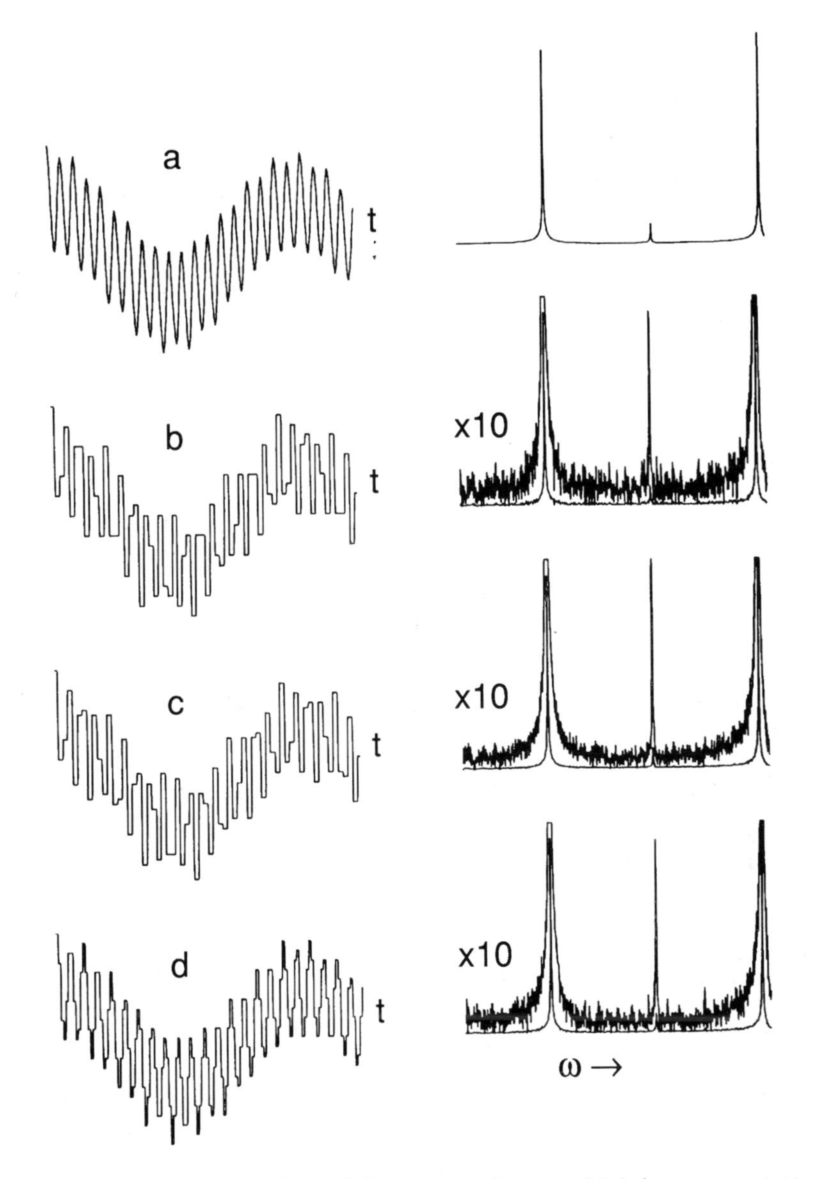

Figure 5.4 Time-domain (left) and frequency-domain (right) representations of a noiseless time-domain analog signal (a) consisting of a sum of three damped sinusoids of different frequency, sampled every (T/N) seconds (b and c) or every $(T/2N)$ seconds (d), for the same total acquisition period, T. The quantization noise evident in spectrum (b) is reduced by a factor of $\sqrt{2}$, either by increasing the ADC word length by 1 bit (c) or by doubling the time-domain sampling rate (d). Oversampling effectively spreads the quantization noise over twice the spectral bandwidth. (Figure provided by G. M. Alber)

5.1.4 Precision in determination of peak position, width, and height as a function of peak shape, signal-to-noise ratio, and digital resolution

From the above discussion, it is beginning to be clear that the quality of a spectrum is limited by the type and magnitude of the spectral noise and by the discrete nature of the data. In this section, we shall quantitate those relations, for several common spectral line shapes.

For a given line shape, the principal parameters, x_i, of a spectral peak (see Figure 5.5) are: peak center frequency, ν_0; peak full width, $\Delta\nu$, at half-maximum height; and peak maximum magnitude, A. When random (vertical) noise, $n(\nu)$, is present, the present problem is to predict the precision, $P(x_i)$, with which each of these parameters can be determined, when only a *single* data set is available. For example, it would be desirable to know the precision of a single mass measurement in FT ion cyclotron resonance mass spectrometry, without having to repeat the measurement many times.

For scaling purposes, it is convenient to define a "reduced" frequency, u, measured from the center of the peak, and scaled in multiples of absorption-mode peak width at half-maximum peak height.

$$u = \frac{\lambda}{\Delta\nu}(\nu - \nu_0) \qquad (5.6)$$

in which λ is a line width scaling factor. Various previously discussed spectral line shapes may then be expressed as follows:

$$\text{Absorption-mode Lorentzian} = \frac{A}{1+u^2}\ ; \qquad \lambda = 2 \qquad (5.7)$$

$$\text{Magnitude-mode Lorentzian} = \frac{A}{\sqrt{1+u^2}}\ ; \qquad \lambda = 2\sqrt{3} \qquad (5.8)$$

$$\text{Absorption-mode Gaussian} = A\exp(-u^2)\ ; \qquad \lambda = \sqrt{4\log_e 2} \qquad (5.9)$$

$$\text{Absorption-mode sinc} = A\,\frac{\sin \pi u}{\pi u}\ ; \qquad \lambda = 3.791 \qquad (5.10)$$

$$\text{Magnitude-mode sinc} = \sqrt{2}\,A\,\frac{\sqrt{1-\cos u}}{|u|}\ ; \qquad \lambda = 7.582 \qquad (5.11)$$

in which A is the maximum peak height (at $\nu = \nu_0$) for each peak shape.

The precision for each peak parameter can be defined relative to its standard deviation, σ. σ represents the root-mean-square deviation from the average value that would be obtained from many measurements (see Appendix A).

$$\text{Amplitude precision:} \qquad P(A) = \frac{A}{\sigma(A)} \qquad (5.12)$$

$$\text{Peak center precision:} \qquad P(\nu_0) = \frac{\Delta\nu}{\sigma(\nu_0)} \qquad (5.13)$$

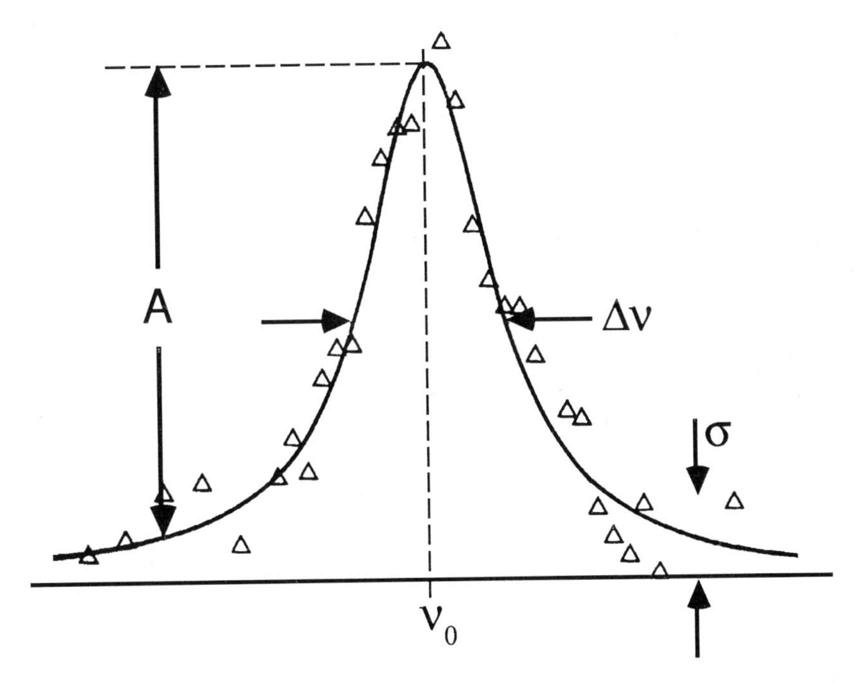

Figure 5.5 Non-linear least squares fit (solid line) of a theoretical absorption-mode Lorentzian spectral line to a Lorentzian to which ordinate white noise of root-mean-square deviation, $\sigma(A)$, has been added. The spectral parameters determined from the fit are: peak center frequency, ν_0; full width, $\Delta\nu$, at half-maximum peak height; and amplitude, A. (From L. Chen *et al.*, *Chemometrics and Intelligent Lab. Systems* **1986**, *1*, 51–58.)

Peak width precision: $$P(\Delta\nu) = \frac{\Delta\nu}{\sigma(\Delta\nu)} \qquad (5.14)$$

The amplitude precision is scaled according to the absolute peak height, A (Eq. 5.12). [It may be easier to think of the *imprecision*, $1/P(A) = \sigma(A)/A$, in which the absolute imprecision in amplitude, $\sigma(A)$, is divided by the absolute amplitude, A, to give the relative imprecision, $\sigma(A)/A$.] However, the absolute precision for either width or peak position clearly depends upon the absolute line width, and is therefore scaled according to $\Delta\nu$ (Eqs. 5.13,14).

In equations 5.12 to 5.14, the following assumptions apply:

(i) Each parameter is determined by *least-squares* fit of the experimental data to a proposed line shape.

(ii) *Gaussian-distributed* (in both absorption and dispersion spectra) *white noise* is present in the *ordinate only* (i.e., the frequency scale is perfectly known).

(iii) The line shape is known.

(iv) The noise is independent of the signal (i.e., "detector-limited").

Under the above conditions, it can be shown that the precision, $P(x_i)$, for the parameter, $x_i = A$, ν_0, or $\Delta\nu$, is predictable from the observed absorption-mode peak height-to-noise ratio, SNR,

$$P(x_i) = c(x_i)\ SNR\ \sqrt{K} \tag{5.15}$$

in which

$$K = \text{number of data points per line width} \tag{5.16}$$

and $c(x_i)$ is a constant which depends upon the line shape. Eq. 5.15 can be derived (see Further Reading) from an equally weighted non-linear least squares fit to the absorption-mode spectrum $A(\nu)$, whose set of normal equations coefficients form a matrix, H, whose elements are

$$H_{ij} = \sum_{m=1}^{N} \left(\frac{\partial A(\nu)}{\partial x_i}\right)_m \left(\frac{\partial A(\nu)}{\partial x_j}\right)_m \tag{5.17}$$

$$= \frac{K}{\lambda}\, G_{ij} \tag{5.18}$$

in which N is the number of data points. In the limit that the discrete intervals, Δu, are sufficiently small,

$$G_{ij} = \sum_{m=1}^{N} \left(\frac{\partial A(\nu)}{\partial x_i}\right)_m \left(\frac{\partial A(\nu)}{\partial x_j}\right)_m \Delta u$$

$$G_{ij} = \int_{-\infty}^{\infty} \left(\frac{\partial A(\nu)}{\partial x_i}\right) \left(\frac{\partial A(\nu)}{\partial x_j}\right) du \tag{5.19}$$

Equation 5.15 can then be derived from the general relations (Eqs. 5.12 to 5.14) between the standard deviation for each parameter and the root-mean-square spectral (vertical) noise, n,

$$\sigma(x_i) = [(H^{-1})_{ii}] \cdot n \tag{5.20}$$

For a two-parameter fit to the spectrum (i.e, best-fit values for ν_0 and A, assuming that $\Delta\nu$ is precisely known), $c(A)$ and $c(\nu_0)$ from Eq. 5.15 become (see Problems)

$$c(A) = \frac{1}{\sqrt{\lambda\,(G_{11})^{-1}}} \tag{5.21a}$$

$$c(\nu_0) = \frac{\Delta\nu}{A\sqrt{\lambda\,(G_{22})^{-1}}} \tag{5.21b}$$

in which λ is defined in Eqs. 5.7-5.11. Once the matrix elements of G have been computed in Eq. 5.19, the inverse matrix, G^{-1}, can be computed, and $c(A)$ and $c(\nu_0)$ can then be generated from Eqs. 5.21.

For a three-parameter fit (i.e., best-fit for A, ν_0, and $\Delta\nu$), one must invert the 3x3 G-matrix defined by

$$G = \begin{bmatrix} \int_{-\infty}^{\infty}\left(\frac{\partial A(\nu)}{\partial A}\right)\left(\frac{\partial A(\nu)}{\partial A}\right)du & \int_{-\infty}^{\infty}\left(\frac{\partial A(\nu)}{\partial A}\right)\left(\frac{\partial A(\nu)}{\partial \nu_0}\right)du & \int_{-\infty}^{\infty}\left(\frac{\partial A(\nu)}{\partial A}\right)\left(\frac{\partial A(\nu)}{\partial \Delta\nu}\right)du \\ \int_{-\infty}^{\infty}\left(\frac{\partial A(\nu)}{\partial \nu_0}\right)\left(\frac{\partial A(\nu)}{\partial A}\right)du & \int_{-\infty}^{\infty}\left(\frac{\partial A(\nu)}{\partial \nu_0}\right)\left(\frac{\partial A(\nu)}{\partial \nu_0}\right)du & \int_{-\infty}^{\infty}\left(\frac{\partial A(\nu)}{\partial \nu_0}\right)\left(\frac{\partial A(\nu)}{\partial \Delta\nu}\right)du \\ \int_{-\infty}^{\infty}\left(\frac{\partial A(\nu)}{\partial \Delta\nu}\right)\left(\frac{\partial A(\nu)}{\partial A}\right)du & \int_{-\infty}^{\infty}\left(\frac{\partial A(\nu)}{\partial \Delta\nu}\right)\left(\frac{\partial A(\nu)}{\partial \nu_0}\right)du & \int_{-\infty}^{\infty}\left(\frac{\partial A(\nu)}{\partial \Delta\nu}\right)\left(\frac{\partial A(\nu)}{\partial \Delta\nu}\right)du \end{bmatrix} \tag{5.22}$$

$c(A)$ and $c(\nu_0)$ are then obtained from Eqs. 5.21 as before, and $c(\Delta\nu)$ is found from

$$c(\Delta\nu) = \frac{\Delta\nu}{A\sqrt{\lambda\,(G_{33})^{-1}}} \tag{5.23}$$

For each of the five line shapes of Eqs 5.7 to 5.11, the approximate G-matrix elements shown in Eq. 5.19 have been evaluated analytically. Following inversion of that G-matrix, explicit values for $c(A)$, $c(\nu_0)$, and $c(\Delta\nu)$ generated from Eqs. 5.21 and 5.23 are listed in Table 5.3.

Table 5.3 lists the dependence of precision in measurement of spectral peak height, width, and position, as a function of absorption-mode peak height-to-noise ratio SNR and number of data points per line width, K, according to Equation 5.15. All entries in the table were calculated analytically from Eqs. 5.15 to 5.23. (For the absorption-mode and magnitude-mode sinc line shapes, the G_{33} integral in Equation 5.23 becomes infinite; thus, only the two-parameter fit precision can be evaluated for the sinc line shapes.)

There are two ways to estimate the precision of an equally weighted non-linear least-squares fit of a particular curve to a given digitized spectrum. First, the standard deviation for each spectral parameter (amplitude, width, and position) can be estimated from Eq. 5.15, based on a *single* data set (Table 5.3).

Table 5.3 Theoretical slope (Eq. 5.15) of a plot of precision *vs.* $SNR \cdot \sqrt{K}$. Except as noted, all values are for a three-parameter fit (A, ν_0, and $\Delta\nu$) to the spectral data.

Line Shape	Display Mode	A	ν_0	$\Delta\nu$
Lorentzian	Absorption-Mode	0.627 (0.886)[a]	1.253 (1.253)[a]	0.443
	Magnitude-Mode	0.550 (0.952)[a]	1.166 (1.166)[a]	0.337
Gaussian	Absorption-Mode	0.708 (0.868)[a]	1.445 (1.445)[a]	0.614
Sinc	Absorption-Mode	(0.910)[a]	(1.992)[a]	
	Magnitude-Mode	(0.910)[a]	(1.992)[a]	

[a] Two-parameter fit, assuming that $\Delta\nu$ is precisely known (A three-parameter fit for the sinc lineshape is not feasible—see text.)

Alternatively, one could compute the precision directly from the standard deviation of the best-fit values for peak position, amplitude, and width for each of a *large number* of independent discrete spectra. For example, Figure 5.6 compares the results of this "brute-force" procedure with the theoretical predictions from Eq. 5.15 for an absorption-mode Lorentzian peak. The scatter in the plots is due to the limited number (30) of data sets. For other absorption-mode and magnitude-mode spectral line shapes, plots of precision vs. $SNR \cdot \sqrt{K}$ would resemble those in Figure 5.6, except for different slope (see Table 5.3).

As seen from the first column in Table 5.3, the precision in determination of peak *height* is always higher (i.e., better) if the line width is known in advance. However, when the peak shape is symmetrical about its center, the precision in determination of peak *position* is the same whether or not the peak width is precisely known, as seen from the second column of Table 5.3.

The significance of Eq. 5.15 is that *there is an absolute limit to the precision with which we may use non-linear least squares fits to determine any spectral parameter*, even if the line shape is perfectly known. No amount of apodization can change that precision—apodization can affect only the *appearance* of the spectrum, not its fundamental *information* content. We shall discuss various spectral estimation techniques in Chapter 6.

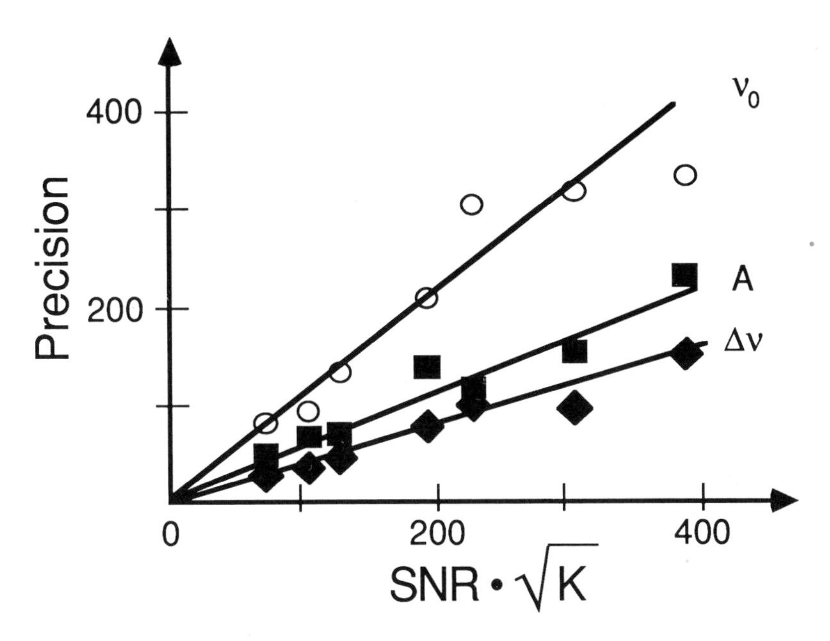

Figure 5.6 Plots of precision in peak position (O), amplitude (■), and line width (◆) vs. the product of spectral peak height-to-noise ratio, *SNR*, and the square root of the number of data points, *K*, per line width. Each straight line represents the predicted slope (see Table 5.3) for that parameter, for an absorption-mode Lorentzian peak shape. The data points were computed directly from the standard deviation of best-fit values from 30 independent simulated spectra of the type shown in Figure 5.3. (Adapted from Chen *et al.*, *Chemometrics and Intelligent Lab. Systems* **1986**, *1*, 51-58.)

5.1.5 Effect of apodization on spectral signal-to-noise ratio and resolution

As we shall discuss in greater length in Chapter 6, apodization can only *decrease* the precision of spectral information available by computer analysis (whether by FT or other spectral estimator methods). For example, multiplication of a time-domain discrete signal by the decreasing exponential weight factor, $\exp(-t/\tau)$, *preferentially* reduces the noise (which is present at equal rms magnitude at all t-values) more than signal (whose magnitude decreases with time). However, *some* of the signal is also suppressed by the weighting operation, and therefore the overall information content of the spectrum is reduced.

Because spectral peak height-to-noise ratio, *SNR*, is the quotient of *two* spectral parameters, it is possible to find a time-domain weight function which reduces noise more than signal, and which thereby optimizes signal-to-noise *ratio*. For example, for a signal of the form,

$$S(t) = A_0 \exp(-t/T_2) \tag{5.24}$$

and a time-domain decreasing exponential weight function,

$$W(t) = \exp(-t/\tau) \tag{5.25}$$

the optimal choice of τ is the so-called "matched filter", namely

$$\tau = T_2 \quad \text{("matched filter")} \tag{5.26}$$

as shown in Figure 5.7 (see Problems).

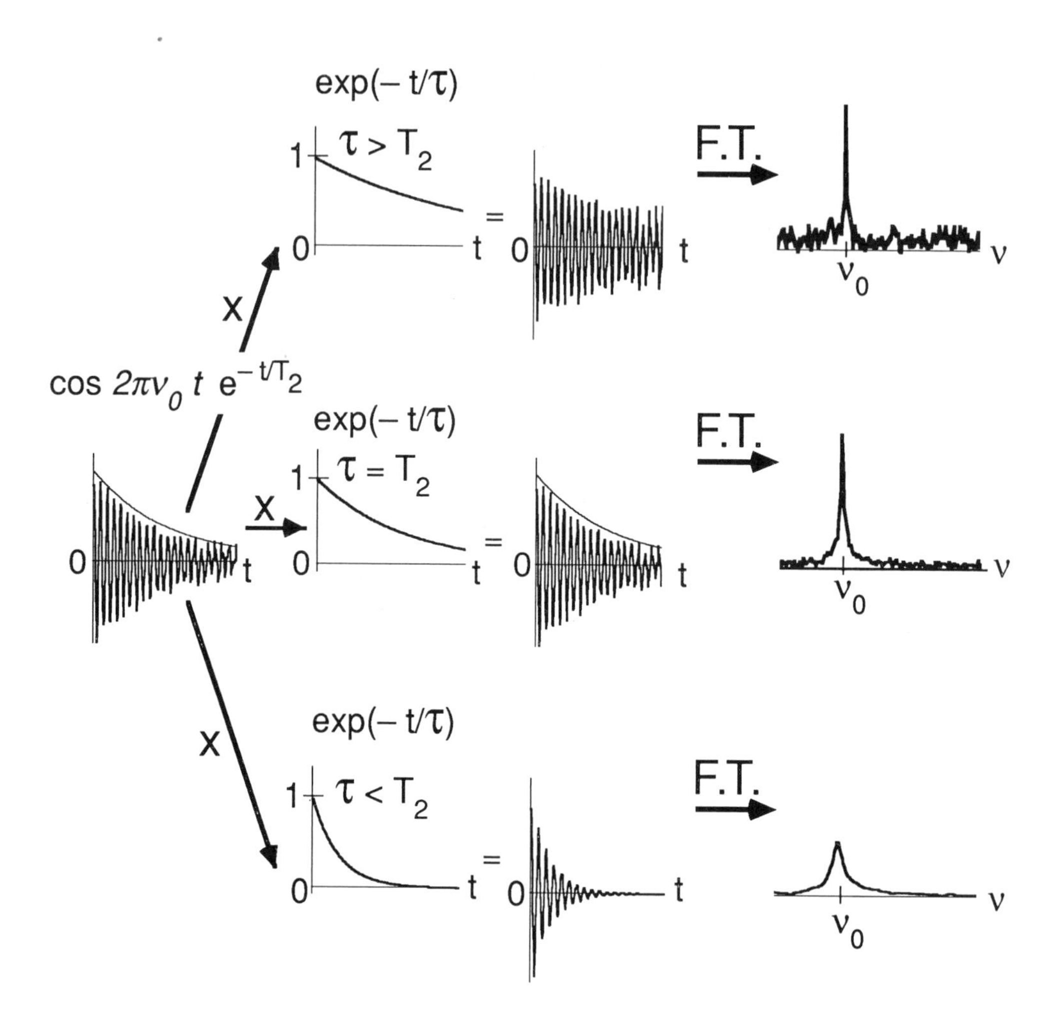

Figure 5.7 Effect of time-domain decreasing exponential apodization (Eq. 5.25) on frequency-domain peak-height-to-noise ratio, for three choices of exponential damping constant, τ. The optimum peak-height-to-noise ratio for this apodization function is for $\tau = T_2$ (i.e., the "matched filter" condition of Eq. 5.26). Note that a decrease in τ results in increased peak-height-to-noise ratio at the cost of increased spectral peak width (see text).

As noted earlier, frequency-domain peak *area* is a better measure than peak *height*, for determination of the relative numbers of oscillators (e.g., ions in ICR, magnetic nuclei in NMR, molecules in IR) of various frequencies. The time-domain weight function, $\exp[-(t/T_2)^a]$, with $a > 1$, can be shown to give better frequency-domain peak-area-to-noise ratio than the "matched filter" of Eq. 5.26, but at the cost of requiring more precise (advance) knowledge of the magnitude of T_2 (see Weiss *et al.* in Further Reading). A treatment much like that which led to Eq. 5.15 shows that the ultimate precision in determination of peak area by non-linear least-squares methods is related to the time-domain initial signal-to-noise ratio and the number of time-domain data points (see Liang & Marshall in Further Reading).

Unfortunately, if the apparent precision in one parameter (in this case, peak height-to-noise ratio) is enhanced by apodization, then at least as much information must be lost in some other way, since apodization cannot increase the overall spectral information content. In this example, spectral *resolution* is reduced by the apodization, which broadens the spectral half-maximum peak width from $(2/T_2)$ rad s^{-1} to $[(2/T_2) + (2/\tau)]$, as seen in Figure 5.7.

Similarly (see Problems), the reader can show that an increasing exponential weight function, $\exp(+t/\tau)$, can increase spectral *resolution* at the cost of reduced peak height-to-noise ratio. Again, apodization does not increase the overall spectral information content, as will be more evident when *non*-FT spectral estimators are used to determined spectral peak positions and line widths (see Chapter 6).

5.2 Noise as a spectral source: Fourier analysis of random motions

5.2.1 Random noise: autocorrelation function and power spectrum

By now the reader has come to appreciate that a linear oscillator (see weight-on-a-spring in Chapter 1) can be made to vibrate by application of some sort of perturbation whose motions contain frequency components near the "natural" vibrational frequency of the spring. Up to now, we have considered only *coherent* driving forces [e.g., sinusoidal or Dirichlet (rectangular pulse) excitation], so that all of the driven springs in any given region of space oscillate in phase.

In this section, we wish to determine the frequency components of *incoherent* radiation sources or molecular motions. It will then become evident that a randomly time-varying electric (or magnetic) field can have frequency components that also induce transitions between energy levels (quantum mechanical language) or drive electrons on springs near their "natural" frequencies (classical language) even in the absence of externally applied radiation, thus explaining the T_1 processes that account for the "natural" line width of Lorentzian spectral lines (see Chapter 7.3.2 and Chapter 8.3.2). We first present a general formula for computing the frequency components of any time-varying random process, and then show how some particular types of random motion (e.g., chemical exchange, rotational diffusion, translational diffusion) are related in a natural way to spectral line-broadening in magnetic resonance and other types of spectroscopy and scattering experiments.

By a random or stochastic process, we mean a process in which the amplitude, $f(t)$, of the instantaneous molecular linear or angular position (or electromagnetic field) depends on time in a way not completely definite, but with some sort of *average* variation, to be defined more precisely below. For example, the three time-domain traces in Figure 5.8 might represent the instantaneous positions of three different molecules, each undergoing a one-dimensional random walk as a function of time, or three different time-domain electronic detector readings in the absence of a signal. For such a case, the *average* ("mean") position of a molecule away from the origin is zero. Nevertheless, the noise level at a given instant is clearly non-zero, and we need a way to quantitate its magnitude. One might think to take the *absolute-value* of the noise, but the absolute-value operation can be mathematically inconvenient (e.g., for computation of definite integrals). The most useful representation of noise magnitude turns out to be its *root-mean-square* value, as we shall soon see.

Suppose that we record an arbitrary *random* displacement, $f(t)$, for a specified time interval, T. As for a *coherent* time-domain signal, we can express $f(t)$ as an infinite series (Eq. 5.27a) of sinusoidal oscillations having component amplitudes, a_n and b_n, at corresponding frequencies, ω_n (compare to Eq. 3.9):

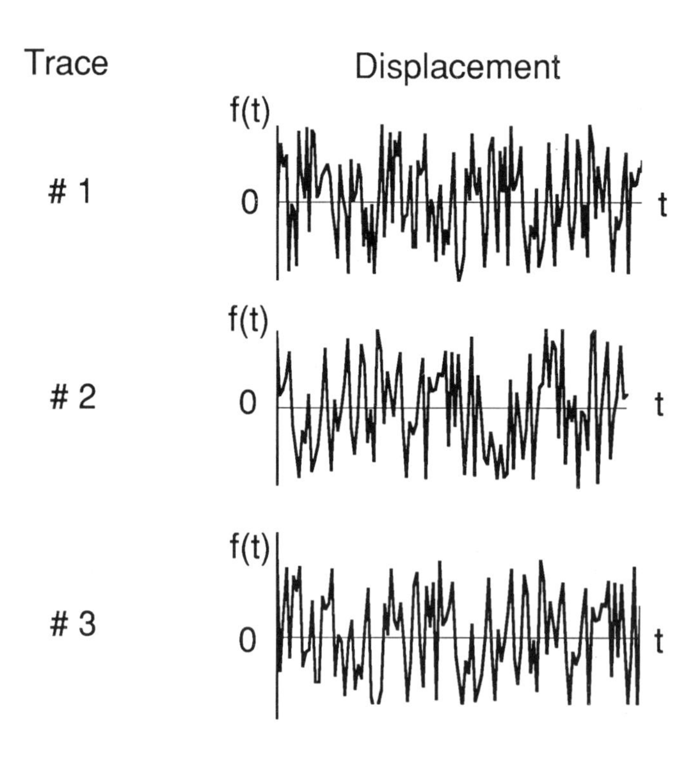

Figure 5.8 Hypothetical recordings of position as a function of time for three molecules undergoing one-dimensional translational diffusion. The behavior is different for different *molecules*. Moreover, the frequency components of the motion vary with *time* (sometimes the oscillations are slow and sometimes fast), so that meaningful frequency analysis of the motion must involve a *time-average* and average over different particles *(ensemble-average)*.

$$f(t) = \sum_{n=0}^{\infty} \left(a_n \cos \omega_n t + b_n \sin \omega_n t \right) \tag{5.27a}$$

in which

$$\omega_n = \frac{2\pi n}{T} \tag{5.27b}$$

As for the undriven (Chapter 1.2) or driven (Chapter 1.3) weight-on-a-spring displacement, we have broken down the response into components that vary with time as either cosine or sine oscillations, but with one major difference. In the preceding examples, the motion or oscillation was *coherent* ; that is, all of the particles (or electromagnetic waves) in one region of space were oscillating with the same phase, so we could predict the group ("ensemble") behavior simply by knowing the behavior of any one particle (or wave). Now, however, the magnitude of the random signal varies from time to time and from one particle to another (Figure 5.8), so that the observed experimental result is based on some sort of average (over time and over particles) of the behavior of individual particles (waves).

If we attempt a simple time-average of $f(t)$ in Eqs. 5.27, however, averaged over the observation period from time zero to time T, the time-average, $\langle f(t) \rangle$, goes to zero because the average of either $\cos(2\pi n\, t/T)$ or $\sin(2\pi n\, t/T)$ goes to zero over that interval:

$$\frac{1}{T} \int_0^T \cos\left(\frac{2\pi n\, t}{T}\right) dt = \frac{1}{T} \int_0^T \sin\left(\frac{2\pi n\, t}{T}\right) dt = 0 \tag{5.28}$$

a_n and b_n in the present analysis represent the point-by-point amplitudes of the cosine and sine component oscillations at a series of equally spaced discrete spectral frequencies (Eq. 5.27b), just as for the coherent time-domain signals treated in Chapter 3. The new feature is that although a_n and b_n for any one particle at any one time are nonzero, the *time-average* values of a_n and b_n are zero, and we can no longer obtain a_n and b_n directly from observing the *average* behavior of an ensemble of particles.

However, if we conduct an experiment that measures the *square* of $f(t)$, such as a measurement of radiation *intensity* (rather than radiation *field* value), then the time-averaged frequency-components of the resulting spectrum can be determined from (see Problems)

$$\frac{1}{T} \int_0^T \cos^2\left(\frac{2\pi n\, t}{T}\right) dt = \frac{1}{T} \int_0^T \sin^2\left(\frac{2\pi n\, t}{T}\right) dt = \frac{1}{2} \tag{5.29}$$

so that the *time-average* value of $[f(t)]^2$, again denoted by brackets, is given by

$$\langle [f(t)]^2 \rangle = \sum_{n=0}^{\infty} \left(\frac{a_n{}^2 + b_n{}^2}{2} \right) = \text{constant for any one particle} \tag{5.30}$$

Finally, when we take an additional "ensemble"-average over the various particles (or successive time-domain acquisitions), we obtain the noise *power spectrum* of the random motion, $P(\omega)$:

$$P(\omega_n) = \overline{\left(\frac{a_n{}^2 + b_n{}^2}{2}\right)} \tag{5.31}$$

from which the time-averaged, ensemble-averaged squared $f(t)$ becomes

$$\overline{< [f(t)]^2 >} = \sum_{n=0}^{\infty} P(\omega_n) \tag{5.32}$$

in which the superscript line denotes an ensemble-average. $P(\omega_n)$ thus represents the time-averaged, ensemble-averaged frequency component (at frequency ω_n) of the squared amplitude of the random motion of interest. What remains is to determine the form of $P(\omega_n)$ corresponding to a suitable model for the random motion process.

From the pictures in Figure 5.8, it is clear that we cannot predict with certainty the value of $f(t)$ at a particular instant for a particular particle. Nevertheless, if we know something about the *rate* of the random fluctuations, then we can say that if $f(t)$ has a given value at a particular time for a given particle, then $f(t)$ will not have changed very much for a sufficiently short "lag time", τ, afterward (i.e., the value at time, $t + \tau$, will be "correlated" with the value at time, t). It is also intuitively obvious that if we wait a sufficiently long time, we shall no longer be able to predict the value of $f(t + \tau)$ based its prior value, $f(t)$: i.e., $f(t + \tau)$ will have become totally "uncorrelated" with $f(t)$ for sufficiently long τ. Therefore, since the length of time for $f(t + \tau)$ to become uncorrelated with $f(t)$ has something to do with the characteristic rate constant (e.g., diffusion constant in the translational diffusion case) for the random process, we might expect to learn about the frequency components of the random process from study of the ensemble-averaged "autocorrelation function," $g(\tau)$, previously encountered in Chapter 2.2.3 (see also Appendix D):

$$g(\tau) = \overline{< f(t) f(t + \tau)>} \tag{5.33}$$

in which (as before) the brackets denote a time-average and the bar denotes an ensemble-average over all particles undergoing the random process. A remarkable property of the autocorrelation function appears when we express $f(t)$ and $f(t + \tau)$ in terms of their frequency components defined by Eq. 5.27a):

$$g(\tau) = \frac{1}{T} \int_0^T \left(\sum_{n=0}^{\infty} a_n \cos \omega_n t + b_n \sin \omega_n t \right) \left(\sum_{n=0}^{\infty} a_n \cos \omega_n (t+\tau) + b_n \sin \omega_n (t+\tau) \right) dt$$

$$= \sum_{n=0}^{\infty} \overline{\left(\frac{a_n{}^2 + b_n{}^2}{2} \right)} \cos \omega_n \tau \tag{5.34}$$

$$= \sum_{n=0}^{\infty} P(\omega_n) \cos \omega_n \tau \tag{5.35}$$

The demonstration of Eq. 5.34 is left as an exercise (see Problems), and Eq. 5.35 then follows immediately from Eq. 5.31.

Replacing the sum in Eq. 5.35 by an integral, we obtain

$$\overline{< f(t)\ f(t + \tau) >} = g(\tau) = \int_0^{\infty} P(\omega) \cos \omega\tau \ d\omega \tag{5.36}$$

which is more typically expressed as the equivalent inverse Fourier transform

$$P(\omega) = \frac{1}{2\pi} \int_0^{\infty} g(\tau) \cos \omega\tau \ d\tau \tag{5.37}$$

Equation 5.37 can be thought of as a recipe for finding the relative amount of power at any given frequency, resulting from random particle or electromagnetic wave motion having the correlation function, $g(\tau)$. Thus, as soon as we work out the form of $g(\tau)$ for a particular type of motion (see Problems), we will immediately be able to compute the power spectrum as a function of frequency from Eq. 5.37.

For all of the specific random processes that we will consider, the correlation function, $g(\tau)$, decreases exponentially with time.

$$g(\tau) = g(0) \exp(-\tau/\tau_c); \quad 0 \leq \tau < \infty; \quad \tau_c = \text{"correlation time"} \tag{5.38}$$

in which the *correlation time*, τ_c , is a measure of the time it takes for $f(t + \tau)$ to become uncorrelated with its earlier value, $f(t)$, and $g(0)$ is the mean-square instantaneous amplitude of the random motion (see below). For example, if the random process is a first-order radioactive decay, then the number of undecayed atoms decreases exponentially with time, and that number is a measure of the "correlation" of the final state of the system with its initial state. For that case, the correlation time, τ_c , is simply the reciprocal of the first-order reaction rate constant (see Problems). As other examples, translational (or rotational) diffusion may be described by a correlation function which represents the fraction of molecules which have not translated (or rotated) by a characteristic distance (or angle) during the correlation time, τ_c (see Marshall, *Biophysical Chemistry*, Chapter 21).

When the correlation function is of the form given by Eq. 5.38, we can compute the power spectrum easily from Eq. 5.37, and our previous knowledge that the Fourier transform of an exponential is a Lorentzian (see Appendix D):

$$P(\omega) = \frac{1}{2\pi}\ g(0) \int_0^{\infty} \exp(-\tau/\tau_c) \cos \omega\tau \ d\tau$$

$$= \frac{1}{2\pi}\ g(0) \frac{\tau_c}{1 + \omega^2 \tau_c{}^2} \tag{5.39}$$

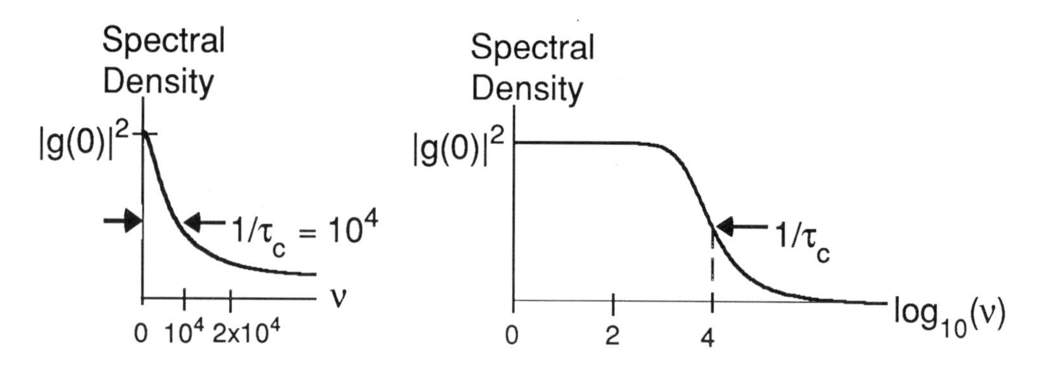

Figure 5.9 Power spectrum (spectral density) as a function of frequency [linear scale (left) or log scale (right)] for a random process described by an exponentially decaying correlation function with correlation time, $\tau_c = 10^{-4}$ s. As seen most clearly on the semi-log plot, the noise power spectrum is almost "white" (i.e., constant in magnitude) from zero frequency up to frequencies of the order of $(1/\tau_c)$. [The spectral density, $P(\omega)$, of Eq. 5.37 is simply a Lorentzian curve, whose form in the right-hand plot is distorted by the use of a log scale for frequency.]

The power spectrum ("spectral density"), $P(\omega)$, of a randomly fluctuating motion thus takes the familiar Lorentzian form, and is illustrated in Figure 5.9 as a function of ω and as a function of $\log_{10}(\omega)$. The power spectrum is flat from zero frequency up to frequencies of the order of $(1/\tau_c)$, and falls off rapidly near $(1/\tau_c)$. The "correlation time", τ_c, is thus a measure of the shortest time in which the particle position or orientation angle (or electromagnetic field) mean-square value can change significantly. Alternatively, since $(1/\tau_c)$ is effectively the highest (angular) frequency generated by that time-domain noise fluctuation, $(1/\tau_c)$ may also be thought of as the frequency-domain "bandwidth" of that noise (since random noise always extends down to zero frequency at the low-frequency end of the band).

Figure 5.9 shows that such a random process effectively acts as a broadband source of frequencies up to about $(1/\tau_c)$ s^{-1}. Thus, when the random motion produces a change in electric (or magnetic) field strength, the random motion can act as a broadband *source* of electromagnetic radiation over that range of frequencies. If the "natural" frequency of the sample of interest happens to fall at a frequency less than about $(1/\tau_c)$ s^{-1}, such a random process will drive that system at its natural frequency (classical language) so as to cause transitions between the corresponding energy levels (quantum mechanical language) *even in the absence of an external radiation source.* Since those transitions occur at a rate that defines the "lifetime", T_1, of the population difference between the two energy levels involved, we can now account for the "natural line width", $(2/T_1)$, of several types of spectral absorption lines, in terms of microscopic motion of the particles in the sample. It remains only to show how the spectral line width for a sample (related to the rate of transitions between the two energy levels) is related to the correlation time, τ_c, for the particle random motion, and then (see Problems) how that τ_c is determined by some particular types of random motion.

For a given random process, there is a well-defined mean-square value for the instantaneous amplitude of the motion:

$$\overline{<|f(t)|^2>} = \text{constant} \tag{5.40}$$

But $\overline{<|f(t)|^2>}$ in Eq. 5.40 is just the value of the correlation function, $g(\tau)$, for $\tau = 0$. Recalling Eqs. 5.33 and 5.36, we thus reach the immediate important result,

$$\overline{<|f(t)|^2>} = g(0) = \int_0^\infty P(\omega)\, d\omega \tag{5.41}$$

or,

$$\int_0^\infty P(\omega)\, d\omega = \text{area under a plot of } P(\omega) \text{ vs. } \omega = \text{constant} \tag{5.42}$$

independent of the correlation time, τ_c, for that process. The significance of Eq. 5.42 is evident from Figure 5.10, which shows plots of spectral density versus frequency (log scale) for three different choices of correlation time for a given random process. Since the *area* under each curve must be the same (Eq. 5.42), whereas the cut-off frequency $(1/\tau_c)$ varies from plot to plot, the *height* of each curve must also vary as shown in the Figure. If we now specify that this random process produces a fluctuating electric (or magnetic) field, then the plots show the relative power spectrum of radiation produced by the process.

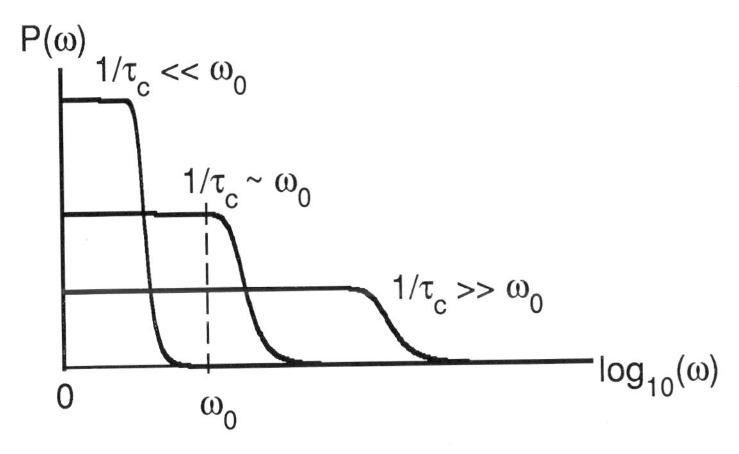

Figure 5.10 Spectral density, $P(\omega)$, as a function of frequency (log scale), for random electromagnetic field fluctuations characterized by three different correlation times. The quantity of interest (see text) is the spectral density at the frequency, ω_0, of the natural vibration of some oscillator in a sample exposed to the fluctuating field. For either long (topmost) curve) or short (lowermost curve) correlation time, τ_c, the spectral density at ω_0 is small; the spectral density at ω_0 is a maximum when $(1/\tau_c) \approx \omega_0$. These curves account for dependence of the induced transition rate constant $(1/T_1)$ on correlation time as shown in Fig. 5.11.

Figure 5.10 allows us to estimate the transition rate induced by randomly fluctuating external electric or magnetic fields. If ω_0 is the "natural" frequency for a particular type of motion (e.g., NMR precession or IR vibration), then the rate constant, $(1/T_1)$, for induced (radiationless) transitions will be proportional to the amount of power reaching the sample at frequency ω_0. For example, when motion is slow and the correlation time is very long $[(1/\tau_c) << \omega_0]$, very little power reaches the sample at frequency ω_0, so that relatively infrequent transitions are induced and $(1/T_1)$ is small. In contrast, when $(1/\tau_c) \approx \omega_0$, there is maximal spectral density at ω_0, and the rate of induced transitions is also a maximum. Lastly, when motion is very fast and the correlation time (Eq. 5.38) is very short $[(1/\tau_c) >> \omega_0]$, the spectral density at ω_0 is again small and $(1/T_1)$ is small. From the consequences of Eq. 5.42 shown in Fig. 5.10, it is clear that the rate constant for induced transitions (in the absence of externally applied radiation), $(1/T_1)$, must show the dependence on correlation time illustrated in Figure 5.11, with a maximum spectral line-broadening (i.e., minimum T_1) when the random motion has an inverse correlation time of the order of the natural frequency of the system, namely, $(1/\tau_c) \approx \omega_0$. Figure 5.11 thus accounts for natural spectral line width, $\Delta\omega = (2/T_1)$, in terms of the characteristic correlation time for change in the mean-square magnitude of a randomly varying local electromagnetic field. The reader is referred to the Problems and Further Reading for evaluation of the correlation function and correlation time for each of several types of random processes (e.g., chemical exchange, rotational diffusion, translational diffusion) that produce corresponding randomly fluctuating electromagnetic fields. Armed with those results, one can then proceed in reverse order to determine characteristic rate constants (e.g., chemical reaction rate constant; rotational or translational diffusion constant) from measured spectral line widths (see Chapter 8 and Marshall, *Biophysical Chemistry*, Chapter 21, in Further Reading).

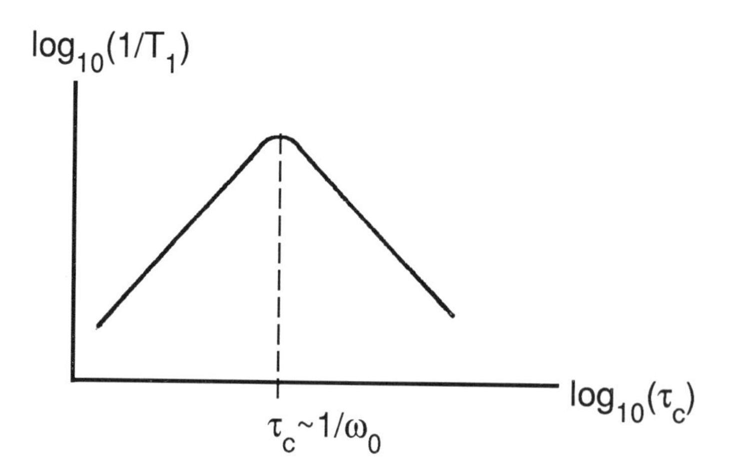

Figure 5.11 Plot of transition probability, $(1/T_1)$, (log scale) versus correlation time, τ_c, (log scale) for a particular random process. The Figure shows that the transition probability is a maximum when $(1/\tau_c) \approx \omega_0$, as inferred from Figure 5.10. These transitions occur in the *absence* of externally applied radiation, and thus define the "natural" spectral line width for the sample in question (see text).

5.2.2 Pseudorandom noise: shift register and Hadamard S-matrix sequences

In the preceding section, we showed that truly random noise can serve as a radiation source, with relatively flat power over a frequency-domain bandwidth from zero to $(1/\tau_c)$, where τ_c is the "correlation time" which characterizes that random process. A potential advantage of a random noise radiation source over (say) a short time-domain spike is that the random noise root-mean-square magnitude can be much smaller (i.e., much lower time-domain dynamic range) than the magnitude of a short rectangular pulse (see Chapter 2.1.2). Unfortunately, truly random noise is not reproducible from one data acquisition to the next, so that signal-averaging would require Fourier transformation of each time-domain data set and subsequent co-addition in the *frequency*- domain, a much slower process than direct co-addition of *time*- domain signals.

Fortunately, it is possible to construct *pseudo-random* sequences, in which the magnitude (or phase, or width, or time-delay) of successive time-domain pulses varies so as to give the same sort of power spectrum as that from truly random noise, but reproducible from one excitation to the next. Moreover, electrical engineers have been clever enough to design algorithms which can generate such sequences without having to store the entire pattern in a computer memory.

Shift register sequences

A shift register is a device which replaces a given binary number (1 or 0) with another number obtained by shifting the digits of a binary word as shown in Figure 5.12. If the binary digits from two of the locations in such a device are added and then placed in the first location of the register after a shift, repetition of such a process generates a *maximal-length pseudo-random sequence* of $(2^n - 1)$ n- digit binary numbers (n is an integer) as illustrated in Figure 5.12 for the $(2^3 - 1) = 7$ periods generated from a 3-stage shift register. Similarly (see Problems), a 4-stage register can produce $(2^4 - 1) = 15$ periods of four digits each.

The "maximal-length" sequences resulting from the algorithm of Figure 5.12 have several useful properties. First, every possible n- digit binary period (except for all 0's) appears exactly once during the sequence. In each period of $p = 2^n - 1$ bits, the number of 1's (namely, 2^{n-1}) exceeds the number of 0's by exactly one. Second, for each "run" of consecutive 0's within a given period, there is a run of 1's of equal length. Third, if we lay all of the periods side by side to form a single long sequence, and designate the digits of that sequence by $a_0, a_1, a_2, \cdots$, then the discrete autocorrelation function of the sequence takes the form,

$$g(k) = \frac{1}{2^n - 1} \sum_{i=1}^{2^n - 1} a_i \, a_{i-k}$$

$$= \frac{2^{n-1}}{2^n - 1} \quad \text{for } k = 0, \pm(2^n - 1), \pm 2(2^n - 1), \cdots \qquad (5.43a)$$

$$= \frac{2^{n-2}}{2^n - 1} \quad \text{for all other } k\text{-values} \qquad (5.43b)$$

In other words, the maximal-length sequence "looks" random, in the sense that it contains approximately equal numbers of 0's and 1's and no preferred "runs" of a given number of 0's or 1's in a row. However, the sequence as a whole is *periodic*, because it repeats every $(2^n - 1)$ periods as shown by Eqs. 5.43. By comparison, remember that the autocorrelation function of a sinusoid is another sinusoid. Thus, the maximal length sequence possesses both "random" and "periodic" properties—hence "pseudo-random". It is also precisely reproducible from one experiment to another. Finally, it is possible to generate a long maximal-length sequence from a relatively small number of memory locations (e.g., 32,767-bit sequence from a 15-bit shift register).

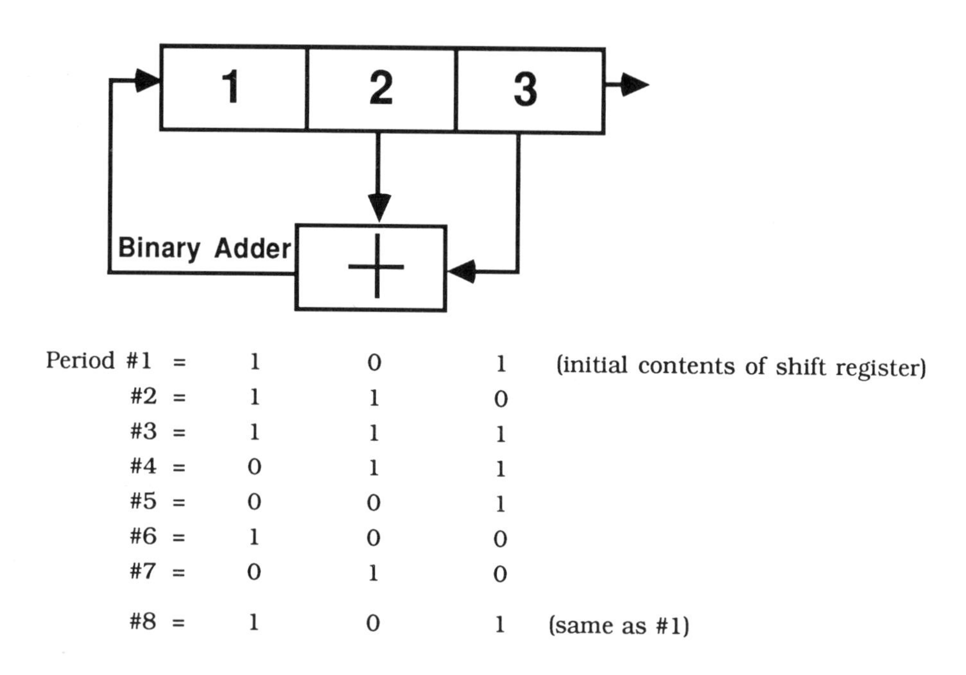

Period #1 =	1	0	1	(initial contents of shift register)
#2 =	1	1	0	
#3 =	1	1	1	
#4 =	0	1	1	
#5 =	0	0	1	
#6 =	1	0	0	
#7 =	0	1	0	
#8 =	1	0	1	(same as #1)

Figure 5.12 Shift register circuit for generation of a pseudo-random sequence of binary numbers. Beginning from any non-zero initial "period" (sequence) of 0's and 1's (e.g., 1 0 1 in this example), the contents of shift register locations 2 and 3 are added (see Table) to give a third (one significant digit) binary number (see Addition Table below). The original number in location 3 is discarded, the numbers originally in locations 1 and 2 are shifted to locations 2 and 3, and the new number is placed in location 1. The resulting shift register contents now form a second period of 0's and 1's, and the process is repeated to generate $2^3 - 1 = 7$ such sequences. Note that every possible combination of 0's and 1's (except for 0,0,0) appears exactly once during the sequence of 7 periods.

Addition Table:

$0 + 0 = 0$
$0 + 1 = 1$
$1 + 0 = 1$
$1 + 1 = 10$, whose last digit = 0

Hadamard sequences

We previously encountered the Hadamard H- and S-matrix codes in Chapter 4.1.3.1. In this section, we note that the rows of 0's and 1's of the Hadamard S-matrix (or, preferably, the rows of −1's and 1's of its inverse matrix) can be treated as "periods" of a pseudo-random sequence. Such a sequence can be used to "code" a time-domain sequence of voltage pulses of opposite sign to excite a spectrum with approximately constant excitation power over a bandwidth given by the reciprocal of the spacing between successive time-domain pulses. Used in this way, the Hadamard scheme can be thought of as "sampling" of the time-domain response to an impulse (δ-function) excitation. Hadamard decoding (as in Chapter 4.1.3.1) can then be used to recover the desired time-domain response to an impulse excitation. The advantage of such a scheme, as noted earlier, is that one can indirectly determine the response to a true impulse excitation, without having to produce the high excitation power needed for direct use of an impulse excitation. There is even a generating scheme analogous to the shift-register example of Figure 5.10 for producing the desired sequence.

We shall encounter the shift register principle again when we discuss "linear prediction" methods (Chapter 6.3), in which future time-domain data points are predicted from appropriate linear combinations of prior time-domain data points.

Further Reading

L. Chen, C. E. Cottrell, & A. G. Marshall, *Chemometrics and Intelligent Laboratory Systems* **1986**, *1*, 51-58. (Spectral precision in peak frequency, height, and width as a function of peak shape, signal-to-noise ratio, and number of data points per line width)

J. W. Cooper, *Anal. Chem.* **1978**, *50*, 801A-812A. (Good treatment of FT problems arising from discrete data)

M. A. Delsuc & J. Y. Lallemand, *J. Magn. Reson.* **1986**, *69*, 504-507. (Oversampling in FT/NMR spectroscopy)

W. H. Flygare, in *Fourier, Hadamard, and Hilbert Transforms in Chemistry*, Ed. A. G. Marshall, (Plenum, New York, 1982), pp. 207-270. (Describes 1-bit digitization of high-frequency signals, esp. pp. 224-225)

D. A. Hanna, *Amer. Soc. for Mass Spectrometry 33rd Conf. on Mass Spectrometry & Allied Topics*, San Diego, CA, May, 1985, pp. 435-436. (Short but lucid treatment of magnitude-mode noise and how to measure it by fitting a Rayleigh distribution to an experimental spectral peak-free baseline)

Z. Liang, Ph.D. thesis, The Ohio State University, **1990**. (Spectral peak area precision as a function of signal-to-noise ratio and number of data points per line width; relation between time-domain and frequency-domain precision)

G. H. Weiss, J. E. Kiefer, and J. A. Ferretti, *Chemometrics and Intelligent Laboratory Systems* **1988**, *4*, 223-229. (Effect of apodization on peak-area-to-noise estimation)

J. D. Winefordner, R. Avni, T. L. Chester, J. J. Fitzgerald, L. P. Hart, D. J. Johnson, and F. W. Plankey, *Spectrochimica Acta* **1976**, *31B*, 1-19. (Comparison of signal-to-noise ratios for single-channel and multichannel spectroscopy)

Problems

5.1 The Poisson distribution, $P_N(m)$, gives the probability that m desired events (e.g., radioactive decays, electrons crossing a junction in an electrical circuit, etc.) will occur in N trials, in which the probability of success in any one trial is much less than 50%. $\overline{m}$ is the average number of counts in a given N-trial experiment (e.g., $\overline{m}$ of N unstable nuclei decaying in a given observation period).

$$P_N(m) = \frac{\overline{m}}{m!}\exp(-\overline{m}) \tag{5.1.1}$$

(a) Plot $P_N(m)$ *vs.* m, for $\overline{m} = 5$. Note the asymmetrical shape of the curve.

(b) Demonstrate the most important property of the Poisson distribution, namely that

$$\sqrt{<(m - \overline{m})^2>} = \sqrt{\overline{m}} \tag{5.1.2}$$

In other words, the root-mean-square imprecision in any counting experiment is simply the square root of the average number of counts.

Hint: Average value of a discrete function, $f(m)$, is defined as

$$<f(m)> = \sum_{m=0}^{N} f(m)\, P_N(m) \tag{5.1.3}$$

where, in this case, $N \rightarrow \infty$.

5.2 An analog-to-digital converter (ADC) produces a quantized signal, $f(t_n)$ which differs from the true analog signal, $f(t)$, by the quantization error,

$$e(t_n) = f(t_n) - f(t) \tag{5.2.1}$$

Each discrete sample is the (integer) number of counts produced by the ADC, scaled to match the dynamic range of the signal: e.g., an analog dynamic range of $\pm 5\ V$ would require that a 10-bit ADC produce 512 counts for a 5-volt positive signal.

The error, $e(t_n)$, results when the ADC rounds off each voltage value to within half of one count. The ADC error, δ (in counts), ranges from $-0.5 \le \delta < +0.5$, corresponding to a probability density function, $P(\delta)$, given by:

$$P(\delta) = \begin{cases} 1 & -0.5 \le \delta < +0.5 \\ 0 & \text{otherwise} \end{cases} \tag{5.2.2}$$

(a) Assume that the quantization error is random and uniformly distributed according to Eq. 5.2.2, and show that its mean and variance are given by:

$$\overline{\delta} = \int_{-\infty}^{\infty} \delta\, P(\delta)\, d\delta = 0 \qquad (5.2.3a)$$

$$\sigma(\delta)^2 = \int_{-\infty}^{\infty} (\delta - \overline{\delta})^2\, P(\delta)\, d\delta = \frac{1}{12} \qquad (5.2.3b)$$

The mean and variance given here are in counts and counts-squared units. In order to convert the variance into physical units, an appropriate multiplying factor must be used. Thus, if one count corresponds to V_0 volts, then the variance would be $V_0^2/12$.

(b) The full range of an ADC is usually a power of 2. Obtain an expression for the variance for an $(n + 1)$-bit ADC with full-scale reponse of $\pm V$ volts. You should find that the variance (i.e., mean–square quantization error) decreases exponentially with increasing n.

(c) If the signal variance (i.e., ratio of signal power to noise variance) is denoted as $\sigma^2{}_S$, then obtain an expression for signal-to-noise ratio, SNR,

$$\text{SNR} = 10 \log_{10}\left(\frac{\sigma^2{}_S}{[\sigma(\delta)]^2}\right)$$

in decibels, as a function of n, V, and σ_S. Note that SNR increases by ~6 dB for each additional ADC bit. Thus, high-quality audio-frequency music recording and playback with a signal-to-noise ratio of ~90-96 dB requires n = 16-bit quantization (and, of course, careful matching of the input maximum voltage to the full-scale response of the ADC).

5.3 For each of the spectral line shapes of Eqs. 5.7 to 5.11, evaluate the slope, $c(x_i)$, of a plot of precision versus the product of peak height-to-noise ratio SNR and the square root of the number of data points, K, per line width. In other words, compute the G-matrix of Eq. 5.22, and invert it to obtain the matrix elements, $(G^{-1})_{11}$, $(G^{-1})_{22}$, and $(G^{-1})_{33}$, needed for Eq. 5.20. Check your results against the entries in Table 5.3.

5.4 In section 5.1.3, we tacitly (and incorrectly) assumed that magnitude-mode noise could be simulated simply by adding Gaussian random noise to a theoretical magnitude-mode spectrum. The correct way to describe magnitude-mode noise is to introduce Gaussian random noise separately into the independent absorption and dispersion spectra. Therefore, let x and y represent the (ordinate) noise values in the absorption and dispersion spectra, respectively. Then suppose that x and y are Gauss-distributed (see Figure); e.g., that the probability, $P(x)\,dx$, that the absorption-mode ordinate noise at any given frequency in a particular spectrum falls between x and $x + dx$ is given by,

$$P(x)\,dx = \frac{1}{\sqrt{2\pi\sigma}} \exp\left(\frac{-x^2}{2\,\sigma^2}\right) dx \qquad (5.4.1)$$

and that the dispersion-mode ordinate noise follows a normal distribution with the same root-mean-square deviation,

$$P(y)dy = \frac{1}{\sqrt{2\pi\sigma}} \exp\left(\frac{-y^2}{2\,\sigma^2}\right) dy \qquad (5.4.2)$$

(a) Solve to find the probability, $P(r)\,dr$, that the magnitude-mode noise falls between r and $r + dr$, where (see Figure)

$$r^2 = x^2 + y^2 \qquad (5.4.3)$$

Equation 5.4.3 defines a curve of constant r along which the magnitude-mode noise is constant.

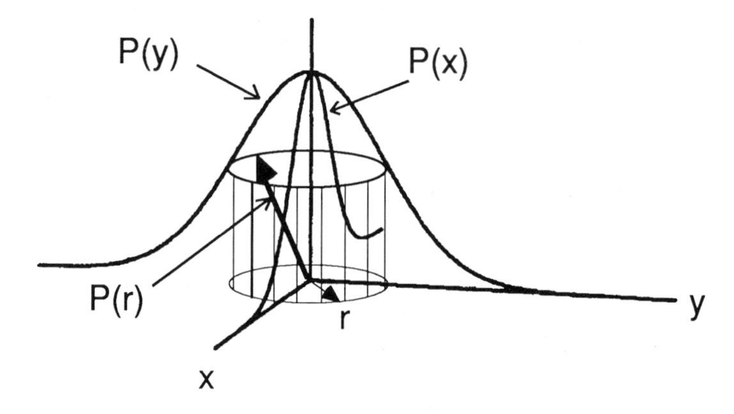

(b) Sketch a plot of $P(r)$ vs. r (note that only positive r-values are possible). Magnitude-mode noise may be quantitated by fitting a magnitude-mode spectral baseline segment to such a plot (see Hanna reference listed in Further Reading).

[Note: Although the Rayleigh distribution, $P(r)\,dr$, of this problem accurately describes magnitude-mode spectral noise in signal-free spectral segments (i.e., the spectral "baseline"), magnitude-mode noise in the vicinity of a spectral peak (i.e., signal-to-noise ratio ≥ 3 or so) is better described by a Gaussian distribution. Thus, although (detector-limited) magnitude-mode and absorption-mode noise have different distributions, their effect on the precision of determining spectral peak height (or width or amplitude) appears to be about the same.]

5.5 Suppose that a time-domain signal is of the form,

$$f(t) = \cos(\omega_0 t) \exp(-t/T_2), \quad 0 \leq t \leq \infty \qquad (5.5.1)$$

Next, multiply $f(t)$ by the *SNR* "enhancement" weight function (see Fig. 5.7),

$$w(t) = \exp(-t/\tau), \quad \tau > 0 \qquad (5.5.2)$$

(a) Obtain analytical expressions for the absorption-mode spectral peak maximum height and full peak width at half-maximum peak height, for the FT of the product function, $f(t)\, w(t)$.

(b) Plot absorption-mode resolution $(\omega_0/\Delta\omega)$ as a function of τ/T_2, and show that resolution is maximal for $\tau \to \infty$ (i.e., unit weight function).

(c) Since time-domain noise is expected to have uniform (root-mean-square) magnitude for any value of t, you may assume that the frequency-domain noise at any frequency value is (on the average) proportional to the square root of the area under $w(t)$ $vs.$ t, for $0 \le t \le \infty$. With that assumption, obtain an analytical expression for frequency-domain noise as a function of τ.

(d) Finally, obtain an analytical expression for absorption-mode peak height-to-noise ratio as a function of τ/T_2 for the absorption spectrum obtained by FT of $f(t)\, w(t)$. Set the first derivative of your expression equal to zero and solve for τ/T_2 to find the maximum peak height-to-noise ratio. You should find a maximum at the "matched filter" condition, $\tau = T_2$.

5.6 Confirm Eq. 5.34 (also known as the Wiener-Khintchine theorem).

5.7 The time-average value of a randomly fluctuating time-domain function, $f(t)$, is zero (Eq. 5.28). Evaluate the time-average value of $[f(t)]^2$, when $f(t)$ is represented by a series of cosine and sine functions. In other words, show that Eq. 5.30 follows from Eqs. 5.27

5.8 In this problem, you will determine the correlation function for chemical exchange between two sites of equal population.

$$A \underset{k}{\overset{k}{\rightleftharpoons}} B \tag{5.8.1}$$

Let $P_A(t)$ and $P_B(t)$ be the probabilities of finding the molecule at site A or B at time, t. Then, from the rules of ordinary chemical reaction kinetics,

$$\frac{d\, P_A(t)}{dt} = -k\, P_A(t) + k\, P_B(t)$$

and

$$\frac{d\, P_B(t)}{dt} = k\, P_A(t) - k\, P_B(t) \tag{5.8.2}$$

Since the forward and reverse rate constants are equal, the probability, $P_A(0)$ or $P_B(0)$, of finding a molecule initially at site A or site B at time zero is:

$$P_A(0) = P_B(0) = \frac{1}{2} \tag{5.8.3}$$

Next, confirm that the following equations represent the *conditional* probability, P(final, τ; initial, 0), that a molecule at a given initial site at time, 0, will be found at the indicated final site at time, τ:

$$P(\mathrm{A}, \tau; \mathrm{A}, 0) = \frac{1}{2}\,[1 + \exp(-2k\tau)] \tag{5.8.4a}$$

$$P(\mathrm{B}, \tau; \mathrm{A}, 0) = \frac{1}{2}\,[1 - \exp(-2k\tau)] \tag{5.8.4b}$$

$$P(\mathrm{A}, \tau; \mathrm{B}, 0) = \frac{1}{2}\,[1 - \exp(-2k\tau)] \tag{5.8.4c}$$

$$P(\mathrm{B}, \tau; \mathrm{B}, 0) = \frac{1}{2}\,[1 + \exp(-2k\tau)] \tag{5.8.4d}$$

Suppose that the value of (say) the magnetic field, B, is different at the two sites:

$$B_A = +b \tag{5.8.5a}$$

$$B_B = -b \tag{5.8.5b}$$

You are now in a position to compute the correlation function, $g(\tau)$, for fluctuations in magnetic field strength resulting from random jumps between sites A and B. Mathematically, you need simply ensemble-average (i.e., sum, with appropriate initial probability weighting) over all (in this case, two) initial sites at time zero, and then ensemble-average (i.e., sum, with appropriate conditional probability weighting) over all (in this case, two) final sites at time, τ. [The time-average indicated by the brackets in Eq. 5.8.6 has been taken into account by writing the rate law, Eq. 5.8.2]

$$g(\tau) = \overline{< B(0)\; B(\tau) >} \tag{5.8.6}$$

$$= \sum_{\text{initial sites}} \; \sum_{\text{final sites}} P_{\text{initial}}\, B_{\text{initial}}\, P(B_{\text{final}}, \tau;\, B_{\text{initial}}, 0)$$

Finally, you are asked to substitute from Eqs. 5.8.3, 5.8.4, and 5.8.5 into Eq. 5.8.6 to obtain the final result:

$$g(\tau) = \overline{< B(0)\; B(\tau) >} = b^2 \exp(-\tau/\tau_c) \tag{5.8.7a}$$

in which $$\frac{1}{\tau_c} = 2k \tag{5.8.7b}$$

Although derived for a particular random process, the correlation function you have computed in Eqs. 5.8.7 is quite general. $g(\tau)$ for any commonly encountered random process will generally consist of the product of a mean-square magnitude for the fluctuating quantity of interest and an exponential time-decay characterized by a "correlation time", τ_c, which is the reciprocal of some characteristic rate constant for the random process This particular example is widely used to describe chemical exchange in NMR.

5.9 Now consider another two-site exchange, for which the forward and reverse rate constants are unequal:

$$\mathrm{A} \underset{k_{-1}}{\overset{k_1}{\rightleftharpoons}} \mathrm{B} \tag{5.9.1}$$

Again suppose that the magnetic field strengths at sites A and B are different, and compute the correlation function and its associated correlation time) for fluctuations in magnetic field strength as in the previous problem.

Hint #1: For the initial condition that $[A] = [A]_0$ and $[B] = 0$ at time zero, the concentrations of A and B evolve as:

$$[A] = [A]_0 - \frac{k_1 [A]_0}{k_1 + k_{-1}} \left(1 - \exp[-(k_1 + k_{-1})\, t]\right) \tag{5.9.2a}$$

$$[B] = \frac{k_1 [A]_0}{k_1 + k_{-1}} \left(1 - \exp[-(k_1 + k_{-1})\, t]\right) \tag{5.9.2b}$$

Hint #2: The subsequent algebra may be simplified by letting:

$$B_A = \frac{k_1 + k_{-1}}{k_{-1}}\, b \tag{5.9.3a}$$

and

$$B_B = \frac{k_1 + k_{-1}}{k_1}\, (-b) \tag{5.9.3b}$$

5.10 As a final exercise in correlation functions, compute the correlation function and its associated correlation time for three-site chemical exchange between three kinetically equivalent sites (A, B., and C), at which the magnetic field values are $+b$, 0, and $-b$, respectively. This example is one model for the random jumps of a methyl group between each of its three potential wells, in NMR relaxation time analysis.

$$A \underset{k}{\overset{k}{\rightleftharpoons}} B, \quad B \underset{k}{\overset{k}{\rightleftharpoons}} C, \quad A \underset{k}{\overset{k}{\rightleftharpoons}} C \tag{5.10.1}$$

Hint: For the initial condition that $[A] = [A]_0$; and $[B] = [C] = 0$, the concentrations at the three sites are:

$$[A] = [A]_0 \left(\frac{1}{3} + \frac{2}{3} \exp(-3k\,t)\right) \tag{5.10.2a}$$

$$[B] = [A]_0 \left(\frac{1}{3} - \frac{2}{3} \exp(-3k\,t)\right) \tag{5.10.2b}$$

$$[C] = [A]_0 \left(\frac{1}{3} - \frac{2}{3} \exp(-3k\,t)\right) \tag{5.10.2c}$$

5.11 Show that the following shift register circuit gives a maximum-length sequence which is the reverse of that in Figure 5.12, for the same initial shift register contents: 1, 0, 1.

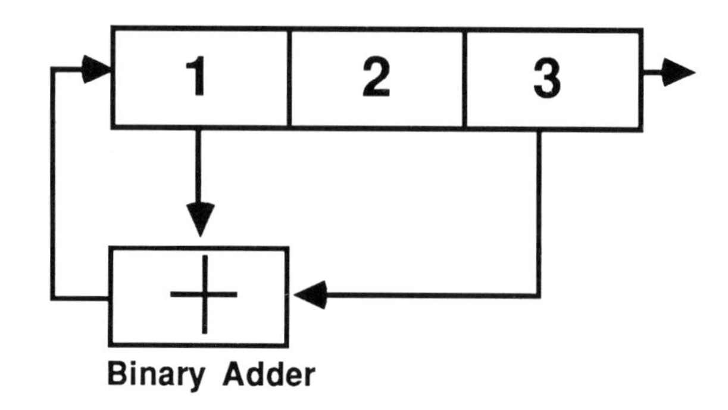

5.12 Show that the following shift register circuit does not generate a maximal-length sequence, for the initial shift register contents: 1, 0, 1

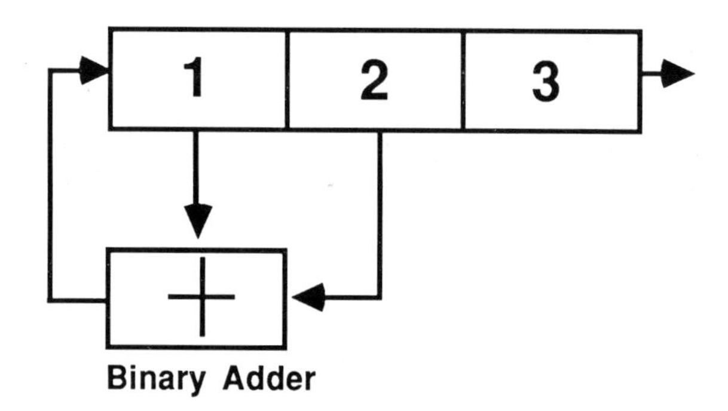

Solutions to Problems

5.1 From the definition of average value,

$$(< (m - \overline{m})^2 > = \sum_{m=0}^{N} (m - \overline{m})^2 \frac{\overline{m}}{m!} \exp(-\overline{m})$$

$$= \sum_{m=0}^{N} \left(m^2 \frac{\overline{m}}{m!} \exp(-\overline{m}) - 2m\,\overline{m} \frac{\overline{m}}{m!} \exp(-\overline{m}) + \overline{m}^2 \frac{\overline{m}}{m!} \exp(-\overline{m}) \right)$$

The rest of the solution is simply to expand the summation for each of the three terms, and collect like terms. The result should be

$$= \overline{m} \ \exp(-\overline{m}) \left(1 + \overline{m} + \frac{\overline{m}^2}{2!} + \frac{\overline{m}^3}{3!} + \cdots \right)$$

$$= \overline{m} \ \exp(-\overline{m}) \ \exp(+\overline{m}) = \overline{m} \qquad \textbf{Q.E.D.}$$

5.2 (a) $\overline{\delta} = \int_{-0.5}^{+0.5} \delta \, d\delta = 0$

$$\sigma(\delta)^2 = \int_{-0.5}^{0.5} \delta^2 \, P(\delta) \, d\delta = \frac{(1/2)^3}{3} - \frac{(-1/2)^3}{3} = \frac{1}{12}$$

(b) $\sigma(\delta) = \frac{1}{12} \left(\frac{V}{2^n}\right) = \frac{V^2 \, 2^{-2n}}{12}$

(c) $\mathrm{SNR} = 10 \log_{10} \left(\frac{\sigma^2{}_S}{[\sigma(\delta)]^2}\right) = 6.02 \, n + 10.8 - 20 \log_{10} \left(\frac{V}{\sigma_S}\right)$

For more detail, see: A. V. Oppenheim and R. W. Schafer, *Discrete-Time Signal Processing*, Prentice-Hall, Englewood Cliffs, NJ, 1989, pp. 119-123.

5.3 See Chen *et al.* reference in Further Reading.

5.4 See Hanna reference in Further Reading.

5.5 First, note that the product, $f(t) \, w(t)$,

$$f(t) \, w(t) = \cos(\omega_0 \, t) \exp\left(-t \left(\frac{1}{\tau} + \frac{1}{T_2}\right)\right)$$

is of the same functional form as $f(t)$, except that T_2 is replaced by T^*,

$$\frac{1}{T^*} = \left(\frac{1}{\tau} + \frac{1}{T_2}\right) = \frac{\tau + T_2}{\tau T_2}$$

Thus, the FT of $f(t) \, w(t)$ is a Lorentzian whose absorption-mode peak width is $(2/T^*)$ and whose maximum peak height is T^*.

(a,b) Resolution $= \frac{\omega_0}{\Delta\omega} = \frac{2\omega_0}{\left(\frac{1}{\tau} + \frac{1}{T_2}\right)} = \frac{2\omega_0 \, \tau T_2}{\tau + T_2}$

Resolution is clearly a maximum, $2\omega_0 T_2$, for $\tau \to \infty$ (i.e., unit weight function).

(c) The area under the weight function is simply,

$$\int_0^{\infty} \exp(-t/\tau)\, dt = \tau$$

(d) Therefore,

$$\text{Peak height-to-noise ratio} = T^*/\sqrt{\tau}$$

$$= \frac{\tau T_2}{\sqrt{\tau}\,(\tau + T_2)} = \frac{\sqrt{\tau}\, T_2}{\tau + T_2}$$

The maximum of this function can be found by setting the first derivative with respect to τ equal to zero and solving for τ. The result is that $\tau = T_2$.

5.6 First, apply the trigonometric identities,

$$\cos(a + b) = \cos a \cos b - \sin a \sin b$$

$$\sin(a + b) = \sin a \cos b + \cos a \sin b$$

to the terms in the second sum. Then expand both sums and apply the orthogonality properties:

$$\int_0^T \sin \omega_n t \sin \omega_m t\, dt = \int_0^T \cos \omega_n t \cos \omega_m t\, dt = 0,\ n \neq m;$$

$$\int_0^T \sin \omega_n t \cos \omega_m t\, dt = 0$$

The desired result is then rapidly obtained.

5.7 Use the same trigonometric identities as in the previous problem, plus Eq. 5.29. Eq. 5.30 then follows directly.

5.8 The equilibrium probabilities, $P(\text{A},0)$ and $P(\text{B},0)$, of finding the molecule at site A or site B are obtained by letting $\tau \to \infty$ in Eqs. 5.8.4:

$$P(\text{A},0) = P(\text{B},0) = 1/2$$

Eq. 5.8.6 may then be evaluated as:

$$g(\tau) = P(A,0)\, B_A\, P(B_A, \tau\, ;\, B_A, 0)\, B_A$$

$$+ P(A,0)\, B_A\, P(B_B, \tau\, ;\, B_A, 0)\, B_B$$

$$+ P(B,0)\, B_B\, P(B_A, \tau\, ;\, B_B, 0)\, B_A$$

$$+ P(B,0)\, B_B\, P(B_B, \tau\, ;\, B_B, 0)\, B_B$$

$$= \frac{1}{4}\Big(b^2\, [1 + \exp(-2k\, t)]\Big) + \frac{1}{4}\Big(-b^2\, [1 - \exp(-2k\, t)]\Big)$$

$$+ \frac{1}{4}\Big(-b^2\, [1 - \exp(-2k\, t)]\Big) + \frac{1}{4}\Big(b^2\, [1 + \exp(-2k\, t)]\Big)$$

$$= b^2 \exp(-2k\, t) \qquad \textbf{Q.E.D.}$$

5.9 The solution to this problem is very similar to that for Problem 5.8. Starting from

$$P(A,0) = \frac{k_{-1}}{k_1 + k_{-1}} \quad \text{and} \quad P(B,0) = \frac{k_1}{k_1 + k_{-1}}$$

and the hint, one should be able to obtain

$$g(\tau) = B_A\, B_B \exp(-\tau / \tau_c)\, ; \text{ in which } \frac{1}{\tau_c} = k_1 + k_{-1}$$

5.10 As in the previous case, the solution is based on the conditional probabilities corresponding to the chemical rate equations for the system:

$$\begin{aligned}
P(A,\tau\, ;\, A,0) &= \frac{1}{3} + \frac{2}{3} \exp(-3k\, t) \\
P(B,\tau\, ;\, A,0) &= \frac{1}{3} - \frac{1}{3} \exp(-3k\, t) \\
P(C,\tau\, ;\, A,0) &= \frac{1}{3} - \frac{1}{3} \exp(-3k\, t) \\
P(A,\tau\, ;\, B,0) &= \frac{1}{3} - \frac{1}{3} \exp(-3k\, t) \\
P(B,\tau\, ;\, B,0) &= \frac{1}{3} + \frac{2}{3} \exp(-3k\, t) \qquad (5.10.3) \\
P(C,\tau\, ;\, B,0) &= \frac{1}{3} - \frac{1}{3} \exp(-3k\, t) \\
P(A,\tau\, ;\, C,0) &= \frac{1}{3} - \frac{1}{3} \exp(-3k\, t) \\
P(B,\tau\, ;\, C,0) &= \frac{1}{3} - \frac{1}{3} \exp(-3k\, t) \\
P(C,\tau\, ;\, C,0) &= \frac{1}{3} + \frac{2}{3} \exp(-3k\, t)
\end{aligned}$$

and the equilibrium probabilities,

$$P(A,0) = P(B,0) = P(C,0) = \frac{1}{3} \tag{5.10.4}$$

This time there are nine terms in the computation of $g(\tau)$ from Eq. 5.8.6:

$$\begin{aligned} g(\tau) = {} & P(A,0)\, B_A\; P(B_A,\tau\,;\, B_A,0)\, B_A \\ & + P(A,0)\, B_A\; P(B_B,\tau\,;\, B_A,0)\, B_B \\ & + P(A,0)\, B_A\; P(B_C,\tau\,;\, B_A,0)\, B_C \\ & + P(B,0)\, B_B\; P(B_A,\tau\,;\, B_B,0)\, B_A \\ & + P(B,0)\, B_B\; P(B_B,\tau\,;\, B_B,0)\, B_B \\ & + P(B,0)\, B_B\; P(B_C,\tau\,;\, B_B,0)\, B_C \\ & + P(C,0)\, B_B\; P(B_A,\tau\,;\, B_B,0)\, B_A \\ & + P(C,0)\, B_B\; P(B_B,\tau\,;\, B_B,0)\, B_B \\ & + P(C,0)\, B_B\; P(B_C,\tau\,;\, B_B,0)\, B_C \end{aligned} \tag{5.10.5}$$

Substitution of Eqs. 5.10.3 and 5.10.4 into Eqs. 5.10.5 yields the desired result,

$$g(\tau) = \frac{2}{3} b^2 \exp(-3k\,t)$$

5.11 By the same procedure used to generated Figure 5.12,

Period #1 =	1	0	1	(initial contents of shift register)
#2 =	0	1	0	
#3 =	0	0	1	
#4 =	1	0	0	
#5 =	1	1	1	
#6 =	1	1	1	
#7 =	0	1	1	
#8 =	1	0	1	(same as #1)

5.12 As in the preceding problem,

Period #1 =	1	0	1	(initial contents of shift register)
#2 =	1	1	0	
#3 =	0	1	1	
#4 =	1	0	1	(same as #1)

In other words, the sequence repeats on every 4th rather than every 8th period; it can therefore not be a maximal length sequence.

Chapter 6

Non-FT methods for proceeding from time- to frequency-domain

6.1 What's wrong with Fourier transforms?

Throughout this book, we have concentrated on the Fourier transformation as *the* method for converting a time-domain waveform into a frequency-domain spectrum. The Fourier transform is in fact ideally suited for finding the frequency-domain spectrum of a *noiseless* time-domain *continuous* signal of *infinite* duration. However, experimental time-domain signals are "corrupted" in three ways.

First, experimental time-domain signals are of *finite* duration. Thus, as we noted in Chapter 2, Fourier transformation of a truncated time-domain signal leads to undesirable frequency-domain "wiggles" ("Gibbs oscillations") which make it hard to observe a small peak in the vicinity of a large peak. If the time-domain acquisition is delayed by more than one dwell period after excitation, such wiggles cannot be removed by deconvolution and phase correction.

Second, the Fourier transform operation distributes time-domain *noise* (including isolated "spikes") uniformly throughout the frequency-domain (see Figure 5.1). Truncation and noise thus effectively limit the certainty with which we can establish the peak frequencies (and widths, magnitudes, and phases) in the frequency-domain spectrum of a given time-domain signal.

Third, the *discrete* nature of a spectrum obtained from discrete sampling of a time-domain continuous signal further limits the spectral information content, as discussed in section 5.1.4.

Moreover, the discrete Fourier transform is inherently designed to convert *equally-spaced* time-domain data into *equally-spaced* frequency-domain data. However, such sampling is not necessarily optimal for two reasons: (a) the time-domain signal-to-noise ratio decreases during the observation, so that one might prefer to acquire more data points per unit time at earlier stages of the acquisition where the signal-to-noise ratio is highest; and (b) certain spectra have an inherently non-linear frequency scale (e.g., FT/ICR, for which mass and frequency are inversely related), so that a spectrum with inherently non-uniform discrete frequency spacing might be preferable.

Most of the above drawbacks can be traced to our (tacit) assumption that the time-domain and frequency-domain representations of a signal are "deterministically" connected, just as in classical physics, in which future properties (e.g., position, momentum, angular momentum, etc.) of a particle can be predicted perfectly and uniquely from its present properties. However, just as the fundamental postulate of quantum mechanics states that one can at best know the precise *probability* (but not a *unique* value) of a particular future measurement of a physical property, the spectral representation of a given time-domain noisy (classical) signal is also knowable only to within some computable *probability*.

Specifically, because sampling, time-domain truncation, and the presence of noise "corrupt" our ability to predict deterministically a *unique* frequency-domain discrete spectrum from its corresponding time-domain discrete signal, there can be *many* spectra which could equally well represent a given time-domain data set. The question then becomes how to choose the (discrete) spectrum which "best" represents the observed (discrete) time-domain signal. The short answer is that "best" depends on one's choice of judging criteria, and there is no uniformly accepted single criterion of "best". Therefore, in the remainder of this chapter, we shall outline two of the most generally applied classes of "spectral estimation" methods, along with some discussion of how those methods might be generalized. Finally, one should be aware (see below) that spectral estimation methods tend to be named for their specific algorithms (e.g., "linear prediction", "autoregression", "maximum-entropy", "maximum likelihood", "all-poles", "all-zeroes", etc.), even when two or more algorithms can be shown to give equivalent results.

Finally, this chapter, more than any other in this book, is necessarily replete with new jargon. Although many spectroscopic applications are based on only one class of estimators (namely, the "all-poles" method discussed below), the reader needs to understand how that method fits into the general approach. Of the many classes of spectral estimators, our intent is to present enough of the mathematics to enable the reader to use at least some of the available methods and to acquire sufficient vocabulary to be able to learn about the others from the standard literature.

In general, FT or one of the non-FT methods may offer superior precision in estimating signal parameters, depending on the signal-to-noise ratio, number of data points, and one's degree of prior knowledge about the functional form of the time-domain signal. For example, Figure 6.1 (see Eq. 6.56) shows that autoregression is superior to FT analysis when signal-to-noise ratio is very high (say, >1,000) *and* the number of data points is relatively small (say, <1,000).

6.2 New ways to represent an analog spectrum: the (continuous) transfer function and (continuous) Laplace transform for a mass-on-a-spring

The class of non-FT "spectral estimation" methods known as "linear prediction" is based on the assumption that the "natural" motions of a spectroscopic "system" can be described by a specified linear differential equation. We therefore begin this section by (re-)analyzing the particular (second-order) linear differential equation which corresponds to our familiar mass-on-a-spring model. Because we no longer expect a Fourier transform relationship between time- and frequency-domain data, we should not be surprised to find that a different transform (the "Laplace" transform) turns out to be more suitable. Finally, we need to change notation somewhat, in order to match the literature jargon on this subject, which comes mainly from electrical engineering sources.

In the new vocabulary, Figure 6.2 shows a schematic "system" (in this case, our usual mass-on-a-spring model) subjected to an "input" (i.e., "excitation" in our previous jargon), leading to an "output" (i.e., time-domain "signal"). If we represent the output by a time-domain function, $f(t)$, and the input by an excitation function, $e(t)$, then for a mass-on-a-spring system, the recipe that relates the input to the output is given by a particular second-order linear differential equation (Eq. 1.17):

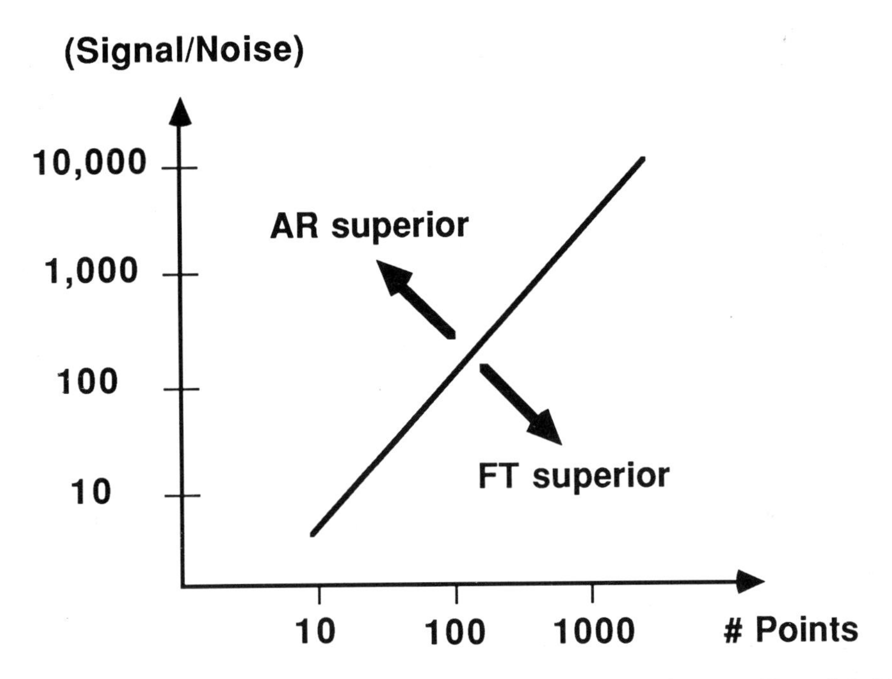

Figure 6.1 Method producing the "best" spectral resolution (Eqs. 6.56 and 6.57), according to time-domain initial signal-to-noise ratio (log scale) and number of time-domain data points (log scale). Autoregression is superior to Fourier analysis for all grid points *above* the plotted line (namely, high signal-to-noise ratio and small number of data points) and inferior for all grid points *below* the line.

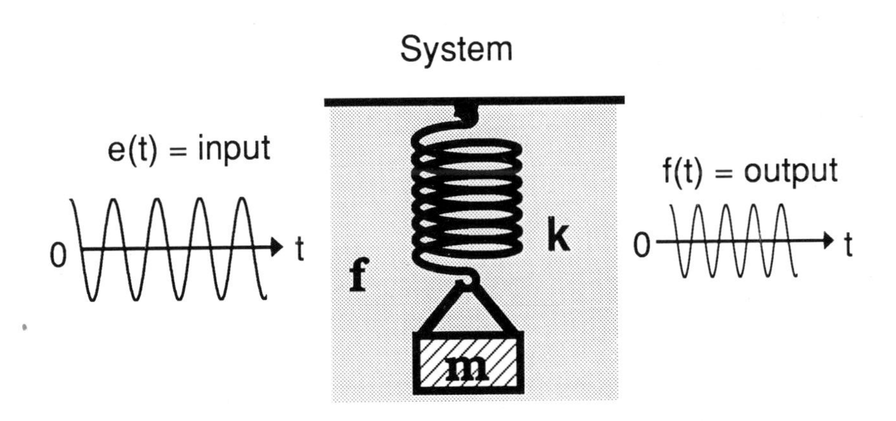

Figure 6.2 Schematic diagram of the relation between the time-domain signal "output" obtained by exciting a mass-on-a-spring ("system") by an arbitrary time-domain excitation waveform ("input").

$$e(t) = m \frac{d^2 f(t)}{dt^2} + \mathrm{f} \frac{d f(t)}{dt} + k\, f(t)\,; \quad f(t) \text{ is real} \tag{6.1}$$

If two functions are equal, then their Fourier transforms are also equal. However (see Problem 2.19), the Fourier transform of $d\,[f(t)]/dt$ is simply $i\,\omega\,F(\omega)$. Thus, we may take the Fourier transforms of both sides of Eq. 6.1 and apply Problem 2.19 to yield

$$E(\omega) = m\,(i\,\omega)^2\,F(\omega) + \mathrm{f}\,(i\,\omega\,)\,F(\omega) + k\,F(\omega) \tag{6.2}$$

It is useful to define the "transfer function", $H(\omega)$, as the output divided by the input (compare to Eq. 2.21):

$$H(\omega) = \frac{F(\omega)}{E(\omega)} \tag{6.3}$$

Since the Fourier transform of a δ-function excitation is unity [i.e., $E(\omega) = 1$ in Eq. 6.3], the "transfer function", $H(\omega)$, can be thought of as the system response to an (idealized) impulse excitation. Eqs. 6.2 and 6.3 may be combined to obtain

$$H(\omega) = \frac{1}{m\,(i\,\omega)^2 + \mathrm{f}\,(i\,\omega) + k} \tag{6.4a}$$

$$= \frac{1}{(k - m\,\omega^2) + i\,\mathrm{f}\omega} \tag{6.4b}$$

$$= \frac{k - m\,\omega^2}{(k - m\,\omega^2)^2 + \mathrm{f}^2\omega^2} - i\,\frac{\mathrm{f}\,\omega}{(k - m\,\omega^2)^2 + \mathrm{f}^2\omega^2} \tag{6.4c}$$

Up to now, we have simply re-derived a previously known result (Eq. 1.32) in a somewhat different way. The real and imaginary parts of the "transfer function", $H(\omega)$, are now recognized as the familiar dispersion and absorption spectra for a damped harmonic oscillator.

Furthermore, at this stage (see right-hand-side of Eq. 6.4a), it is useful to consider H to be a function of $(i\,\omega)$:

$$H(i\omega) = \frac{1}{m\,(i\,\omega)^2 + \mathrm{f}\,(i\,\omega) + k} \tag{6.4d}$$

Since we know that the real and imaginary parts of $H(i\omega)$ are the dispersion and absorption spectra, it follows that

$$H(i\,\omega) \cdot H^*(i\,\omega) = \frac{1}{(k - m\,\omega^2)^2 + (\mathrm{f}\,\omega)^2} \tag{6.5a}$$

$$= |M(\omega)|^2 = P(\omega) \tag{6.5b}$$

in which $M(\omega)$ and $P(\omega)$ are the familiar magnitude and power spectra (see Eqs. 1.24, 1.25).

Therefore, the natural (angular) frequency, ω_0, and damping constant, τ,

$$\frac{1}{\tau} = \frac{f}{2m} \qquad (1.14a)$$

$$\omega_0 = \sqrt{\frac{k}{m}} \qquad (1.10)$$

could now be extracted from the power spectrum obtained by plotting $H(i\omega) \cdot H^*(i\omega)$ *vs.* ω.

The key new feature of the above treatment is that the (Fourier) transform converts the original *differential equation* 6.1 into the simpler *power series* appearing in the denominator of Eq. 6.4. Readers familiar with differential equations will recognize that the denominator of Eq. 6.4 is the "characteristic equation" of the original differential Eq. 6.1. The "natural" frequencies obtained by solving the original differential equation may therefore be obtained by finding the roots of that equation.

Unfortunately, the power series is expressed in terms of a complex variable, $(i\omega)$. Although we could solve for the (complex) roots of the denominator of Eq. 6.4a, it is preferable to define a new kind of transform,

$$H(s) = \int_{-\infty}^{\infty} f(t)\,\exp(-st)\,dt \qquad \text{(Laplace transform)} \qquad (6.6a)$$

in which $s = \sigma + i\omega$, where σ (s^{-1}) and ω (rad s^{-1}) are real (6.6b)

The Laplace transform is simply a Fourier transform, in which the "frequency" is mathematically complex. Perhaps unsurprisingly, all of the major properties of Fourier transforms also hold for Laplace transforms: shift theorem, convolution theorem, transforms of derivatives and integrals, etc. Thus, we may apply the Laplace transform to our original "system" differential equation 6.1 to obtain the Laplace analogs of Eq. 6.2, 6.3, and 6.4:

$$E(s) = m s^2 F(s) + f s F(s) + k F(s) \qquad (6.7)$$

$$H(s) = \frac{F(s)}{E(s)} \qquad (6.8)$$

and $$H(s) = \frac{1}{m s^2 + f s + k} \qquad (6.9)$$

Next, we solve for the (complex) roots of the quadratic expression in the denominator of Eq. 6.9:

$$m s^2 + f s + k = 0 \qquad (6.10)$$

to obtain

$$s = \frac{-f \pm \sqrt{f^2 - 4mk}}{2m} = -\left(\frac{1}{\tau}\right) \pm \sqrt{\left(\frac{1}{\tau}\right)^2 - \omega_0^2} \qquad (6.11)$$

in which τ and ω_0 are again defined by Eqs. 1.14a and 1.10.

For the case of physical interest, namely $\omega_0 >> (1/\tau)$, the argument of the square root in Eq. 6.11 is negative, and Eq. 6.11 can be rewritten as

$$s_1 = -\left(\frac{1}{\tau}\right) + i\sqrt{\omega_0^2 - \left(\frac{1}{\tau}\right)^2} \tag{6.12a}$$

$$s_{-1} = -\left(\frac{1}{\tau}\right) - i\sqrt{\omega_0^2 - \left(\frac{1}{\tau}\right)^2} \tag{6.12b}$$

In other words, the purely mathematical manipulation of setting the denominator of $H(s)$ equal to zero yields a mathematically complex "frequency", s, whose real and imaginary parts are the desired damping rate constant, $(1/\tau)$, and the two natural frequencies, $\pm[\omega_0^2 - (1/\tau)^2]^{1/2}$ for a mass-on-a-spring displaced by a sudden impulse [compare to Eqs. 1.14a,b, resulting from the solution of the "characteristic" Eq. 6.1 with $e(t) = \delta(t)$]. (Although the negative-frequency component of a causal real time-domain function is not observed directly, it can nevertheless affect the positive-frequency spectrum, as discussed in Sec. 4.3.1.)

By combining Eqs. 6.12 and 6.9, we can express the transfer function, $H(s)$, produced by the Laplace transform of Eq. 6.1, as

$$H(s) = \frac{1}{(s - s_1)(s - s_{-1})} \tag{6.13}$$

Thus, $H(s)$ approaches infinity as s approaches s_1 or s_{-1}. Figure 6.3 compares the Laplace and Fourier complex-variable representations of the *same* frequency-domain spectrum. In the more familiar Fourier spectrum, $H(\omega)$, each "natural" *frequency*, ω_0, of the system is obtained from the peak *position*, and the time-domain damping rate constant, $(-1/\tau)$, is obtained from the *width* of that peak. In the Laplace spectrum, $H(s)$, the "peaks" appear as infinite spikes (see Eq. 6.13), and the natural frequencies and damping rate constants are simply the *projections* of those spikes onto the real (σ) and imaginary ($i\omega$) axes. Alternatively, ω_0 and τ can be determined from a plot of the power spectrum, $H(i\omega)\cdot H^*(i\omega)$ *vs.* ω as from Eqs. 6.5.

It is left as a (short) exercise for the reader to show that Eq. 6.13 can be rewritten in the form,

$$H(s) = \frac{c}{s - s_1} - \frac{c}{s - s_{-1}} \tag{6.14}$$

in which $c = 1/\{2i\ [(\omega_0^2 - (1/\tau)^2]^{1/2}\}$.

When a mathematically complex function can be written as a "Laurent" series of fractions as in Eq. 6.14, any value of s for which one of the denominators goes to zero ($s = s_1$ or $s = s_{-1}$ in this case) is called a "pole". We have therefore shown that the assumption that a given system can be represented by a damped harmonic oscillator model is equivalent to solving for the (two) "poles" of the transfer function, $H(s)$. The spectral information (namely, peak frequency and width) can then be found from Eqs. 6.12.

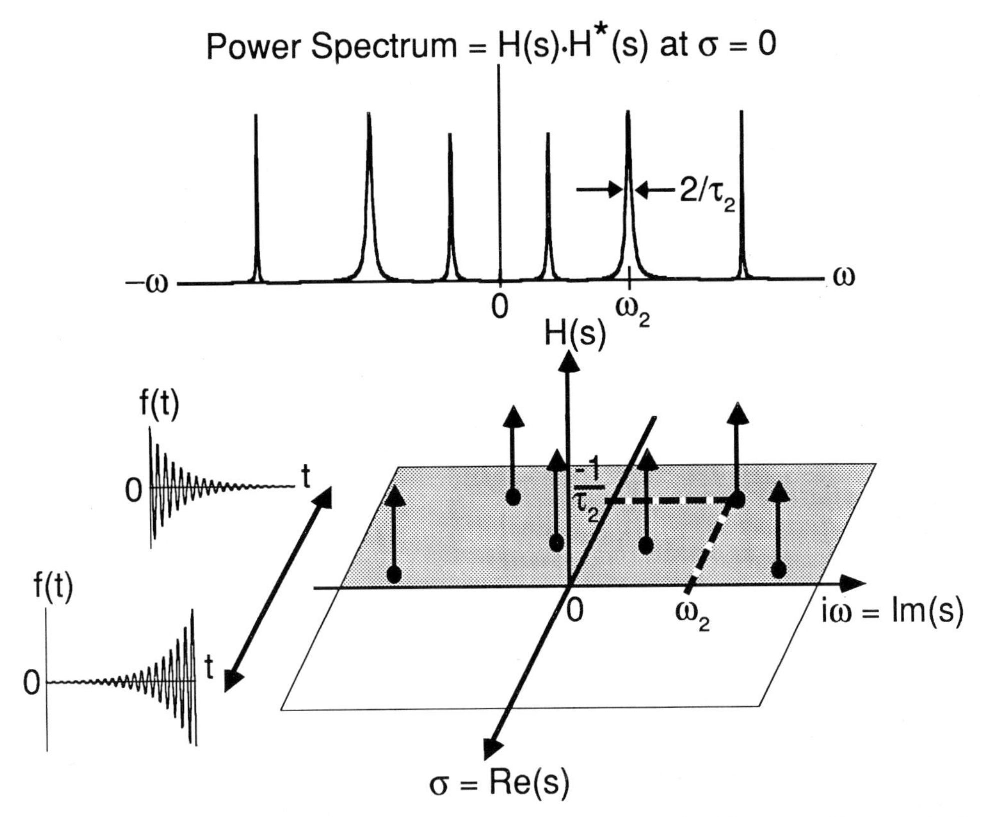

Figure 6.3 Spectral representations obtained from the (mathematically complex) Laplace transform (see Eq. 6.6), for a system of three damped harmonic oscillators (e.g., frictionally damped masses-on-springs) of different natural frequency and damping constant. Top: The power spectrum, evaluated from $H(s) \cdot H^*(s)$ at $\sigma = 0$, is the same as that obtained by Fourier transformation. Bottom: The Laplace spectrum yields the natural frequency and damping rate constant from the real and imaginary coordinates of each δ-function.

For a system consisting of n independent masses-on-springs, Eq. 6.13 can be generalized (see Problems) to the form,

$$H(s) = \prod_{k=1}^{n} \frac{1}{(s - s_k)(s - s_{-k})} \tag{6.15}$$

Eq. 6.15 can be rearranged to the form,

$$H(s) = \sum_{k=1}^{n} \left(\frac{c_k}{s - s_k} - \frac{c_k}{s - s_{-k}} \right) \tag{6.16}$$

whose "poles" (namely, $s = s_k$ and s_{-k}) again yield the desired spectral frequencies, line widths, and phases. Armed with those results, one can then plot the transfer function, $H(\omega)$, which is the desired frequency-domain spectrum (compare to Eqs. 6.4 for a single-peak spectrum). In the next section, we will examine various "linear prediction" methods for extracting the spectral peak frequencies and widths from a *discrete* time-domain response to a given excitation. The "autoregression" method (section 6.3.1) is equivalent to modeling a system as a collection of damped harmonic oscillators, and is therefore sometimes known as the "all-poles" method, because it is equivalent to solving for the poles of the continuous transfer function as shown above.

A final reason for including the Laplace transform approach is that it is used extensively in the engineering texts which treat these sorts of problems. Therefore, the reader who wishes to pursue this topic further will need to be familiar with Laplace transform techniques.

6.3 Difference equation, discrete transfer function, and discrete z-transform for the mass-on-a-spring

In the previous section, we showed that the differential equation corresponding to a mass-on-a-spring model leads to an "all-poles" method for finding the frequencies, line widths, and phases of the frequency-domain spectrum corresponding to a time-domain *analog* signal. The *assumption* that the system follows that particular differential equation provides the basis for the "all-poles" method.

In this section, we extend the *analog* idea of the previous section to a *discrete* time-domain signal for the same system (i.e., damped harmonic oscillator). We begin by proceeding from an analog *differential* equation to a discrete *difference* equation:

$$\left.\frac{df(t)}{dt}\right|_{\text{at } t = t_n} = \frac{f(t_n + \Delta t) - f(t_n)}{\Delta t} \quad \text{as } \Delta t \to 0 \tag{6.17}$$

The key to linear prediction methods is that a discrete *difference* equation can also be expressed as a *recursion* formula, which allows for the computation ("prediction") of the n'th point of a discrete time-domain function (i.e., a series of equally-spaced time-domain data points) from one or more of its previous values. We will apply a discrete Fourier (or Laplace) transform to the difference equation to obtain a discrete transfer function from which we may either obtain a discrete power spectrum (FT approach) or obtain the natural frequencies and relaxation times of the system from the poles of the discrete transfer function produced by means of a discrete Laplace transform. [The discrete version of the Laplace transform is more conveniently expressed as the so-called "z-transform" (see below).] Among several possible algorithms for extraction of the natural frequencies and relaxation times, we will demonstrate only the "autocorrelation" method, because it is widely used and because we have already had some experience with such functions in Chapters 2.2.3 and 5.2.1. A schematic diagram relating the analysis of continuous and discrete time-domain functions is shown in Figure 6.4.

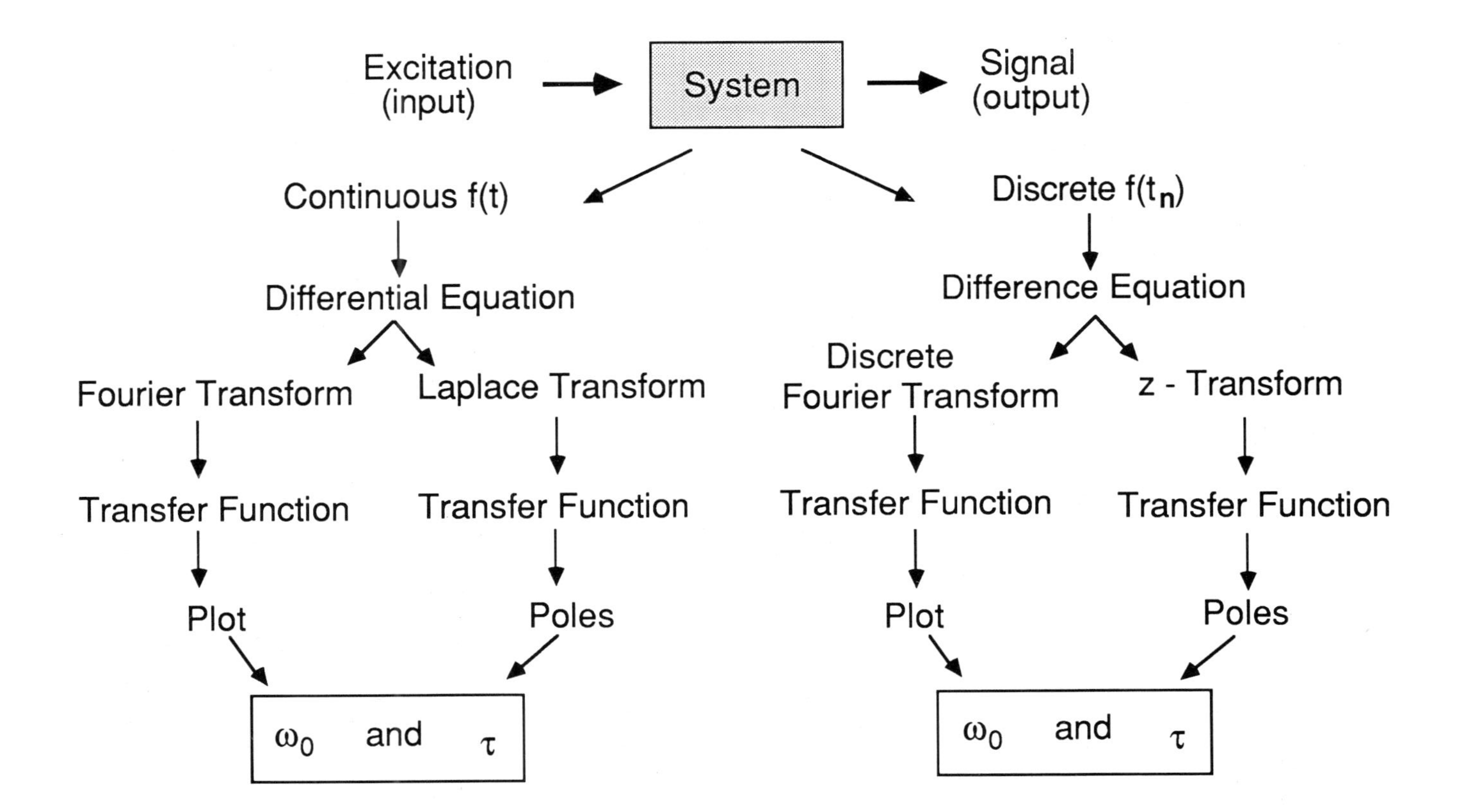

Figure 6.4 Relation between methods for obtaining the power spectrum and the natural frequencies and relaxation times of a system consisting of damped oscillators described by a continuous (left) or discrete (right) time-domain signal.

From Eq. 6.17, the reader should confirm that the second derivative of a time-domain function becomes:

$$\left.\frac{d^2 f(t)}{dt^2}\right|_{\text{at } t = t_n} = \frac{f(t_n + 2\Delta t) - 2f(t_n + \Delta t) + f(t_n)}{(\Delta t)^2} \quad \text{as } \Delta t \to 0 \tag{6.18}$$

Thus, a differential equation, such as

$$\frac{d^2 f(t)}{dt^2} = 0 \tag{6.19}$$

becomes a difference equation, in this case,

$$f(t_n + 2\Delta t) - 2f(t_n + \Delta t) + f(t_n) = 0 \tag{6.20}$$

Next, if we express t_n as

$$t_n = n\,\Delta t\ ,\quad 0 \leq n \leq N \tag{6.21}$$

then the difference equation 6.20 (corresponding to the differential equation 6.19) can be expressed as the "recursion" formula, Eq. 6.22:

$$f(t_n) - 2f(t_{n-1}) + f(t_{n-2}) = 0 \tag{6.22}$$

In other words, for a system described by the second-order differential equation 6.19, a given time-domain discrete value, $f(t_n)$, may be computed ("predicted") from its two immediately preceding values, $f(t_{n-1})$ and $f(t_{n-2})$.

Since our favorite mass-on-a-spring problem is characterized by a second-order linear differential equation, we should not be surprised to find that the corresponding recursion equation is of the form,

$$f(t_n) + a_1 f(t_{n-1}) + a_2 f(t_{n-2}) = 0 \tag{6.23}$$

At this stage, we can save some effort by recalling our previously derived solution to the mass-on-a-spring problem, expressed in terms of the natural frequency, ω_0, and the relaxation time, τ:

$$f(t_n) = A \exp(-n\,\Delta t/\tau)\,\sin(\omega_0\, n\,\Delta t + \varphi) \tag{6.24}$$

in which

$$\frac{1}{\tau} = \frac{f}{2m} \tag{1.14a}$$

$$\text{and } \omega_0 = \sqrt{\frac{k}{m}} \tag{1.10}$$

We may now evaluate the coefficients, a_1 and a_2, by substituting $f(t_{n-1})$ and $f(t_{n-2})$

$$f(t_{n-1}) = A \exp[(n-1)\Delta t/\tau] \sin[\omega_0 (n-1)\Delta t + \varphi] \tag{6.25a}$$

$$f(t_{n-2}) = A \exp[(n-2)\Delta t/\tau] \sin[\omega_0 (n-2)\Delta t + \varphi] \tag{6.25b}$$

into Eq. 6.23 to obtain (left as an exercise)

$$a_1 = -2 \exp(-\Delta t/\tau) \cos \omega_0 \Delta t \tag{6.26a}$$

$$a_2 = \exp(-2\Delta t/\tau) \tag{6.26b}$$

The desired recursion formula for a single mass-on-a-spring thus becomes,

$$f(t_n) - 2 \exp(-\Delta t/\tau) \cos \omega_0 \Delta t \; f(t_{n-1}) + \exp(-2\Delta t/\tau) f(t_{n-2}) = 0 \tag{6.27}$$

Proceeding as we did before for a continuous $f(t)$, we next subject the system to a time-varying discrete excitation function ("input" sequence), $e(t_n)$. The difference equation 6.27 then takes the form

$$f(t_n) + a_1 f(t_{n-1}) + a_2 f(t_{n-2}) = e(t_n) \tag{6.28}$$

Just as we applied a continuous Fourier transform to the differential equation of a continuous time-domain signal resulting from a δ-function excitation, we now apply a discrete Fourier transform to the difference equation of the discrete time-domain signal, $f(t_n)$, excited by a discrete excitation,

$$e(t_n) = 1 \text{ for } n = 0$$

$$e(t_n) = 0 \text{ for } n \geq 1 \tag{6.29}$$

The reader is left to show that the (discrete) transfer function resulting from discrete Fourier transformation of Eq. 6.28 is

$$H(\omega_k) = \frac{1}{1 + a_1 \exp(-i\omega_k \Delta t) + a_2 \exp(-2i\omega_k \Delta t)} \tag{6.30a}$$

in which $\omega_k = \frac{2\pi k}{T}$, and $T = N\Delta t$; $0 \leq \omega_k \leq \frac{N\pi}{T} = \omega_{\text{Nyquist}}$ (6.30b)

and we have made use of the discrete version of the shift theorem (Chapter 2.4.1): namely, if the discrete Fourier transform of the discrete time-domain sequence,

$$f(t_0), f(t_1), f(t_2), \cdots, f(t_{N-1}) \tag{6.31a}$$

is $F(0), F(2\pi/T), F(4\pi/T), \cdots, F[2(N-1)\pi/T]$ (6.31b)

then the Fourier transform of the time-domain sequence which has been shifted by Δt,

$$f(t_0 - \Delta t), f(t_1 - \Delta t), f(t_2 - \Delta t), \cdots, f(t_{n-1} - \Delta t) \tag{6.32a}$$

is

$$F(0), e^{(-i\Delta t/T)} F(2\pi/T), e^{(-i2\Delta t/T)} F(4\pi/T), \cdots, e^{[-i(N-1)\Delta t/T]} F[2(N-1)\pi/T] \tag{6.32b}$$

It is left as an exercise (see Problems) to show that, as for the continuous time-domain signals of the preceding section, $H(e^{-i\omega_k}) \cdot H^*(e^{-i\omega_k})$ is the discrete power spectrum.

As in the preceding section, we can define a complex "frequency", s_k,

$$s_k = \sigma + i\omega_k \tag{6.33}$$

and use its corresponding discrete Laplace transform to obtain a complex discrete transfer function, the roots of whose denominator (i.e., the "poles" of the transfer function) again yield the natural frequencies and relaxation times of the system. We begin by taking the discrete Laplace transform of the recursion Eq. 6.27, to obtain

$$\sum_0^\infty f(t_n)\exp(-s_k t_n) + a_1 \sum_0^\infty f(t_{n-1})\exp(-s_k t_{n-1})$$

$$+ a_2 \sum_0^\infty f(t_{n-2})\exp(-s_k t_{n-2}) = \sum_0^\infty e(t_n)\exp(-s_k t_n) \tag{6.34}$$

from which the transfer function, $H(s_k)$,

$$H(s_k) = \frac{F(s_k)}{E(s_k)} \tag{6.35}$$

can be evaluated as for a δ-function excitation as for the continuous-variable case to yield

$$H(s_k) = \frac{1}{1 + a_1 \exp(-s_k \Delta t) + a_2 \exp(-2 s_k \Delta t)} \tag{6.36}$$

Since s_k appears in Eq. 6.36 in the form, $\exp(-s_k)$, it is convenient (and, in electrical engineering texts, standard) to define a new variable, z_k,

$$z_k = \exp(s_k \Delta t) \tag{6.37}$$

so that $H(s_k)$ may be expressed in the more compact form,

$$H(s_k) = \frac{1}{1 + a_1 z_k^{-1} + a_2 z_k^{-2}} = H(z_k) \tag{6.38}$$

When expressed in terms of the variable, z_k, the discrete Laplace transform is known as the "z-transform". Also, z^{-1} is often called the "unit delay operator", since it corresponds to a time shift of $-\Delta t$ (see Eqs. 6.31, 6.32 and Chapter 5.2.2).

The roots of the polynomial in z in the denominator of Eq. 6.38 correspond to the poles of the transfer function expressed in terms of z_n, as in the case of the Laplace transform. Just as the *continuous* power spectrum was obtained (see Figure 6.3) from the *continuous* transfer function, $H(s)$ evaluated at $\sigma = 0$, the *discrete* power spectrum is obtained from the *discrete* transfer function, $H(z_k)$ evaluated at $z_k = \exp(i\,2\pi\nu_k \Delta t)$:

$$P(z_k) = H(z_k)\, H^*(z_k) \tag{6.39a}$$

The power spectrum, $P(\nu_k)$, with ν in Hz, is related to $P(z_k)$ by the scale factor, Δt (see Chapter 3.6 and Appendix D):

$$P(\nu_k) = \Delta t\ P(z_k)\ \text{, evaluated at } z_k = \exp(i\,2\pi\nu_k\,\Delta t) \tag{6.39b}$$

For a continuous time-domain function, we showed that the roots of the polynomial of the denominator of the transfer function (expressed in terms of s) yield the natural frequencies and damping constants for the system. For the discrete case, the roots of the polynomial of the discrete transfer function (expressed in terms of z) also yield the natural frequencies and damping constants of the system. In the case of a real time domain function, this polynomial has two roots (for each natural frequency) corresponding to the positive and negative angular frequencies (i.e., 2 poles) of that signal component.

For a system characterized by p different natural frequencies subjected to a δ-function input, the 2-term time-domain recursion formula is replaced by a $2p$-term formula,

$$f(t_n) = -\sum_{k=1}^{2p} a_k\, f(t_{n-k}) + G\,\delta_{n\,1}, \tag{6.40}$$

in which $\delta_{n\,1} = 1$ for $n = 1$ and zero otherwise.

The corresponding transfer function, $H(z_k)$, for a system with p different natural frequencies has a denominator "characteristic" polynomial of degree $2p$ (i.e., $2p$ "poles", half of which correspond to negative "image" frequencies):

$$H(z_k) = \frac{G}{1 + \sum_{k=1}^{2p} a_k\, z^k} \tag{6.41}$$

The factor, G, is called the "gain" of the system, because it determines the magnitude of the output, $f(t_n)$ for a given input (excitation), $e(t_n)$.

Example

Suppose that a discrete Laplace transformation of a particular experimental time-domain response to an impulse excitation yields the transfer function,

$$H(z_k) = \frac{1}{1 - 92\, z_k^{-1} + 0.17\, z_k^{-2} + 0.90\, z_k^{-3} + 0.85\, z_k^{-4}} \tag{6.42a}$$

As shown in Figure 6.5, we could use $H(z_k)$ directly to obtain the power spectrum $P(\nu) = \Delta t\,[H(z_k)\, H^*(z_k)]$ evaluated at $z_k = \exp(i\,2\pi\nu_k\,\Delta t)$ vs. ν_k. Alternatively, if we could factor the denominator of the transfer function of Eq. 6.42a,

$$H(z_k) = \frac{1}{(1 - 1.83\, z_k^{-1} + 0.90\, z_k^{-2})\ (1 + 0.91\, z_k^{-1} + 0.94\, z_k^{-2})} \tag{6.42b}$$

then we could compute the natural frequencies, ω_1 and ω_2, and damping constants, $\sigma_1 = 1/\tau_1$ and $\sigma_2 = 1/\tau_2$, from the (positive-frequency) z-roots shown in Eq. 6.42b,

$$z_1 = \exp(-s_1) \text{ and } z_2 = \exp(-s_2) \tag{6.42c}$$

to obtain the corresponding s-values, and thence the desired σ– and ω–values.

$$s_1 = \sigma_1 + i\omega_1 = -0.05 + i\,0.25\left(\frac{N\pi}{T}\right)$$

$$s_2 = \sigma_2 + i\omega_2 = -0.03 + i\,2.07\left(\frac{N\pi}{T}\right) \tag{6.42d}$$

A shorter procedure is to obtain each damping constant and natural frequency from the corresponding a_1 and a_2 values (see Eq. 6.26) in each quadratic expression in the denominator of Eq. 6.42b.

In summary, we have shown that we can extract natural frequencies and relaxation times, provided that we can determine the coefficients, a_k, in the transfer function of Eq. 6.41. In the next section, we present one (of several) methods for determining the a_k coefficients.

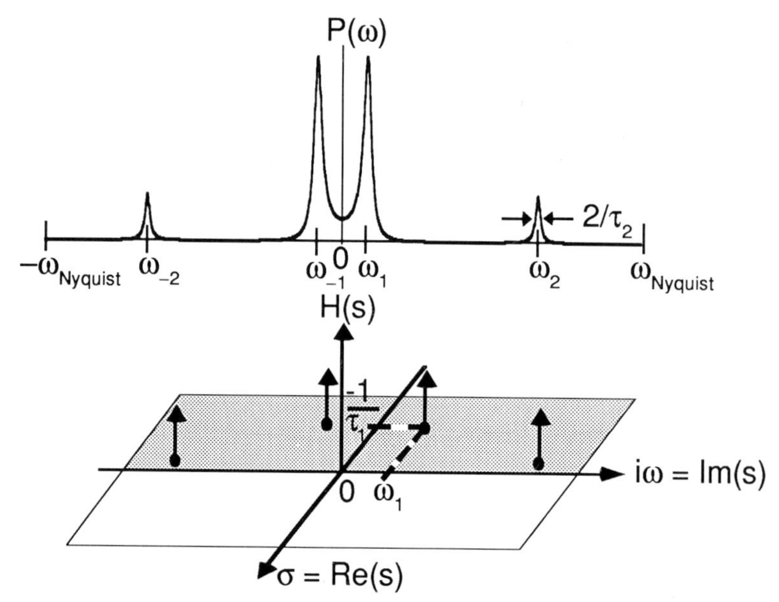

Figure 6.5 Extraction of natural frequencies (ω_1 and ω_2) and time-domain damping constants, τ_1 and τ_2, from the discrete transfer function, $H(s_k)$ of Eq. 6.36, produced by (discrete) z-transform of the discrete time-domain response to a δ-function excitation of a system consisting of two damped oscillators of different frequency. The τ's and ω's may be recovered either from the (discrete) power spectrum, Eq. 6.39b (top diagram), or by solving for the "poles" (shown as δ-functions in the bottom diagram) of $H(z_k)$ obtained from the z-roots (Eq. 6.42c) obtained by setting the denominator of $H(z_k)$ of Eq. 6.38 equal to zero.

6.3.1 Linear prediction: the autoregression (AR) model

The autoregression model is based on the assumption that the system is defined by the differential equation of one or more damped harmonic oscillators (Eq. 6.1), with corresponding discrete recursion formula (Eqs. 6.27 and 6.40). That model leads to an "all-poles" solution (Eqs. 6.13, 6.14 for a continuous time-domain signal, or Eq. 6.38 for a discrete signal).

Given a particular time-domain discrete data set, $f(t_n)$, the present problem is to determine the recursion formula (Eq. 6.40) "gain" factor, G, and "predictor" coefficients, a_k, which in turn yield the natural frequencies and damping constants. Evaluation of the a_k coefficients generally begins from the following least-squares approach.

6.3.1.1 *Least squares criterion*

If we first suppose that the excitation (input), $e(t_n)$, is unknown, then we may begin by ignoring $e(t_n)$ in Eq. 6.28. In other words, we imagine that the signal at time t_n, $f(t_n)$, can be predicted from a suitable linear combination of $2p$ preceding *signal* values alone: $f(t_{n-1})$, $f(t_{n-2})$, etc., in which p is the (assumed) number of distinct natural (positive) frequencies (we shall return to that point later). The mean-square deviation between predicted and actual time-domain signal values is then computed and minimized. Since it can be shown that the input and output power must be the same for the mass-on-a-spring system, any remaining non-zero mean-square deviation arising from the excitation in Eq. 6.40 can be interpreted as G^2 (for impulse or δ-function excitation) or noise variance, σ^2 (if noise is the excitation). Therefore, if the experimentally measured discrete N-point time-domain signal at time, t_n, is $f(t_n)$, and the estimated value computed (predicted) from $2p$ prior values is $\boldsymbol{f}(t_n)$, then

$$\boldsymbol{f}(t_n) = -\sum_{k=1}^{2p} a_k\, f(t_{n-k}) \tag{6.43}$$

The difference (deviation, error, "residual") between the *actual* value, $f(t_n)$, and the *predicted* value, $\boldsymbol{f}(t_n)$ is

$$f(t_n) - \boldsymbol{f}(t_n) = f(t_n) + \sum_{k=1}^{2p} a_k\, f(t_{n-k}) \tag{6.44}$$

We seek to minimize the *total* squared error, E, obtained from successive prediction of each time-domain data point from its preceding $2p$ values. The range of n (i.e., the number time-domain data points chosen for analysis) is determined by the particular regression method (see below).

$$E = \sum_n \left(f(t_n) + \sum_{k=1}^{2p} a_k\, f(t_{n-k}) \right)^2 \tag{6.45}$$

The minimum of E is found by setting its derivative with respect to a_k equal to zero:

$$\frac{\partial E}{\partial a_k} = 0, \qquad 1 \le k \le 2p \tag{6.46}$$

From Eqs. 6.45 and 6.46, one can readily obtain (see Problems) the following set of $2p$ so-called "normal" equations (i.e., one for each value of k):

$$\sum_{k=1}^{2p} a_k \sum_n f(t_{n-k})\, f(t_{n-i}) = -\sum_n f(t_n) f(t_{n-i}) \qquad 1 \le k \le 2p \tag{6.47a}$$

or

$$\sum_{k=1}^{2p} a_k \sum_n f(t_{n-k+i})\, f(t_n) = -\sum_n f(t_{n+i}) f(t_n) \qquad 1 \le k \le 2p \tag{6.47b}$$

Note that the summations over n in Eq. 6.47b are in fact discrete correlation functions (compare, e.g., Eq. 5.40). Therefore, we shall pursue the solution of Eq. 6.47b by an autocorrelation method (see below). Although far from optimal when compared to several more recent alternatives, the autocorrelation method is representative of the autoregression family of spectral estimator methods.

6.3.1.2 *Autocorrelation method*

For an infinitely long (i.e., $-\infty < n < +\infty$) time-domain discrete stationary signal (i.e., a signal whose characteristics do not depend on the choice of starting instant for the measurement), the normal equation can be rewritten in the form,

$$\sum_{k=1}^{2p} a_k\, R(i-k) = -R(i) \tag{6.48a}$$

in which

$$R(i) = \sum_{n=-\infty}^{\infty} f(t_n) f(t_{n+i}) \tag{6.48b}$$

is the (discrete) *autocorrelation function* for the (discrete) time-domain signal, $f(t_n)$. $R(i-k)$ is another correlation function, which is a function of two indices (i and k), and which may thus be expressed in matrix form as the *autocorrelation matrix*,

$$R(i-k) = \sum_{n=-\infty}^{\infty} f(t_{n-k+i})\, f(t_n) \tag{6.48c}$$

Since the experimental signal, $f(t_n)$, is defined only over a finite time interval, the autocorrelation function must be truncated to some smaller number of data points, N:

$$R(i) = \sum_{n=0}^{N-i-1} f(t_n)\, f(t_{n+i}) \tag{6.49}$$

Both $R(i)$ and $R(i-k)$ can then be computed readily from the (discrete) time-domain signal, $f(t_n)$. It is important to recognize that N will in general be much smaller than the total number experimental time-domain data points. In matrix notation (shown as boldface—see Appendix A), the a_k and $R(i)$ sequences may be expressed as column vectors, and Eq. 6.48a may be written:

$$\mathbf{R_{i-k}} \cdot \mathbf{a_k} = -\mathbf{R_i} \,, \tag{6.50a}$$

or,

$$\begin{bmatrix} R_0 & R_1 & R_2 & R_3 & \cdots & R_{2p-1} \\ R_1 & R_0 & R_1 & R_2 & \cdots & R_{2p-2} \\ R_2 & R_1 & R_0 & R_1 & \cdots & R_{2p-3} \\ R_3 & R_2 & R_1 & R_0 & \cdots & R_{2p-4} \\ \vdots & \vdots & \vdots & \vdots & & \vdots \\ R_{2p-1} & R_{2p-2} & R_{2p-3} & R_{2p-4} & \cdots & R_0 \end{bmatrix} \cdot \begin{bmatrix} a_1 \\ a_2 \\ a_3 \\ a_4 \\ \vdots \\ a_{2p} \end{bmatrix} = - \begin{bmatrix} R_1 \\ R_2 \\ R_3 \\ R_4 \\ \vdots \\ R_{2p} \end{bmatrix} \tag{6.50b}$$

Eqs. 6.50 are known as the "Yule-Walker" equations, which express $\mathbf{R_i}$ in terms of $\mathbf{R_{i-k}}$ and $\mathbf{a_k}$. We seek the *inverse* equation, in which the desired $\mathbf{a_k}$ are expressed in terms of $\mathbf{R_i}$ and the *inverse* autocorrelation matrix, $\mathbf{R_{i-k}}^{-1}$:

$$\mathbf{a_k} = \mathbf{R_{i-k}}^{-1} \mathbf{R_i} \tag{6.51}$$

The great advantage of the autocorrelation method is that the autocorrelation matrix, $\mathbf{R_{i-k}}$, is a symmetric "Toeplitz" matrix (i.e., a matrix in which all elements along any given diagonal are equal). The special "Toeplitz" property of the $\mathbf{R_{i-k}}$ matrix allows for the use of a computationally efficient "Levinson" matrix inversion algorithm to obtain $\mathbf{R_{i-k}}^{-1}$. The autoregression/autocorrelation algorithm may be found in the Kay reference at the end of this chapter.

We have thus succeeded in determining the a_k coefficients which provide the best fit (by the criterion of minimum squared deviation) between predicted and measured time-domain signal values. The system natural frequencies and damping constants can in turn be obtained from the a_k's by solving the characteristic equation appearing as the denominator of the discrete transfer function (see example of Eqs. 6.42). The remaining deviation, E, in Eq. 6.45 is assigned to the "gain" factor, G, which is computed from

$$G^2 = R(0) + \sum_{k=1}^{p} a_k R(k) \tag{6.52}$$

Alternatively, we may combine Eqs. 6.50 and 6.52 to give a single matrix equation,

$$\begin{bmatrix} R_0 & R_1 & R_2 & R_3 & \cdots & R_{2p} \\ R_1 & R_0 & R_1 & R_2 & \cdots & R_{2p-1} \\ R_2 & R_1 & R_0 & R_1 & \cdots & R_{2p-2} \\ R_3 & R_2 & R_1 & R_0 & \cdots & R_{2p-3} \\ \vdots & \vdots & \vdots & \vdots & & \vdots \\ R_{2p} & R_{2p-1} & R_{2p-2} & R_{2p-3} & \cdots & R_0 \end{bmatrix} \cdot \begin{bmatrix} 1 \\ a_1 \\ a_2 \\ a_3 \\ \vdots \\ a_{2p} \end{bmatrix} = \begin{bmatrix} G^2 \\ 0 \\ 0 \\ 0 \\ \vdots \\ 0 \end{bmatrix} \qquad (6.53)$$

The use of Eq. 6.53 to determine the "gain", G, natural frequency and damping constant for a single damped oscillator from a 3-point time-domain data set is given in the Problems. For random excitation, G^2 can be replaced by σ^2 in Eq. 6.53.

How does one choose the proper number of poles and the proper number of time-domain data points?

We have shown that a real time-domain signal arising from excitation of p different natural (positive) frequencies has $2p$ "poles" (half of which represent negative-frequency spectral "peaks" as discussed in section 4.3.1). The selection of the number of poles (also known as the "model order") in autoregression spectral estimation is critical, as shown in Figure 6.6. Too low an order results in an over-smoothed estimate which may obscure small peaks, whereas too large an order results in the appearance of spurious peaks (which may be comparable in magnitude to true peaks). Similarly, the number of time-domain data points (above the minimum number, $N = 2p$), also affects the quality of the result (see Figure 6.7). The most common statistical criterion for choosing model order (proposed by Akaike) is termed the "final predicted error", and estimates the model order as that for which the following function is a minimum:

$$\text{Final Predicted Error} = \frac{N + 2p + 1}{N - 2p + 1} \cdot \sigma^2 \qquad (6.53)$$

in which N is the number of time-domain data points chosen for analysis, p is the number of (positive-frequency) poles, and G^2 is the squared "gain" [replaced by σ^2, the white noise variance, when the input (excitation) is a white noise source]. As illustrated in Figure 6.6, Eq. 6.53 shows that the final error is reduced by choosing N larger than the minimum number ($N = 2p$) needed to determine the frequencies of p different oscillators. However, if N is too large, then the computational time required becomes unreasonably high (see Table 6.1), and the required random access memory size also becomes unreasonably large. Therefore, autoregression is generally used when the number of time-domain data points is relatively small (e.g., < 1,024).

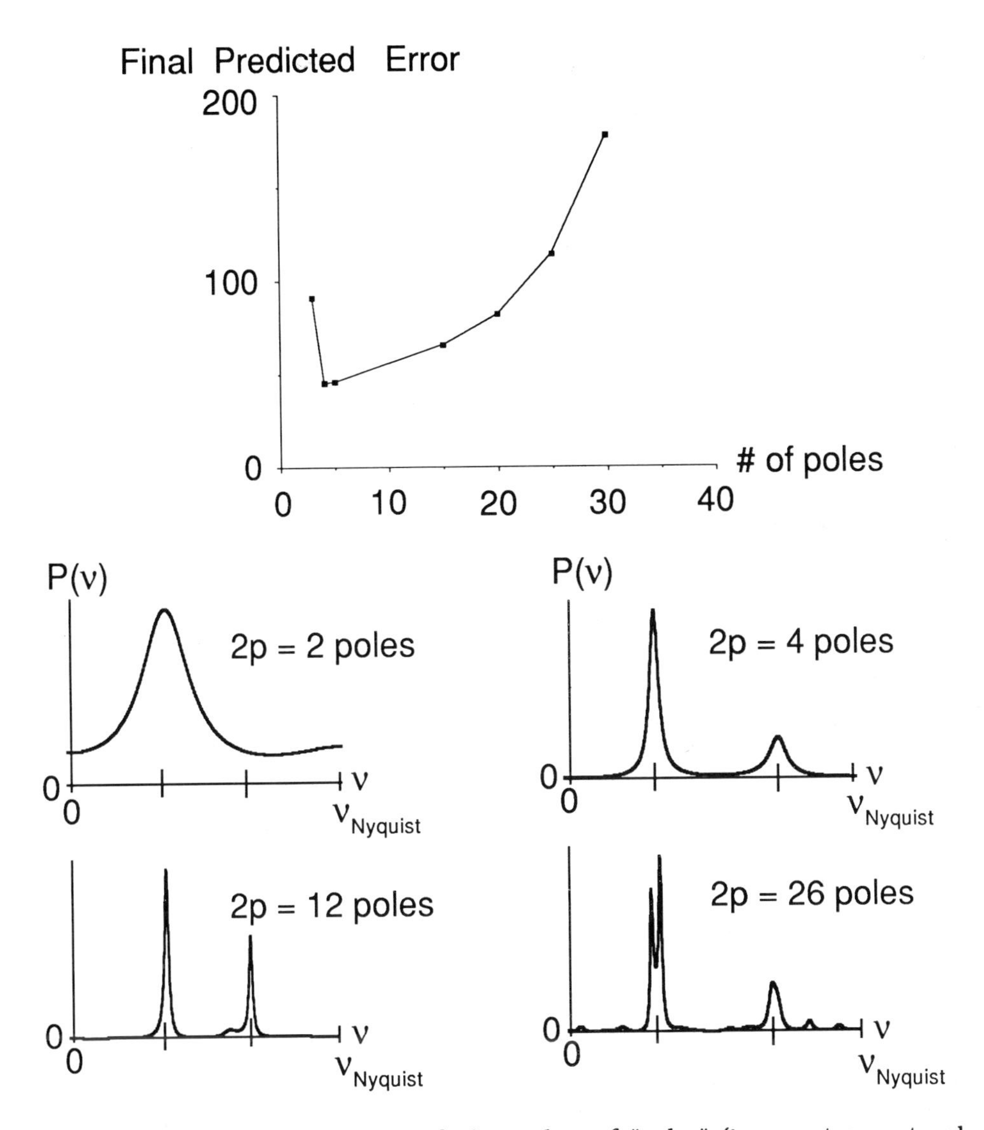

Figure 6.6 Effect of the pre-specified number of "poles" (i.e., system natural frequencies, or number of spectral peaks) on spectra obtained by the autoregression (AR) autocorrelation method, for a 40-point time-domain signal consisting of two undamped sinusoids of relative magnitudes 1.0 (signal-to-noise ratio = 5.0) and 0.7 at frequencies of 0.3 ν_{Nyquist} and 0.7 ν_{Nyquist}. With too few "poles", the AR spectrum fails to resolve all of the peaks. With too many "poles", the AR spectrum attempts to account for noise by adding spurious peaks. The optimum number of poles can be predicted from the minimum of a plot of "final predicted error" (Eq. 6.53) *vs.* number of poles shown in the uppermost diagram. As expected, the "best" spectrum is obtained when the pre-specified number of poles ($2p = 4$) matches the true number of natural frequencies of the system ($p = 2$ in this case).

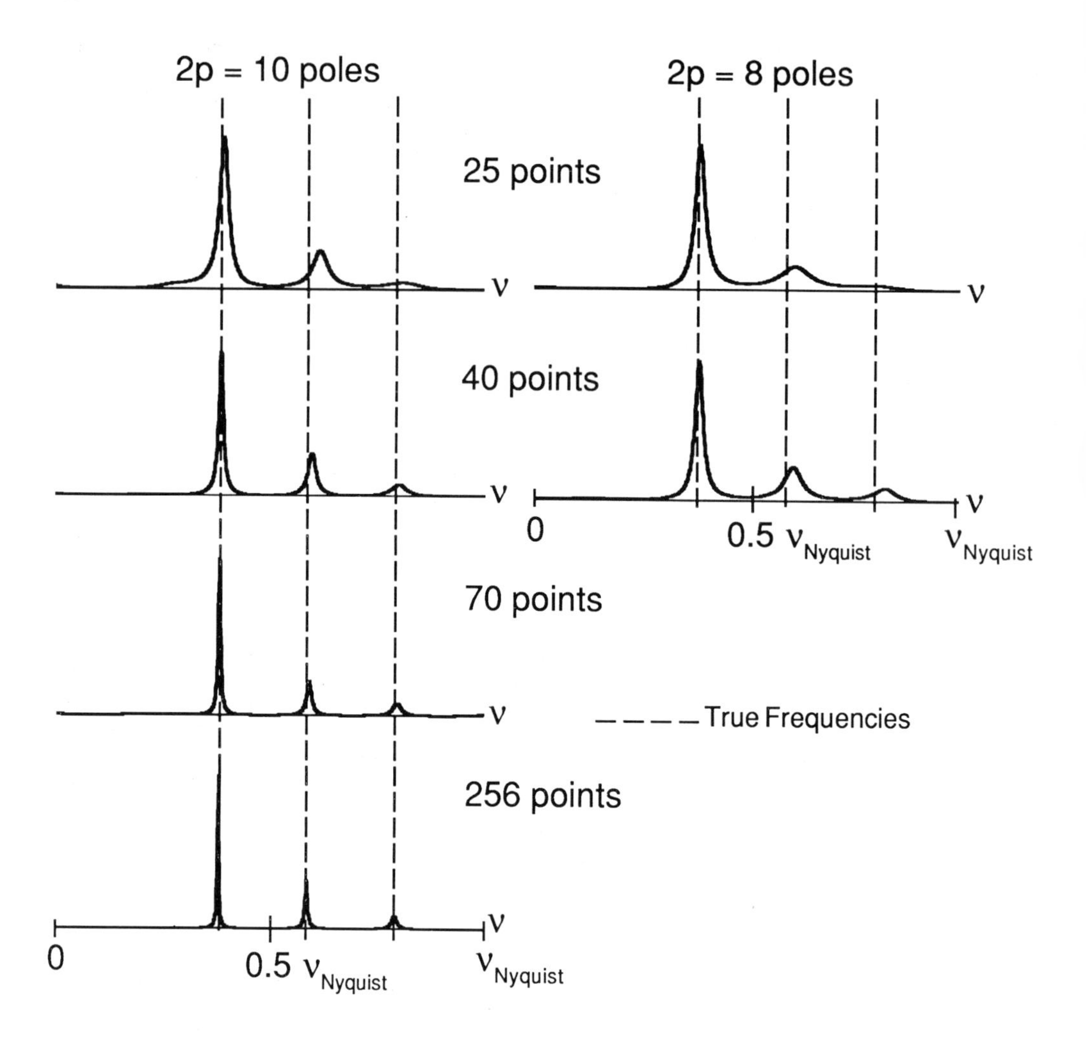

Figure 6.7 Effect of number of time-domain data points used for AR autocorrelation spectral estimates for the same discrete time-domain signal consisting of $p = 3$ undamped oscillators of frequency, 0.4 ν_{Nyquist}, 0.6 ν_{Nyquist}, and 0.8 ν_{Nyquist}, with relative time-domain amplitude 1.0 (signal-to-noise ratio = 4), 0.8, and 0.5 (to give relative power spectral magnitudes of 1.0, 0.64, and 0.25). The spectral quality (i.e., resolution and peak center accuracy) clearly improves as the number of data points increases, but at the cost of significantly increased computation time (see Table 6.1). The error in peak position is even more obvious when fewer poles are included (compare $2p = 8$ to $2p = 10$ poles spectra).

Table 6.1 Relative computational time in seconds (IBM PS-2/30 with 8087 math coprocessor) for AR spectral estimation, for the stated number of positive-frequency poles and number of data points, for a time-domain signal consisting of the sum of four exponentially damped noisy sinusoids.

Number of poles	Number of data points		
	50	150	256
3	2.74	3.10	4.16
7	3.29	3.92	4.93
10	3.43	4.27	5.67
13	3.59	4.80	6.43
15	3.94	5.39	7.13

Precision

A very important drawback to the autoregression (AR) model is its sensitivity to noise, which has not been included in our original signal model (Eq. 6.28). Thus, autoregression methods can be expected to work well only when the time-domain signal-to-noise ratio is high. Stated another way, the AR method cannot distinguish between spectral signal and noise "peaks", and it will therefore continue to find more and more "peaks" as the pre-specified number of "poles" increases, as illustrated in Figure 6.6. Marple has shown analytically that the minimum separation, $\nu_2 - \nu_1$, at which two AR-generated power spectral peaks can be resolved (see Figure 6.8) is given approximately by

$$\nu_2 - \nu_1 = \frac{1.03}{2p\,\Delta t\ [snr\,(2p + 1)]^{0.31}} \quad \text{in Hz} \tag{6.54}$$

in which $2p$ is the number of poles, and snr is the time-domain signal-to-noise ratio [in linear (as opposed to dB) units]. For comparison, the linewidth, $\Delta\nu$, at half-maximum peak height for a power spectrum obtained by discrete Fourier transformation of an N-point time-domain undamped sinusoidal signal is given by

$$\Delta\nu = \frac{0.86}{T} \tag{6.55}$$

in which $T = N\,\Delta t$ is the time-domain acquisition period for a signal sampled every Δt seconds.

Figure 6.1 was constructed by comparing the precision computed from either Eq. 6.54 or 6.55, assuming that all of the N time-domain data points were used in both cases. Finally, it is worth noting that for very short-length data sets, the apparent *frequency* of an AR spectral peak can shift if the time-domain signal *phase* is non-zero; that problem can be greatly reduced by increasing the number of time-domain data points, just as FT spectral precision increases with increasing data acquisition period.

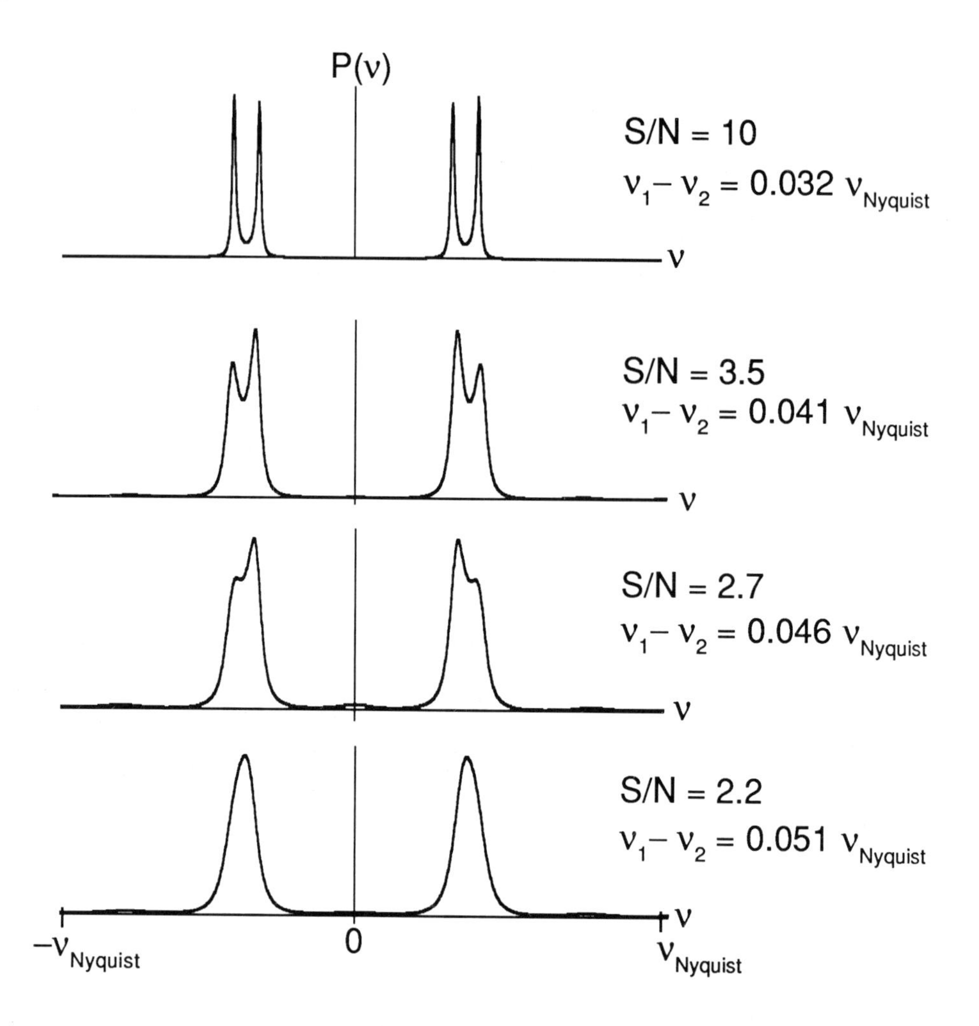

Figure 6.8 Effect of signal-to-noise ratio (S/N) and frequency separation on autoregression (AR) spectra. Autoregression (AR) $2p = 8$ pole spectra were computed from time-domain discrete 256-point signals consisting of the sum of two undamped equal-magnitude sinusoids of indicated signal-to-noise ratio (S/N) and differing in frequency by $(\nu_2 - \nu_1) = 0.04\ \nu_{\text{Nyquist}}$. For each indicated S/N-value, the minimum resolvable peak separation computed from Eq. 6.54 is indicated for each spectrum. For example, two peaks separated by 0.04 ν_{Nyquist} are predicted (and are observed) to be resolved at S/N = 10 but not at S/N = 2.2.

AR versus FT

Consider the problem of resolving two equal-magnitude spectral peaks of different frequency. The theoretical minimum frequency separation to produce a barely resolved autoregression (power) spectrum of $2p = 2 \cdot 2 = 4$ poles is given by Eq. 6.54, and the corresponding FT (power) spectrum requires a frequency difference of one line width, as given by Eq. 6.55. Since resolution is defined as the *ratio*, $\nu_1/(\nu_2 - \nu_1)$, we may divide the reciprocal of Eq. 6.54 by the reciprocal of Eq. 6.55 to obtain a measure of the relative resolution of AR compared to FT:

$$\text{Relative resolution} = \frac{4\,\Delta t\ (snr \cdot 5)^{0.31}}{1.03} \cdot \frac{0.86}{N\,\Delta t} = \frac{5.5\ (snr)^{0.31}}{N} \tag{6.56}$$

If the relative resolution is greater than 1, AR is superior to FT; and conversely for relative resolution less than 1. Thus, the two methods give the same resolution when Eq. 6.56 is set equal to 1 and all N time-domain data are used in both cases.

$$\frac{5.5\ (snr)^{0.31}}{N} = 1\text{ , which may be rearranged to give}$$

$$\log_{10} snr = -\,2.39 + 3.22 \log_{10} N \tag{6.57}$$

Equation 6.57 is the straight line plotted in Figure 6.1. In Figure 6.9, AR and FT spectra are compared for two simulated time-domain data sets corresponding to locations on either side of the line in Figure 6.1. As expected, AR is superior to FT (note the narrower peaks without Gibbs oscillations for the AR spectrum) when the signal-to-noise ratio is high (100) and the number of data points is small (20). Conversely, FT is superior to AR (note the more accurate relative peak heights in the FT spectrum) when the signal-to-noise ratio is low (2.0) and the number of data points is large (256). The Figure also shows that modified autoregression methods (see below) can significantly enhance the quality of AR spectra; nevertheless, Eq. 6.57 still provides a good measure of the relative resolution of AR and FT methods.

Peak magnitudes:

It is important to note that the relative peak heights in an AR power spectrum do not necessarily match those for the "true" power spectrum which would be obtained by discrete FT in the absence of noise. Although the relative *areas* of the AR power spectral peaks are correct, the relative peak *heights* are not. In fact, for high signal-to-noise ratio, *snr*, and/or a large (pre-specified) number, $2p$, of poles, the AR power spectral peak height, $P_{AR}(\omega_0)$, is actually proportional to the *square* of the "true" power spectral peak height, $P(\omega_0)$:

$$P_{AR}(\omega_0) = G^2\,\Delta t \left(1 + 2p\,\frac{P(\omega_0)}{\sigma_N{}^2}\right)\left(1 + (2p+1)\,\frac{P(\omega_0)}{\sigma_N{}^2}\right) \tag{6.58a}$$

$$\lim_{(2p\ snr) >> 1} P_{AR}(\omega_0) = (2p)^2\,G^2\,\Delta t\ (snr)^2 \tag{6.58b}$$

in which $\sigma_N{}^2$ is the (white) noise variance, and $snr = P(\omega_0)/\sigma_N{}^2$.

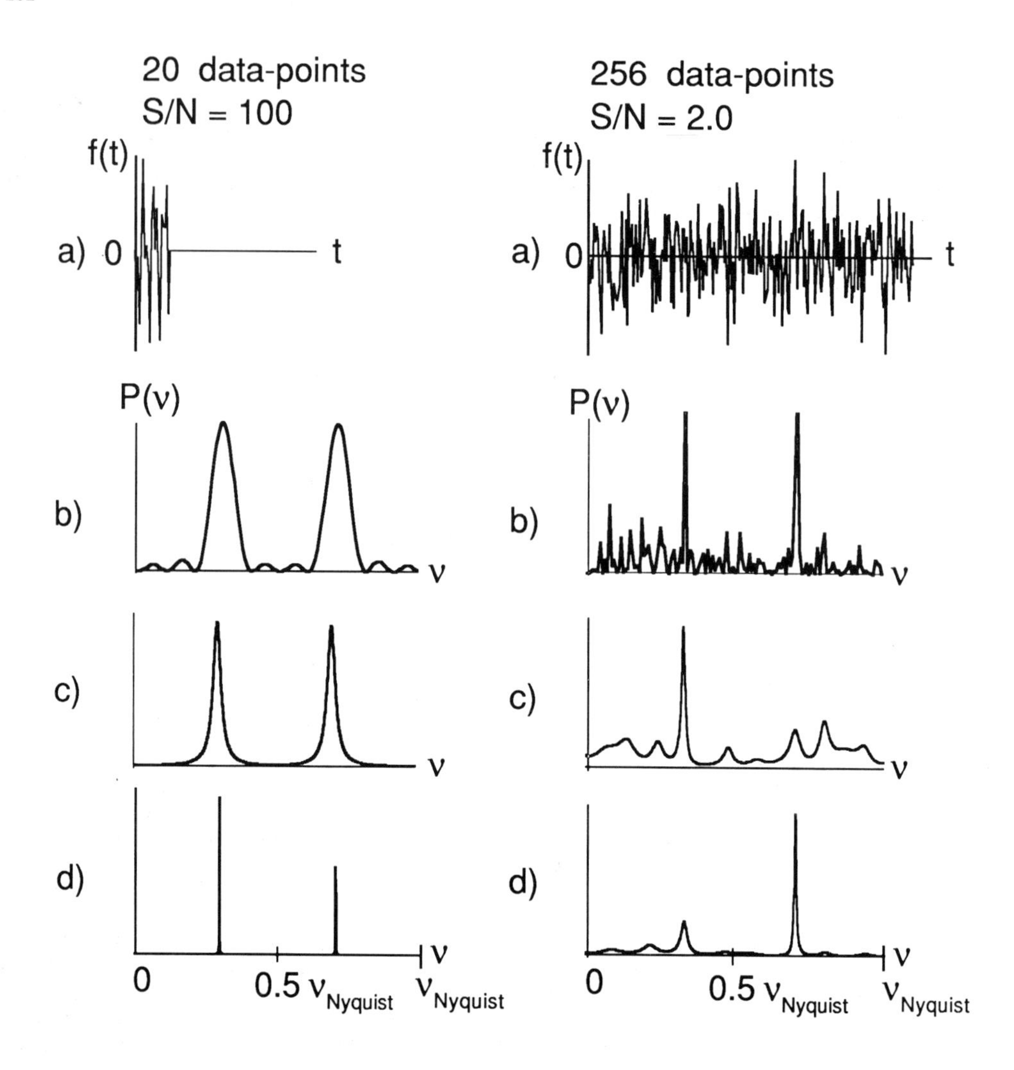

Figure 6.9 Direct comparison of unapodized Fourier transform (b) and autoregression [(c) and (d)] data reduction of the same time-domain discrete data sets (a) consisting of two undamped sine waves of frequency 0.32 ν_{Nyquist} and 0.70 ν_{Nyquist}, of indicated signal-to-noise ratio and number of time-domain data points. The left-hand results correspond to the upper left region of Figure 6.1 (AR predicted superior), and the right-hand results to the lower right region of Figure 6.1 (FT predicted superior). Left-hand non-FT spectra were computed from autoregression autocorrelation with $2p$ = 6 poles (c) and from the Burg autoregression method with $2p$ = 4 poles (d). Right-hand non-FT spectra were computed from autoregression autocorrelation with $2p$ = 28 poles (c) and from the autocorrelation moving average method (see below) with $2p$ = 18 poles and 2 zeroes (d).

Effect of time-domain signal phase

The "true" power (or magnitude) spectrum is independent of time-domain signal phase. However, the power spectrum computed by means of linear prediction methods from the transfer function [namely, $H(z_k)\,H^*(z_k)$], *does* depend on phase, particularly when the number of time-domain data points is small, as shown in Figure 6.10.

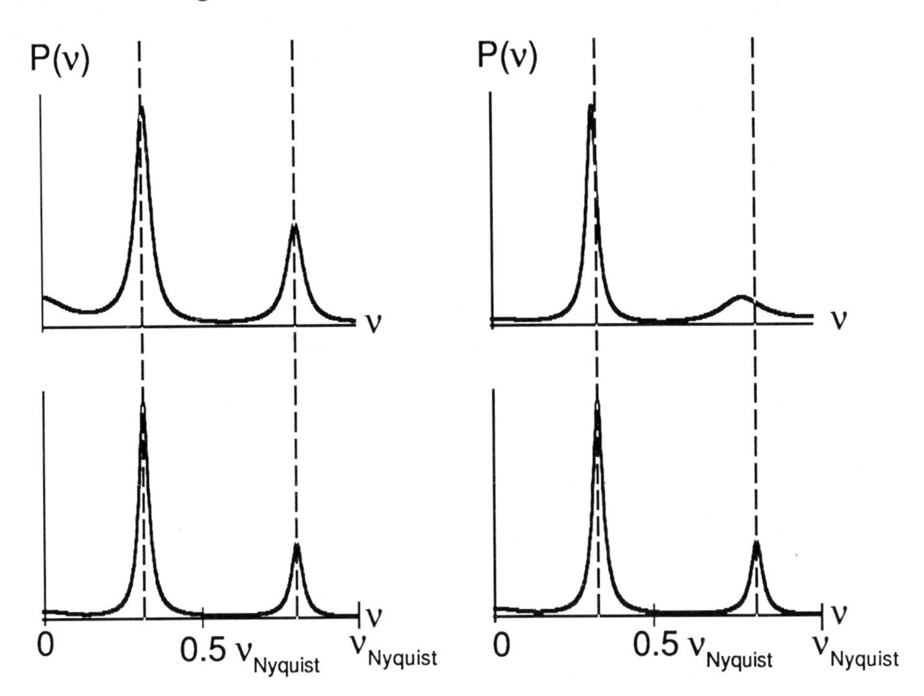

Figure 6.10 Effect of time-domain signal phase on the spectrum estimated by the AR autocorrelation method ($2p = 6$ poles) for a time-domain discrete data set consisting of two undamped sinusoids of relative magnitude 1.0 and 0.7 and frequency, $\nu_1 = 0.32\ \nu_{\text{Nyquist}}$ and $\nu_2 = 0.8\ \nu_{\text{Nyquist}}$. Left: both sinusoids have phase, $\varphi_1 = \varphi_2 = 0°$ (i.e., pure sine waves). Right: $\varphi_1 = 45°$; $\varphi_2 = 10°$. Top spectra: 60 time-domain data points. Bottom spectra: 250 time-domain data points. Note that the peak frequency error decreases as the number of time-domain data points used in AR spectral estimation increases.

Other linear prediction algorithms

One problem with the AR autocorrelation method is that it occasionally produces solutions which correspond to exponentially *increasing* rather than exponentially *decreasing* time-domain behavior (i.e., $\sigma > 0$ rather than $\sigma < 0$ in Eq. 6.6.b, corresponding to peaks on the bottom (rather than top) half of Figure 6.3). This "instability" problem is most severe when signal-to-noise ratio is low and/or the number of data points is small. We are therefore led to seek more "robust" methods in the AR family.

If, instead of the autocorrelation function, we had used the "covariance" function (i.e., the "covariance method), a similar error minimization procedure would have resulted. However, in the covariance matrix no longer has the convenient Toeplitz property, and the matrix inversion step corresponding to Eq. 6.51 is much lengthier. Some of the best autoregression methods are nevertheless based on the covariance approach (e.g., modified covariance method, Burg method, etc.); the interested reader is referred to Further Reading at the end of this chapter for further discussion. An example of the improvement offered by the Burg AR method is illustrated in Figure 6.9.

6.3.2 Moving-average (MA), autoregression/moving average (ARMA) methods

In view of Figures 6.1 and 6.9, it is logical to wonder how one could improve spectral resolution by linear prediction autoregression (AR) methods when the signal-to-noise ratio is low and the number of data points is small. As noted in the preceding section, an inherent defect of AR methods is their failure to take in account *noise* in the time-domain response (output). If one tries to account for noise by increasing the number of *poles* in the AR model, spurious peaks may appear in the spectrum.

A better approach would appear to be to modify the AR model to include at the outset (white) noise, N_n (with variance, $\sigma_N{}^2$): i.e., replace $f(t_n)$ by $f(t_n) + N_n$ in Eq. 6.40, in which N_n denotes the noise value at time, t_n. The resulting power spectrum, $P(z_n)$, can then be derived by the methods of the preceding section, to give:

$$P(z_n) = \frac{G^2}{A(z_n)\,A^*(z_n)} + \sigma_N{}^2 = \frac{G^2 + \sigma_N{}^2\,A(z_n)\,A^*(z_n)}{A(z_n)\,A^*(z_n)} \tag{6.59a}$$

in which

$$A(z_n) = 1 + \sum_{k=1}^{2p} a_k z_k{}^{-1} \tag{6.59b}$$

Eq. 6.59 is a special case of what is known as an "autoregression moving average" (ARMA) model. The general form of the ARMA model is given by the recursion relation, Eq. 6.60,

$$f(t_n) = -\sum_{k=1}^{p} a_k\, f(t_{n-k}) + G \sum_{k=0}^{q} b_k\; e(t_{n-k})\ ;\quad b_0 = 1 \tag{6.60}$$

Eq. 6.60 states that the signal (output) at time, t_n, is a linear combination not only of previous time-domain *signal* (output) data points (as in AR) but also of previous *excitation* (input) values. Its transfer function can be derived as in the preceding section:

$$H(z_k) = G\,\frac{1 + \sum\limits_{k=1}^{2q} b_k z_k{}^{-1}}{1 + \sum\limits_{k=1}^{2p} a_k z_k{}^{-1}} \tag{6.61}$$

The new transfer function contains not only "poles" (i.e., z-values for which the denominator goes to zero), but also "zeroes" (i.e., z-values for which the numerator goes to zero). The ARMA model is therefore also known as the "all poles and zeroes" model. Finally, an "all-zeroes" model (corresponding to a transfer function given by Eq. 6.61 with unit denominator) is also known as a "moving average" (MA) or "all-zeroes" model. Extraction of the system natural frequencies and damping constants from Eq. 6.61 is tedious and will not be discussed here. The point of this paragraph is to show that there are alternatives to autoregression, which offer improved performance at the cost of additional computational complexity and reduced speed. Figure 6.11 documents the improvement of ARMA over AR for noisy spectra from relatively small data sets.

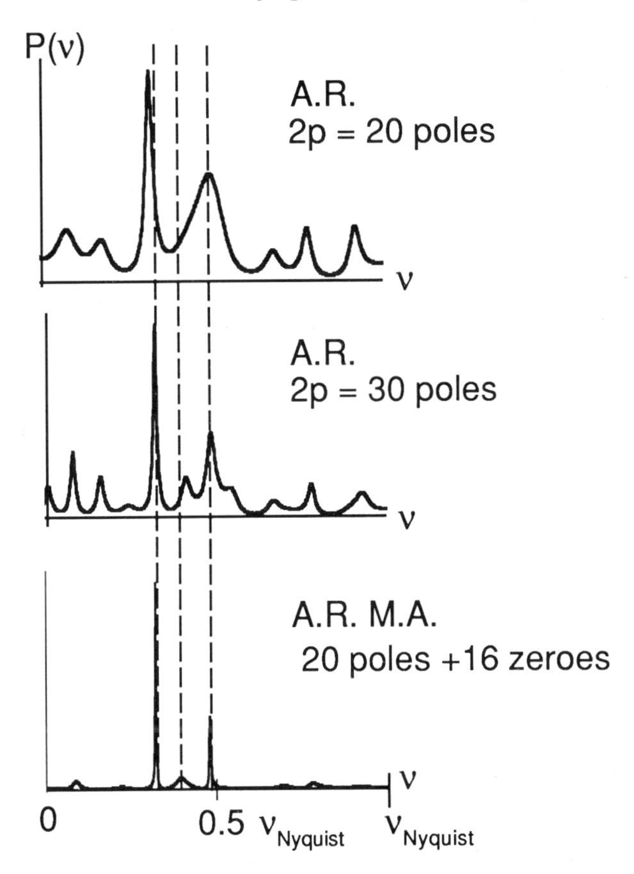

Figure 6.11 Power spectra computed by "all-poles" (top and middle), and "all poles and zeroes" (bottom) spectral estimation methods for the same time-domain 256-point signal consisting of a sum of three undamped sinusoids (frequencies, 0.32 ν_{Nyquist}, 0.40 ν_{Nyquist}, and 0.48 ν_{Nyquist}) of relative magnitude 1.0 (signal-to-noise ratio = 1.5), 0.7, and 1.0. Because of the low signal-to-noise ratio, none of the methods gives a particularly accurate representation of the relative peak magnitudes, but the ARMA spectrum is clearly superior in distinguishing the three peaks of interest from noise.

6.4 Maximum entropy methods

We began this chapter by noting that FT spectral estimation is based on a *causal* relation between time- and frequency-domain signals. Linear prediction methods are based on assumed prior knowledge of the form of the system "transfer function", which is in turn based on the form of the differential equation which describes the number and types of natural motions of a system. Maximum entropy methods (MEM) are based on estimating the "most probable" spectrum which matches the time-domain data, subjected to specified constraints.

In order to apply MEM, we first establish a criterion for deciding how well a proposed discrete spectrum matches its experimental time-domain discrete data set, to provide a "constraint" that any spectral estimate must satisfy. Next, we introduce the concept of a "most probable" result, by connecting that idea to other problems already familiar to chemists (namely, coin-tossing and the Boltzmann distribution). We then relate the concepts of entropy and probability. All of these aspects are combined to produce MEM.

6.4.1 Agreement between trial and actual time-domain data sets

We begin (see Figure 6.12) by letting

$$f(t_n),\ n = 0, 1, 2, \cdots, N-1 \tag{6.62}$$

represent a discrete N-point time-domain signal. Next, let

$$G(\omega_m),\ m = 0, 1, 2, \cdots, M;\quad M \geq N-1 \tag{6.63}$$

represent a discrete trial spectrum (computed by as yet unspecified non-FT means). Next, compute the (inverse) Fourier transform of $G(\omega_m)$ to obtain the discrete time-domain trial signal, $g(t_n)$. We agree to judge the quality of the trial spectrum, $G(\omega_m)$, according to the magnitude of the sum, C, of the squared differences between the trial and actual time-domain data sets:

$$C = \sum_{n=0}^{N-1} \frac{[f(t_n) - g(t_n)]^2}{(\sigma_n{}^2)} \tag{6.64}$$

in which σ_n represents the standard deviation,

$$\sigma_n = \frac{1}{J-1} \sum_{j=1}^{J} \sqrt{[f(t_{nj}) - \langle f(t_n)\rangle]^2)} \tag{6.65}$$

corresponding to a series of $j = 1, 2, \cdots, J$ measurements of the n'th time-domain data point, $f(t_{nj})$ whose average value is $\langle f(t_n)\rangle$. If, as is the case for detector-limited noise, σ_n is the same for all n (i.e., noise is independent of frequency and/or peak heights), then the usual way to measure σ is to determine the root-mean-square time-averaged noise either from all N time-domain data points of a separate time-domain acquisition in the absence of signal (i.e., noise only) or from (inverse FT of) a selected peak-free segment of the spectral baseline.

The value of C is a measure of the disagreement between the trial and actual data sets. For example, $C = 0$ corresponds to the spectrum whose inverse FT fits perfectly to the actual time-domain data set. For $C > 0$, we can consult a standard χ^2 (chi-square) table, to establish the "confidence" with which two data sets agree. For example, the confidence limit is >99% for a trial spectrum for which $C \leq (N + 3.29N)$. (Some authors prefer the criterion, $C < N$.) Armed with a criterion for judging various possible spectra, we next proceed to choose the "best" spectrum among several having the same C-value.

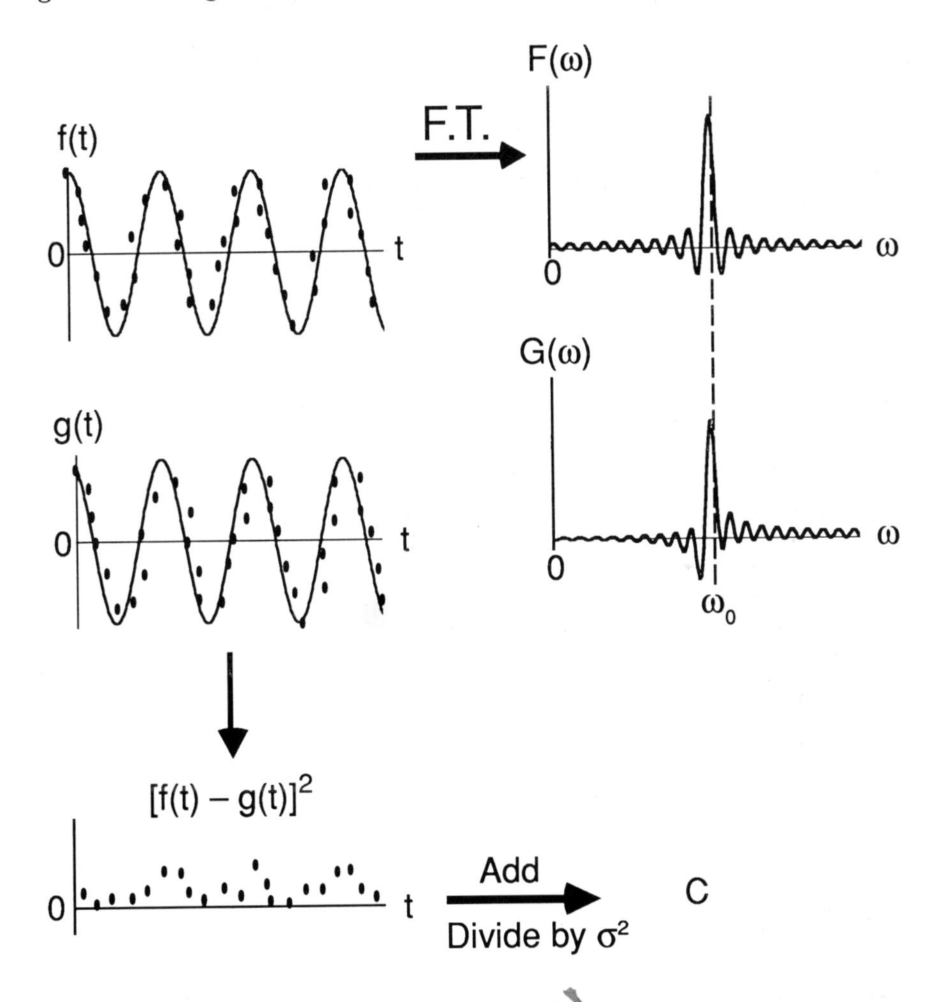

Figure 6.12 Evaluation of a trial discrete spectrum, $G(\omega)$, designed to represent a given time-domain discrete signal, $f(t)$. Upper left: $f(t)$. Upper right: frequency-domain spectrum, $F(\omega)$, obtained by Fourier transformation of $f(t)$. Lower right: $G(\omega)$. Lower left: time-domain discrete signal, $g(t)$, obtained by inverse Fourier transformation of $G(\omega)$. The quality of $G(\omega)$ is judged by the agreement (Eq. 6.64) between $f(t)$ and $g(t)$.

Although computation of the C-value in Eq. 6.64 affords a means for comparing trial and experimental data, the C-criterion alone is (perhaps surprisingly) not sufficient to establish the "best" spectrum. For example, one might at first intuitively infer that the spectrum produced by direct FT of the actual time-domain data should be the "best" spectrum, because the inverse FT of that spectrum yields the original time-domain data set *exactly* (i.e., $C = 0$ in Eq. 6.64). In the remainder of this chapter, we shall try to expand our intuition to show that we need a second criterion which will turn out to be maximization of the entropy of the trial spectrum.

6.4.2 Probability peaking: most probable result

Faced with uncertainty in choice of the "best" spectrum, we should not be surprised to find that the "best" answer is a matter of probability. In general, the *probability*, $P(m)$ of a particular result, m, is defined by:

$$P(m) = \frac{\text{Number of ways of obtaining the result } m}{\text{Total number of ways of obtaining all possible results}} \tag{6.66}$$

For example, in one throw of two (honest) dice, there is only one way to roll a 2 (namely, 1,1), two ways to roll a 3 (1,2; 2;1), and so on. From Eq. 6.66, the reader can easily show that the odds of rolling a 7 are thus, 6/(1+2+3+4+5+6+5+4+3+2+1) = 6/36 = 1/6. The number, 7, is also the most probable result, because there are more ways (6) of obtaining it than of any other number between 2 and 12.

Example: Coin tosses

An MEM spectrum (see below) is the *most probable* spectrum which matches the experimental time-domain data to within some previously specified criterion (e.g., a particular maximum C-value in Eq. 6.64). However, we first need to show that only the *most probable* result matters, so that we can ignore all other possible outcomes. Therefore, consider first the simpler problem illustrated in Figure 6.13. The relative probability of obtaining m heads after N tosses of a single coin is given by:

$$P_N(m) = \frac{N!}{m!\,(N-m)!}\left(\frac{1}{2}\right)^N \tag{6.67}$$

When the number of tosses is relatively small (say, $N < 10$), the number of heads from a particular experiment can differ significantly from the most probable result, namely, $N/2 = 5$ heads. However, as the number of tosses becomes large, we will be very likely to observe a result very close (percentage-wise) to the most probable result (i.e., $N/2$ heads in a particular N-toss experiment). Thus, for large N, we can safely ignore all but the *most probable* result.

The result shown in Figure 6.13 is a direct consequence of the *number of ways* of producing a given result (in this case, the relative number of heads after N coin tosses): e.g., there is only one way that *all* of the tosses can be heads, but there is a huge number of ways that *half* of the tosses can be heads. For large N, we are therefore almost certain to obtain $\sim N/2$ heads.

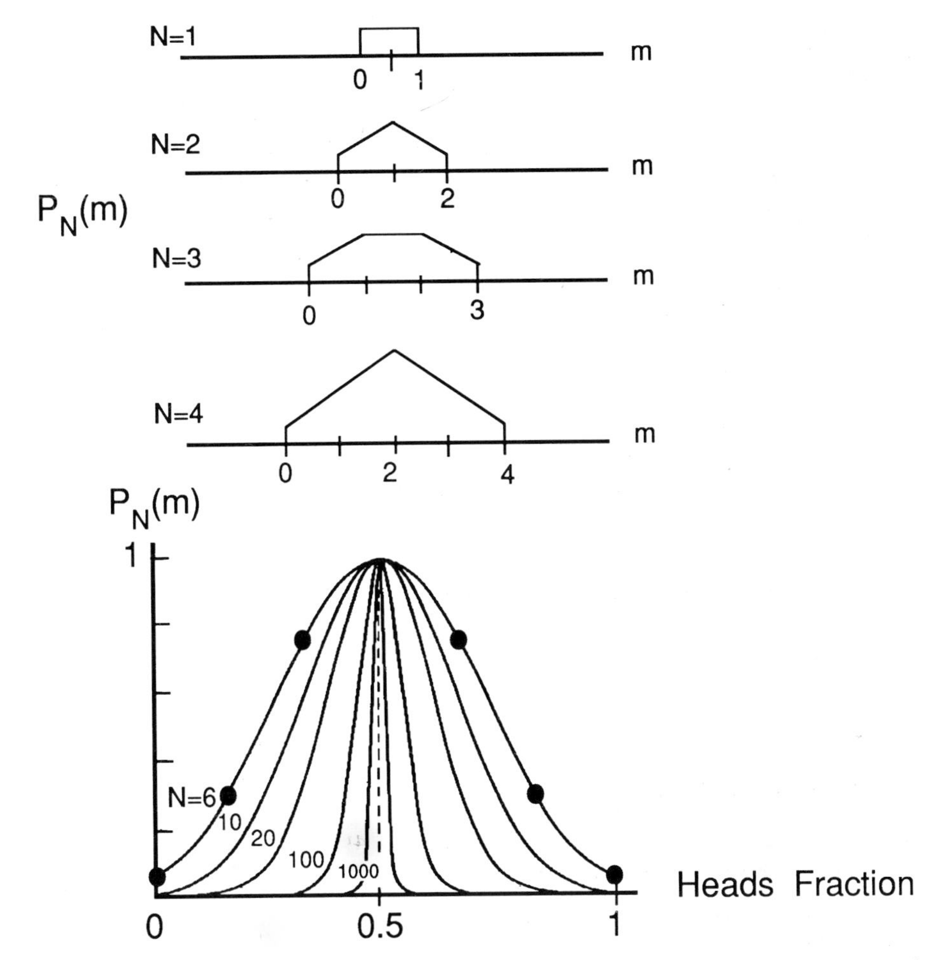

Figure 6.13 "Peaking" of the relative probability of obtaining a given fraction of tosses as heads, after N tosses of a single coin. As the number of tosses increases, the jagged shape (top diagram) becomes a smooth curve (bottom diagram), and one is more and more likely to obtain a result very close to the most probable result (namely, 0.5).

Example: Sorting particles into energy levels: Boltzmann distribution

In statistical mechanics, the fundamental problem is how to sort a fixed (large) number of particles, N_0, into a known distribution of energy levels, with N_i particles in energy level, E_i, for a given total energy, E_0, of all of the particles taken together:

$$\sum_i N_i = N_0 \tag{6.68a}$$

$$\sum_i N_i E_i = E_0 \tag{6.68b}$$

The number of ways of putting N_1 particles into level E_1, N_2 into E_2, etc., is given by

$$\frac{N_0!}{N_1!\,N_2!\,N_3!\cdots N_i!\cdots}$$

The *probability*, P, for that particular arrangement is simply the above factor divided by the total number of ways, N_{total}, of assigning the particles to the various energy levels.

$$P = \frac{\left(\dfrac{N_0!}{N_1!\,N_2!\,N_3!\cdots N_i!\cdots}\right)}{N_{total}} \tag{6.69}$$

To find the most probable distribution, we need to find the maximum value of P, by setting its derivatives with respect to the various N_i equal to zero. It is therefore more convenient to find the maximum value of $\log_e P$,

$$\log_e P = \log_e\left(\frac{N_0!}{N_1!\,N_2!\,N_3!\cdots N_i!\cdots}\right) - \log_e N_{total} \tag{6.70}$$

The way in which the "constraints" (in this case, a fixed total number of particles and fixed total energy) are taken into account is known as the method of "Lagrange multipliers", a common technique for this sort of problem. The constraints (Eqs. 6.68) are first rewritten in the form,

$$\sum_i N_i - N_0 = 0 \tag{6.71a}$$

$$\sum_i N_i E_i - E_0 = 0 \tag{6.71b}$$

If $\log_e P$ is to be a maximum, then it should seem reasonable that we can add any multiple (e.g., the Lagrange multipliers, λ_1 and λ_2) of the left-hand sides of Eqs. 6.71a and 6.71b (since both quantities are equal to zero) to $\log_e P$, and then compute the maximum of the combined expression (remember that the derivatives of the constants, N_0 and E_0, are zero):

$$\frac{\partial}{\partial N_i}\log_e P + \lambda_1 \frac{\partial}{\partial N_i}\sum_i N_i + \lambda_2 \frac{\partial}{\partial N_i}\sum_i N_i E_i = 0 \tag{6.72}$$

Eq. 6.72 is an example of what is known as the calculus of variations. After some algebraic manipulation (see, e.g., Davidson reference, p. 244), Eq. 6.72 leads to the Boltzmann distribution for the probability, P_i, that a particle is found in the (non-degenerate) energy level E_i.

$$P_i = \frac{N_i}{N_0} = \frac{\exp(-E_i/kT)}{\sum_i \exp(-E_i/kT)} \tag{6.73}$$

As for the coin-tossing example, it turns out that the Boltzmann distribution is not just the *most probable* distribution of particles among energy levels; it is so probable that it is essentially the *only* distribution that one need consider when the number of particles is large (e.g., $\sim 10^{23}$).

6.4.3 Entropy, probability, and spectral "smoothness"

The key feature of the Boltzmann example is the use of Lagrange multipliers to determine the maximum of a probability function subject to two previously specified *constraints* (Eqs. 6.68). In this section, we first establish the relation between entropy and spectral probability, and again use a Lagrange multiplier method to maximize that entropy subject to a previously specified constraint (e.g., a pre-specified maximum C-value in Eq. 6.64), to obtain the "most probable" spectrum.

We begin by representing a proposed (discrete) power spectrum by a probability distribution,

$P(\omega_j)$ = Probability that there is a power spectral peak of amplitude, $[G(\omega_j)]^2$, at frequency, ω_j

$$= \frac{[G(\omega_j)]^2}{\sum_j [G(\omega_j)]^2} \tag{6.74}$$

We desire a spectrum which suppresses uninformative peaks arising from (e.g.) Gibbs oscillations (for truncated time-domain signals) or noise—in other words, we seek the "smoothest" spectrum (i.e., "flattest" probability distribution) which matches a given experimental time-domain discrete data set to within (say) a maximum C-value in Eq. 6.64.

It is natural to associate "smoothness" with entropy. For example, thermodynamic entropy is a measure of "disorder". For a collection of particles in a container, entropy is maximized when the particles are distributed randomly in position (i.e., with a probability distribution which is "flat" as a function of spatial position). For a Boltzmann (i.e., "most probable") distribution, it is possible to relate thermodynamic entropy, S, for N particles to a probability distribution,

$$S(P_1, P_2, \cdots) = -\sum_{j=1}^{N} P_j \log_e P_j \tag{6.75}$$

in which P_j is the probability of finding a particle in a particular (non-degenerate) energy level.

By analogy, we define *spectral* "entropy", S, according to

$$S[P(\omega_1), P(\omega_2), \cdots, P(\omega_{N-1})] = -\sum_{j=0}^{N-1} P(\omega_j) \log_e P(\omega_j)] \tag{6.76}$$

An MEM algorithm

In order to maximize S subject to the (rewritten) constraint Eq. 6.77a,

$$\sum_{n=0}^{N-1} \frac{[f(t_n) - g(t_n)]^2}{\sigma_n^2} - C = 0 \tag{6.77a}$$

we use the method of Lagrange multipliers as for the Boltzmann example. Specifically, we could maximize the sum, H, of S and the product of λ' (a Lagrange multiplier) and Eq. 6.77a,

$$H = S + \lambda' \left(\sum_{n=0}^{N-1} \frac{[f(t_n) - g(t_n)]^2}{\sigma_n^2} - C \right) \tag{6.78}$$

in which C represents a pre-specified difference (e.g., $C = N$) between the actual time-domain data set and the time-domain data set obtained by inverse FT of the proposed discrete spectrum. However, if the left-hand side of Eq. 6.77a is zero, then its square is also equal to zero. A particularly convenient MEM algorithm is therefore based on introducing a second constraint equation,

$$\left(\sum_{n=0}^{N-1} \frac{[f(t_n) - g(t_n)]^2}{\sigma_n^2} - C \right)^2 = 0 \tag{6.77b}$$

to give the final function to be optimized:

$$H = S + \lambda_1 D \ P(\omega_j) - \lambda_2 [D \ P(\omega_j)]^2 \tag{6.79a}$$

in which

$$D = \left(\sum_{n=0}^{N-1} \frac{[f(t_n) - g(t_n)]^2}{\sigma_n^2} - C \right) \tag{6.79b}$$

For a given λ_1 and λ_2, we seek a maximum for H. Therefore, we set the partial derivative of H with respect to each power spectral value equal to zero:

$$\frac{\partial H}{\partial P(\omega_j)} = 0 \tag{6.80}$$

In vector notation,

$$\begin{pmatrix} \frac{\partial H}{\partial P(\omega_0)} \\ \frac{\partial H}{\partial P(\omega_1)} \\ \vdots \\ \frac{\partial H}{\partial P(\omega_{N-1})} \end{pmatrix} = \begin{pmatrix} \frac{\partial S}{\partial P(\omega_0)} \\ \frac{\partial S}{\partial P(\omega_1)} \\ \vdots \\ \frac{\partial S}{\partial P(\omega_{N-1})} \end{pmatrix} + \lambda_1 \begin{pmatrix} \frac{\partial D}{\partial P(\omega_0)} \\ \frac{\partial D}{\partial P(\omega_1)} \\ \vdots \\ \frac{\partial D}{\partial P(\omega_{N-1})} \end{pmatrix} - 2D\lambda_2 \begin{pmatrix} \frac{\partial D}{\partial P(\omega_0)} \\ \frac{\partial D}{\partial P(\omega_1)} \\ \vdots \\ \frac{\partial D}{\partial P(\omega_{N-1})} \end{pmatrix} \qquad (6.81)$$

and arbitrary values of λ_1 and (positive-valued) λ_2. Each term on the right-hand side of Eq. 6.81 is then evaluated analytically, e.g.,

$$\frac{\partial S}{\partial P(\omega_j)} = -\log_e [P(\omega_j) - 1] \qquad (6.82)$$

The derivation of Eq. 6.83 for $\partial D / \partial P(\omega_j)$ is left as an exercise (see Problems).

$$\frac{\partial D}{\partial P(\omega_j)} = \frac{-2}{N\sigma^2} \sum_{n=0}^{N-1} \big(f(t_n) - g(t_n)\big) \exp(i\, 2\pi j\, n / N) \qquad (6.83)$$

The (iterative) computation begins from a first-guess power spectrum which is maximally flat:

$$P(\omega_j) = \frac{1}{N}, \quad i = 0, 1, 2, \cdots, N-1 \qquad (6.84)$$

Eq. 6.84 is then substituted into Eqs. 6.82 and 6.83, which are in turn substituted into Eq. 6.81 to yield a full set of derivatives, $\partial H / \partial P(\omega_i)$. λ_1 and λ_2 are then varied systematically until all of the $\partial H / \partial P(\omega_j)$ approach zero, to establish a local maximum for H. The final values of λ_1 and λ_2 are then substituted into (e.g.) Eq. 6.85

$$\frac{\partial H}{\partial P(\omega_2)} = 0 = \frac{\partial S}{\partial P(\omega_2)} + \lambda_1 \frac{\partial D}{\partial P(\omega_2)} - 2D\lambda_2 \frac{\partial D}{\partial P(\omega_2)} \qquad (6.85)$$

to obtain a new value for each $P(\omega_j)$, in this case, $P(\omega_2)$.

At this stage, D is computed in order to test whether or not the constraint equation is satisfied (i.e., $C \leq N$). If not, then the new $P(\omega_j)$ are used to define a new trial spectrum, and the above procedure repeated until the final spectrum matches the time-domain data to within the desired agreement factor.

In the above process, entropy is maximized from the $\partial S / \partial P(\omega_j)$ terms in Eq. 6.81, and the constraint which forces the spectrum to match the experimental data is provided by the $\partial D / \partial P(\omega_j)$ terms in Eq. 6.81. The final spectrum represents a compromise between the two requirements (see Figure 6.14).

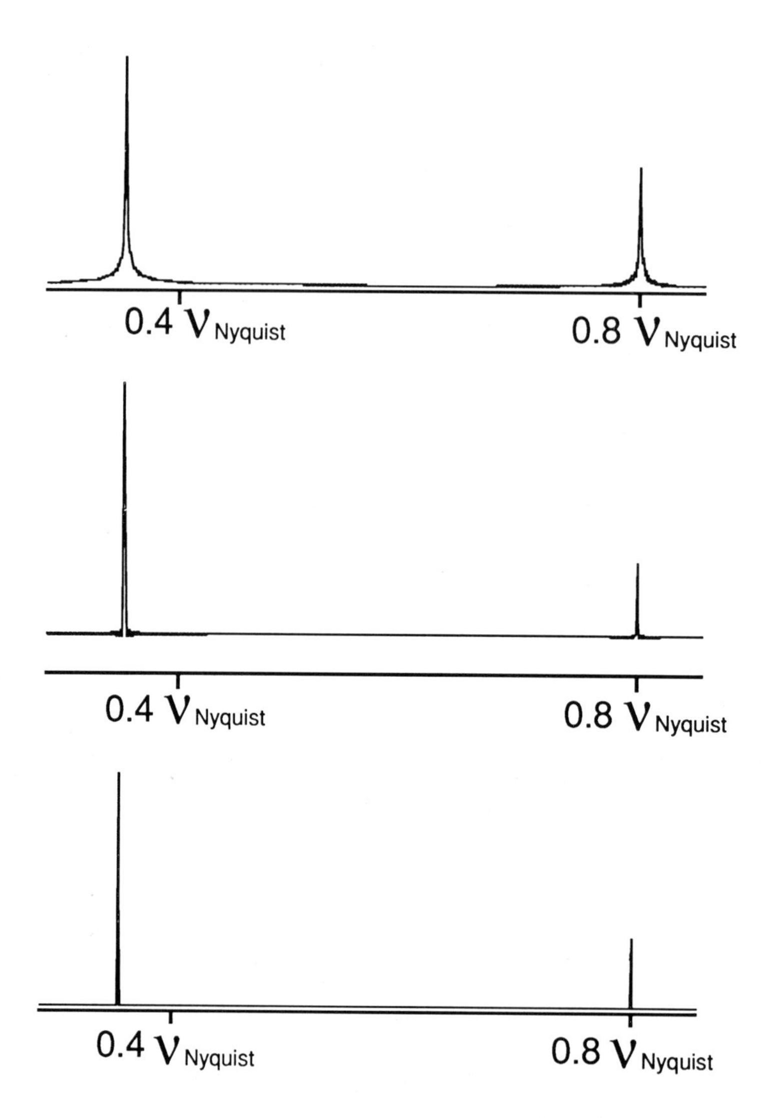

Figure 6.14 Phased absorption-mode Fourier transform spectrum (top) and MEM magnitude-mode spectra (middle, bottom) obtained from the same time-domain 512-point discrete signal (zero-filled to N = 1,024 points), consisting of two damped sinusoids of relative magnitudes 1.0 and 0.5, frequencies 0.35 and 0.8 ν_{Nyquist}, and exponential damping constant, τ = acquisition period = $256/\nu_{\text{Nyquist}}$. The two MEM spectra correspond to agreement factors (Eq. 6.64) of C = 2876 (middle) and C = 1026 ≅ N (bottom). Note the absence of Gibbs oscillations in the lowermost spectrum, and the higher baseline in the middle spectrum (see text). The computational time for each MEM spectrum was approximately 10,000 times longer than that for the FT spectrum.

As evident from Figure 6.14, MEM does good job of eliminating Gibbs oscillations, and can give narrower (but not necessarily more precise) spectral peaks. However, the appearance of an MEM spectrum is highly sensitive to the match between the assumed and actual level of noise in the experimental data. If the assumed noise level is comparable to the actual noise level (lowermost spectrum in Figure 6.14), then the MEM spectrum is similar to that obtained by FFT, except that the MEM spectrum is narrower and Gibbs oscillations are absent from the MEM spectrum. However, if the assumed noise level is greater than the actual noise level (middle spectrum in Figure 6.14), then MEM can "over-smooth" the spectrum and can actually suppress small true peaks, because the MEM spectral baseline (i.e., the MEM-inferred spectral noise level) is significantly above zero. Therefore, just as auto-regression methods work best when the true number of spectral *peaks* is known, MEM works best when the true *noise level* is known (see next section).

Because MEM can require repeated iterations (~100), each of which (for the illustrated algorithm) requires FFT at each local maximum stage, the MEM computation is *much* slower than FFT for large data sets (N > 1,024 time-domain data points), as seen from Table 6.2. Finally, although the above-described MEM algorithm is relatively fast computationally, other (computationally more complex) algorithms can more accurately locate the maximum of H in Eq. 6.78. For example, Bayesian analysis uses the MEM principle but also allows for an assumed functional form for the time-domain signal (see Bretthorst reference in Further Reading).

It can be shown that the relative peak magnitudes of an MEM spectrum are maximally uncorrelated with each other. As a result, MEM is used in spectroscopy mainly for applications (e.g., astronomical imaging) in which the various spectral positions and magnitudes are truly independent. However, in most "chemical" spectra (NMR, IR, mass spectra), at least some peak positions and magnitudes can be highly correlated (e.g., J-coupled multiplets in NMR, isotope peaks in mass spectrometry), and MEM methods may require additional modification.

Table 6.2 Typical relative approximate computational times for various spectral estimation methods. The actual computational time varies according to the particular algorithm, the spectral peak distribution and noise, and the relative multiplication and addition speeds of a given computer.

Spectral Estimation Method	Relative Computational Time for N time-domain data points
FT	N^2 (non-FFT) or $N \log_e N$ (FFT)
AR Autocorrelation	N^2 or N^3, depending on algorithm
MEM	~100 $N \log_e N$

6.4.4 Relation between MEM and autoregression methods

The maximum-entropy spectral estimation method of the preceding section was based on maximizing spectral entropy subject to a maximum least-squares deviation (constraint) between actual and computed time-domain data. In this section, we note that MEM spectra can be based on other constraints. Consider, for example, the constraint that the experimental and computed time-domain *autocorrelation function* (rather than the time-domain data set itself) agree (as judged by least-squares difference) to within a specified limit with the autocorrelation function computed (by inverse FT) from the estimated power spectrum. Burg has shown that the MEM spectrum resulting from such an autocorrelation constraint is identical to that obtained from the autoregression autocorrelation method described in Chapter 6.3.1., provided that the noise is Gaussian-distributed, and the time- and frequency-domain data points are equally spaced. Burg's "maximum entropy" method differs from the procedure described here in that the entropy of a probability density function of the *time-domain* discrete signal (rather than the *frequency-domain* discrete spectrum) is maximized; the Burg method leads to different formulas and a reconstruction algorithm with different properties (see Skilling reference in Further Reading).

An interesting feature of the MEM method is that it may be applied to "incomplete" data sets. For example, MEM has been used to estimate the spectrum from an "exponentially sampled" time-domain NMR free-induction decay (i.e., an ordinary time-domain signal in which data are sampled less and less frequently as time proceeds). Thus, spectral analysis is not limited (as in FT analysis) to equally-spaced data (see Barna *et al.* reference in Further Reading).

Finally, it is worth noting that the MEM method yields the "best" spectrum (i.e., the most probable spectrum whose individual spectral magnitudes are uncorrelated with each other except as required by the data) whose inverse FT matches the true time-domain discrete data to within a pre-specified agreement factor, *if nothing is known about the functional form of the time-domain signal.* However, if the functional form of $f(t)$ is known, then any of several MEM-based or other non-FT methods may yield a "better" spectrum for a particular data set. One of the most generally useful methods is Bayesian analysis, which combines the MEM criterion with assumed ("prior") knowledge of the functional form of $f(t)$, as described in the Bretthorst reference in Further Reading. Unlike autoregression (Figure 6.1), the precision of Bayesian analysis is superior to that from FT analysis, for a sufficiently large number of time-domain data points, N.

Further Reading

G. L. Bretthorst, *Bayesian Spectrum Analysis and Parameter Estimation*, in *Lecture Notes in Statistics*, Vol. 48 (Spring-Verlag, NY, 1988). (Comprehensive description of Bayesian analysis)

M. G. Bellanger, *Adaptive digital filters and signal analysis*, Dekker, NY, **1987**. (Introduction to signal modeling)

H. J. Blinchikoff & A. I. Zverev, *Filtering in the time and frequency domains*, J. Wiley & Sons, New York, **1987**. (Very good introduction to Laplace and z-transforms)

J. P. Burg, *Modern Spectrum Analysis*, I.E.E.E. Press, NY, **1978**. (MEM applied to signal theory.)

N. Davidson, *Statistical Mechanics*, McGraw-Hill, NY, **1962**. (Old but excellent treatment of statistical mechanics)

S. F. Gull & G. J. Daniell, *Nature* **1978**, *272*, 686-690. (MEM applied to radioastronomy--one of the earliest MEM physical applications)

S. F. Gull & P. J. Hore, *J. Mag. Res.* **1985**, *62*, 561-567. (Good presentation of the parameters involved in MEM)

E. T. Jaynes, *Phys. Rev.*, **1957**, *106(4)*, 620-630. (Detailed discussion of MEM)

E. T. Jaynes, *I.E.E.E. Proc.* **1982**, *70(9)*, 939-952. (Good treatment of MEM limitations.)

S. M. Kay & S. L. Marple Jr., *Proc. I.E.E.E.* , **1981**, 69(11), 1380-1419. (Very good introduction to linear prediction formalism)

S. M. Kay, *Modern spectral estimation - Theory and Application*, Prentice-Hall, Englewood Cliffs, NJ, **1987**. (Very good explanations of AR, ARMA, and related methods; a software package with examples is also available)

J. Makhoul, *Proc. I.E.E.E.* **1975**, 63(4), 561-580. (Very good introduction to linear prediction formalism)

S. L. Marple Jr., *Digital spectral analysis with applications*, Prentice-Hall, Englewood Cliffs, NJ, **1987**. (More practical than the previous book for discussion of properties of AR, ARMA, and related spectra)

C. E. Shannon, *Bell System Techn. J.* **1948**, *27*, pp. 379-423 and 623-656. (Justification of the connection between entropy and probability)

S. Sibisi, *Nature* **1983**, *301*, 134-136. (MEM applied to FT/NMR)

J. Skilling & S. F. Gull **1984**, SIAM Amer. Math Soc. proc. *Appl. Math. 14*, 167-189. (Discusses MEM applied to discrete time-domain or frequency-domain data sets)

D. Van Ormondt & K. Nederveen, *Chem. Phys. Lett.* **1981**, *82(3)*, 443-446. (Discussion of various ways to set the constraints in MEM)

Problems

6.1 Evaluate the transfer function, H, of a system characterized by the following differential equation describing the time-dependent response, $x(t)$, of a single damped mass-on-a-spring to an impulse excitation, $\delta(t)$:

$$m\frac{d^2x}{dt^2} + f\frac{dx}{dt} + kx = \delta(t)$$

(a) by use of *Fourier transform* methods

(b) by use of *Laplace transform* methods

Express your result in the form of Eq. 6.14, and evaluate the constant, c.

Solve for the "poles" of the system (i.e., s_1 and s_{-1} in Eq. 6.14).

(c) What is the physical meaning of the "poles"?

(d) Show that the product $H(i\omega)\, H^*(i\omega)$ corresponds to the power spectrum of the system (compare with Eq. 1.30).

6.2 Since the function $|H(i\omega)|$ corresponds, by definition, to the ratio of the output to the input, then $|H(i\omega)|$ can be considered to be the output (response) to an excitation (input) of unit magnitude:

$$|H(i\omega)| = \frac{\text{output magnitude}}{\text{input magnitude} = 1}$$

Thus, experimental determination of $|H(i\omega)|$ provides a direct measurement of the *gain* of a system. However, when dynamic range is large, it is convenient to introduce a logarithmic scale for $|H(i\omega)|$:

(a) Show that the expression $-\log_e H(i\omega)$ can be written in the following form:

$$-\log_e H(i\omega) = \alpha(\omega) - i\,\theta(\omega)\text{, in which}$$

$$\alpha(\omega) = -\log_e |H(i\omega)| \quad \text{and} \quad \theta(\omega) = \arctan\left(\frac{Im\,[H(i\omega)]}{Re\,[H(i\omega)]}\right)$$

$\alpha(\omega)$ is known as the *attenuation* of the system (in nepers). Attenutation is more commonly expressed in decibels (dB): $A(\omega) = -20 \log_{10}|H(i\omega)|$.

(b) Determine the conversion factor between neper and dB.

(c) Suppose that the transfer function of a system is given by the following expression:

$$H(s) = \frac{0.5}{s^3 + 2s^2 + 2s + 1}$$

Show that the poles of $H(s)$ are $s = -1$ and $s = -0.5 \pm i\sqrt{3}/2$, and evaluate $H(s)$ at $s = i\omega$.

(d) Evaluate the magnitude spectrum, $|H(i\omega)|$.

(e) From the previous definitions and results, determine and plot the attenuation and phase spectra, $A(\omega)$ *vs.* ω and $\theta(\omega)$ *vs.* ω.

(f) Compute $A(\omega)$ and $\theta(\omega)$ at a particular frequency, say, $\omega = 1$ rad s^{-1}.

6.3 This problem serves as a model for extraction of spectral peak frequencies from a time-domain signal consisting of a sum of undamped sinusoids, based on so-called "Pisarenko decomposition".

(a) By use of trigonometric identities, show that:

$$\sin n\,\Omega = 2 \cos \Omega\, \sin (n-1)\Omega - \sin (n-2)\Omega \qquad (6.3.1)$$

(b) Rewrite the previous equation in the form of a difference equation in x_n: i.e., let $x_n = \sin(n\,\Omega + \phi)$, and express its z-transform in the following format [i.e., determine $S(z)$ and $D(z)$ in Eq. 6.3.2].

$$S(z) \cdot [\text{characteristic polynomial}(z)] = D(z) \qquad (6.3.2)$$

(c) Show that the roots, z_1 and z_2, of that characteristic polynomial have the form: $z_1 = \exp(i\,\omega\,\Delta t)$, and $z_2 = z_1^*$.

(d) Show that the corresponding frequencies, ν_1 and ν_2 (in Hz), are given by

$$\nu_i = \arctan\left(\frac{Im\,(z_i)}{Re\,(z_i)}\right) / \, 2\pi\,\Delta t$$

6.4 In this problem, you will determine the recursion equation coefficients, a_k, for a system consisting of a single damped harmonic oscillator (i.e., a system described by a second-order differential equation with constant coefficients).

(a) Starting from the general *real* time-domain response (Eq. 6.24) of a single damped harmonic oscillator to an impulse excitation, show that the coefficients, a_1 and a_2, of Eqs. 6.23 are given by Eq. 6.26.

(b) Repeat part (a) for the general *complex* time-domain response,

$$\boldsymbol{f}(t_n) = \exp(i\,\omega_0\, n\,\Delta t + \varphi)$$

6.5 In the text, we evaluated the transfer function for a *single* damped harmonic oscillator. In this problem, we extend the method to a system of damped harmonic oscillators of p different natural frequencies ($p \geq 2$) which may be described by the difference equation,

$$f(t_n) = -\sum_{m=1}^{2p} a_m\, f(t_{n-m}) + \sum_{k=0}^{2q} b_k\, e(t_{n-k})$$

(a) evaluate the transfer function, $H(e^{i\omega})$, of such a system by use of the *Fourier* transform and then by means of the *z-transform*. Show that $H(e^{i\omega})$ may be expressed as the quotient, $H(e^{i\omega}) = Z(e^{i\omega})/P(e^{i\omega})$, of an all-zeroes polynomial, $Z(e^{i\omega})$ and an all-poles polynomial $P(e^{i\omega})$.

(b) Show that $P(e^{i\omega})$ can be rewritten as:

$$P(\boldsymbol{f}) = c \prod_{n=1}^{M} \left(1 - \exp[-\,i\,2\pi\,\Delta t\,(\boldsymbol{f} - \boldsymbol{f}_k)]\right)$$

in which c is a constant and $\boldsymbol{f}$ is a complex number.

(c) Suppose that $P(\boldsymbol{f})$ has only pure-imaginary roots: e.g., $\boldsymbol{f}_k = (i\,\alpha_k / 2\pi\,\Delta t)$, in which $\alpha_k > 0$. Verify that $P(\boldsymbol{f})$ is given by:

$$P(f) = 1 - \exp[-i2\pi\,\Delta t\,(f - i\alpha_k/2\pi\,\Delta t)] = 1 - \exp(-\alpha_k)\,z^{-1}$$

which corresponds to the solution of a first-order difference equation (i.e., a massless weight-on-a-spring), with $\alpha_k = 1/\tau_k$.

(d) Conversely, suppose that $P(f)$ has a complex root: e.g., $f_k = (i\alpha_k + \omega_k)/2\pi\,\Delta t$, $(\alpha_k, \omega_k > 0)$. Show that $f_k = (i\alpha_k - \omega_k)/2\pi\,\Delta t$ is also a solution. Combine the two solutions to show that

$$P(f) = \left(1 - \exp\{-i2\pi\,\Delta t\,[f - (i\alpha_k + \omega_k)]/2\pi\,\Delta t\,\}\right)$$
$$\times \left(1 - \exp\{-i2\pi\,\Delta t\,[f - (i\alpha_k - \omega_k)]/2\pi\,\Delta t\,\}\right)$$
$$= 1 - 2\exp(-\alpha_k)\cos\omega_k\,z^{-1} + \exp(-2\alpha_k)\,z^{-2}.$$

Compare the preceding expression (for damped oscillators) with the Pisarenko decomposition result of the previous problem (for undamped oscillators).

(e) A system of damped oscillators of 2 or more natural frequencies can be modeled by a set of second-order differential equations which in turn leads to a $2p$'th order difference equation. Show that any system based on a difference equation of higher than second order ($p \geq 2$) can be broken down into a series of first and second order difference equations. Your result shows why the autoregressive model is suitable for extraction of spectral peak frequencies and line widths from a time-domain signal consisting of a sum of an arbitrary number of damped sinusoids.

6.6 By comparing Eq. 6.40 and Eq. 6.44, show that

$$f(t_n) = -\sum_{k=1}^{2p} a_k\,f(t_{n-k}) + e_n \tag{6.6.1}$$

Now compare Eqs. 6.40 and Eq. 6.6.1 to show that

$$e_n = G\,e(t_n)\,,$$

namely, that the difference between the actual and recursion-calculated value of $f(t_n)$ corresponds to the input, $e(t_n)$, weighted by the "gain" factor, G .

6.7 Complete the derivation of the "normal" equations which form the basis for the autoregression autocorrelation spectral estimation method; i.e., obtain Eq. 6.47a from Eq. 6.45.

6.8 In this problem, the reader uses the "Yule-Walker" equation to evaluate the "predictor estimator" recursion formula autoregression coefficients, a_k, from the autocorrelation function of a discrete three-point time-domain signal.

Consider a time domain signal defined by three data-points: $f(t_0)$, $f(t_1)$, $f(t_2)$.

(a) Write out the autocorrelation function coefficients: $R(0)$, $R(1)$, and $R(2)$ of Eq. 6.49 in terms of $f(t_0)$, $f(t_1)$, $f(t_2)$.

(b) Given that $R(0) = 5.75$, $R(1) = 4.10$, and $R(2) = 0.50$, calculate the "prediction estimator" coefficients, a_1 and a_2, and the gain, G, of the transfer function of the system.

(c) Write the difference equation for this system (see Eqs. 6.28 and 6.40).

(d) Using Eqs 6.26, obtain the (positive and negative) natural frequency and damping constant of the oscillator described by the previous difference equation.

6.9 Derive Eq. 6.83 for $\partial D / \partial P(\omega_j)$.

Solutions to Problems

6.1 (a) See Eqs. 6.1 - 6.4.

(b) See Eqs. 6.6 - 6.11.

(c) The root of the denominator corresponds to the solution of the characteristic differential equation of the system; it therefore yields the normal modes of the system.

(d) See Eqs. 6.4 - 6.5.

6.2 (a) and (b) Express a complex number as the product of its magnitude and phase, i.e., $[H = |H| \exp(i\psi)]$. You should then be able to show that 1 neper = 0.115 decibel. (c) Multiply $H(s)$ by $(s + 1)$; similarly for the other pair of solutions. (d) $|H(i\omega)| = 0.5/\sqrt{1 + \omega^6}$. (f) $A(\omega) \cong 9$ dB and $\theta(\omega) = 0$ at $\omega = 1$ rad s^{-1}.

6.3 (a) Write $\sin(n-2)\Omega$ as $\sin(n-1-1)\Omega$ and then use Eq. A.25. (b) For initial conditions, $x_{-1} = \sin(-\Omega + \varphi)$ and $x_{-2} = \sin(-2\Omega + \varphi)$ and the result that if $y_{-1}(n) = y(n-1)$ then:

$$Y_{-1}(z) = \sum_{n=0}^{\infty} y_{-1}(n) z^{-n} = z^{-1}[Y(z)] + y(-1)$$

$$[X(z)] = \frac{\sin\varphi - \sin(-\Omega + \varphi) z^{-1}}{1 - (2\cos\Omega) z^{-1} + z^{-2}}$$

Eq. 6.3.2 follows directly, from which (c) and (d) are readily obtained.

6.4 First calculate the *z-transform* of $f(t_n)$. Then separate its real and imaginary parts:

$$\mathrm{Re}[F(z)] = \frac{1 - (e^{-\alpha} \cos \omega_0 \Delta t)\, z^{-1}}{1 - (2\, e^{-\alpha} \cos \omega_0 \Delta t)\, z^{-1} + e^{-2\alpha} z^{-2}}$$

$$\mathrm{Im}[F(z)] = \frac{1 - (e^{-\alpha} \sin \omega_0 \Delta t)\, z^{-1}}{1 - (2\, e^{-\alpha} \cos \omega_0 \Delta t)\, z^{-1} + e^{-2\alpha} z^{-2}}$$

(For a complete treatment, see the Bellanger reference, pp. 15-17.)

6.5 (a) $$H(z) = Z(z)/P(z) = \frac{1 + \sum_{k=1}^{2q} b_k z^{-k}}{1 + \sum_{m=1}^{2p} a_m z^{-m}}$$

The evaluation of $H(z)$ at $z = e^{-i\omega}$ gives $H(e^{-i\omega})$.

(b) From the equation for the sum of terms in a geometric series, one may readily calculate the *z-transform* of the complex exponential, $y(n) = e^{-in\omega_0}$, namely, $Y(z) = \frac{1}{1 - e^{-i\omega_0} z^{-1}}$. Since p damped harmonic oscillators will be characterized by $2p$ natural frequencies (half of which are negative), the linear property of the *z-transform* leads at once to the desired result.

(c) $H(z) = 1 - \exp(-\alpha_k)\, z^{-1}$

(d) If $P(e^{-if})$ has a complex solution, then $P(e^{-if})$ is a polynomial of at least degree 2. It may thus be analyzed into polynomials of degree 2 with complex conjugate solutions. Note the "damping" factor which now appears now in $P(z)$. (For a complete treatment, see the Bellanger reference, pp. 15-23.).

(e) This problem may be solved by use of the partial-fraction expansion technique.

6.6 We have modeled the signal according to Eq. 6.40, and then determined its various parameters with the least-squares method, assuming that the input was unknown. By comparison to Eq. 6.44:

$$e_n = f(t_n) - \boldsymbol{f}(t_n) = f(t_n) + \sum_{k=1}^{2p} a_k f(t_{n-k})$$

and Eq. 6.40, it should be clear that the input signal will result in the corresponding response signal, $f(t_n)$, only if $G u_n = e_n$; i.e., the signal is proportional to the error of the system. (For more detailed explanation, see Makhoul reference.)

6.7 Expand Eq. 6.45 and use Eq. 6.46 to obtain Eq. 6.47a. Mathematically, Eqs. 6.47a and 6.47b differ by a shift in index, i. Physically, that shift corresponds to a time lag, $t_{n+i} - t_n$. That is why we needed to recast Eq. 6.47a as Eq. 6.47b in order to connect the autocorrelation function with the difference equation method.

6.8 (a) See Eq. 6.48b.

(b) $a_1 = -1.32$; $a_2 = 0.85$; $G^2 = 0.76$.

(c,d) Check with example given in the text (Eqs. 6.42).

6.9 It is necessary to inverse Fourier transform the trial spectrum in order to calculate $\dfrac{\partial D}{\partial P(\omega_j)}$. From there on, the route to Eq. 6.83 is straightforward:

$$D = \sum_{n=0}^{N-1} \frac{\left(f(t_n) - \sum_j P(\omega_j)\, e^{inj/N} \right)^2}{\sigma_n{}^2} - C$$

Chapter 7

Fourier transform ion cyclotron resonance mass spectrometry

7.1 Natural motions of an ion in a static electromagnetic trap

We have chosen Fourier transform ion cyclotron resonance (FT/ICR) mass spectrometry as our first spectroscopic application of FT methods, because the natural motions involved can be treated wholly classically (i.e., without any need to resort to quantum mechanics).

Although Fourier transform methods have been applied both to all-electric ion traps and to time-of-flight mass spectrometry, the principal embodiment of Fourier transform mass spectrometry (FTMS) is ICR mass spectrometry. We therefore begin this chapter by analyzing the natural motions of an ion trapped by static electric and magnetic fields (Chapter 7.1). We then show how a coherent signal is generated by an oscillating (or rotating) radiofrequency electric field, damped by ion-molecule collisions, and detected by the image current induced on opposed detector plates (Chapter 7.2). Mass resolution and mass calibration are described in Chapter 7.3, followed by a discussion of various complications (e.g., space charge, non-uniform static and rf electric field, non-uniform magnetic field, etc.). ICR aspects unique to the FT approach are discussed in Chapter 7.4, followed by applications in Chapter 7.5.

7.1.1 Cyclotron motion: ICR orbital frequency, radius, velocity, and energy

An ion moving in the presence of spatially uniform electric and magnetic fields, $\boldsymbol{E}$ and $\boldsymbol{B}$, is subjected to the "Lorentz" force given (in S.I. units) by Eq. 7.1,

$$\text{Force} = \text{mass} \cdot \text{acceleration} = m \frac{d\boldsymbol{v}}{dt} = q\boldsymbol{E} + q\boldsymbol{v} \times \boldsymbol{B} \tag{7.1}$$

in which m, q, and $\boldsymbol{v}$ are ionic mass, charge, and velocity, and the vector cross product term means that the direction of the magnetic component of the Lorentz force is perpendicular to the plane determined by $\boldsymbol{v}$ and $\boldsymbol{B}$ (see Figure 7.1).

We shall first consider Eq. 7.1 in the absence of the electric field, $\boldsymbol{E}$. As shown in Figure 7.1, it is convenient to assign the magnetic field direction along the z-axis of a Cartesian coordinate frame: $\boldsymbol{B} = B_0 \boldsymbol{k}$. Since angular acceleration, $dv/dt = v_{xy}{}^2/r$, Eq. 7.1 then becomes

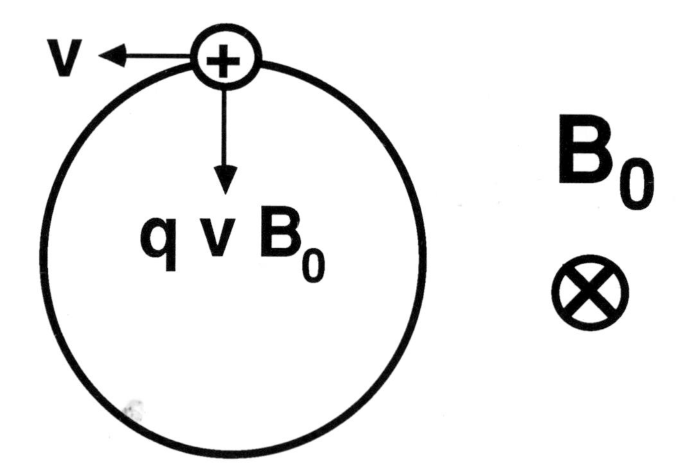

Figure 7.1 Origin of ion cyclotron motion. The path of an ion moving in the plane of the paper is bent into a circle (see text) by the inward-directed Lorentz magnetic force produced by a magnetic field directed perpendicular to the plane of the paper.

$$\frac{m\,v_{xy}{}^2}{r} = q\,v_{xy}\,B_0 \tag{7.2}$$

in which $v_{xy} = \sqrt{v_x{}^2 + v_y{}^2}$ is the ion velocity in the xy plane, and r is the radial distance measured from the ion to the center of its cyclotron orbit (see below). Since angular velocity, ω, about the z-axis is defined by

$$\omega = \frac{v_{xy}}{r} \tag{7.3}$$

Eq. 7.2 can thus be expressed as:

$$m\,\omega^2\,r = q\,B_0\,\omega\,r, \text{ or simply}$$

$$\boxed{\omega_c = \frac{qB_0}{m}} \quad \text{(S.I. units)} \tag{7.4}$$

Equation 7.4 is the celebrated "cyclotron" equation, in which the "ion cyclotron frequency" is denoted as ω_c. The remarkable result of Eq. 7.4 is that all ions of a given mass-to-charge ratio, m/q, have the *same* ICR frequency, *independent of their initial velocity.* It is this property that makes the ICR phenomenon so useful for mass spectrometry, because translational energy "focussing" is not needed for precise determination of m/q.

Several useful conclusions follow directly from Eq. 7.4. First, at a representative static magnetic field value of 3.0 tesla, ICR frequencies for ions formed from typical molecules range from a few kHz to a few MHz (see Figure 7.2).

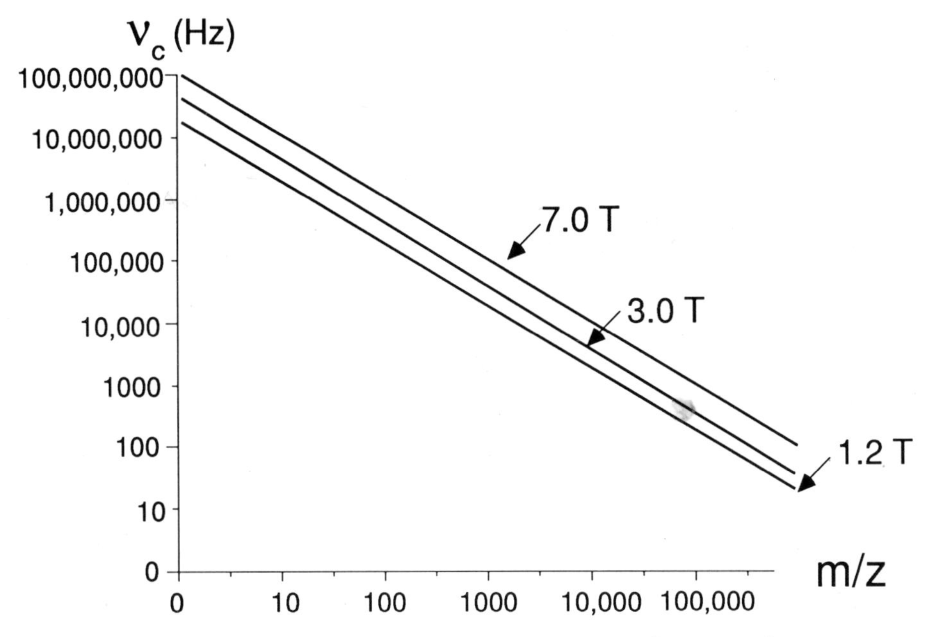

Figure 7.2 ICR orbital frequency, $\nu_c = \omega_c/2\pi$, in Hz, as a function of ionic mass-to-charge ratio, m/z, in atomic mass units (u) per multiple (z) of electronic charge, at each of three common magnetic field strengths: 1.2, 3.0, and 7.0 tesla. Note that ICR frequencies for ions in the usual "chemical" mass range (~15–10,000 u) typically lie between a few kHz and a few MHz.

Second, from the first derivative of Eq. 7.4 with respect to m,

$$\frac{d\omega_c}{dm} = \frac{-qB_0}{m^2} = -\frac{\omega_c}{m} \tag{7.5}$$

we obtain the useful relation,

$$\boxed{\frac{\omega_c}{d\omega_c} = -\frac{m}{dm}} \tag{7.6}$$

In other words, frequency resolution and mass resolution in ICR mass spectroscopy are the same (except for a minus sign). We shall return to derive mass resolution from frequency resolution, once we have established the form of the detected FT/ICR time-domain signal in Chapter 7.2.

We can rearrange Eq. 7.2 to yield the *ion cyclotron orbital radius*,

$$\boxed{r = \frac{m v_{xy}}{qB_0}} \tag{7.7}$$

of an ion of velocity, v_{xy}. For example, the average *x-y* translational energy of an ion in equilibrium with its surroundings at temperature, T (in K), is given by:

$$\frac{m\langle v_{xy}^2\rangle}{2} \cong kT \qquad (7.8)$$

in which k is the Boltzmann constant. Solving Eq. 7.8 for v_{xy}, and substituting back into Eq. 7.7, we obtain

$$r = \frac{1}{qB_0}\sqrt{\frac{mkT}{2}} \qquad (7.9)$$

The reader can quickly confirm that at room temperature, a typical singly charged ion of m =100 u in a magnetic field of 3 tesla (i.e., 30,000 Gauss) has an ICR orbital radius of ~0.04 mm. Even a singly charged ion of mass 10,000 u has an ICR orbital radius of only ~0.4 mm. Thus, ions of thermal energy formed from all but the largest molecules are confined by the magnetic field to conveniently small orbital radii for ICR excitation and detection (see Figure 7.3).

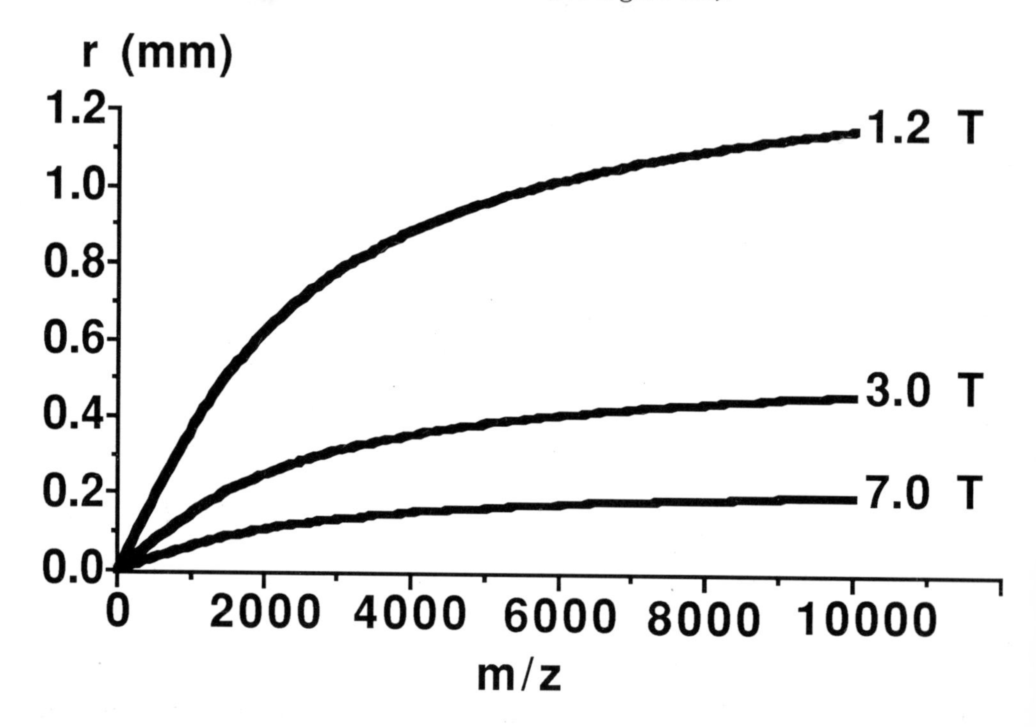

Figure 7.3 ICR orbital radius, r (Eq. 7.9), versus ionic mass-to-charge ratio, m/z (in u per multiple of electronic charge) at each of three typical magnetic field strengths: 1.2, 3.0, and 7.0 tesla. Note that even relatively heavy ions are confined to conveniently small-radius orbits by such magnetic fields.

Conversely, we can compute the *velocity* and *translational energy* of an ion excited (by as yet unspecified means) to a larger orbital radius. From Eq. 7.7,

$$v_{xy} = \frac{qB_0 r}{m} \qquad (7.10)$$

From Eq. 7.10, a singly charged ion of 100 u mass excited to an ICR orbital radius of 1 cm in a magnetic field of 3 tesla has a translational velocity, v_{xy} = 2.89 x 10^4 meters/second, corresponding to a translational energy,

$$\text{Kinetic energy} = \frac{m\, v_{xy}^{\,2}}{2} = \frac{q^2\, B_0^2\, r^2}{2\, m} \tag{7.11}$$

of 434 eV. Thus, ions can be "heated" to high translational energy even in a relatively small container, and then induced to break into smaller fragments by collision with neutral gas molecules (see below). The dependence of ion kinetic energy on ICR orbital radius and magnetic field strength is shown in Figure 7.4.

Finally, it is instructive to note that the excited ion of 100 u of the preceding paragraph will travel a distance of ~30 kilometers during a 1-second observation period! Since the accuracy of any mass measurement increases with ion path length, it is immediately clear that ICR offers inherently much higher mass resolution than that from a "beam" instrument of ordinary laboratory dimensions.

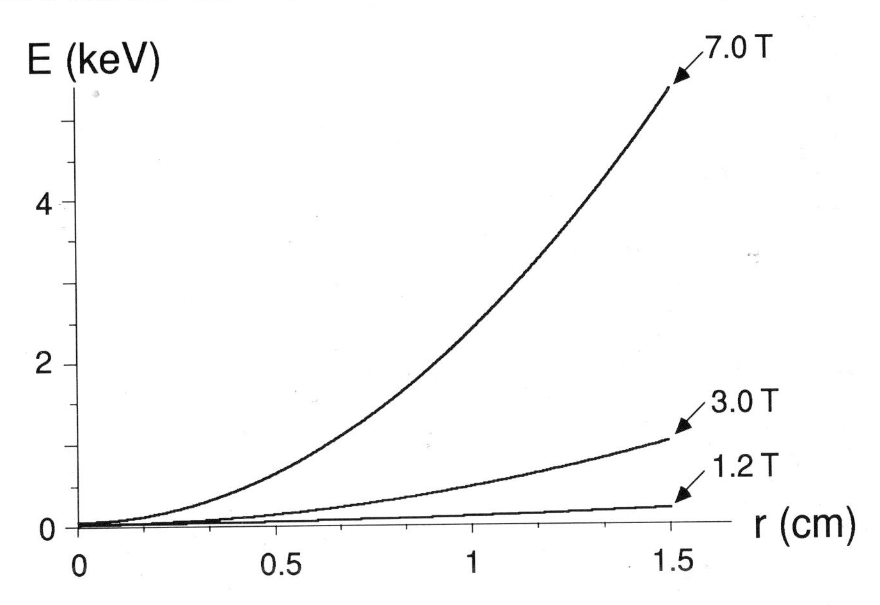

Figure 7.4 Ion translational energy (Eq. 7.11) as a function of ICR orbital radius, at each of three common magnetic field strengths: 1.2, 3.0, and 7.0 tesla. Note that ions can be accelerated to relatively high (≥ 1 keV) energy (as in collision-induced dissociation experiments) while still confined to relatively small (~1 cm) orbital radius (compare to previous Figure).

7.1.2 Trapping oscillation

A static magnetic field applied along the z-direction effectively confines ions in the x- and y-directions according to the cyclotron motion described in the preceding section. However, ions are still free to escape in the z-direction, parallel to the magnetic field. In order to prevent such escape, it is usual to apply a small (~1 volt) electrostatic potential to each of two "trapping" electrodes positioned at $z = \pm d/2$ away from the center of the ion trap. To a good approximation (see below), the electrostatic potential, $V(z)$, between two opposed flat electrodes (each at applied voltage, $+V_T$, with other electrodes grounded) varies quadratically with z-distance (near the center of the trap):

$$V(z) \cong \frac{V_T}{2} + \frac{k' z^2}{2}, \quad k' \text{ is a constant} \tag{7.12}$$

The electric field, $E(z)$, is obtained from the negative gradient (in this case, the first z-derivative) of the electrostatic potential:

$$E(z) = -\frac{dV(z)}{dz} = -k' z \tag{7.13}$$

Finally, the (restoring) force, $F(z)$, on an ion trapped between the two electrodes is given by

$$F(z) = m\frac{d^2 z}{dt^2} = -q k' z \tag{7.14}$$

The reader can immediately recognize Equation 7.14 as the familiar harmonic oscillator equation. Thus, ions trapped in a quadratic z-potential must oscillate back and forth along the z-direction at a natural "trapping" frequency,

$$\omega_T = \sqrt{\frac{k' q}{m}} \tag{7.15}$$

For a given geometrical arrangement of charged electrodes which produce an approximately quadratic z-potential, k' in Eq. 7.12 can be computed analytically. For example, a two-dimensional "quadrupolar" potential [which is well-approximated near the center of a cubic ion trap (Chapter 7.2.3) from which the plates at $y = \pm d/2$ have been removed] produces $V(z)$ with k' given by

$$k' = \frac{4V_T}{d^2} \tag{7.16}$$

From Eq. 7.15, the ion trapping frequency is then found to be

$$\omega_T = 2\pi\nu_T = \frac{2}{d}\sqrt{\frac{q V_T}{m}} \tag{7.17}$$

For a one-inch electrode separation ($d = 0.0254$ m) and a trapping electrode potential, $V_T = 1$ V, an ion of m/z = 100 has a trapping frequency of ~12.3 kHz, compared to its cyclotron frequency of ~461 kHz at 3.0 tesla. In general, the trapping frequency is much smaller than the ICR orbital frequency (see Problems), as shown in Figure 7.5.

7.1.3 Magnetron motion

A third natural "magnetron" motion results from the relatively slow mass-independent precession of an ion along a path of constant electrostatic potential, as shown in Figure 7.6. Magnetron motion arises in a natural way as one of two solutions to the equations of (transverse) motion of an ion in static electric and magnetic fields (see Problems). Although magnetron motion can be excited, either intentionally or as an unintentional consequence of cyclotron excitation (see below), it can generally be ignored in most FT/ICR applications.

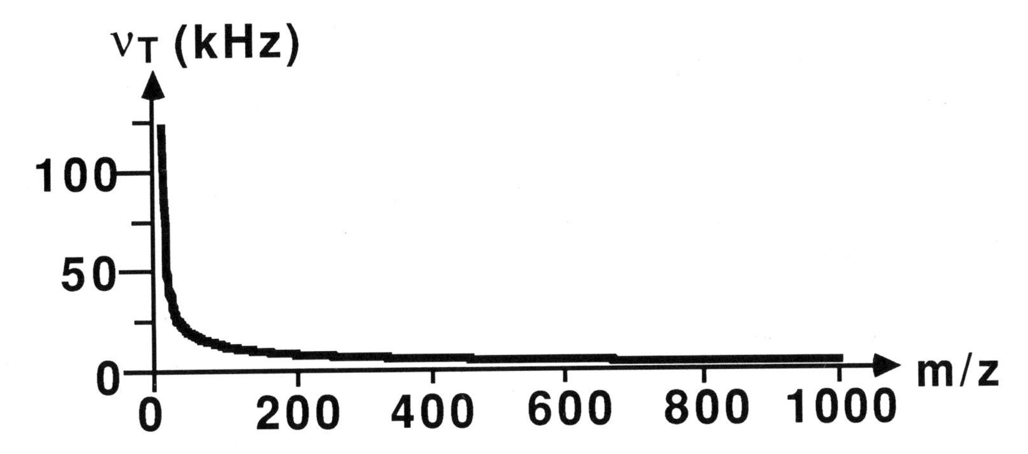

Figure 7.5 Trapping frequency, ν_T (Eq. 7.17), as a function of ionic mass-to-charge ratio, m/z, in u per multiple of electronic charge, for ions in a hollow cubic trap whose z-electrodes are separated by 2.54 cm and held at 1 V each. Note that the trapping frequency is typically much smaller than the corresponding ICR orbital frequency for that m/z ratio (compare to Figure 7.2).

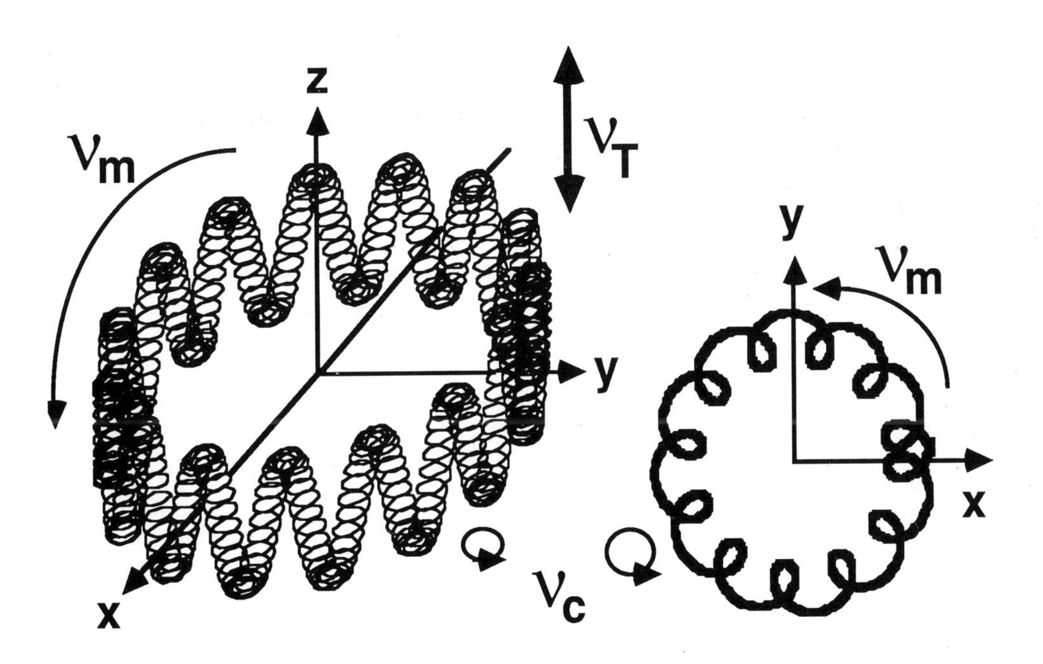

Figure 7.6 Schematic diagram of all three natural motions of an ion trapped by uniform magnetic and quadrupolar electric static fields: ω_c (cyclotron), ω_T (trap), and ω_m (magnetron). The magnetron motion is circular about a guiding center that follows a contour of constant electric potential. (Figure provided by M. Wang)

7.2 Static electromagnetic ion traps

It is important to recognize that ICR orbital motion does not by itself generate an observable electrical signal. At its instant of formation (e.g., by electron bombardment of a neutral gas) or arrival in the ion trap, the *phase* of the each ion's orbital motion is random—i.e., an ion may start its cyclotron motion at any point around the circle shown in the left diagram of Figure 7.7. Thus, any charge induced in either of two opposed detector plates will be balanced, on the average, by an equal and opposite charge induced by an ion whose phase is 180° different (i.e., an ion located on the "far" side of the same orbit).

7.2.1 Excitation of a coherent time-domain ICR signal

In order to create a signal on the detector plates, an ion packet whose cyclotron orbits are initially centered on the z-axis must be made *spatially coherent*, by moving the ion packet *off-center*. Spatial coherence is created by applying a oscillating *resonant* ($\omega = \omega_c$) *phase-coherent* electric field excitation, $\mathbf{E}(t)$, of the form,

$$\mathbf{E}(t) = E_0 \cos \omega_c t \; \mathbf{j} \tag{7.18}$$

As noted in Chapter 4.3, the linearly-polarized electric field of Eq. 7.18 can be analyzed into two counter-rotating components, $\mathbf{E}_L(t)$ and $\mathbf{E}_R(t)$,

$$\mathbf{E}(t) = \mathbf{E}_L(t) + \mathbf{E}_R(t) \tag{7.19a}$$

$$\mathbf{E}_L(t) = \frac{E_0}{2} \cos \omega t \; \mathbf{j} + \frac{E_0}{2} \sin \omega t \; \mathbf{i} \tag{7.19b}$$

$$\mathbf{E}_R(t) = \frac{E_0}{2} \cos \omega t \; \mathbf{j} - \frac{E_0}{2} \sin \omega t \; \mathbf{i} \tag{7.19c}$$

The radiofrequency electric field component rotating in the same sense as (and at the same frequency—i.e., "in resonance with") the ion of interest will push that ion continuously forward in its orbit, as shown in the right diagram of Figure 7.7. (The electric field component rotating in the opposite sense as the ion is so far off-resonance as to have virtually no significant effect.)

As the ion speeds up, its radius increases linearly with time, and the ion absorbs power according to the dot product:

$$A(t) = q \, \mathbf{E}(t) \cdot \mathbf{v}_{xy} \tag{7.20}$$

For a positive (or negative) ion (initially at rest) subjected to circularly-polarized on-resonance excitation (Eq. 7.19c or 7.19b) for a period, T_{excite}, the instantaneous rate of power absorption in the absence of any time-domain damping during the excitation is given by (see Problems):

$$\boxed{A(T_{excite}) = \frac{E_0^2 \, q^2 \, T_{excite}}{4m}} \tag{7.21}$$

Integration of Eq. 7.21 from time zero to time, T_{excite}, yields the total energy absorbed during the excitation period. If we further assume that all of the absorbed energy is converted into kinetic energy, then

$$\frac{m\,\omega_c^2\,r^2}{2} = \int_0^{T_{excite}} A(t)\;dt$$

$$= \frac{q^2\,E_0^2\,(T_{excite})^2}{8\,m} \qquad (7.22)$$

Substituting for $\omega_c = qB_0/m$ from Eq. 7.4, we can solve for r:

$$\boxed{r = \frac{E_0\,T_{excite}}{2\,B_0}} \qquad (7.23)$$

For example, an ion in a magnetic field of 3.0 tesla can be excited to a radius of 1.67 cm in 1.0 ms by a constant rf electric field of amplitude, 1 V/cm. Thus, ions can be excited to detectable ICR orbital radius by a relatively small rf electric field.

A delightful feature of Eq. 7.23 is that the orbital radius of the excited ion is *independent of* m/q ! Thus, all ions of a given m/q-range can be excited to the *same* ICR orbital radius, by application of an rf electric field whose magnitude is *constant with frequency*. In Section 7.4.2, we shall consider various practical ways to produce such optimal flat-power excitation.

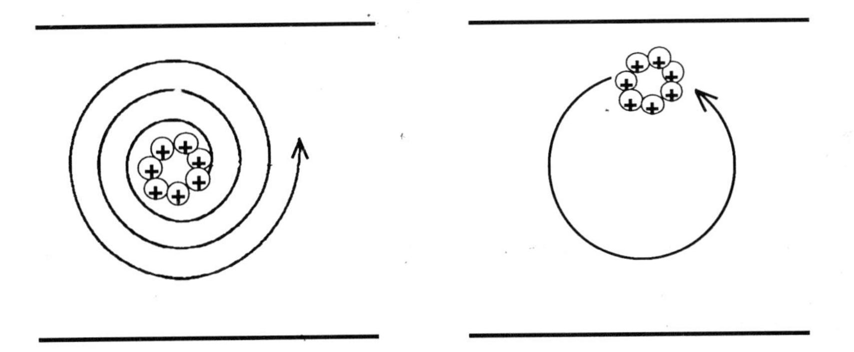

Figure 7.7 Incoherent ion cyclotron orbital motion (left) is converted to coherent (and therefore detectable) motion (right) by the application of a rotating electric field which rotates in the same sense and at the ICR frequency of the ions of that m/q value. (See text).

7.2.2 Detection of an ICR signal

Figure 7.7 suggests several methods for detecting ions from their excited ICR orbital motion. First, one could excite ions of a given m/q until they strike the detector plate(s), and measure the charge deposited on the plate(s). An ICR "omegatron" mass spectrometer based on that principle was developed ~1950, and it found some application for residual gas analysis of relatively low-mass species.

Alternatively, one could measure the power absorbed from the detector circuit during the excitation process (left-hand diagram of Figure 7.7), as the excitation frequency (or applied magnetic field strength) is slowly swept across the ICR frequency range (and thus the mass range) of interest. As in NMR (see Chapter 8), it proved experimentally simpler (and thus historically earlier) to vary the magnetic field strength at constant excite/detect frequency (marginal oscillator detector) than to vary the excite/detect frequency at constant magnetic field strength (with a capacitive bridge detector in ICR). The magnetic field-scanning ICR instrument based on power-absorption detection was developed in the mid-1960's, and about 40 were in use by the mid-1970's. A handful of capacitive-bridge detector instruments remains in use today.

Finally, one could first excite ions to an orbital radius somewhat smaller than the separation between the detector plates (left-hand diagram in Figure 7.7), and then turn off the excitation power and observe the "image current" resulting from the alternating charge induced in the detector plates by the coherent ICR orbital motion (right-hand diagram in Figure 7.7). That oscillating current can be converted to an oscillating voltage by passing the current through an impedance, and then amplified and digitized to give a time-domain discrete voltage signal, $f(t)$, of the form,

$$f(t) \propto \sum_{i=1}^{M} N_i \exp(-t/\tau_i) \cos(\omega_i t + \varphi_i) \qquad (7.24)$$

in which N_i, τ_i, ω_i, and ϕ_i are the number, time-domain exponential damping constant, ICR frequency, and initial phase of ions of M different m/q values. Image current detection, developed by M. B. Comisarow in 1973, is employed in all FT/ICR mass spectrometers. In the next sections, we shall consider in more detail the generation, detection, and features of image-current FT/ICR signals.

7.2.3 ICR ion traps (cubic, cylindrical, hyperbolic, orthorhombic, screened), and their effect on mass resolution and upper mass limit

From the preceding discussion, it is clear that an FT/ICR mass spectrometer requires at least one pair of opposed electrodes to excite coherent ICR orbital motion, one pair (which may be the same pair) of opposed electrodes to detect that motion, and some means for trapping the ions so that they do not escape along the magnetic field- (z-) axis. Because the detected signal voltage is many orders of magnitude smaller than the excitation voltage, it is experimentally convenient to separate the excitation and detection events in time as well as in space, by locating two pairs of orthogonal excitation and detection electrodes at $\pm a/2$ on the x-axis and $\pm b/2$ on the y-axis. Typically, $a = b$.

Although it is in principle possible to trap ions by use of spatially inhomogeneous magnetic field (namely, the "magnetic mirror" effect employed in plasma fusion experiments), it is more practical to add another pair of "trap" electrodes at $\pm d/2$ on the z-axis. Each "trap" electrode is charged to V_T volts of the same charge sign as the ions of interest (e.g., positive trap voltage to trap positive ions).

Several static electromagnetic ion trap designs for FT/ICR are shown in Figure 7.8. In each case, ions are trapped by applying a d.c. voltage, V_T, to each "trap" electrode, with the remaining plates electrically grounded. For example, a cubic trap, with static voltage, V_T, applied to the two electrodes at $z = \pm a/2$ and the remaining four electrodes at ground potential, produces an approximately *quadrupolar* static electrostatic potential, V_Q, near the center of the trap:

$$V_Q\,(x, y, z) = \frac{V_T}{3} - \frac{\alpha V_T x^2}{a^2} - \frac{\alpha V_T y^2}{a^2} + \frac{2\alpha V_T z^2}{a^2} \tag{7.25a}$$

or

$$V_Q\,(r, z) = \frac{V_T}{3} - \frac{\alpha V_T r^2}{a^2} + \frac{2\alpha V_T z^2}{a^2} \tag{7.25b}$$

in which $\alpha = 1.386$. [Note that the potential at the center of the cubic trap, i. e., $x = y = z = 0$, is simply the average of the potentials applied to the six plates: $(V_T + V_T + 0 + 0 + 0 + 0)/6 = V_T/3$.] [Traps of other symmetrical shapes also produce an approximately quadrupolar potential near the center of the trap, but with different coefficients relating $V_Q\,(x, y, z)$ to V_T.]

At first glance, the quadrupolar electrostatic field would appear to make the ICR problem much more complicated, because such a potential produces a *radial electric field*, $E\,(r)$,

$$E\,(r) = -\frac{\partial V_Q\,(r, z)}{\partial r} = \frac{2\alpha V_T\, r}{a^2} = E_0\, r, \tag{7.26}$$

in which $E_0 = \dfrac{2\alpha V_T}{a^2}$ for the cubic trap case illustrated here.

The radial electric field acting on the ion produces an *outward*-directed electric force which opposes the *inward*-directed Lorentz magnetic force from the applied magnetic field. Equation 7.4 must therefore be modified accordingly —for simplicity, the center of the ICR orbit is taken to lie on the z-axis:

Radial Force = Inward Force + Outward Force

$$m\omega^2 r = qB_0\omega r - qE_0 r \tag{7.27}$$

Several important conclusions follow from Eq. 7.27. First, when we solve again for the "natural" frequency of the motion, we find that there are now *two* "natural" frequencies (see Problems):

$$\omega_0 = \frac{qB_0 + \sqrt{q^2 B_0^2 - 4mqE_0}}{2m} \quad \text{(Cyclotron frequency)} \tag{7.28a}$$

$$\omega_m = \frac{qB_0 - \sqrt{q^2 B_0^2 - 4mqE_0}}{2m} \quad \text{(Magnetron frequency)} \tag{7.28b}$$

In the limit of zero trapping voltage, $E_0 \to 0$, and Eq. 7.28a gives $\omega_0 = qB_0/m$, which we recognize as the "cyclotron" orbital frequency of Eq. 7.4. Eq. 7.28b represents "magnetron" motion (Figure 7.6), in which the ion precesses around a surface of constant electrostatic potential. Because the radial electric force is typically much weaker than the (radial) magnetic force in Eq. 7.27, the magnetron frequency of Eq. 7.28b is typically much smaller than the cyclotron orbital frequency of Eq. 7.28a (see Problems).

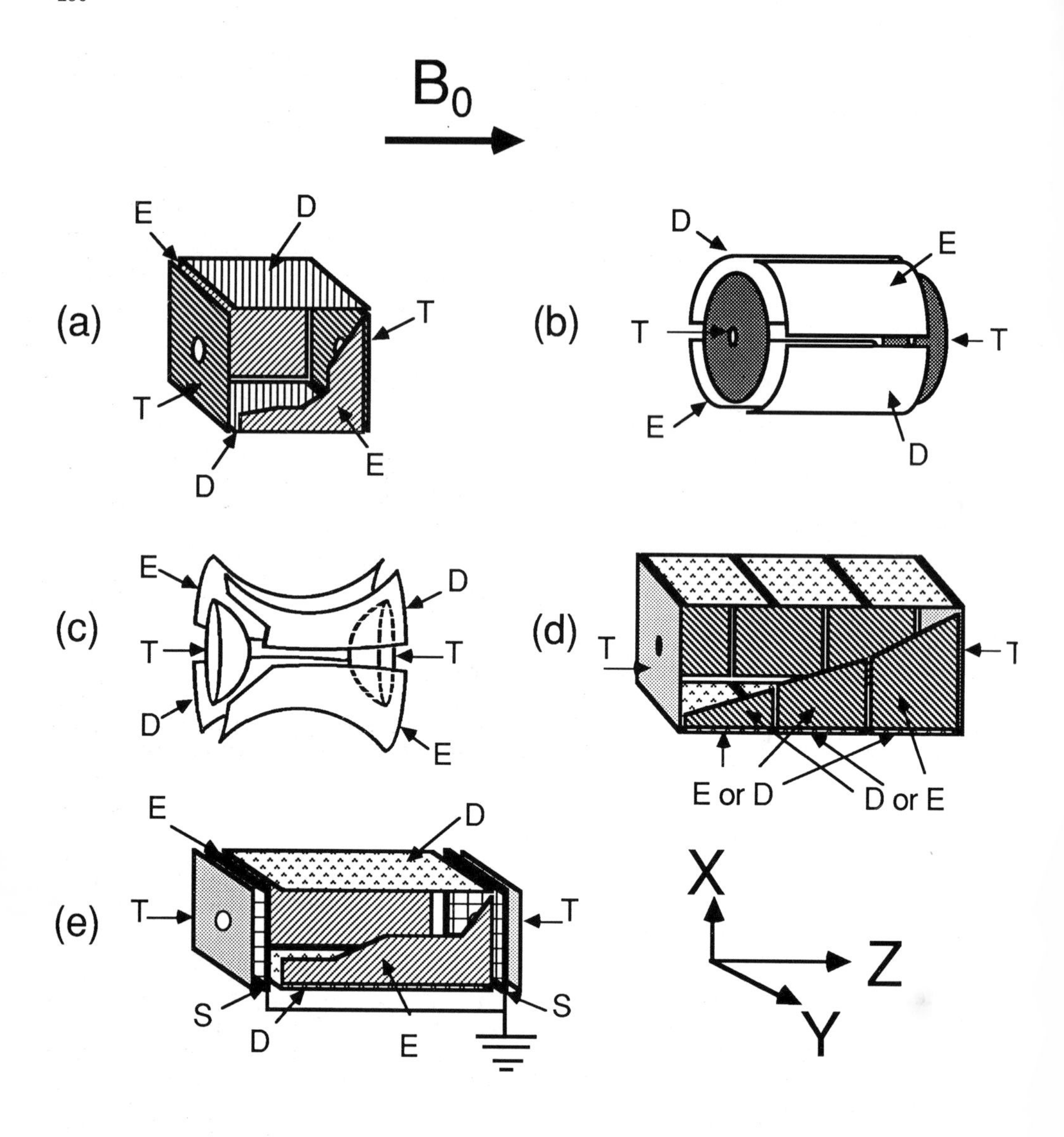

Figure 7.8 Static electromagnetic ion traps of various geometry. (a) Cubic; (b) Cylindrical; (c) Hyperbolic; (d) Multi-section; (e) Screened. E, D, T, and S denote excite, detect, trap, and screen electrodes. Traps (a) and (b) approximate the quadrupolar electrostatic potential of trap (c). Traps (d) and (e) are designed to reduce the electric field within the trap. See text for further discussion of the relative advantages of each design.

Second, the imposition of a trapping potential *lowers* the ICR orbital frequency (i.e., shifts ω_0 to *lower* frequency) because the radial electric field has the same effect as a decrease in magnetic field strength (see Eq. 7.27 and Problems). Eq. 7.28a is the basis for mass calibration in FT/ICR (see Chapter 7.4.1).

Third, even though the electric radial force increases with increasing ICR orbital radius, the factor, r, cancels from each term in Eq. 7.27, so that *ICR orbital frequency is independent of ICR orbital radius* in a quadrupolar electrostatic field. Therefore, we might expect highest mass resolution for the hyperbolic trap (Figure 7.8c), because ions anywhere within the region bounded by its electrodes should have the same ICR frequency.

The final consequence of Eq. 7.27 is somewhat more subtle. In order to prevent ions from escaping along the z-axis, we have applied an electric field along the z-direction. However, Gauss's law of electrostatics requires that if electric field lines enter a closed region (in our case, the ion trap), then electric field lines must also leave that region if the bounded volume is to be electrically neutral. Thus, a static voltage applied only to the trap electrodes to generate an electric field pointing *inward* from $\pm a/2$ on the z-axis, must necessarily generate an electric field directed radially *outward* as in the special quadrupolar example of Eq. 7.25. The importance of the outward-directed radial electric field is that it limits the highest mass ion which can be contained in the trap, as we now show.

We begin from our previously computed ICR orbital radius for a "thermal" ion in the absence of any applied electric field (Eq. 7.9). Since we now recognize that an ion trap must contain "side" electrodes in order to provide for excitation and detection of the ICR signal, it is clear that the ultimate upper mass limit is the mass at which the ICR radius reaches the radius of the trap. Rewriting Eq. 7.9,

$$m_{\text{upper}} = \frac{2\,q^2 B_0^2\, r^2}{k\,T} \tag{7.29}$$

we find that at 3.0 tesla, the upper mass limit for a singly charged ion would be ~10.8 million atomic mass units in a trap of one-inch cross-sectional diameter, if we could ignore the radial *electric* field.

However, the radial electric field produced by the trap electrodes effectively reduces the strength of the magnetic field and therefore (see Eq. 7.29) further reduces the upper mass limit. The effect is shown in Figure 7.9 at three common magnetic field strengths. For example, the upper mass limit for singly charged thermal ions at 3.0 tesla is ~40,000 amu in a 1" (0.0254 m) cross-sectional diameter quadrupolar trap at 1 V trapping potential. The actual upper mass limit will of course be much smaller, because ions must begin with an ICR orbital radius much smaller than that of the trap in order that a detectable coherent signal be generated by exciting their cyclotron orbital motion to a larger radius for detection.

The remaining traps in Figure 7.8d,e are designed to reduce the electric field to *zero* in the detection region, rather than to shape the electric field to the *quadrupolar* form of Eq. 7.25. The screened trap (Figure 7.8e) should significantly extend the ICR upper mass limit, by reducing the radial electric field magnitude by a factor of ~30.

Although the discussion evolving from Figure 7.8 was limited to *static* electric fields, a similar problem arises from the curved *radiofrequency* electric field generated by finite-size transmitter electrodes. When an rf voltage is applied between the two excitation (transmitter) electrodes in the traps of Figure 7.8, a component of that rf electric field "leaks" into the z-direction, and can push ions out of the trap in the z-direction ("z-ejection"). Fortunately, the rf electric field can be effectively flattened (and its z-component virtually eliminated) by adding a small number of "guard rings" as in the original omegatron.

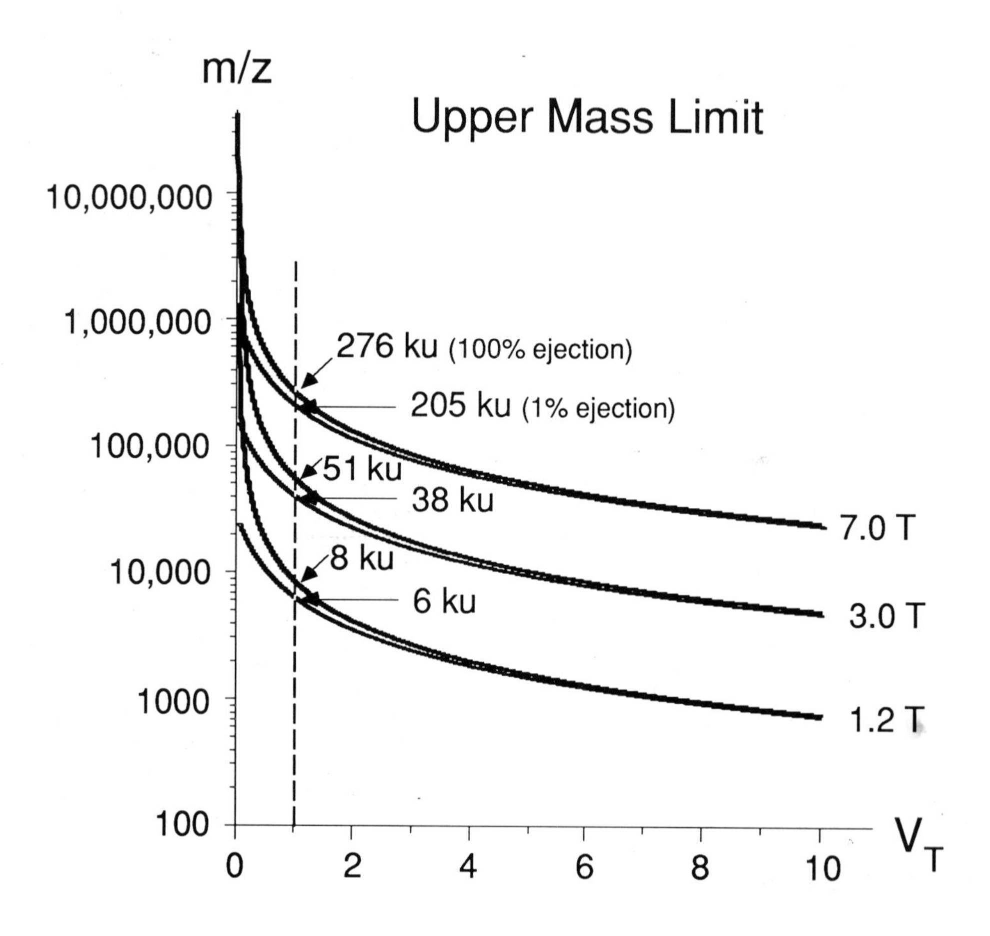

Figure 7.9 Upper mass limit for a singly charged ion in a Penning trap (i.e., homogeneous static magnetic field plus quadrupolar electrostatic potential), as a function of trapping voltage, V_T, for each of three common magnetic field strengths. At $V_T = 1$ V, each upper arrow denotes the highest m/z for which an ion has a stable orbit, independent of ion initial velocity (temperature); each lower arrow denotes the highest m/z for which the radius of stable ICR orbits for 99% of thermal 25 °C ions fit within the 1" separation between the detector electrodes.

Finally, as noted in Chapter 4.3.2, *quadrature excitation* can be realized in ICR by use of *both* pairs of side electrodes for excitation, with a 90° phase difference between the two channels [e.g., single-frequency excitation with cos ωt on one pair of plates and sin ωt on the other pair]. Similarly, *quadrature detection* can be realized either by independent detection on both pairs of side electrodes or (in heterodyne-mode) by dividing the signal from one pair of detector electrodes in half and phase-shifting one of the two signals by 90° before (complex) FFT.

7.2.4 Ion formation at high pressure with detection at low pressure: dual-trap; external injection of ions

As we shall learn in the next section, ion-molecule collisions and reactions broaden FT/ICR mass spectral peaks and thereby reduce mass resolution. Therefore, it is desirable to conduct ICR *detection* at *low pressure* ($\leq 10^{-8}$ torr). However, it may be more convenient or necessary (as in fast atom bombardment or gas chromatography) to *form* ions at *high pressure* ($\geq 10^{-6}$ torr). The two requirements are incompatible with any of the single-trap designs of Figure 7.8.

There are two general ways to combine high-pressure ion formation with low-pressure ICR excitation/detection, as shown in Figure 7.10. One method (Figure 7.10, top) is to connect two single-section ion traps together along the magnetic field (z) axis, with a small (~1-3 mm) hole in the gas conductance plate separating the two traps. Ions formed in the high-pressure trap can be pushed along the z-axis to the low-pressure trap by applying suitable d.c. potentials to the trap electrodes. Neutrals enter the low-pressure trap only slowly through the small hole. In the dual-trap design, a pressure differential of ~100-fold can be maintained between two 10 cm diameter chambers connected by a 2 mm diameter hole.

Alternatively (Figure 7.10, bottom), it is sometimes easier to form ions in a chamber far removed (in z-direction) from the center of the (usually super-conducting solenoidal) magnet. Ions may then be extracted through a small hole and accelerated toward the ion trap by use of charged acceleration lenses. Unfortunately, unless all ions are aimed precisely on-axis, the large magnetic field gradients between the ion source and the ion trap act as a "magnetic mirror" to reflect ions away from the trap. The magnetic mirror effect may be defeated, either by accelerating the ions so strongly that they penetrate the gradient (with subsequent deceleration just before the ions reach the trap), or by keeping the ions close to the z-axis with long quadrupolar rods or other ion-focussing lenses as the ions pass through the magnetic field gradient. An additional complication of either dual-trap ion transfer (see Giancaspro & Verdun in Further Reading) or external ion injection is that ions of different m/z ratio arrive at different times. At this writing, ion trap design continues to be in an active state of development in FT/ICR mass spectrometry.

7.3 Damping of the time-domain ICR signal

7.3.1 Types of damping: homogeneous vs. inhomogeneous

We know from Chapter 4.2. that the frequency-domain peak width for a time-domain exponentially damped sinusoidal signal (e.g., the time-domain ICR signal of Eq. 7.24) is inversely related to the time-domain exponential damping constant. In this section, we point out that time-domain signals can be exponentially damped in either of two very general ways: *homogeneous* and *inhomogeneous*. The two line-broadening mechanisms are nicely illustrated in ICR by a system of two ions of different initial phase, as shown in Figure 7.11.

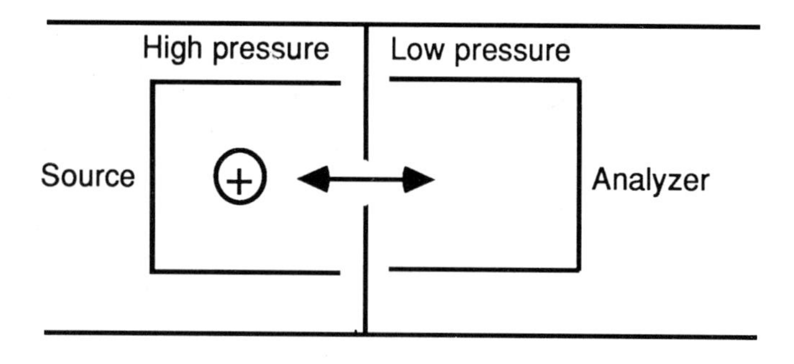

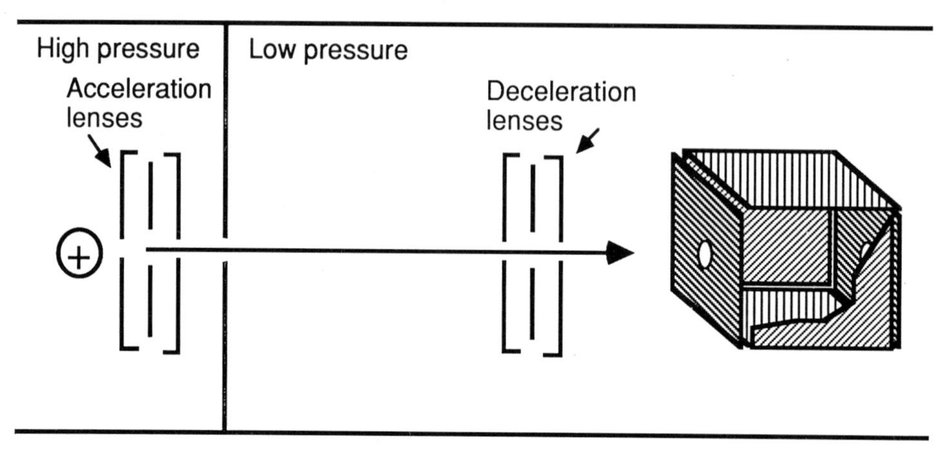

Figure 7.10 Two methods for producing ions in a high-pressure "source", with subsequent excitation/detection in a low-pressure "analyzer". Left: Dual-trap, in which ions can move freely between two regions separated by a small aperture, but neutrals cannot. Right: External "source" with electric-field-induced acceleration/deceleration and/or focussing to keep the ions on-axis as they move toward the ion trap in the center of a solenoidal magnet.

7.3.2 Damping ("relaxation") mechanisms

Excitation of an initially incoherent ensemble of ions of different initial position, speed, and phase produces a spatially localized (and therefore phase-coherent) ion packet. The ICR signal resulting from the detector electrode oscillating charge induced by that orbiting packet can decrease *inhomogeneously* by loss of *phase-coherence.* Inhomogeneous relaxation in turn occurs when different ions in the packet have different ICR frequencies. For example, ions located in the center of the trap at their instant of formation will undergo trapping oscillations of relatively small z-amplitude, whereas ions initially formed near the trap electrodes will oscillate over almost the whole z-distance between the trap electrodes. During the detection period, the trapping motion effectively averages out differences in electric or magnetic field over z-positions ranging from zero to the trapping oscillation z-amplitude. For example, mass resolution

>500,000 has been observed in a magnet for which the magnetic field inhomogeneity was about 1 part in 40,000. However, ions of different z-amplitude will nevertheless experience different average effective magnetic field strength, if the radial electric field strength varies significantly with z. Similarly, ions of different initial speed (and therefore different initial ICR orbital radius) will experience different average electric and magnetic field—radial inhomogeneity is most significant for high-temperature and/or high-mass ions of large initial (pre-excitation) ICR orbital radius. (In general, electric field inhomogeneity poses a greater problem than magnetic field homogeneity, at least for superconducting solenoidal magnets.) Inhomogeneous line-broadening can therefore be reduced by reducing electric and magnetic field inhomogeneity.

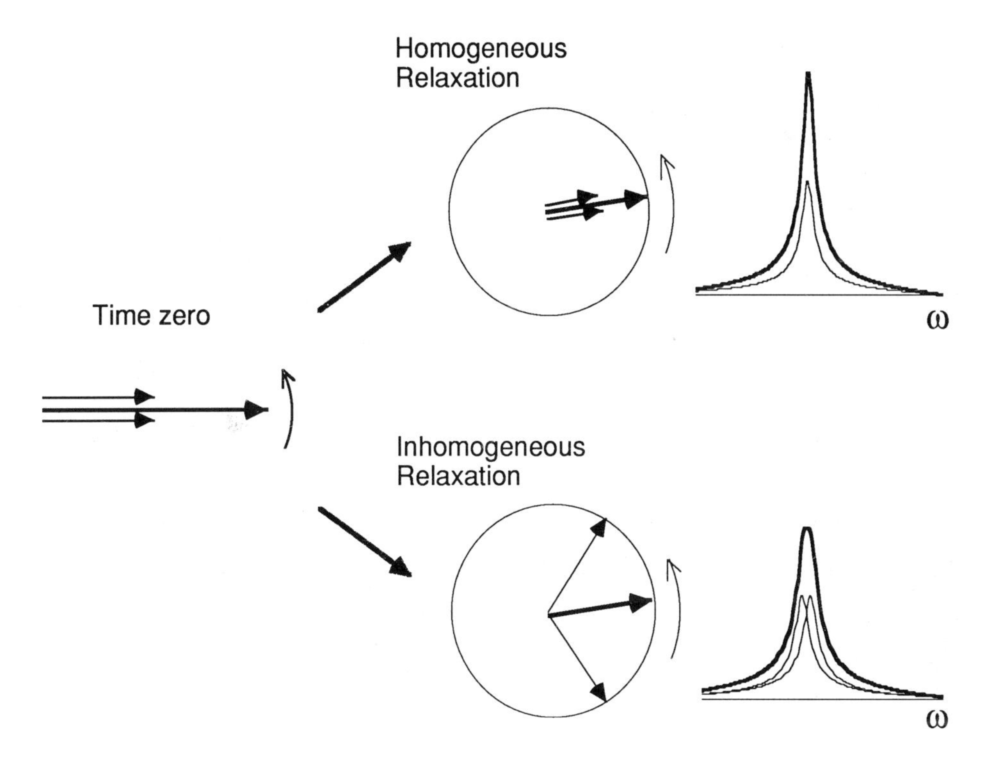

Figure 7.11 Homogeneous and inhomogeneous relaxation, illustrated by the ICR behavior of two excited ion packets of different initial phase at time zero. (The magnetic field points into the plane of the paper.) Each ion packet is represented by an electric monopole vector. The ICR signal is proportional to the vector sum of the two monopoles, represented as the radius of the solid circle in each case. In homogeneous relaxation, both packets decrease in magnitude without change in phase angle. In inhomogeneous relaxation, the ICR frequencies of the two ion packets differ slightly, and the ICR signal therefore decreases with time as the phases of the two packets diverge—even in the absence of collisions which reduce the magnitudes of the two monopoles. The corresponding frequency-domain spectra are shown at the right. [Adapted in part from M. B. Comisarow in *Lecture Notes in Chemistry*, Vol. 31, *Ion Cyclotron Resonance Spectrometry II*, Eds. H. Hartmann and K.-P. Wanczek (Springer-Verlag, Berlin, 1982)]

Homogeneous ICR relaxation for a packet of ions of identical m/z and identical ICR orbital frequency is the result of an exponential decrease in the *magnitude* of the electric (rotating) monopole signal formed from the vector sum of the electric monopoles of its constituent ions. If we neglect ion leakage out of the trap, homogeneous relaxation can arise only from loss of ICR signal due to ion-neutral or ion-ion collisions. If each ion-neutral collision results in a chemical reaction, or if low-mass ions collide with high-mass neutrals so that each ion-neutral collision effectively randomizes the phase of the ion, the effect of each ion-neutral collision is to remove that ion from the coherent packet. In that case, the time-domain signal for $N(t)$ ions of a given m/z value takes the form,

$$f(t) \propto N(t) \cos \omega_c t = N_0 \exp(-t/\tau) \cos \omega_c t \quad (7.30a)$$

in which N_0 is the number of ions at the beginning of the detection period, and

$$\frac{1}{\tau} = \xi = \frac{m_{ion}\ m_{neutral}}{m_{ion} + m_{neutral}} \nu_{collision} = \text{"reduced" collision frequency} \quad (7.30b)$$

in which $\nu_{collision}$ is the number of ion-molecule collisions per second. When $m_{ion} << m_{neutral}$, the observed ICR orbital frequency (see Eqs. 7.30) is simply ω_c.

$$\omega_{observed} = \omega_c \quad \text{(velocity randomized on each collision)} \quad (7.30c)$$

However, if $m_{ion} >> m_{neutral}$ (e.g., a large peptide ion colliding with helium background gas molecules), then each collision slows the ion but doesn't change the ion *direction* very much. Thus, the coherent ion packet spirals down to zero ICR orbital radius, even though the *number* of coherent ions remains essentially the same during the detection period. In this limit, the appropriate model is an equation of motion in which ion velocity is *frictionally* damped (i.e., reduced to zero in many small steps) according to the equation of motion,

$$m \frac{d\mathbf{v}}{dt} = q\mathbf{E}(t) + q(\mathbf{v} \times \mathbf{B}) - \xi \mathbf{v} \quad (7.31a)$$

Equation 7.31a is simply our familiar damped harmonic oscillator, for which the "natural" frequency is

$$\omega_{observed} = \sqrt{\omega_c^2 - \xi^2} \quad \text{(frictional loss of ion velocity)} \quad (7.31b)$$

Evidence for the validity of the model of Eq. 7.31a is McLafferty *et al.*'s experimental observation that high-mass ions can be re-excited after a detection period, because the ions have relaxed back to the center of the trap. It is interesting to note that the observed ICR frequency differs slightly (compare Eqs. 7.30c and 7.31b) for low-mass and high-mass ions because of their different relaxation mechanism.

Finally, it is worth noting that the ion-molecule collision mechanism itself varies with ion speed. At low ion speed (e.g., thermal ions with ICR orbital radius much less than 1 mm, at 3 tesla), an ion induces an electric dipole moment in the neutral. The $(1/r_0^4)$ ion/induced-dipole "Langevin" potential (r_0 here is the distance between ion and neutral) causes the ion to spiral toward the neutral until collision occurs. In the Langevin limit, ion-neutral collision rate is essentially independent of ion speed—a faster-moving ion encounters more neutrals, but must approach a given neutral more closely in order to spiral toward it for a collision. However, at high ion speed (corresponding to excited ICR ions with ICR orbital radius >2 mm at 3 tesla), the collision mechanism is better described by a hard-sphere model, for which the collision frequency increases with increasing ion speed. In other words, ICR spectral line width is essentially independent of ICR orbital radius until the radius exceeds a few mm, beyond which ICR line width increases with increasing ICR orbital radius. Thus, at a given pressure, ICR frequency (or mass) resolution is in principle highest at low ICR orbital radius (if we ignore signal-to-noise ratio, which increases with ICR radius).

7.3.3 Coulomb (space-charge) ion-ion repulsions: ICR frequency shift and mass spectral peak broadening

When more than ~10,000 ions are present in the trap, it becomes necessary to consider the static and dynamic effects of ion-ion repulsions. It can be shown that in a spatially uniform static electromagnetic field, Coulomb repulsions between ions do not affect the ICR orbital frequency of the center of mass of the ion packet. However, in the spatially *non-uniform* electromagnetic field of a typical ion trap, Coulomb repulsions can shift and broaden FT/ICR mass spectral peaks, by pushing like-charge ions apart into regions of different applied external electric or magnetic field.

One should also consider ICR signal relaxation due to ion-ion collisions. Unfortunately, there is no simple analytical treatment for either the static or dynamic effect of multiple-ion Coulomb interactions, and one must resort to computer-intensive trajectory calculations of each of 10,000 or more ions in a trap of given geometry. Experimentally, it is observed that ICR mass spectral signal-to-noise ratio increases with number of ions up to a limit of about 10^5 ions (at 3 tesla), beyond which the spectral peaks broaden and distort because of ion-ion Coulomb repulsions and close encounters. Supercomputers offer a means for future systematic exploration and characterization of these effects.

7.4 FT aspects of ICR

7.4.1 Mass resolution, mass accuracy, and mass calibration

The cyclotron resonance Eq. 7.4 gives a direct relation between the experimentally measurable ICR frequency and the most important molecular property (i.e., mass, from which chemical formula may be inferred) available from an ICR experiment. Thus, ultra-precise estimation of the true (continuous) FT/ICR spectral peak position (frequency) from the discrete spectrum is much more important in FT/ICR than in other forms of FT spectroscopy. We have previously discussed the effects of noise and *discrete* sampling (Chapter 5.1, Chapter 6). In this section, we discuss the effect of *analog* peak broadening on spectral resolution.

We have already shown (Eq. 7.6) that ICR frequency resolution and mass resolution are the same (except for sign), over a sufficiently small range of frequency or mass, $\Delta\omega$ or Δm:

$$\frac{\omega}{\Delta\omega} = -\frac{m}{\Delta m} \tag{7.32}$$

In FT/ICR mass spectrometry, $\Delta\omega$ is usually defined (Chapter 4.2.1) to be the frequency-domain full spectral peak width at half-maximum peak height, since a valley just begins to appear between two peaks of equal height and shape when they are separated by slightly more than $\Delta\omega$.

Since the frequency of an FT/ICR mass spectral peak is approximately qB_0/m, mass resolution in Eq. 7.32 can be expressed as

$$\boxed{\frac{m}{\Delta m} = -\frac{qB_0}{m\,\Delta\omega}} \tag{7.33}$$

It is useful to evaluate ICR mass resolution for either $T >> \tau$ (high-pressure limit) or $T << \tau$, where T is the time-domain acquisition period and τ is the time-domain exponential damping constant (relaxation time—Eq. 7.30b). The results (see Problems) are shown in Table 7.1 for typical FT/ICR spectral line shapes.

Table 7.1 FT/ICR mass resolution formulas (SI units).

$\frac{m}{\Delta m}$	Low-pressure ($T << \tau$)	High-pressure ($T >> \tau$)
Absorption-mode	$\frac{0.264\, q B_0 T}{m}$	$\frac{q B_0 \tau}{2m}$
Magnitude-mode	$\frac{0.132\, q B_0 T}{m}$	$\frac{q B_0 \tau}{2\sqrt{3}\, m}$

In the low-pressure limit, FT/ICR mass spectral peak width, $\Delta\omega$, is independent of m/q. Even in the high-pressure limit, $\Delta\omega = 2/\tau$, the peak width is determined by the ion-neutral collision frequency, which does not vary strongly with m/q. Thus, as m/q increases, the peak *widths* remain relatively constant, but the peaks are *closer together*, since ICR frequency varies inversely with m/q. That is why (see Eq. 7.33 and Table 1) ICR mass resolution (at constant B_0, for a given T and τ) varies *inversely* with m/q, as shown in Figure 7.12.

It is worth noting that Figure 7.12 applies to other magnet-based mass separators (e.g., "magnetic sector" mass spectrometers). However, in magnetic sector mass spectrometry (or in the original ICR mass spectrometers based on marginal oscillator detection at constant frequency) it is usual to vary the magnetic field strength in order to bring ions of a given m/q into focus (or resonance). In that case, mass resolution follows one of the dashed lines in Figure 7.12. Thus, FT/ICR achieves the highest mass resolution possible for magnet-based mass analysis, by operating at maximum magnetic field strength throughout the mass range. (A fixed-field magnet also produces a spatially more homogeneous field.)

In the absence of electric space charge and trapping potentials, measurement of the frequency of a *single* ion of known mass would serve to calibrate the magnetic field strength, from which the m/q values of all other ions in the mass spectrum could be computed from Eq. 7.4. However, the introduction of trapping potential (see Problems) and space charge changes the ICR mass/frequency relation to the form shown in Eq. 7.28a, from which Eq. 7.34 can be derived. A and B are constants obtained by fitting a particular set of ICR mass spectral peak frequencies for ions of at least two known m/q values to Eq. 7.34.

$$m/q = \frac{A}{\nu_0} + \frac{B}{\nu_0^2} \tag{7.34}$$

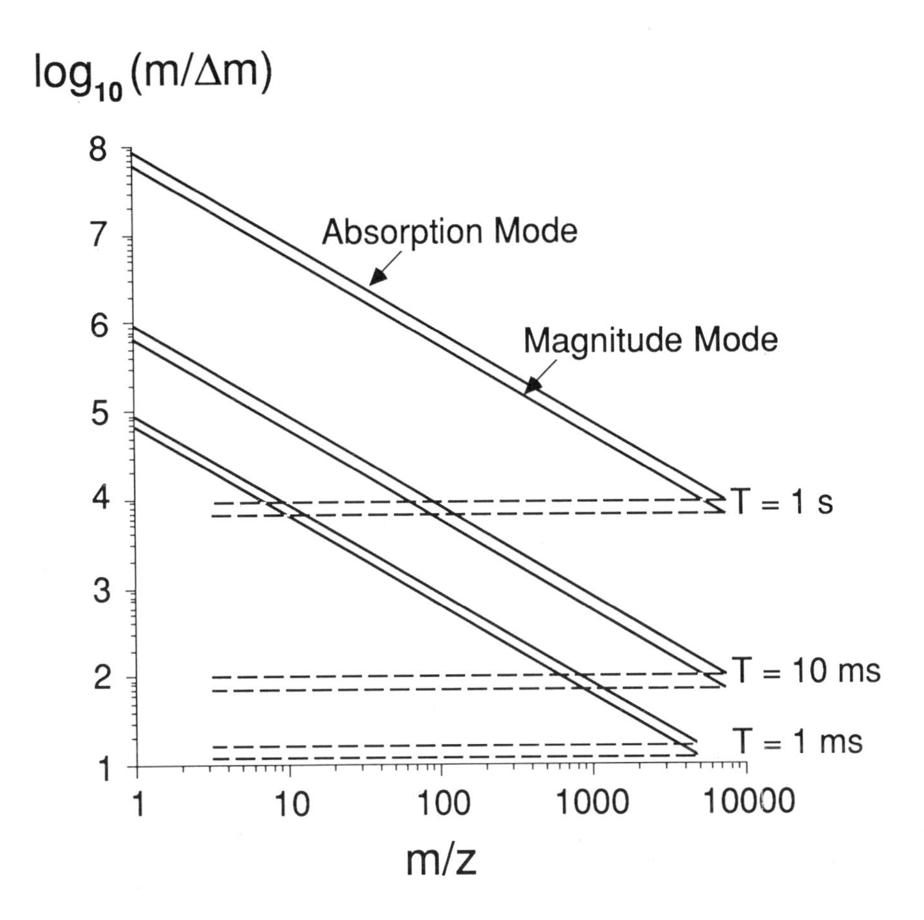

Figure 7.12 FT/ICR theoretical mass resolution at 3.0 tesla as a function of mass-to-charge ratio, m/z, in atomic mass units per electronic charge, for the indicated line shapes, in the low-pressure limit (i.e., no ion-neutral collisions during the detection period), for each of three time-domain acquisition periods, T. Although mass resolution decreases with increasing m/z, mass resolution can be spectacularly high if the time-domain acquisition period is long enough. The dashed lines indicate mass resolution for magnetic field-swept instruments (see text).

7.4.2 Pulsed single-frequency, frequency-sweep, and stored waveform inverse FT (SWIFT) excitation

The first and simplest excitation waveform for FT/ICR spectroscopy was a single-frequency rectangular pulse ("impulse", "burst") excitation. The frequency-domain spectrum of that time-domain waveform is the "sinc" function described in Chapter 2.1.2. Of course, an actual experimental pulse will exhibit a finite (typically exponential) rise and fall period when the power is turned on and off. However, the magnitude-mode spectrum is qualitatively similar and only slightly broadened compared to a hypothetical rectangular pulse (see Problems), as shown in Figure 7.13a,b.

The problem with pulsed excitation is obvious from Figure 7.13a,b: namely, that the excitation magnitude varies significantly with frequency. Since ICR signal strength is proportional to ICR orbital radius, which is in turn proportional to excitation magnitude, we seek a flat excitation magnitude spectrum in order to produce a mass spectrum whose peak areas accurately reflect the relative abundances of ions of different m/q. Comisarow therefore introduced frequency-sweep ("chirp") excitation (Chapter 2.1.6) to excite ICR signals over a wide m/q range with approximately flat frequency-domain magnitude (Figure 7.13c).

It is possible to excite ICR signals by irradiation with random or pseudo-random time-domain noise. Unfortunately, the resulting frequency-domain excitation magnitude spectrum is also noise (i.e., random in magnitude). The magnitude spectrum of a random noise source approaches a smooth Gaussian after a very large number of accumulated signals, but is still not flat. The spectrum from pseudo-random noise can be made flat, but only by accumulating a series of N suitably incremented pseudorandom signals to define excitation power at N frequency-domain values. Since N in FT/ICR can be 128K or greater, pseudo-random noise excitation is generally not feasible.

The preferred excitation method for FT/ICR is the stored waveform inverse FT method discussed in Chapter 4.2.3. It achieves the flattest and most frequency-selective excitation magnitude spectrum that is possible for a given time-domain excitation period. SWIFT excitation profiles for the three most general FT/ICR experiments are shown in Figure 7.13d-f.

7.4.3 Direct-mode *vs.* heterodyne-mode experiments; multiple foldover

FT/ICR spectroscopy is unique in that the detected signal may be sampled either directly ("direct-mode") or after passing through a mixer/low-pass-filter ("heterodyne-mode") as shown in Figure 4.22. The choice between the two modes depends upon: (a) the number of available time-domain data points, (b) the maximum available digitizer speed, (c) the lowest m/q value (which determines the highest ICR frequency) to be included in the spectrum, and (d) the time-domain damping constant (i.e., relaxation time, τ, which is inversely related to the gas pressure in the ion trap) for the ICR signal.

For example, *direct-mode* detection at 3.0 tesla of a wide-range positive-ion mass spectrum extending down to H_2O^+ at m/z = 18 requires analog-to-digital conversion (ADC) at a rate of at least $2\nu_c \approx 5.14$ MHz to avoid foldover. Thus, if the available random-access-memory (RAM) for storing the time-domain signal is 64K words, then the maximum time-domain acquisition period is only T = 12.75 ms.

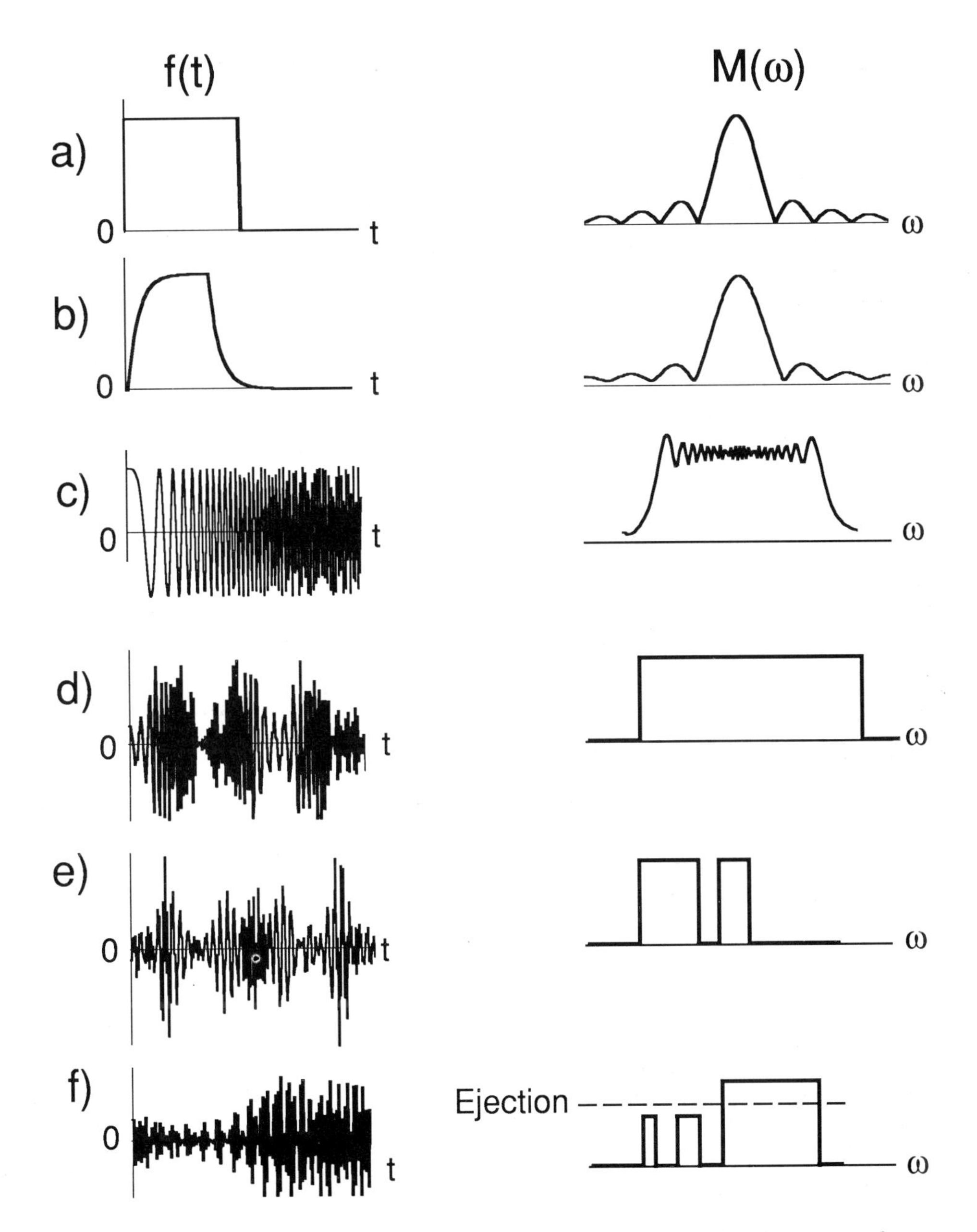

Figure 7.13 Frequency-domain magnitude spectra of several FT/ICR time-domain excitation waveforms. (a) Rectangular pulse; (b) Rectangular pulse with exponential rise and fall; (c) frequency-sweep; (d) SWIFT for broad-band excitation; (e) SWIFT for multiple-ion simultaneous excitation or ejection; (f) SWIFT for MS/MS (see below).

Even in the zero-pressure limit ($T << \tau$), the maximum available magnitude-mode mass resolution (see Table 1) cannot exceed ~27,000. Moreover, mass resolution varies inversely with mass. Thus, mass resolution at m/z = 700 in the same spectrum drops to ~700, or just enough to distinguish between m/z = 700 and 701. Thus, direct-mode detection is suitable for wide-range mass spectral detection, but at the cost of limited mass resolution at the high-mass end of the spectrum.

In order to achieve the highest possible mass resolution, we need to acquire time-domain data for $T \geq 3\tau$. If the available RAM is insufficient for direct-mode detection, then *heterodyne-mode* detection is appropriate. For example, if we seek a (high-pressure limit) mass resolution of ~1,000,000:1 at m/z = 700, then at 3.0 tesla (see Table 1), we need to acquire an undamped signal for $T \geq 3 \times 8.34 \approx 25$ s (requiring a pressure of $<10^{-9}$ torr). If 64K RAM is available, then the appropriate ADC rate would be 65,536/25 ≈ 2600 samples/second, corresponding to a spectral heterodyne-mode bandwidth of ~1300 Hz near the ICR frequency (~66 kHz) of ions of m/z = 700. (Throughout this chapter, m/q denotes mass-to-charge ratio in S.I. units, whereas m/z denotes the same ratio in atomic mass units per electronic charge.)

From the above examples, it should be clear that the choice between direct-mode and heterodyne-mode can be decided from the formulas in Table 7.1 and the following equations:

$$\nu_{\text{sampling}} \cdot T = N = \text{number of time-domain data points} \tag{7.35a}$$

$$\nu_{\text{sampling}} \geq 2\,\nu_c \quad \text{(Nyquist criterion to avoid foldover)} \tag{7.35b}$$

$$T \geq 3\tau \quad \text{(for maximum resolution at a given pressure)} \tag{7.35c}$$

$$\nu_c = \frac{\omega_0}{2\pi} = \frac{qB_0}{2\pi m} = \text{ICR orbital frequency in Hz} \tag{7.35d}$$

In many analytical mass spectrometry applications, one is asked to monitor the relative *amounts* of various *known* species, as in quality control monitoring. The m/q-values (and thus the ICR frequencies) of the ions are therefore known, and we need only determine the relative spectral peak areas to determine the relative numbers of ions of different m/q-values. In such a case, foldover can be tolerated, because we already know the true ICR frequency of each ion in the detected bandwidth, so that we can use Eq. 3.17 to determine the apparent (folded-over) frequencies in the FT spectrum of an undersampled time-domain data set.

An extreme example is illustrated in Figure 7.14, in which the time-domain signal was undersampled by a factor of more than 1,000, to produce a multiply folded-over spectrum. The advantage of multiple-foldover is that it offers significantly enhanced analog and digital resolution, for a given number of time-domain data points—note the resolution of the closely spaced isotopic species at m/z 28 (N_2^+ and CO^+) and at m/z 44 (N_2O^+ and CO_2^+). A similar example (with "only" 8 foldovers) from optical FT/interferometry is given in Chapter 9.3.2.1.

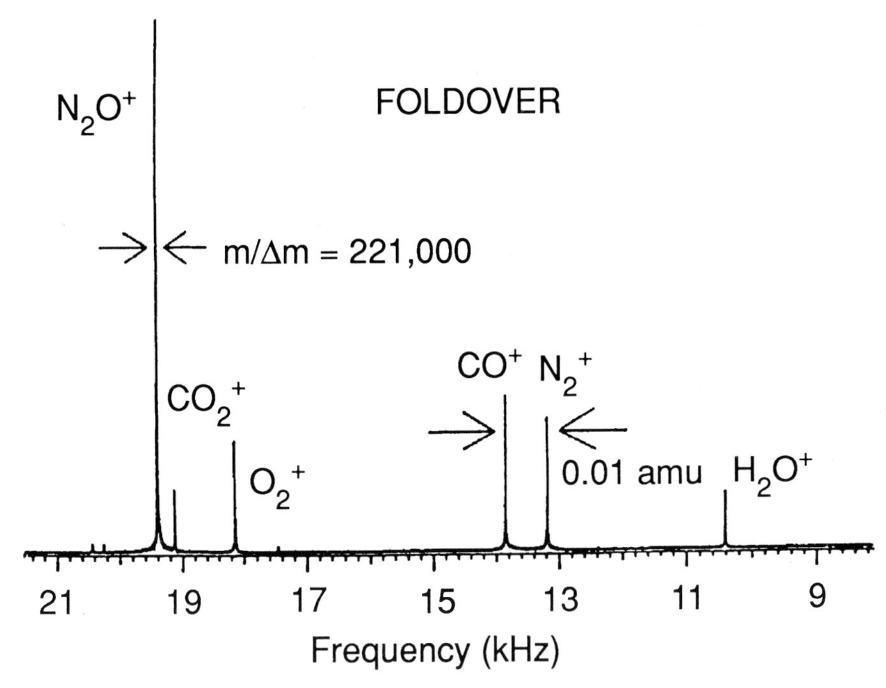

Figure 7.14 High-resolution FT/ICR mass spectrum of respiratory gases mixed with carbon monoxide and nitrous oxide (a general anesthetic). This spectrum was obtained by discrete FT of a time-domain signal direct-mode sampled at a rate corresponding to more than 1,000 foldovers. Note that ultrahigh mass resolution (221,000 at m/z 44) is obtained over a wide (factor of 2.44) range in mass-to-charge ratio. (Adapted, with permission, from M. Wang & A. G. Marshall, *Anal. Chem.* **1988**, *60*, 341-344.)

7.4.4 Phasing of broadband phase-wrapped FT/ICR spectra

Our prior discussion (Chapter 3.5.) of zero- and first-order phasing of Fourier transform spectra suffices for FT/NMR and FT/interferometric spectra, but not for FT/ICR spectra, for two reasons. First, most FT/ICR signals are currently generated by frequency-sweep excitation, for which the phase varies *quadratically* with frequency. Fortunately, since the frequency-sweep parameters (excitation period, swept frequency range, peak-to-peak time-domain amplitude) for any given experiment are known, one may correct for the quadratic phase-dependence by using Eqs. 3.25 to obtain partially phase-corrected discrete real and imaginary spectra:

$$Re\,[\boldsymbol{F}(\omega)]_{\text{final}} = \cos\phi\ Re\,[\boldsymbol{F}(\omega)]_{\text{initial}} - \sin\phi\ Im\,[\boldsymbol{F}(\omega)]_{\text{initial}} \tag{7.36a}$$

$$Im\,[\boldsymbol{F}(\omega)]_{\text{final}} = \sin\phi\ Re\,[\boldsymbol{F}(\omega)]_{\text{initial}} + \cos\phi\ Im\,[\boldsymbol{F}(\omega)]_{\text{initial}} \tag{7.36a}$$

The remaining zero- and first-order phase correction may then be performed as outlined in Chapter 3.5.1.

However, a second (more subtle) problem with FT/ICR phase correction is that signal acquisition (detection) is usually delayed by an interval, T_{delay}, of a millisecond or more after excitation (see Figure 7.15a), in order to allow the transmitter signal [~1–200 $V_{(p-p)}$] to decay sufficiently so as not to interfere with the detected signal (~μV range). As a result, a broadband (0.1–10 MHz) ICR time-domain signal may accumulate many (100–10,000) cycles of phase at the high-frequency end of the spectrum (Figure 7.15b).

If the detected time-domain signal were continuous (analog), then one could correct analytically for even a very steep linear variation of phase with frequency across the spectrum:

$$\phi(\omega) = \phi_0 + 2\pi \nu T_{delay} \tag{7.37}$$

However, the discrete FT (whether with fast algorithm or not) cannot represent phase beyond the range, $-\pi \leq \varphi \leq \pi$. As a result, an attempt to phase-correct the spectrum of Figure 7.15b by substituting $\varphi(\omega)$ from Eq. 7.37 into the usual formula (Eq. 7.36) yields a spectrum (Figure 7.15c) which represents the convolution of the correctly phased spectrum (Figure 7.15h) with the FT (Figure 7.15f) of the rectangular weight function shown in Figure 7.15e.

Stated another way, the inverse FT (Figure 7.15d) of the phase-corrected spectrum (Figure 7.15c) is seen to represent the product of the desired cohererent time-domain signal (Figure 7.15g, which is the same as Figure 7.15a) and the weight function of Figure 7.14e. In this case, we cannot recover the desired spectrum (Figure 7.15h) by deconvolution, since division of time-domain signal (Figure 7.15d) by the time-domain weight function (Figure 7.15e) would result in division by zero. Thus, we have lost our knowledge of the time-domain signal during the delay period, and cannot recover it by FT methods alone. For this problem, autoregression or maximum-entropy methods are needed to produce a phased spectrum without the "Gibbs oscillations" of Figure 7.15c.

7.4.5 Harmonics

In Chapter 4.4.1, we noted that a non-linear relation between excitation magnitude and detected signal magnitude leads to signals at harmonic frequencies (i.e., integer multiples of the "true" signal frequency) and intermodulation frequencies (i.e., arithmetic combinations of integer multiples of the "true" signal frequencies). In FT/ICR, the detector itself is in fact highly linear, in the sense that doubling the number of ions at a given radius will produce a detected signal twice as large. Signals at (odd-integer) harmonic frequencies are nevertheless observed.

The explanation is shown schematically in Figure 7.16, for a cylindrical ion trap geometry. Let the ICR orbital radius of a spatially coherent ion packet be r, and the radius of the cylindrical trap be r_0. When $r \ll r_0$, the ICR signal current induced in the detector electrodes is small, but nearly sinusoidal. However, if the ions are excited to a radius at which they nearly contact the detector electrode, then the ICR signal is constant when the ions are near that electrode, but near-zero otherwise. For the cylindrical detector of Figure 7.16, the frequency-domain FT spectrum (see Problems) of the resultant square-wave modulated signal contains peaks of relative magnitude 1, 1/3, 1/5, 1/7, $\cdots$, at frequencies, ω_c, $3\omega_c$, $5\omega_c$, $7\omega_c$, $\cdots$.

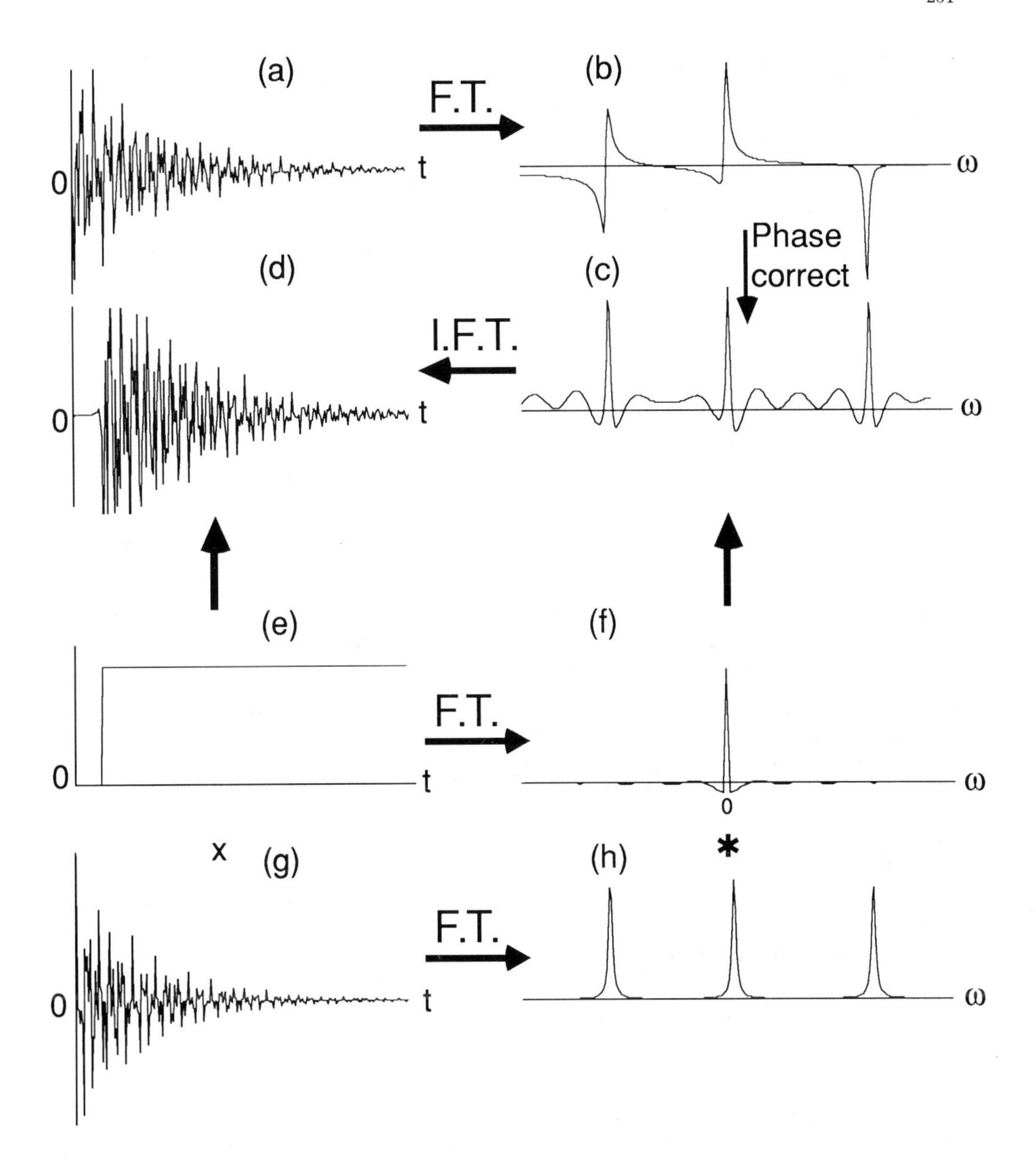

Figure 7.15 Result of linear phase correction of a "phase-wrapped" spectrum obtained by FT of a time-domain data set whose acquisition is delayed by more than one ADC dwell period after the excitation event. Because the discrete FT procedure cannot represent a phase angle larger than $\pm\pi$ radians, spectrum (c) represents the best result in the absence of apodization (see text).

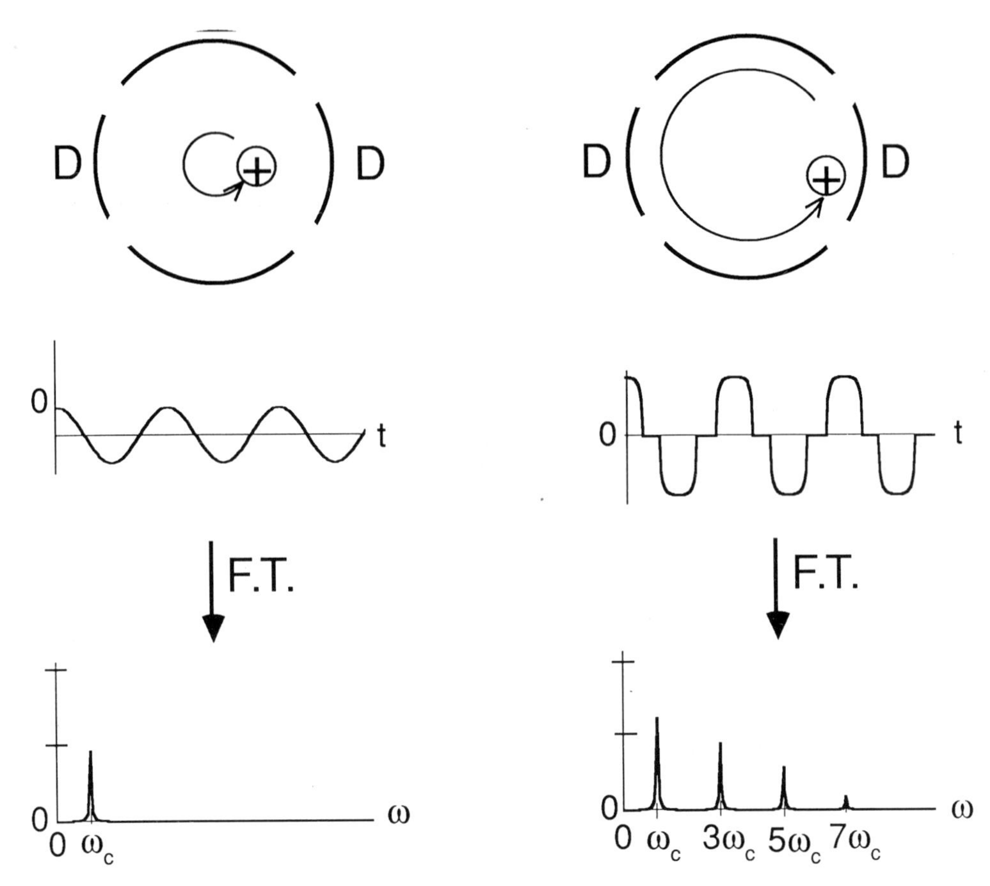

Figure 7.16 Origin of odd-integer harmonic signals in FT/ICR mass spectra. Left: The time-domain ICR signal induced on opposed cylindrical detector electrodes, D, is small but approximately sinusoidal when the ICR orbital radius, r, is much smaller than the radius, r_0, of the detector electrode. The corresponding FT frequency spectrum therefore consists of a single peak at the fundamental ICR orbital frequency, ω_c. Right: As $r \to r_0$, the detected time-domain signal approaches a square wave, whose spectrum contains peaks at the the fundamental and odd harmonic frequencies: ω_c, $3\omega_c$, $5\omega_c$, $\cdots$.

Only odd harmonics are observed, because the ICR signal is detected "differentially": i.e., as the difference between the signals on the two opposed detector electrodes. Although the signal for ions of any given m/q value is non-sinusoidal, detection is still linear, so that no *intermodulation* peaks are observed.

A similar effect results when the time-domain signal voltage exceeds the maximum digitizer range, to give a "clipped" signal, as in Figure 7.17. Although the detection process is non-linear, since the time-domain signal *shape* changes when the signal magnitude increases, differential detection again effectively eliminates signals at even harmonic frequencies ($2\omega_c$, $4\omega_c$, $6\omega_c$, $\cdots$), and only odd harmonics ($3\omega_c$, $5\omega_c$, $7\omega_c$, $\cdots$) are observed (see Problems). In contrast both harmonic (even and odd) and intermodulation distortion are observed in FT/NMR.

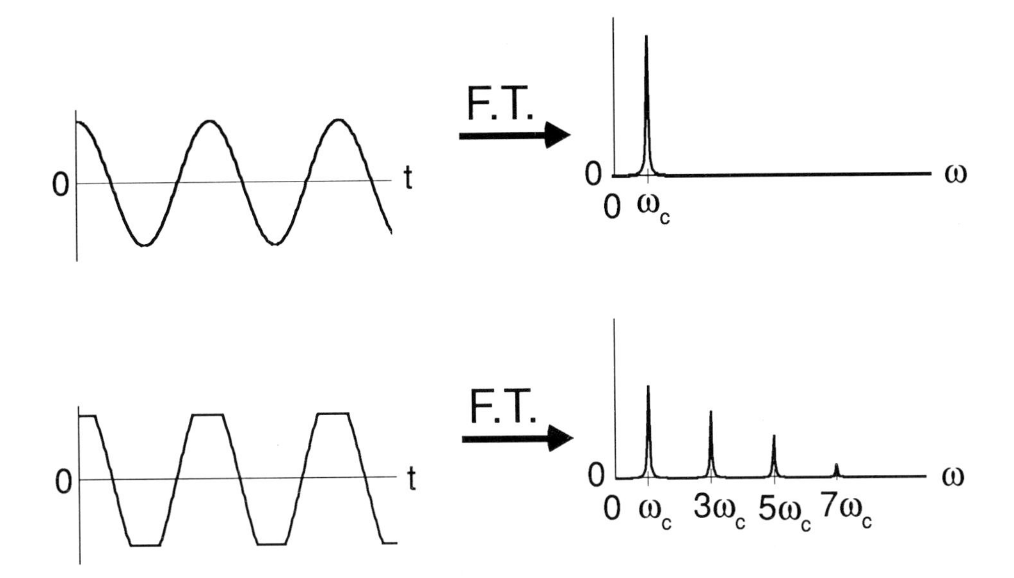

Figure 7.17 Harmonic distortion resulting from "clipping" of a time-domain ICR signal. Top: Fourier transformation of a time-domain signal whose magnitude does not exceed the digitizer range gives a spectrum with peaks only at the fundamental ICR orbital frequencies, ω_c. Bottom: FT of a "clipped" time-domain signal whose magnitude exceeds the digitizer range yields a spectrum with peaks at odd-harmonic frequencies. (See text.)

7.4.6. Quadrature ICR excitation and detection

Quadrature excitation and detection in ICR and NMR differ in two important ways. First, a typical NMR spectrum spans a frequency band which is a small fraction of the Larmor ("natural") NMR frequency, whereas an ICR spectrum can span a range almost as large as the highest ICR frequency in the spectrum. Therefore, all FT/NMR experiments are conducted in *heterodyne* mode, whereas ICR experiments can be performed in either *heterodyne* or *direct* mode. Second, NMR detection is almost always conducted with a *single-axis receiver coil* to yield a *linearly-polarized* detected signal. Although most ICR experiments to date have been performed with a *single pair* of detector plates (to yield a *linearly-polarized* detected signal), ICR detection with *two pairs* of orthogonal detector plates (to yield a *circularly polarized* detected signal) has recently been demonstrated. One interesting and potentially useful feature of dual-channel excitation is that it becomes possible to generate a *circularly polarized* rf electric field excitation. Dual-channel detection with separate digitization channels may also be used for simultaneous acquisition of two signals of different bandwidth (e.g., low-resolution direct-mode detection of a wide-range mass spectrum and high-resolution heterodyne-mode detection of a narrow-range mass spectrum). Dual-channel excitation and detection FT/ICR mass spectrometry are currently under further development.

7.5 FT/ICR features, experiments, and applications

The unique advantages of FT/ICR for mass spectrometry result from particular *combinations* of the following capabilities: (a) multichannel advantage for obtaining the whole mass spectrum at once in ≤ 1 s; (b) ultrahigh mass resolution; (c) ion trapping for extended periods (>1 second); (d) ion formation, reaction, and detection (including MS/MS/· · ·) in a single chamber; (e) operating pressure that is ~1000 times lower than that in most other mass spectrometers; (f) pulsed detection to match pulsed ion formation (e.g., laser desorption); (g) high upper mass limit; (i) variation of experiment type by means of software rather than hardware changes;

Figure 7.18 documents the exponential growth in number of FT/ICR instruments since the first FT/ICR experiment by Comisarow and Marshall in late 1973. Of the more than 90 FT/ICR spectrometers in use worldwide by 1989, about half replaced continuous-wave ICR instruments based on power absorption detection, and the other half represent new academic and industrial users.

In view of the large number of available recent reviews of FT/ICR applications to the identification, structure determination, and reactivity of gas-phase ions (see Further Reading), the remainder of this chapter will be limited to a general introduction to selected FT/ICR experimental techniques.

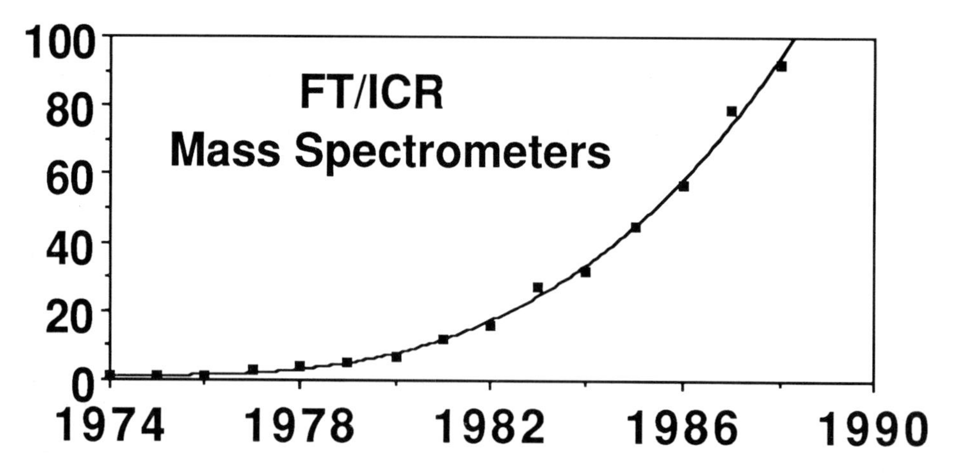

Figure 7.18 Worldwide number of Fourier transform ion cyclotron resonance mass spectrometers, dating from the first FT/ICR experiment in December, 1973 (M. B. Comisarow & A. G. Marshall, *Chem. Phys. Lett.* **1974**, *25*, 282-283).

7.5.1 FT/ICR experimental events

Until the advent of continuous-wave ICR spectroscopy in the mid-1960's, gas-phase ion chemistry was (and continues to be) typically conducted as shown schematically in Figure 7.19. After formation in an ion "source" chamber, ions are accelerated to form a focussed beam. Ions of given energy and m/q can be selected by passing that beam through spatially separated regions of electric and/or magnetic field. Production of ion-molecule collisions and/or reactions requires yet another spatially separated reaction chamber, with ion detection as the last stage. In other words, each stage of the experiment is *spatially separated.*

In contrast, FT/ICR experiments can produce all of the above stages (and, in addition, provide for extended chemical reaction periods by virtue of ion trapping) by use of experiments conducted in a *single* chamber by a series of stages separated in *time* (see Chapter 7.5.2). We therefore begin by analyzing the various temporally separated events of such experiments.

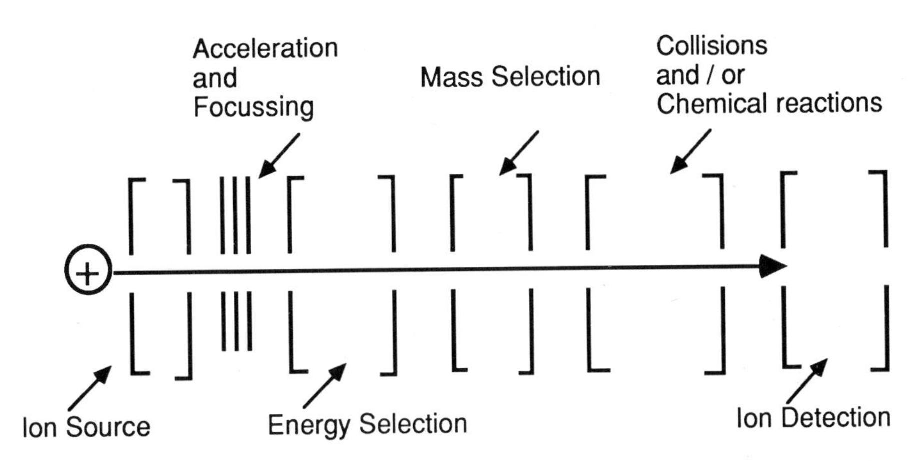

Figure 7.19 Schematic diagram of typical stages of experiments for identifying and characterizing ions and their chemical reaction products. In this diagram (and in most non-ICR experiments), the stages are separated *spatially*. In FT/ICR, the same stages can be achieved by means of a series of *time-separated* events (see below).

Ion formation

The overwhelmingly most common method for ionizing neutral gas-phase molecules is electron ionization ("EI"), namely, bombardment with electrons accelerated to high translational energy (~70 keV) In FT/ICR/MS, the electrons are simply accelerated into the ion trap along the *z* -axis through a small hole in the center of the trapping electrode (Figure 7.8). (Electrons are kept on-axis by their own cyclotron motion about the same applied magnetic field that subsequently confines the ions.) A gas (or vapor from a liquid) can be introduced into a previously evacuated chamber to a pressure of ~10^{-8} torr by means of a controlled-leak valve. Alternatively, a solid sample in a small open capillary tube may be heated until its vapor pressure reaches ~10^{-8} torr. Thus (see Figure 7.20), FT/ICR spectra can be obtained for samples whose volatility is too low for detection by conventional mass spectrometers whose EI ion source pressure must be operated at ~10^{-5} - 10^{-6} torr.

Electron ionization is experimentally simple, but it can lead to extensive fragmentation of the molecular ion, and it requires a gaseous sample. Ions may be produced from *involatile* samples by replacing the electron beam by a beam of accelerated *ions* [e.g., 10 keV Cs^+), *neutrals* (e.g., 5 keV SF_6), or *photons* (e.g., pulsed CO_2 laser at 10.6 μm (infrared); or Nd/YAG laser at 1064 nm (infrared), 532 nm (green), or 266 nm (ultraviolet), at ~10 Mwatt/cm^2 laser power density]. Laser desorption has proved especially suitable for characterization of polymers, as shown in Figure 7.21, and a pulsed Cs^+ beam has proved especially advantageous for peptide sequencing (see below).

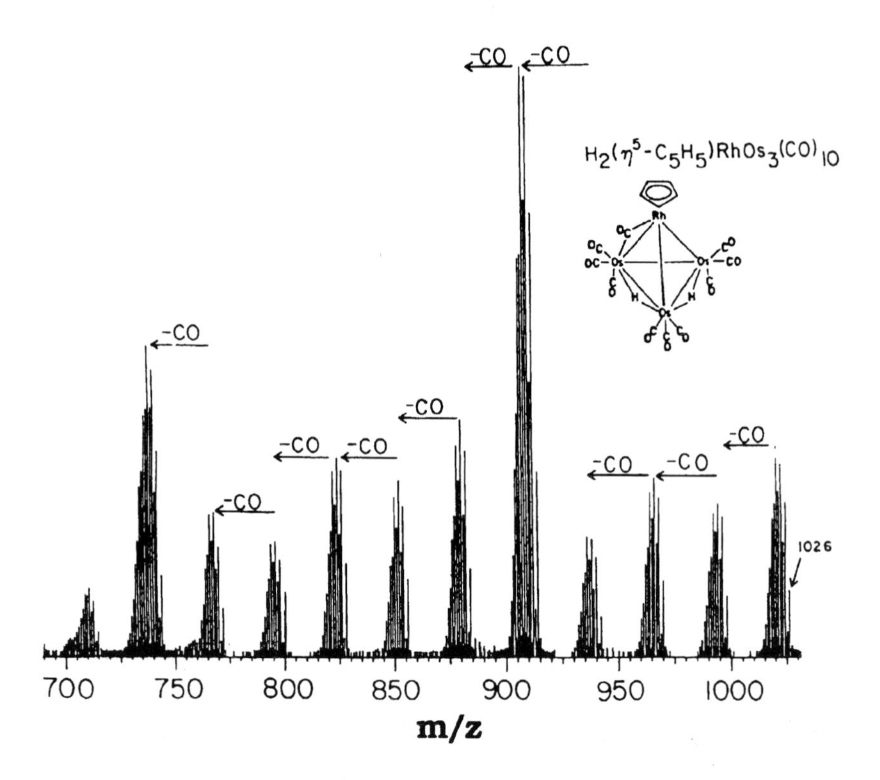

Figure 7.20 FT/ICR mass spectrum (Nicolet FTMS-2000 mass spectrometer operated at 3.0 tesla) of an electron-ionized low-volatility (~10^{-8} torr at ~100 °C) metal cluster compound provided by S. G. Shore. Note the resolution of several isotopes of osmium (e.g., ^{192}Os, ^{190}Os, etc.) for each chemical formula [e.g., the rightmost spectral cluster, $H_2(\eta^5\text{-}C_5H_5)RhOs_3(CO)_{10}^+$]. (The leftmost group of peaks corresponds to $Os_3(CO)_5^+$.)

For other ionization methods (e.g., ^{252}Cf plasma, fast atom bombardment (FAB) with Ar or Xe), the ionization source is typically removed by ~1 meter away from the ion trap along the *z*- (magnetic field-) axis, and ions are injected into the ion trap along the *z*-axis as noted in Chapter 7.2.4.

Ion excitation or ejection

Ion excitation modes were reviewed in Chapter 7.2.1 and 7.4.2. Their use is illustrated schematically in Figure 7.22. Ions irradiated with zero rf amplitude at their ICR frequency may be left unexcited (and therefore unobserved) on or near the *z*-axis of the ion trap. Ions irradiated on-resonance long enough to reach an ICR orbital radius from ~0.1–0.5 of the radius of the trap will generate a coherent ICR signal (until the coherence is destroyed by ion-neutral collisions and reactions). Ions irradiated on-resonance for a sufficiently long period will collide with the side electrodes and be neutralized (i.e., "ejected"). Finally, ions irradiated at a frequency several line widths off-resonance will spiral repeatedly outward and back inward with essentially no net change in final ICR orbital radius.

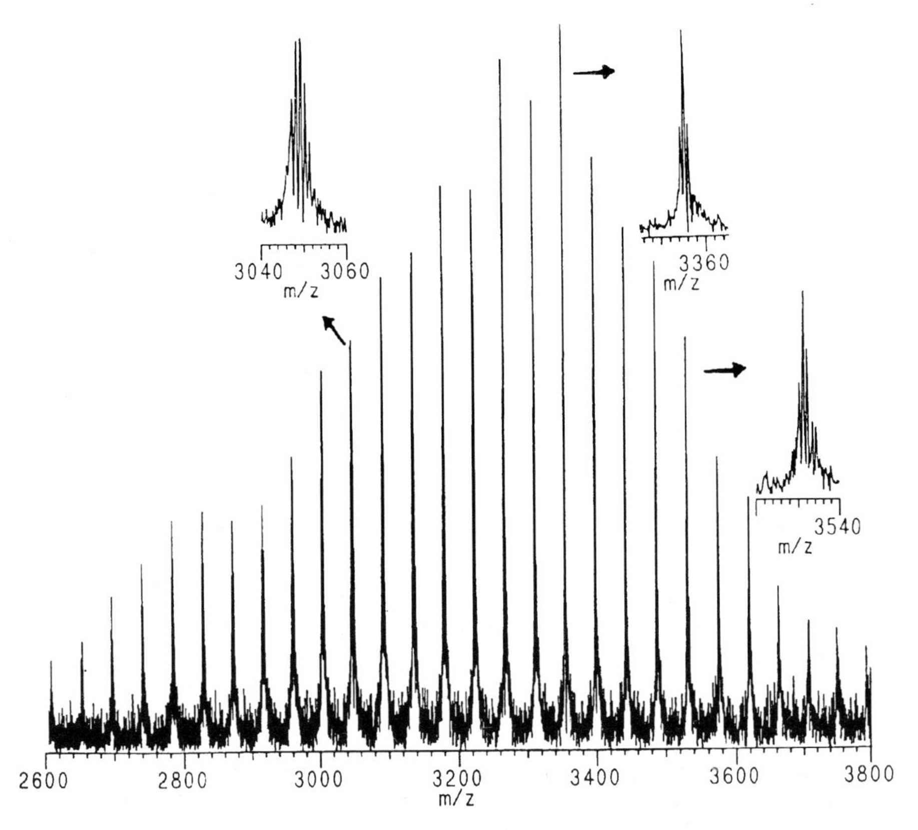

Figure 7.21 Laser desorption (pulsed CO_2 laser at 10.6 μm) FT/ICR mass spectrum (Nicolet FTMS-2000 mass spectrometer operated at 3.0 tesla) of poly(ethylene glycol). The sample was doped with KBr, to produce abundant $(M+K)^+$ ions. The molecular weight distribution of the ions in this mass spectrum accurately reflects the distribution of neutral oligomers in the original polymer. [Spectrum reproduced, with permission, from R. B. Cody, A. Bjarnason, & D. A. Weil, in *Lasers in Mass Spectrometry*, Ed. D. Lubman (Oxford Press, 1989)]

Ion de-excitation

After ions have been excited, but before they have had a chance to collide with neutrals and thereby lose spatial- (and therefore phase-) coherence, the ions may be irradiated a second time. If the phase of that excitation is 180° out of phase with the ICR orbital motion, then ions will be "de-excited" by a force which continuously opposes their forward motion, so that the ions spiral back to their pre-excitation positions on or near the z-axis.

Figure 7.23 shows some of the ion manipulations made possible by various combinations of the ion excitation options of Figure 7.22. The simplest and most generally useful option is *broadband resonant excitation* with flat power, to excite ions from a range of m/q values to the same orbital radius for detection with essentially equal sensitivity. Alternatively, ions of widely separated m/q values

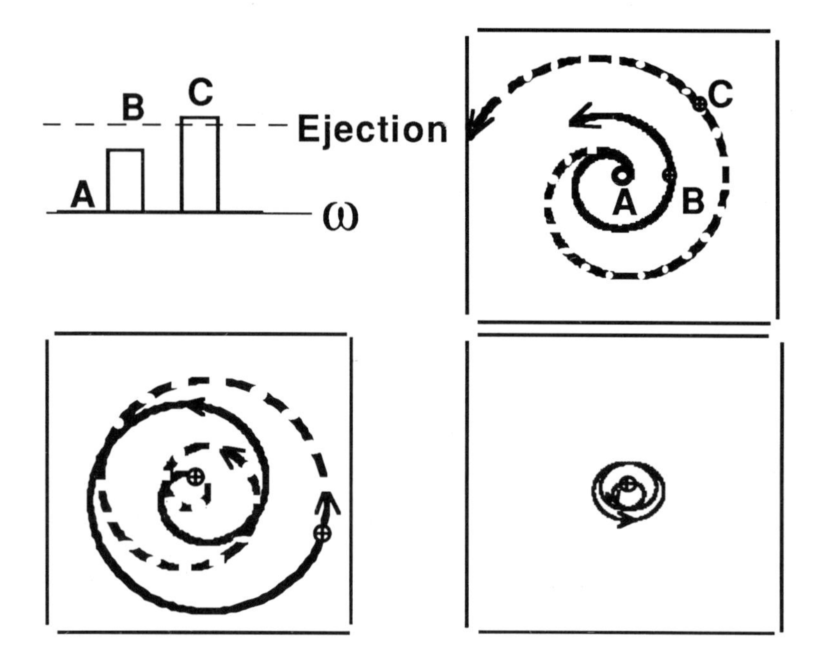

Figure 7.22 ICR excitation options (magnetic field is directed into the plane of the page). On resonance-excitation with zero, medium, or high (A, B, C) rf electric field excitation magnitude and/or duration leaves ions respectively undetectable, detectable, or ejected (upper right). Once excited coherently, ions may be "de-excited" by irradiation with a rotating electric field which is 180° out of phase with the ICR orbital motion (bottom left). Off-resonance irradiation (bottom right) has almost no net effect on the final ICR orbital radius.

may be *excited simultaneously* (as for multiple-ion monitoring) or *ejected simultaneously* (in order to reduce the space charge contribution from unwanted abundant ions), so as to effectively increase the ICR signal dynamic range from its otherwise upper limit of ~1,000:1, for detection of ions of low abundance. One can *eject ions of all but one m/q value* for MS/MS experiments (see below and "notch ejection" Problem). Conversely, ions of a single *m/q* value can be excited, as for collision-induced dissociation experiments (see below).

Ion fragmentation or chemical reaction

Two general ways to identify an ion and/or determine its chemical structure are to analyze the (charged, and therefore ICR-detectable) products formed by (a) fragmentation of the "parent" ion into lower-mass daughter ions (e.g., by collision of translationally excited "parent" ions with a neutral inert gas such as He or Ar, or photodissociation following the absorption of one or more photons by the ion), or (b) chemical reaction of the "parent" ion with neutral molecules. The first process is known as "collision-induced" or "collision-activated" dissociation (CID or CAD), and the degree of fragmentation may be increased by *exciting* (heating) the ions to higher translational energy (i.e., larger ICR orbital radius) before collision. The second process, which typically involves a proton transfer ion-molecule reaction, is called "chemical ionization".

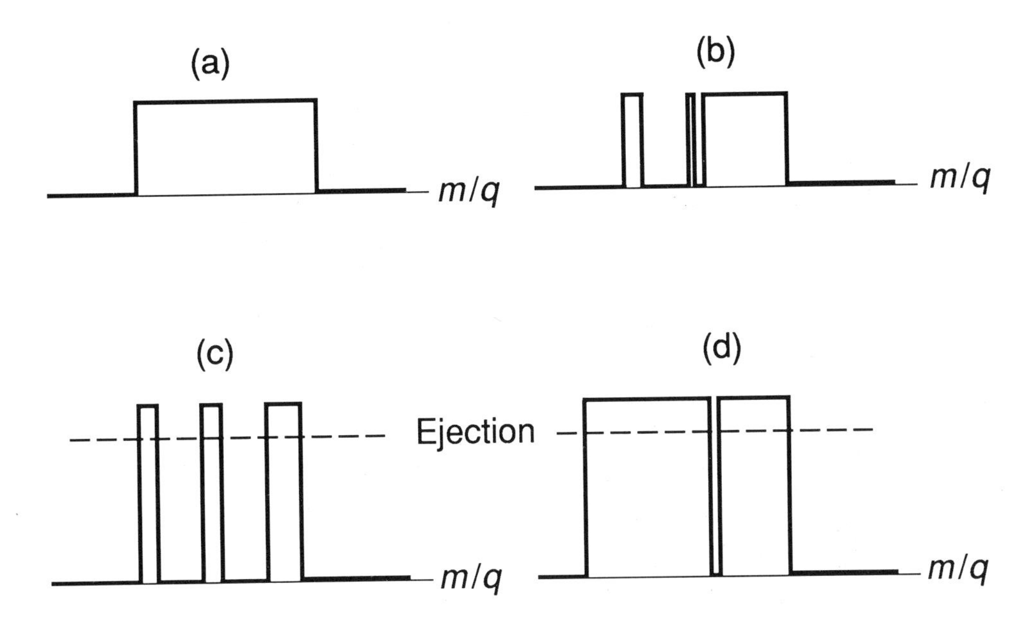

Figure 7.23 ICR excitation modes available by means of SWIFT excitation (Chapter 4.2.3). (a) Broadband excitation of all ions within a specified m/q range (for typical direct-mode or heterodyne-mode detection); (b) Simultaneous excitation of ions of separated by arbitrary m/q differences (multiple-ion monitoring); (c) Simultaneous ejection of ions separated by arbitrary m/q differences (for dynamic range enhancement); (d) Simultaneous ejection of ions of all but one selected m/q value (for MS/MS).

An important advantage of FT/ICR analysis of the above processes is that ions and/or neutrals may be *pre-selected* beforehand. *Ions* of unwanted m/q values may be ejected before collision or reaction as shown in Figure 7.22. The desired (gaseous) *neutral* species may be introduced by rapidly opening and closing a leak valve to produce a *pulse* of collision or reagent gas, which is rapidly pumped away after ~0.1 s collision or reaction period.

Ion detection

Once ions of the desired m/q values have been produced and selectively excited to form a coherent signal, a short delay period (say, 200 μs) is introduced to allow the transmitter signal to die away, and direct- or heterodyne-mode detection is performed.

Ion removal (quench)

At the end of an FT/ICR experimental event sequence, it is desirable to remove *all* ions from the trap, in order to prepare the system for the next experimental event sequence. A quick non-selective way to remove all ions of both charge signs from a static electromagnetic ion trap (Figure 7.8) is to apply a large differential voltage to the two end cap ("trapping") electrodes (say, $+10V$ to one end cap and $-10V$ to the other) for a few milliseconds.

7.5.2 FT/ICR experimental event sequences: the gas-phase chemical laboratory

Figure 7.24 shows a typical FT/ICR experimental event sequence, of the type used to generate the mass spectrum in Figure 7.21. (The spectrum in Figure 7.20 was produced by the same event sequence, but with essentially zero chemical reaction period.) The FT/ICR mass spectrometer thus functions as a complete synthetic and analytical laboratory for gas-phase ion-molecule chemistry.

7.5.2.1 *Normal broad-band or heterodyne detection*

With broadband excitation, zero chemical reaction period, and *direct-mode detection*, the event sequence of Figure 7.24 can yield a wide-range FT/ICR mass spectrum (e.g., Figures 7.20, 7.21) with the full Fourier multichannel advantage (i.e., simultaneous detection of the whole spectrum at once). The multiplex speed advantage is particularly important for gas chromatography/mass spectrometry, in which one or more full-range mass spectra may be acquired during the elution of each chromatograph peak (~1 second) from a capillary GC column.

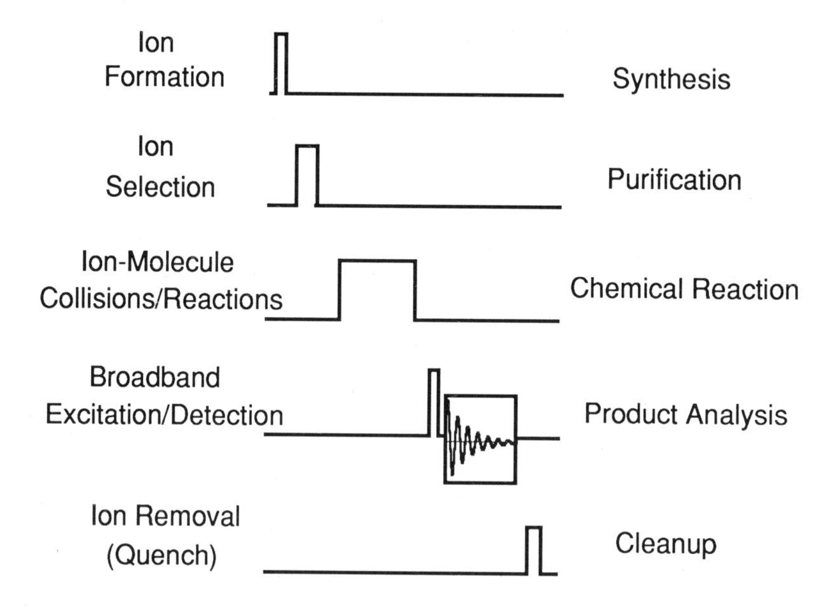

Figure 7.24 Generalized experimental FT/ICR event sequence (left), represented as a series of conventional chemical manipulations (right).

Alternatively, by packing the same number of data points into a smaller bandwidth by *heterodyne detection*, the same event sequence can produce an *ultrahigh-resolution* mass spectrum (see Figure 7.25).

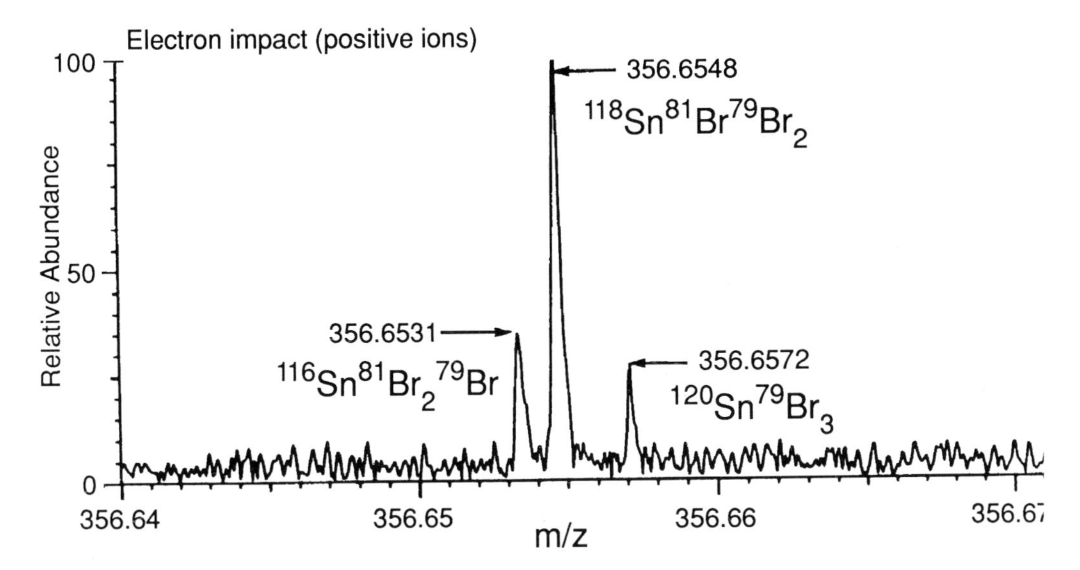

Figure 7.25 Ultrahigh-resolution FT/ICR mass spectrum of the $SnBr_3^+$ isotopic triplet produced from electron ionization of tin bromide. Mass (or frequency) resolution is defined as the ratio of ICR peak frequency to the full line width at half-maximum peak height ($m/\Delta m$ = 833,000 at m/z = 356). Note that ions differing in mass by less than 1 part per million are clearly resolved. (Adapted with permission from: R. B. Cody, J. A. Kinsinger, S. Ghaderi, I. J. Amster, F. W. McLafferty, & C. E. Brown, *Analytica Chimica Acta* **1985**, *178*, 43-66.)

7.5.2.2 *Ion-molecule reaction pathways, rate constants, equilibrium constants, and energetics*

If a chemical reaction period is included in the event sequence of Figure 7.24, the resulting FT/ICR mass spectrum gives the relative concentrations of ionic reactants and ionic products. By varying the reaction period at known partial pressures of neutrals, one may determined the *rate constants* for ion-molecule reactions. Alternatively, if the reaction period is sufficiently long (e.g., 5-20 seconds at 10^{-7} torr) that each ion undergoes many collisions with neutrals, then the *equilibrium constant* for a given reaction may be determined from known partial pressures of neutrals and the relative equilibrium concentrations of ions (from FT/ICR spectral peak relative areas).

For example, consider the relatively simple gas-phase chemical equilibrium involving proton transfer. Suppose that the equilibrium constant, K_A, for addition of a proton to neutral ("reference") molecule, A, is known, and that we wish to determine the equilibrium constant, K_B, for proton addition to neutral molecule, B.

$$A + H^+ \rightleftarrows AH^+ \qquad K_A = \frac{[AH^+]}{[A]\,[H^+]} \tag{7.38a}$$

$$B + H^+ \rightleftarrows BH^+ \qquad K_B = \frac{[BH^+]}{[B]\,[H^+]} \tag{7.38b}$$

For an ionized equilibrium mixture of A and B,

$$A + BH^{+} \rightleftarrows AH^{+} + B \qquad K_{eq} = \frac{[AH^{+}][B]}{[BH^{+}][A]} = \frac{K_A}{K_B} \tag{7.39}$$

Thus, K_{eq} can be determined from experimental measurement of the neutral partial pressures, $[B]$ and $[A]$, and the relative FT/ICR mass spectral peak areas of AH^{+} and BH^{+}. From K_{eq} and K_A, we may then obtain the desired $K_B = K_A / K_{eq}$.

The *gas-phase basicity* and *proton affinity* are defined as the (negative) free energy and enthalpy for the ion-molecule reaction of Eq. 7.38b:

$$-\Delta G_B = RT \log_e K_B = \text{Gas-Phase Basicity;} \; R = \text{gas constant/mole} \tag{7.40a}$$

$$\Delta H_B = \text{Proton Affinity} \tag{7.40b}$$

For example, Figure 7.26 shows the electron ionization FT/ICR mass spectrum of a mixture of two diamines at equal partial pressure (~2.5 x 10^{-7} torr), following a reaction period of 60 s to attain proton transfer equilibrium. The cyclic compound is clearly the stronger base, since its protonated parent ion (m/z = 237) is more abundant than that of the other protonated parent compound (m/z = 145).

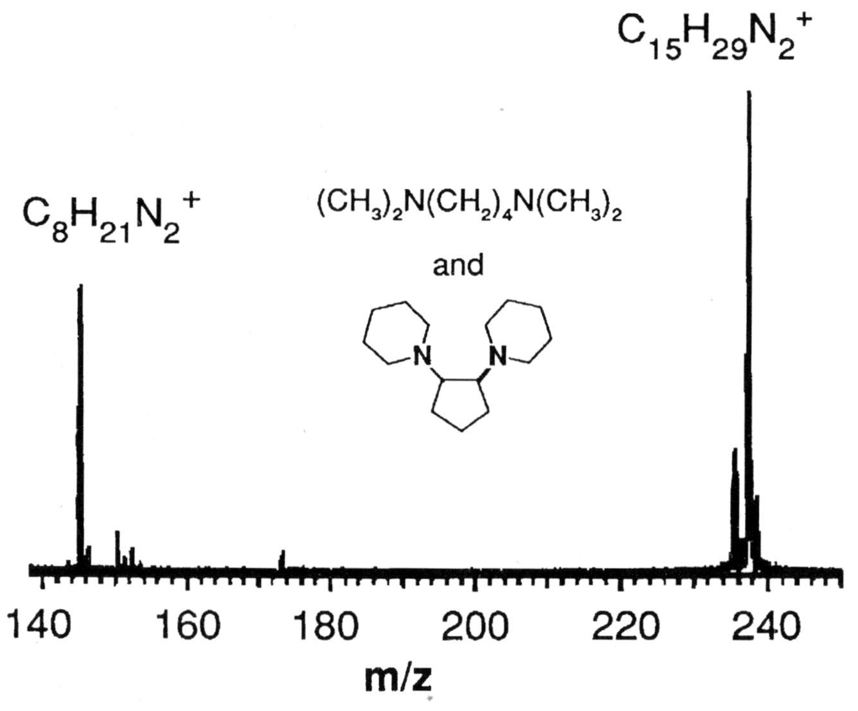

Figure 7.26 Determination of gas-phase basicity from FT/ICR mass spectral peak relative areas for an electron-ionized binary isobaric equilibrium mixture of two diamines at equal (2.5 x 10^{-7} torr) partial pressure (see text). (Spectrum produced by F. R. Tollens, from a sample provided by G. Fraenkel)

From the ratio of the two FT/ICR peak areas and the known neutral partial pressures, the difference in gas-phase basicity (~1.4 kcal mol^{-1} = 5.9 kJ mol^{-1}) can be computed from K_{eq} (Eqs. 7.39, 7.40a). From the results of similar measurements on various pairs of bases, a self-consistent "ladder" of hundreds of gas-phase basicities has been compiled, in large part from ICR data of the type shown in Figure 7.26. At the time the measurement was made,, the cyclic diamine, 1,2-*bis*-(dimethyl-amino)-cyclopentane, shown in Fig. 7.26 had the highest measured gas-phase basicity (247.4 kcal mol^{-1} = 1035 kJ mol^{-1}) of any organic compound.

Similarly, a ladder of gas-phase *acidities* can be compiled from FT/ICR determination of relative concentrations of ionic products of chemical reactions involving the *loss* of a proton: $AH \rightleftarrows A^- + H^+$. Gas-phase basicity and acidity are of more than academic interest. For example, the proton affinity of an organic molecule, M, is often higher than that of its neutral fragments. Thus, an FT/ICR mass spectrum acquired after a short (~1 second) delay period following electron ionization often yields a prominent (and diagnostic) peak at $(M + H)^+$, because the protonated fragment ions transfer their protons to the neutral parent, in a "self chemical ionization" process first demonstrated by FT/ICR by M. L. Gross *et al.* (see Further Reading).

The versatility of FT/ICR mass spectrometry is significantly enhanced by: (a) the use of frequency-selective (e.g., SWIFT) excitation and/or ejection events to select *ions* of a given m/q value for reaction or observation, and (b) the introduction of selected *neutral reagent* gases by means of pulsed valves. One (of the many) latter applications provided by B. Freiser is shown in Figure 7.27. In this remarkable series of experiments, four successive reactions (starting with laser-desorbed Co^+) are performed in an event sequence lasting a few seconds. At each stage, the ion mixture is "purified" by selective ejection of ions of all but the desired m/q-value, and undesired neutrals are pumped away before the pulsed-valve injection of the next desired neutral reagent. These experiments, and the homologous series of experiments starting with laser-desorbed Fe^+, were designed to determine whether $FeCoCH_2^+$ is distinguishable from $CoFeCH_2^+$.

The fastest-growing application of mass spectrometry is the determination of the primary amino acid sequence of a peptide or protein. The usual procedure is to cleave the protein enzymatically into peptides of $\leq$ 2,500 dalton, separate the various peptides chromatographically, and then use a suitable ionization method (e.g., fast atom bombardment or Cs^+ bombardment of a sample dissolved in glycerol) to generate a detectable number of protonated molecular ions, $(M+H)^+$. Although the molecular weight of the protonated peptide helps to establish its amino acid *composition*, it is necessary to break the molecular ion into fragments in order to determine the amino acid primary *sequence*, as shown in Figure 7.28. The desired cleavages may be produced either by collisions with an inert gas (small peptides) or laser irradiation (Figure 7.28) at a frequency designed to break the peptide bond. Each successive cleavage of a peptide may leave the charge on either fragment, to give two (principal) series of fragments ("B" and "Y"). Although it is generally not possible to distinguish isobaric amino acids (e.g., leucine from isoleucine), mass spectrometric peptide sequencing offers major advantages in speed and sensitivity. In favorable cases, as little as 10 picomoles of a protein consisting of >100 amino acids can be fully sequenced in a few days!

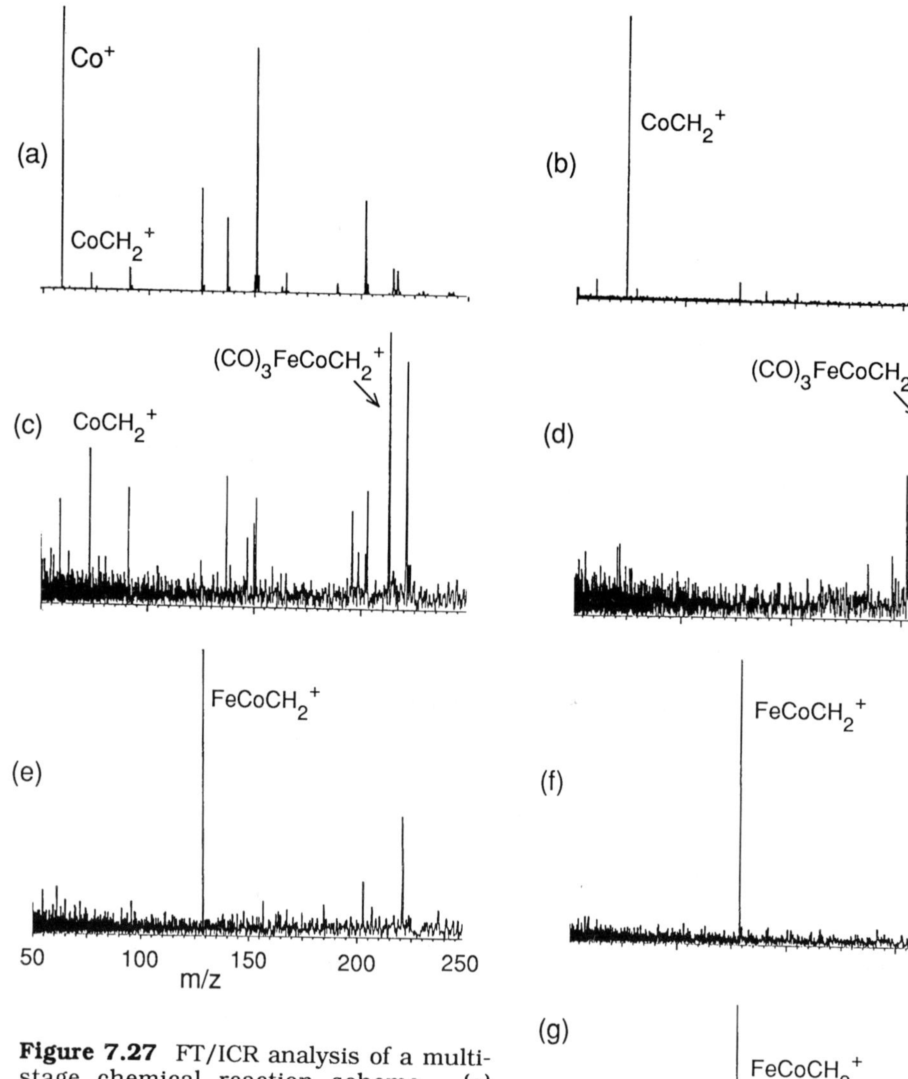

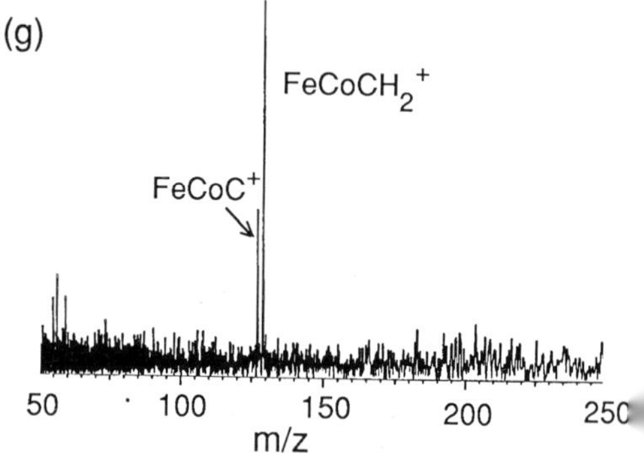

Figure 7.27 FT/ICR analysis of a multi-stage chemical reaction scheme. (a) Laser-desorbed Co^+ reacts with pulsed-valve introduced cycloheptatriene. (b) Ejection of all but $CoCH_2^+$. (c) Reaction of $CoCH_2^+$ with $Fe(CO)_5$ introduced by pulsed-valve. (d) Ejection of low-mass ions. (e) Collision-induced dissociation (CID) of $(CO)_3FeCoCH_2^+$ to form $FeCoCH_2^+$. (f) Ejection of all but $FeCoCH_2^+$. (g) CID of $FeCoCH_2^+$ to give $FeCoC^+$. (Spectra kindly provided by J. Gord and B. S. Freiser).

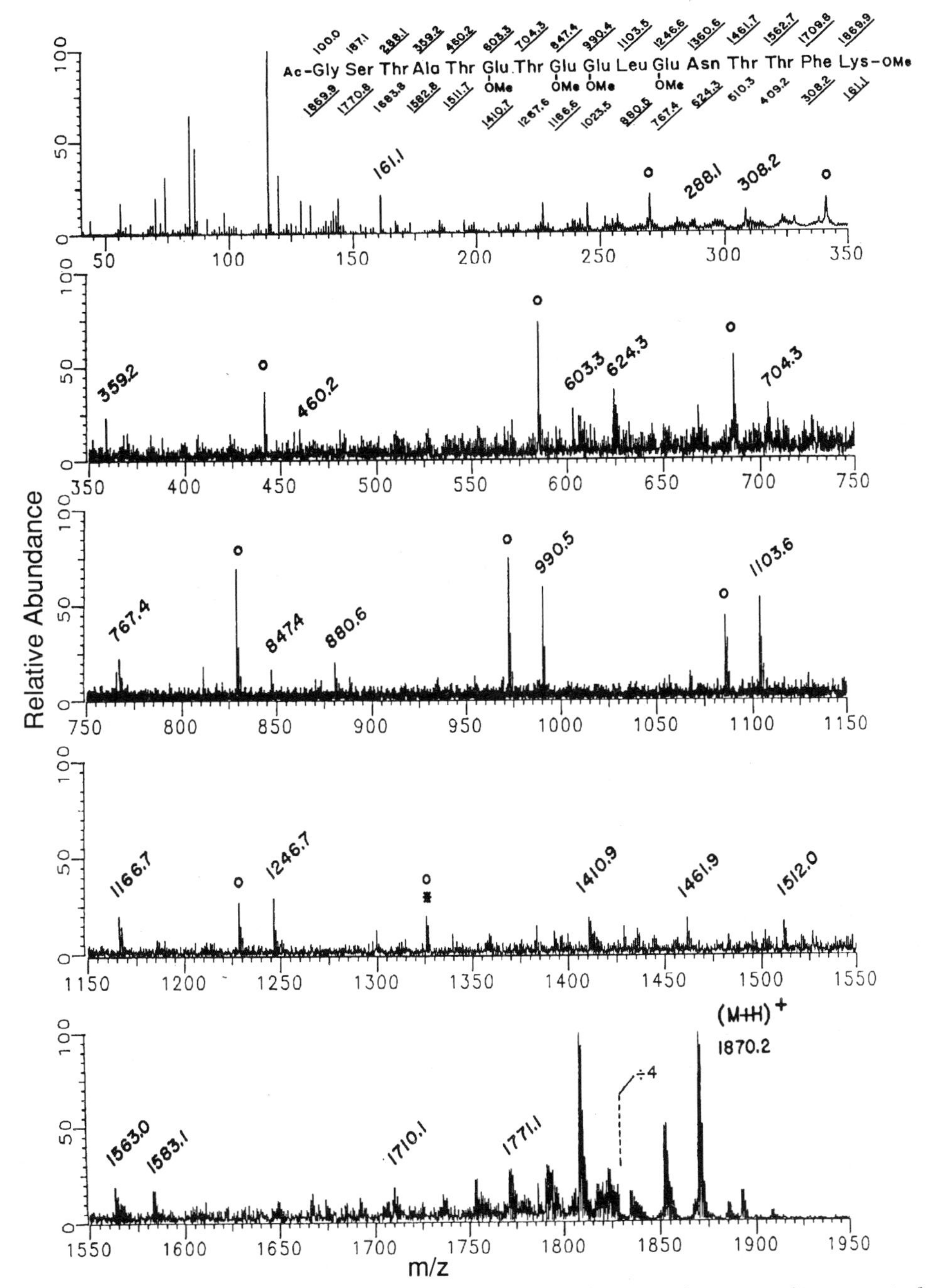

Figure 7.28 FT/ICR mass spectrum of fragments of a hexadecapeptide generated from the endoproteinase Lys-C digest of a 61 kDa calmodulin-dependent phosphodiesterase isozyme. (Spectrum kindly provided by J. Shabanowitz and D. F. Hunt)

Mass spectrometric peptide sequencing is especially important when, as in the example of Figure 7.28, only a tiny amount of peptide is available and/or classical Edman degradation is not available because of a blocked amino-terminus. In that case, wet chemical methods had established only that the carboxy-terminal residue was lysine, and that five acidic amino acid residues were present. To produce Figure 7.28, pseudomolecular $(M+H)^+$ ions produced externally by pulsed Cs^+ ionization of a 10 pmole sample dissolved in glycerol were guided by electrically charged quadrupole rods into the ion trap of a Nicolet FTMS-2000 instrument, and laser-photodissociated to give the illustrated mass spectrum. The masses labeled on the spectrum are the observed masses, whereas the masses labeled on the peptide structure (underlining denotes an observed fragment) are the calculated monoisotopic masses. The FT/ICR mass spectrum of Figure 7.28 is dominated by B-type fragment ions and/or B-fragments resulting from loss of H_2O from threonine-containing fragments (denoted by open circles in the Figure). In this example, the entire polypeptide sequence could be determined from the observed B and Y fragment ions.

FT/ICR mass spectrometry can be used for *synthesis* as well as *analysis*. In example shown in Figure 7.29, methoxide (CH_3O^-) ions were allowed to react with trifluoroacetoxydodecahedrane, to give the observed trifluoroacetate ($CF_3CO_2^-$) product ion:

$$C_{20}H_{19}\text{–}OC(=O)CF_3 + CH_3O^- \longrightarrow C_{20}H_{18} + CH_3OH + {}^-OC(=O)CF_3$$

Although the presumed *neutral* product, dodecahedrene, was not observed directly, the proposed reaction mechanism could be inferred from the following additional facts. First, nucleophilic attack by methoxide at the carbonyl carbon of trifluoroacetoxydodecahedrane could be ruled out, because the same reaction run with $CH_3{}^{18}O^-$ gave solely $CF_3C^{16}O_2^-$, with no $CF_3C^{16}O^{18}O^-$. Second, a major FT/ICR mass spectral peak was observed at m/z = 147, corresponding to the proton-bound adduct, $F_3CCO_2^-(H^{18}OCH_3)$, which would be expected only if methoxide attacks at the C–O bond between trifluoroacetate and dodecahedrane in an elimination reaction. $F_3CCO_2^-(HOCH_3)$ cannot be formed by reaction of $CF_3CO_2^-$ with CH_3OH, because that reaction is sufficiently exothermic that any proton-bound adduct would dissociate promptly after formation. Third, since the reaction of trifluoroacetoxydodecahedrane with HO^- or CH_3O^- (but not $CH_3CH_2O^-$) proceeds to form $CF_3C^{16}O_2^-$, and since the relative basicities of the three reagent anions are known, it can be inferred that ~40 kcal/mole more energy is required to form dodecahedrene from trifluoroacetoxydodecahedrane than to form ethene from ethyl trifluoroacetate. In other words, the above experiments show that dodecahedrene, a molecule not yet synthesized by wet chemistry, exists, and that its carbon-carbon double-bond strain energy is among the highest known.

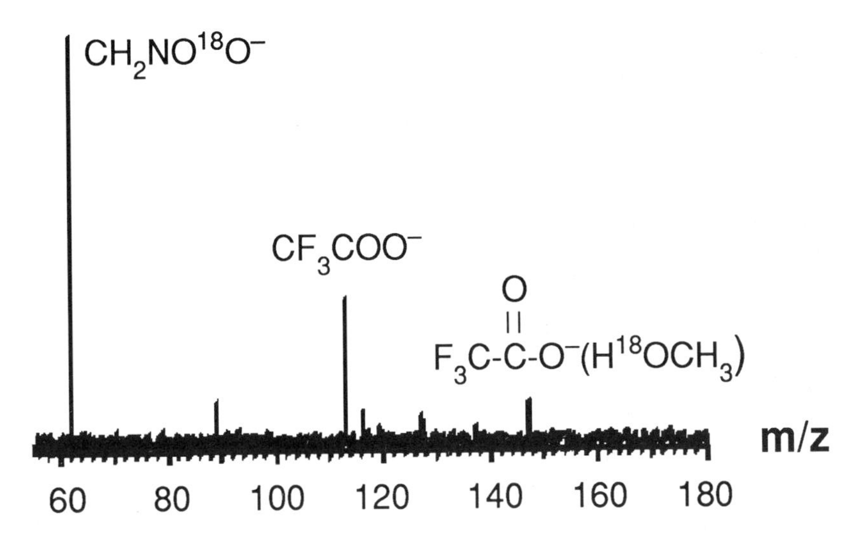

Figure 7.29 FT/ICR mass spectrum (Nicolet FTMS-2000 at 3.0 tesla) of the ionic products of the reaction of neutral trifluoroacetoxydodecahedrane with methoxide ion. ($CH_3NO^{18}O^-$ at m/z = 62 originates from the methyl nitrite reagent needed to generate $CH_3{}^{18}O^-$ for this reaction.) From these and other data, the gas-phase synthesis of dodecahedrene could be inferred (see text). [Reproduced, with permission, from J. A. Kiplinger, F. R. Tollens, A. G. Marshall, T. Kobayashi, D. R. Lagerwall, L. A. Paquette, & J. E. Bartmess, *J. Amer. Chem. Soc.* **1989**, 111 (11).]

7.5.2.3 *Fourier and Hadamard multiplex detection of ion-molecule collision or reaction products: two-dimensional MS/MS*

One of the most powerful methods for structure determination of gas-phase ions is collision-induced dissociation (CID). In this technique, an ion of selected m/q is accelerated to high speed, then allowed to collide with a neutral inert gas, and the structure of the original ion is inferred from the masses of the collision-induced fragments. In its conventional implementation, the technique requires two mass spectrometers (MS/MS): a first mass spectrometer to "filter out" ions of all but the desired m/q, and a second mass spectrometer to scan the m/q values of the product ions. With FT/ICR, the whole experiment can be conducted in a single instrument, by use of the event sequence of Figure 7.24. Moreover, the high mass resolution of FT/ICR provides for determination of the chemical formulas of *daughter* ions as well as parent ions. For example, one can be sure that a loss of ~29 u corresponds to a $\cdot C_2H_5$ group, if the "exact" mass of the loss is sufficiently close to 29.0391 u.

Nevertheless, even the FT/ICR CID experiment detects the daughters from only a *single* parent ion at a time. Thus, it would appear to be necessary to perform N separate CID experiments to obtain the full two-dimensional array of CID patterns from N parent ions. In this section, we shall describe two quite different multiplex methods for introducing the multichannel advantage into the *second* dimension of the CID experiment.

Two-dimensional Hadamard/Fourier transform ICR mass spectrometry

The Hadamard S-matrix encoding scheme of Chapter 4.1.3.1 offers a means for simultaneous detection of the daughter ions formed by CID (or other ion fragmentation methods, such as laser photodissociation) from approximately half of the possible parent ions. The Hadamard MS/MS experiment is shown schematically in Figure 7.30. Consider an ionized mixture containing N parent ions. [In the simple example of Figure 7.30, there is a prominent molecular ion, M^+ (see Figure 7.30a), for each of 11 components of a mixture of thiophene, cyclohexanone, o-xylene, 3-methylanisole, 2-methylphenol, *n*-butylbenzene, 1-fluoro-1-nitrobenzene, 1-fluoro-α-trifluorotoluene, *p*-bromo-fluorobenzene, 1,2,4,-trichlorobenzene, and p-bromoanisole.] Stored-waveform inverse FT (SWIFT) excitation (Figure 7.30b) makes it possible to remove all but the parent ions of interest. The ion mixture is first subjected to irradiation at the ICR frequencies corresponding to $(N + 1)/2 = 6$ of a possible $N = 11$ parent ion m/q-values, thereby "heating" those ions to sufficiently high speed that collisions with neutrals result in extensive fragmentation to yield daughter ions from each of the irradiated parent ions. Subsequent broad-band excitation/detection then yields an FT/ICR mass spectrum from all daughter ions arising from CID of the $(N + 1)/2$ selected parent ions (e.g., Figure 7.30c). At this stage, we have acquired one "observable" ICR mass spectrum, representing a linear combination of the CID products of roughly half of the N possible parent ions.

Next, a different (linearly independent) combination of $(N + 1)/2$ parent ions is selected for CID and FT/ICR excitation/detection, to yield a second "observable" mass spectrum. After N such experiments, each based on a different (Hadamard-encoded) combination of $(N + 1)/2$ parent ions, we will have accumulated N "observable" MS/MS spectra representing N linear combinations of the desired N *unknown* daughter ion mass spectra (one for each parent ion). The Hadamard *encoding* provides for simple *decoding* (see Chapter 4.1.3.1), to yield the final desired daughter ion mass spectrum (e.g., Figure 7.30d) for each of the N original parent ions.

Although the Hadamard experiment takes the same total experimental data acquisition period as one-at-a-time selective excitation of each of N parent ions in turn to give N daughter-ion product spectra, the signal-to-noise ratio of each of the N decoded daughter ion mass spectra is increased by a factor of $\sqrt{(N+1)/2}$, as seen in Figure 7.30d (or a time-saving of a factor of $N/4$ to reach a given signal-to-noise ratio).

Two-dimensional Fourier/Fourier transform ICR mass spectrometry

Figure 7.31 shows a rather different two-dimensional CID experiment modeled after the prior two-dimensional FT/NMR "NOESY" experiment (see next Chapter). The key element of the 2D-FT/ICR experiment is the introduction of an excitation/delay/excitation (P1 $- t_1 -$ P2) sequence followed by a reaction period, τ_m, between ion formation and the final excitation/detection stage (Figure 7.31, top).

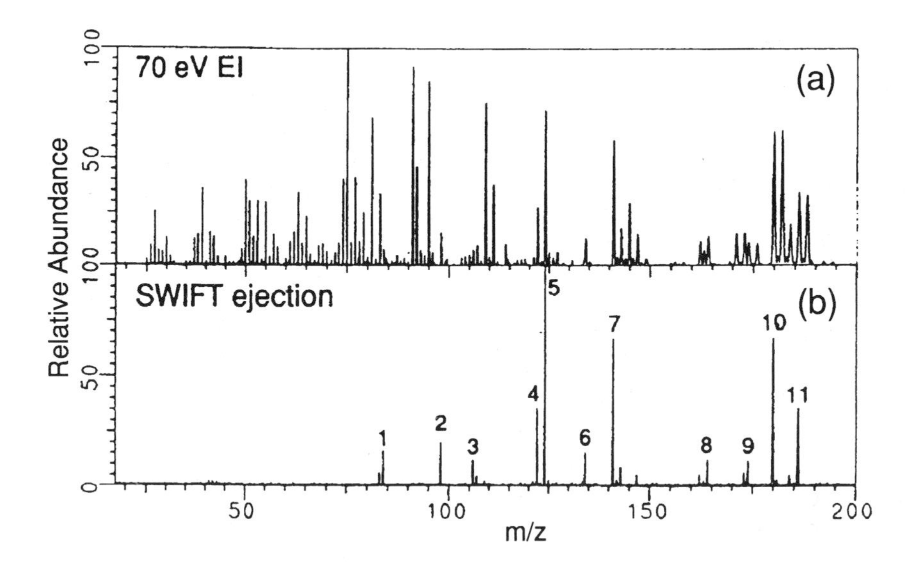

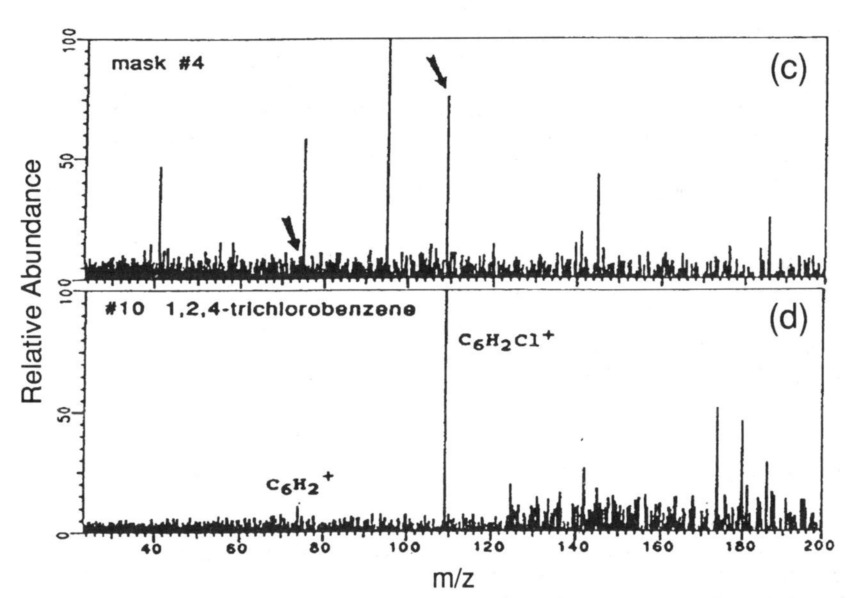

Figure 7.30 Hadamard FT/ICR mass spectra (Nicolet FTMS-2000 at 3.0 tesla). (a) Electron-ionized 11-component mixture (see text). (b) Same system after SWIFT ejection of all but the parent ion of each component. (c) Spectrum obtained from collision-induced dissociation of one of the eleven SWIFT-selected combinations of 6 parent ions. (d) CID mass spectrum of component #10, reconstructed from Hadamard transformation of the 11 CID spectra obtained as in (c). (Adapted, with permission, from E. R. Williams, S. Y. Loh, F. W. McLafferty & R. B. Cody, Jr., *Anal. Chem.* **1989**, *61*, 0000-0000.)

For simplicity, consider parent A^+ ions of a single m/q value, formed along the z- (magnetic field-) axis at time, zero. A^+ ions are first excited by a single-frequency on-resonance rf pulse, P1, to a large ICR orbital radius (say, ~50% of the radius of the ion trap). After a subsequent delay period, t_1, the same A^+ ions are subjected to a second on-resonance single-frequency rf transmitter pulse, P2. If t_1 is exactly k cycles ($k = 0, 1, 2, \cdots$) of the ICR frequency for those ions, then the second rf pulse, P2, will be exactly *in-phase* with the coherent ICR orbital motion, and A^+ ions will be driven coherently outward until they approach the transverse plates of the trap and are lost. On the other hand, if t_1 is exactly $k + (1/2)$ cycles ($k = 0, 1, 2, \cdots$) of the ICR frequency, then P2 will be exactly *180° out-of-phase* with the A^+ ICR orbital motion, and the P2 rf field will oppose the ion motion so as to drive the A^+ ions in a spiral back to their origin on the z-axis of the ion trap. Thus the *number of A^+ ions* available for ion-molecule reactions during the subsequent reaction period, τ_m, is modulated approximately cosinusoidally as a function of the delay period, t_1, between P1 and P2.

Next, the final N-point time-domain signal from the event sequence of Figure 7.31 (top) is Fourier transformed and stored for each of M (equally incremented) delay periods, t_1, between the first pair of rf pulses. (M need not be the same as N.) The modulation of the ICR A^+ spectral magnitude with t_1 is evident in Figure 7.31 (middle). Finally, each M-point "time-domain" data set consisting of the (complex) spectral value at each spectral frequency versus t_1 is Fourier transformed to give the two-dimensional spectrum shown in Figure 7.33 (bottom).

If the A^+ parent ions in the 2D-FT/ICR experiment do not collide or react with neutrals during the above event sequence, and if the modulation in A^+ ICR signal magnitude induced by the time-separated pair of rf pulses is purely sinusoidal, then the 2D-FT/ICR mass spectrum will yield a parent ion mass spectrum located on the *diagonal* of plotted array of Fig. 7.33 (bottom). [In practice, since the modulation is not purely sinusoidal, peaks are also observed at harmonic (i.e., integral multiple) frequencies of the fundamental ICR orbital frequencies.]

However, if parent ions, A^+, are induced to fragment (e.g., by ion-neutral collisions, laser photodissociation, etc.) to form daughter B^+ ions during the additional reaction period, τ_m, then B^+ ions (which are unaffected by the second rf pulse, P2) will be excited (by P3) and detected during the final stage of the event sequence. The *number* of B^+ ions available for detection is proportional to the *number* of A^+ ions available at the beginning of τ_m, and the number of available A^+ ions is in turn modulated by the delay period, t_1. The net result is that the 2D-FT/ICR mass spectrum (Fig. 7.33, bottom) shows *off-diagonal* peaks corresponding to the products of fragmentation or ion-molecule reactions of A^+ to form B^+.

At this writing, it is not yet clear whether the Hadamard/FT or FT/FT approach will prove most useful in practice. The Hadamard approach has the advantage that the number of acquisitions in one of the MS/MS dimensions is simply the number of parent ions to be analyzed, whereas the FT/FT method requires a (generally much larger) number of t_1-values determined by the required digital resolution in the second FT dimension. The Hadamard approach thus requires substantially less data storage, and is moreover free of harmonic distortion. However, the FT/FT method requires no prior knowledge of the parent ion spectrum (other than the specified bandwidth), and its FT/NMR analog has proved very successful. Both methods will benefit from use of SWIFT excitation for optimal m/q selectivity.

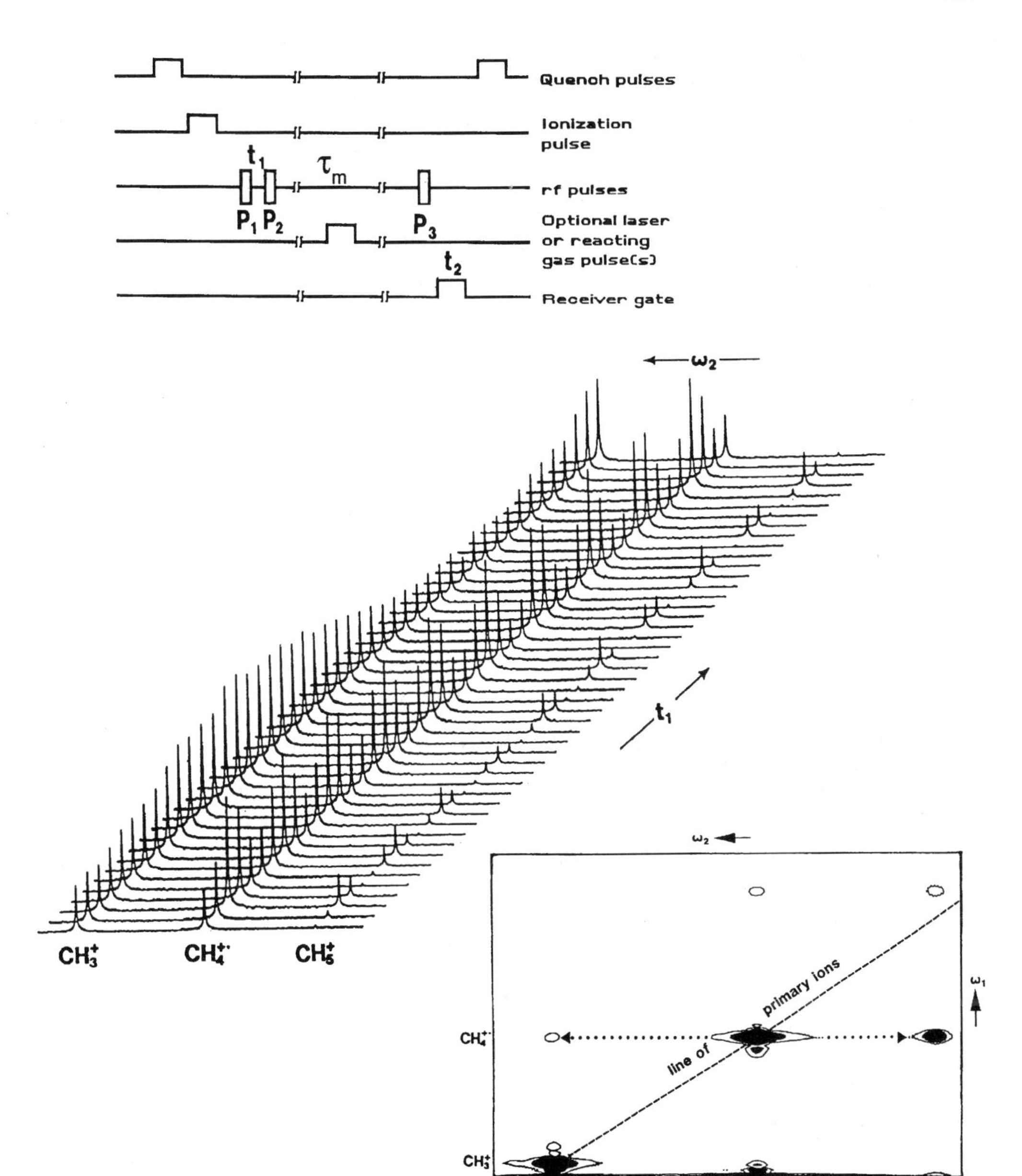

Figure 7.31 Two-dimensional FT/ICR spectroscopy. Top: Experimental event sequence. Middle: Modulation of FT/ICR magnitude-mode spectral peak heights as a function of delay period, t_1, in the event sequence. Bottom: Two-dimensional FT/ICR spectrum (displayed in countours of constant magnitude-mode spectral height) of electron-ionized methane, CH_4. Off-diagonal peaks (see dotted lines) result from the formation of CH_3^+ and CH_5^+ daughter ions from CH_4^+ parent ions. (Adapted, with permission, from P. Pfändler, G. Bodenhausen, J. Rapin, M.-E. Walser, and T. Gäumann, *J. Amer. Chem. Soc.* **1988**, *110*, 5625-5628)

Further Reading

B. Asamoto, *Spectroscopy* **1988**, *3*, 38-46. (Reviews industrial applications)

M. V. Buchanan, Ed. *Fourier Transform Mass Spectrometry: Evolution, Innovation, and Applications, ACS Symp. Series,* **1987**, *359*, 205 pp. (Book with 11 contributed chapters on FT/ICR principles and applications)

R. B. Cody, Jr., A. Bjarnason & D. A. Weil, in D. M. Lubman, Ed., *Lasers in Mass Spectrometry* (Oxford U. Press, NY, 1988), in press. (Good coverage of laser desorption/ionization techniques and applications)

M. B. Comisarow, *J. Chem. Phys.* **1978**, *69*, 4097-4104. (The fundamental reference for ICR signal generation and detection)

M. B. Comisarow, in *Lecture Notes in Chemistry*, Vol. 31, *Ion Cyclotron Resonance Spectrometry II*, Eds. H. Hartmann and K.–P. Wanczek (Springer-Verlag, Berlin, 1982). Good discussion of homogeneous and inhomogeneous relaxation.

B. S. Freiser, in *Techniques for the Study of Ion-Molecule Reactions*, ed. J. M. Farrar and W. Saunders, Jr., Wiley, NY, 1988, Chapter 2. (Overview of FT/ICR techniques, with emphasis on chemical information)

M. L. Gross & D. L. Rempel *Science* **1984**, *226*, 261-268. (Review of early work)

C. D. Hanson, E. L. Kerley & D. H. Russell, D. H., in *Treatise on Analytical Chemistry*, 2nd ed.; Vol. 11; Wiley: NY, 1988; Chapter 2. (Review of FT/ICR techniques and applications)

C. L. Wilkins, A. K. Chowdhur, L. M. Nuwaysir & M. L. Coates, *Mass Spectrom. Rev.* **1989**, *8*, 67-92. (Inclusive review of FT/ICR applications from 1986-88)

A. G. Marshall, *Acc. Chem. Res.* **1985**, *18*, 316-322. (Historical review of the development of FT/ICR mass spectrometry)

A. G. Marshall, M. B. Comisarow & G. Parisod, *J. Chem. Phys.* **1979**, *71*, 4434-4444. (Fundamental relations between relaxation and FT/ICR spectral line shape)

A. G. Marshall, T.-C. L. Wang, L. Chen & T. L. Ricca, *A.C.S. Symp. Ser.*, ed. M. V. Buchanan **1987**, *359*, 21-33. (Review of SWIFT excitation)

A. G. Marshall, *Adv. Mass Spectrom.* **1989**, *11A*, 651-669. (Review of analytical applications)

N. M. M. Nibbering, *Adv. Phys. Org. Chem.* **1988**, *24*, 1-55. (Thorough review of uses of ICR for study of ion-molecule chemistry)

P. Pfändler, G. Bodenhausen, J. Rapin, M.-E. Walser & T. Gäumann, *J. Amer. Chem. Soc.* **1988**, *110*, 5625-5628. (Two-dimensional FT/ICR MS/MS)

C. Giancaspro & F. R. Verdun, *Anal. Chem.* **1986**, *58*, 2097-2099. (Time-of-flight mass discrimination from transfer of ions back and forth in an ICR dual-trap)

K.–P. Wanczek, *Int. J. Mass Spectrom. Ion Proc.* **1984**, *60*, 11-60. (Comprehensive ICR literature review to that date)

E. R. Williams, S. Y. Loh, F. W. McLafferty & R. B. Cody, Jr., *Anal. Chem.* **1989**, *61*, 0000-0000. (Hadamard FT/ICR MS/MS)

Problems

7.1 The shift in ICR orbital frequency due to the "trapping" potential may be computed from Eq. 7.28a. However, since the "trapping shift" is small, you are asked to expand the square root expression in Eq. 7.28a in a Taylor series:

$$\sqrt{1-x} \cong 1 - \frac{x}{2} + \frac{x^2}{4} - \cdots$$

Keeping only the first two terms in the expansion, obtain an expression for the "trapping shift" as a function of V_T, for the quadrupolar electric potential in a cubic trap (Eq. 7.26). You should find that the shift is negative (for positive ions) and independent of m/q.

Finally, compute the "trapping shift", $\partial \nu_T / \partial V_T$, in Hz/volt, for positive ions confined in a cubic trap of 2.54 cm plate-to-plate separation, in a magnetic field of 3.0 tesla.

7.2 Derive Eq. 7.21 for ICR on-resonance power absorption. Hint: Substitute Eq. 7.19c into Eq. 7.1; solve for $\boldsymbol{v}_{xy}$, and substitute for $\boldsymbol{v}_{xy}$ in Eq. 7.20.

7.3 The cyclotron and magnetron "natural" frequencies, ω_0 and ω_m, of Eqs. 7.28 are simply the solutions of the quadratic Eq. 7.27. The "trapping" frequency, ω_T, is given by Eq. 7.15. Compute the unshifted and shifted ion cyclotron orbital frequencies, ν_c and ν_0; the magnetron orbital frequency, ν_m; and the trapping oscillation frequency, ν_T (in Hz) for singly charged positive ions of m/z = 100 and 1000, in a magnetic field of 3.0 tesla, for a cubic trap with $1.0V$ applied to each of the "trapping" electrodes separated by 2.54 cm along the z-axis. Use the three-dimensional quadrupolar potential of Eq. 7.25. Finally, derive an expression for the m/q value at which ω_0 and ω_T are equal, and evaluate that m/q for the above conditions.

7.4 Confirm the expressions for FT/ICR mass resolution in Table 7.1 from the line width at half-maximum peak height for absorption-mode and magnitude-mode sinc and Lorentzian line shapes.

7.5 Show that the FT/ICR mass calibration Eq. 7.34 follows from Eqs. 7.28a and 7.26.

7.6 Evaluate the FT of a rectangular time-domain pulse, $f(t)$ with exponential rise and fall, and plot the FT magnitude-mode spectrum.

$$f(t) = A_0 (1 - \exp(-t/\tau), \quad 0 \le t < T$$

$$f(t) = A_0 \exp(-t/\tau), \qquad T \le t < \infty$$

$$f(t) = 0, \quad t < 0,$$

7.7 Suppose that a coherently excited ion packet of ions of a single m/q ratio reaches an ion cyclotron orbital radius, r, which closely approaches the radius, r_0 of a cylindrical ion trap. The resultant detected ICR signal on one pair of opposed plates (upper right in Figure 7.16) will then approach the waveform shown at the middle right in Figure 7.16.

Assume that the modulation of the waveform at middle right in Figure 7.16 is perfectly rectangular, and compute analytically its FT magnitude-mode spectrum. (You may assume that the number of ions in the coherent packet decreases exponentially with time).

7.8 In this problem, you show that a coherent ICR signal can be "de-excited" by changing the phase of the excitation relative to the phase of the coherent ICR orbital motion.

(a) Suppose that ions of mass-to-charge ratio, m/q, are formed with zero velocity at the origin, in the presence of an applied static magnetic field, $\mathbf{B} = B_0 \mathbf{k}$. At time zero, apply a circularly polarized resonant rf electric field,

$$\mathbf{E} = E_0 \cos(\omega_c t)\,\mathbf{i} + E_0 \sin(\omega_c t)\,\mathbf{j}, \quad 0 \leq t \leq T \tag{7.8.1}$$

At time, T, suppose that the excitation phase changes immediately by 180°:

$$\mathbf{E} = E_0 \cos(\omega_c t + \pi)\,\mathbf{i} + E_0 \sin(\omega_c t + \pi)\,\mathbf{j}$$

$$= -E_0 \cos(\omega_c t)\,\mathbf{i} - E_0 \sin(\omega_c t)\,\mathbf{j}, \quad T \leq t \leq 2T \tag{7.8.2}$$

Write a short algorithm to compute and plot the trajectory (i.e., ionic x- and y-position as a function of time) between $0 \leq t \leq 2T$. (This exercise forms the basis for 2-dimensional FT/ICR Chapter 7.5.2.3.)

(b) A related and useful exercise is to plot the magnitude-mode FT of a frequency-sweep from ω_A to ω_B in T seconds, in which the phase of the frequency-sweep changes suddenly by 180° half-way through the sweep (i.e., at $t = T/2$). You should obtain a magnitude-mode excitation spectrum which produces essentially flat power over the range from ω_A to ω_B, with a sharp "notch" at the frequency of the phase-shift (in this case, $(\omega_A + \omega_B)/2$. Although less uniform in frequency-domain magnitude than a SWIFT excitation, this simple "notch ejection" technique (A. J. Noest & C. W. F. Kort, *Comput. Chem.* **1983**, *7*, 81-86) has proved useful for removal of ions of all but a narrow range of m/q values for MS/MS experiments (see Figure 7.23d and L. J. de Koning, R. H. Fokkens, F. A. Pinske & N. M. M. Nibbering, *Int. J. Mass Spectrom. Ion Proc.* **1987**, *77*, 95-105).

Solutions to Problems

7.1 Begin by combining Eqs. 7.28a and 7.26:

$$\omega_0 = \frac{qB_0 + \sqrt{q^2 B_0^2 - 4mqE_0}}{2m} \quad \text{(Cyclotron frequency)} \qquad (7.28a)$$

$$E_0 = \frac{2\alpha V_T}{a^2} \qquad (7.26)$$

to give

$$\omega_0 = \frac{qB_0 + \sqrt{q^2 B_0^2 - \dfrac{8mq\alpha V_T}{a^2}}}{2m} \qquad (7.1.1)$$

Next, since $\omega_c = \dfrac{qB_0}{m}$ (7.4)

Eq. 7.1.1 may be simplified to the form,

$$\omega_0 = \frac{\omega_c}{2}\left(1 + \sqrt{1 - \frac{8\alpha qV_T}{ma^2 \omega_c^2}}\right) \qquad (7.1.2)$$

Expanding the square root expression in the first two terms of a Taylor series as suggested, we obtain

$$\omega_0 = \frac{\omega_c}{2}\left(1 + 1 - \frac{4\alpha qV_T}{ma^2 \omega_c^2}\right)$$

$$= \omega_c - \frac{2\alpha V_T}{a^2 B_0} \text{ rad s}^{-1} \qquad (7.1.3a)$$

or

$$\nu_0 = \nu_c - \frac{\alpha V_T}{\pi a^2 B_0} \text{ Hz} \qquad (7.1.3b)$$

Substitution of $\alpha = 1.386$, $B_0 = 3.0$ tesla, $V_T = 1.0$ V, and $a = 0.0254$ m into Eq. 7.1.3b yields a trapping shift of ~228 Hz/volt at 3.0 tesla. Note from Eq. 7.1.3 that the trapping shift is always negative and is independent of mass and charge. Finally, note that the magnetron frequency, computed to the same level of approximation, is given by

$$\omega_m = \frac{2\alpha V_T}{a^2 B_0} \text{ rad s}^{-1} \qquad (7.1.4a)$$

or

$$\nu_m = \frac{\alpha V_T}{\pi a^2 B_0} \text{ Hz} \qquad (7.1.4b)$$

Thus, the magnetron frequency has the same magnitude as the trapping shift, and is also independent of m/q.

7.2 See J. L. Beauchamp, *J. Chem. Phys.* **1967**, *46*, 1231-1243.

7.3 $\nu_c = \dfrac{q B_0}{2 \pi m}$ = 460,632 Hz for m/z = 100

= 46,063 Hz for m/z = 1,000

$$\nu_0 = \nu_c - \frac{\alpha V_T}{\pi a^2 B_0} = \nu_c - 228 \text{ Hz (see Problem)}$$

$$\nu_m = \frac{\alpha V_T}{\pi a^2 B_0} = 228 \text{ Hz (see next Problem)}$$

To determine the trapping frequency, we begin by computing the z-component of the electric field from the negative gradient of the three-dimensional quadrupolar potential of Eq. 7.25a:

$$E(z) = -\frac{\partial V_Q(x, y, z)}{\partial z} = -\frac{4 \alpha V_T z}{a^2} \tag{7.3.1}$$

Since force is the product of charge and electric field, the equation for ion motion in the z-direction then becomes,

$$m \frac{d^2 z}{d t^2} = q E(z) = -\frac{4 \alpha q V_T z}{a^2} \tag{7.3.2}$$

Eq. 7.3.2 is clearly of the weight-on-a-spring form,

$$m \frac{d^2 z}{d t^2} = -k z \tag{7.3.3}$$

with $\quad k = \dfrac{4 \alpha q V_T}{a^2}$

Therefore, the trapping frequency, ν_T, is given by

$$\nu_T = \frac{1}{2 \pi} \sqrt{\frac{k}{m}} = \frac{1}{\pi a} \sqrt{\frac{\alpha q V_T}{m}} \tag{7.3.4}$$

Note that the trapping electric field (and thus the trapping frequency) is slightly higher (by a factor of $\sqrt{1.386} \approx 1.18$) for the three-dimensional (Eq. 7.25) than for the two-dimensional quadrupolar potential (Eq. 7.16). For the cubic trap, $\alpha = 1.386$, and

ν_T = 14,492 Hz for m/z = 100;

ν_T = 4,583 Hz for m/z = 1,000

Finally, set the cyclotron orbital frequency (Eq. 7.4) equal to the trapping frequency (Eq. 7.3.4), and solve for m:

$$\frac{qB_0}{m} = \frac{2}{a}\sqrt{\frac{\alpha q V_T}{m}} \tag{7.3.5}$$

$$m = \frac{qB_0^2 a^2}{4\alpha V_T} = 101{,}054 \text{ u at } 3.0 \text{ tesla, } a = 2.54 \text{ cm, } V_T = 1 \text{ V.}$$

7.4 See: A. G. Marshall, M. B. Comisarow & G. Parisod, *J. Chem. Phys.* **1979**, *71*, 4434-4444.

7.5 Again, begin by combining Eqs. 7.28a and 7.26 to obtain Eq. 7.1.2; then rearrange slightly to give

$$\omega_0 - \frac{\omega_c}{2} = \frac{1}{2}\left(\sqrt{\omega_c^2 - \frac{8\alpha q V_T}{m a^2}}\right) \tag{7.5.1}$$

Next, square both sides:

$$\omega_0^2 - \omega_0\omega_c + \frac{\omega_c^2}{4} = \frac{\omega_c^2}{4} - \frac{2\alpha q V_T}{m a^2}$$

$$\text{or} \quad \omega_0^2 = \omega_0\omega_c - \frac{2\alpha q V_T}{m a^2} \tag{7.5.2}$$

Now substitute for $\omega_c = (q B_0/m)$ on the right-hand side of Eq. 7.4.2:

$$\omega_0^2 = \left(\omega_0 B_0 - \frac{2\alpha V_T}{a^2}\right)\frac{q}{m} \tag{7.5.3}$$

Finally, simply solve Eq. 7.4.3 for (m/q):

$$\frac{m}{q} = \frac{\omega_0 B_0 - \dfrac{2\alpha V_T}{a^2}}{\omega_0^2}$$

$$\text{or} \quad \boxed{\frac{m}{q} = \frac{B_0}{\omega_0} - \frac{\dfrac{2\alpha V_T}{a^2}}{\omega_0^2} = \frac{a}{\omega_0} + \frac{b}{\omega_0^2}} \quad \textbf{Q.E.D.} \tag{7.5.4}$$

7.6 See Figure 7.13.

7.7 See Figure 7.16 (lower right).

7.8 (a) See Figure 7.22 (bottom left). For more details, see: A. G. Marshall, T.-C. L. Wang & T. L. Ricca, *Chem. Phys. Lett.* **1984**, *105*, 233-236.

(b)

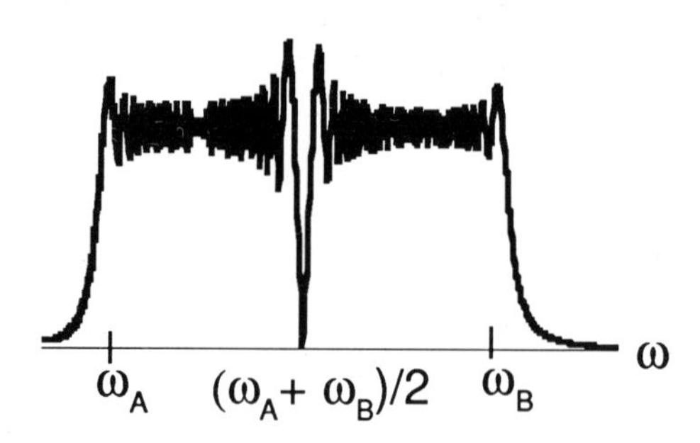

Chapter 8

Fourier transform nuclear magnetic resonance spectroscopy

8.1 Natural motion of a magnetic moment in a magnetic field: Larmor precession

8.1.1 Classical mechanical motion and energy of a magnetic moment in a static magnetic field: laboratory and rotating frame representations

Fourier transform nuclear magnetic resonance (FT/NMR) spectroscopy has been regularly and extensively reviewed and is the sole subject of more than a dozen monographs (see Further Reading). We will therefore limit this chapter to an introduction to some basic principles, followed by some representative nuclear spin manipulations and applications.

The fundamental concepts for a classical mechanical description of NMR are shown in Figure 8.1 and expressed algebraically in Eqs. 8.1 and 8.2. First, a static magnetic field, $\boldsymbol{B}_0$ (in tesla), acting on a classical mechanical bar magnet with magnetic (dipole) moment, μ (in ampere m^{-2}) produces a *torque* (kg m^2 s^{-2}) given by Eq. 8.1.

$$\text{Torque} = \frac{d\boldsymbol{I}}{dt} = \mu \times \boldsymbol{B}_0 \qquad (8.1)$$

in which $\boldsymbol{I}$ is the angular momentum associated with the moving bar magnet.

Second, a classical mechanical magnetic field is generated by charges in motion. Thus, it is natural to find that the magnitude (and direction) of a magnetic (dipole) moment, μ, is proportional to the magnitude of the angular momentum, $\boldsymbol{I}$, of a spinning charge distribution:

$$\mu = \gamma \boldsymbol{I} \qquad (8.2)$$

in which the proportionality constant, γ, is called the *magnetogyric ratio.* Eqs. 8.1 and 8.2 may be combined to yield the equation of motion for a bar magnet in a static magnetic field:

$$\frac{d\mu}{dt} = \gamma\mu \times \boldsymbol{B}_0 \qquad (8.3)$$

Since the "strength" of a bar magnet (i.e., the length of the μ vector) is constant, the motion described by Eq. 8.3 is simply a *precession* of μ at angular velocity, ω_0, about $\boldsymbol{B}_0$, as shown in Figure 8.1 and described algebraically by Eq. 8.4.

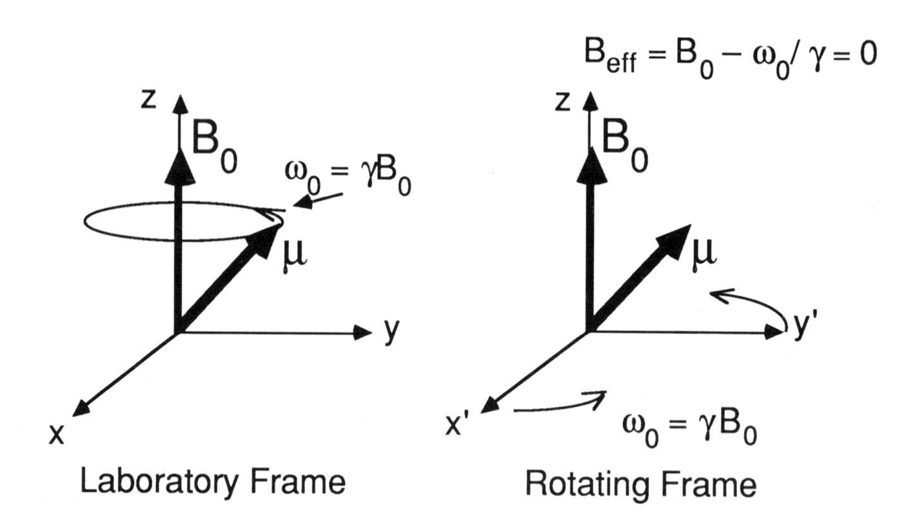

Figure 8.1 Larmor precession of a magnetic moment, μ, about a static magnetic field, $\boldsymbol{B}_0$, oriented along the laboratory-frame z-axis. Left: Magnetic moment rotating at the Larmor frequency about $\boldsymbol{B}_0$ in a static laboratory coordinate frame (x, y, z). Right: Static magnetic moment (i.e., as if no $\boldsymbol{B}_0$ field were present) in a coordinate frame (x', y', z) rotating about the laboratory-frame z-axis at the Larmor frequency.

$$\frac{d\mu}{dt} = \omega_0 \times \mu \tag{8.4}$$

Combining Eqs. 8.3 and 8.4, we find that

$$\boxed{\omega_0 = -\gamma \boldsymbol{B}_0 \ \text{(rad s}^{-1}\text{)}} \tag{8.5a}$$

$$\text{or} \quad \boxed{\nu_0 = \frac{|\omega_0|}{2\pi} = \frac{\gamma |\boldsymbol{B}_0|}{2\pi} \ \text{(Hz)}} = \text{Larmor (precession) frequency} \tag{8.5b}$$

An extremely useful alternative representation of Larmor precession (see Figure 8.1) is a coordinate frame whose x-y plane rotates at the Larmor frequency about the $\boldsymbol{B}_0$ axis (i.e., z-axis). Just as one rider on a merry-go-round appears stationary to another rider on the same merry-go-round, the magnetic moment, μ, is fixed in direction in a coordinate frame rotating at the same angular velocity as μ. However, because the laws of physics must be preserved in any coordinate frame, we conclude that the applied static magnetic field must be zero in the rotating frame—otherwise, the magnetic moment would be forced to precess about the static field. The great value of the famous "rotating frame" representation is that the rapid motion of μ about $\boldsymbol{B}_0$ can be eliminated, thereby vastly simplifying the analysis of further μ motions induced by resonant oscillating magnetic fields (see below).

More generally, the *effective* magnetic field, $\boldsymbol{B}_{\mathrm{eff}}$, in a coordinate frame rotating at frequency, ω, about $\boldsymbol{B}_0$ is given by

$$\boldsymbol{B}_{\mathrm{eff}} = \left(B_0 - \frac{\omega}{\gamma}\right)\mathbf{k} \qquad (8.6)$$

Finally, the *energy* of a classical bar magnet of magnetic moment, μ, oriented at (polar) angle, θ, with respect to a static magnetic field, $\boldsymbol{B}_0$, is given by the dot product (see Figure 8.2):

$$\text{Energy} = -\mu \cdot \boldsymbol{B}_0 = -|\mu|\, B_0 \cos\theta. \qquad (8.7)$$

Thus, the magnetic energy is a maximum or minimum when the magnetic moment is aligned against ($\theta = 180°$) or along ($\theta = 0°$) the applied magnetic field. Alternatively, the magnetic energy is zero when μ is perpendicular ($\theta = 90°$) to $\boldsymbol{B}_0$, so that a magnet may be rotated by arbitrary (azimuthal) angle, ϕ, in the x-y plane without changing its energy. We shall return to these properties when we have considered the differences between classical and quantum mechanical magnetic moments.

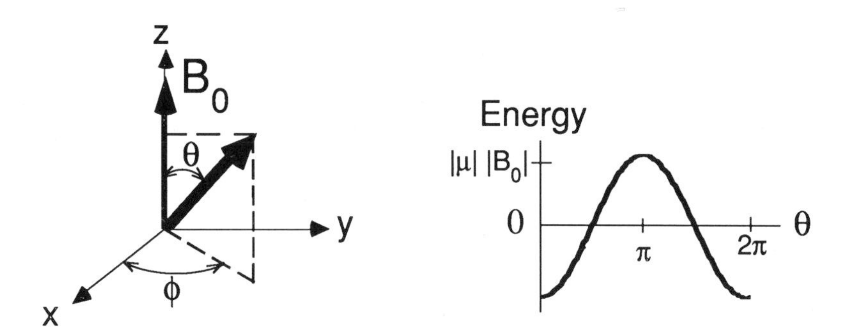

Figure 8.2 Energy of a classical mechanical magnetic moment, μ, as a function of (polar) angle, θ, between μ and an applied magnetic field, $\boldsymbol{B}_0 = B_0\,\mathbf{k}$. Note that the energy is independent of the (azimuthal) angle, ϕ, subtended by μ in the x-y plane.

8.1.2. Quantum mechanical magnetic nuclei

In classical electromagnetic theory, a rotating charge generates a magnetic field whose magnitude is proportional to the rate of rotation (more precisely, the angular momentum) of the rotating charge. Since several atomic *nuclei* (e.g., ^{1}H, ^{13}C, ^{31}P, etc.) have "spin" angular momentum, $\boldsymbol{I}_n$, and are positively charged, they exhibit a *nuclear* magnetic (dipole) moment, μ_n, proportional to $\boldsymbol{I}_n$ as in the classical mechanical case:

$$\mu_n = \gamma_n \boldsymbol{I}_n \qquad (8.8)$$

with proportionality constant, γ_n, the magnetogyric ratio characteristic of that nucleus.

However, quantum mechanical magnetic nuclei differ from classical bar magnets in several important ways. First, a proton (for example) is so small that it would have to spin faster than the speed of light to produce its observed magnetic moment by the classical model of a spinning charge. Thus, we are forced to accept that the proton has a nuclear magnetic moment, even though we cannot account for its magnitude by classical arguments. Second, although both classical and quantum mechanical magnetic moments may point in arbitrary direction with respect to an applied magnetic field, the direction of a quantum mechanical magnetic moment is experimentally *observed* to point only along the magnetic field direction. Third, the z-component of μ_n is quantized (i.e., is observed to have only $2I+1$ values, where I denotes the "nuclear spin quantum number"), as shown in Figure 8.3. Thus, one commonly hears that the magnetic moment of a proton (for which $I = 1/2$, so that the (observable) z-component of μ_n has only two possible values) can point only "up" or "down" with respect to $\mathbf{B}_0$. In fact, a proton magnetic moment can point anywhere, but in a given experiment can only be *observed* to point "up" or "down" along $\mathbf{B}_0$.

A classical bar magnet has a fixed "strength" (i.e., fixed magnetic moment), and therefore only one possible energy for a given orientation of the magnet in a magnetic field (Eq. 8.7). In contrast, an individual quantum mechanical nucleus of spin, I, can have any of $2I+1$ possible z-components of angular momentum, $I(h/2\pi)$, $(I-1)(h/2\pi)$, $\cdots$, $-(I-1)(h/2\pi)$, $-I(h/2\pi)$, and hence $2I+1$ possible energies, E_m.

$$E_m = -\frac{\gamma h m}{2\pi} B_0, \quad m = -I, -I+1, \ldots, I-1, I \tag{8.9}$$

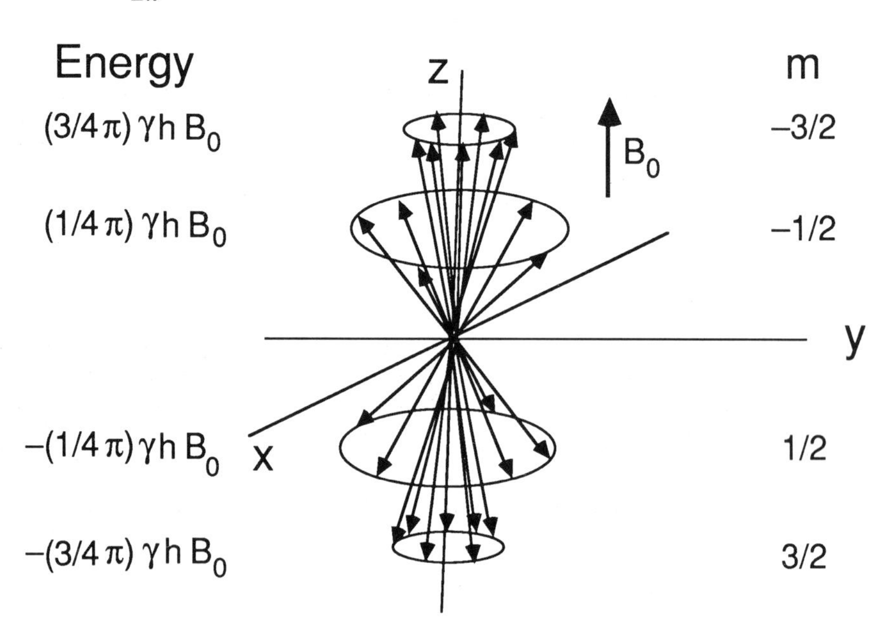

Figure 8.3 Discrete observable energies for a quantum mechanical magnetic nucleus of spin, $I = 3/2$, in a static magnetic field, $B_0 \mathbf{k}$. The $2I+1 = 4$ possible energies are determined by the quantum number, m, of the z-component of nuclear spin angular momentum (see text), and are independent of ϕ.

At thermodynamic equilibrium, the relative probability $P(m)$ that a given nucleus will be observed with magnetic energy, E_m, is given by the Boltzmann factor:

$$P(m) = \frac{\exp\left(\frac{-E_m}{kT}\right)}{\sum_{m=-I}^{I} \exp\left(\frac{-E_m}{kT}\right)} \tag{8.10}$$

in which k is the Boltzmann constant and T is temperature in Kelvin. For typical NMR experiments, $E_m << kT$, so that

$$\exp\left(\frac{-E_m}{kT}\right) \cong 1 - \frac{E_m}{kT} \tag{8.11}$$

An *individual* quantum mechanical magnetic nucleus can be observed to point only along or opposed to $\boldsymbol{B}_0$. However, a *macroscopic ensemble* of quantum mechanical magnetic nuclei will have a net magnetic moment given by the (vector) sum of the magnetic moments of the individual nuclei. For example, for an ensemble of N_0 protons, for which the individual angular momentum z-components are $\pm(h/4\pi)$, Eqs. 8.9 to 8.11 can be combined to yield an equilibrium macroscopic magnetic moment, $\boldsymbol{M}_0 = M_0\,\mathbf{k}$. For spins, $I = 1/2$ (see Problems),

$$\boldsymbol{M}_0 = M_0\,\mathbf{k}\ ; \quad M_0 = N_0\,\gamma^2 \left(\frac{h}{2\pi}\right)^2 \frac{B_0}{4\,k\,T} \tag{8.12}$$

Because the magnetic energy is independent of the azimuthal angle, ϕ, the ϕ-angles of the *individual* nuclear magnetic moments are randomly distributed around the z-axis (see Figure 8.3), so that the resultant net *macroscopic* magnetic moment, $\boldsymbol{M}_0$, lies along the z-axis. Finally, M_0 is proportional to N_0; thus, NMR spectral peak relative areas (see below) provide a direct measure of the relative numbers of nuclei at each Larmor frequency.

8.2 NMR parameters

The enormous practical value of FT/NMR is based on the direct relationship between various spectrally measurable NMR parameters and molecular structure, motions, and chemical reactivity: e.g., chemical bond type, number, and pattern; dihedral angle about a given bond; and intramolecular distances between atoms not directly bonded to each other. Modern FT/NMR experiments can be designed to detect particular chemically bonded groups of atoms (e.g., CH_3) or even to trace out the whole carbon-bonded skeleton of a molecule from a single experiment. In this section, we describe briefly the principal FT/NMR spectral parameters.

8.2.1 Chemical shift: identification of chemical bonds

The Larmor frequencies of some common magnetic nuclei are shown in Figure 8.4. Since FT/NMR spectral peak width is typically < 1 Hz for nuclei of spin, $I = 1/2$, it is clear that chemically different nuclei are readily distinguishable (e.g., ^1H and ^{13}C, whose Larmor frequencies differ by ~375 MHz at 11.75 tesla).

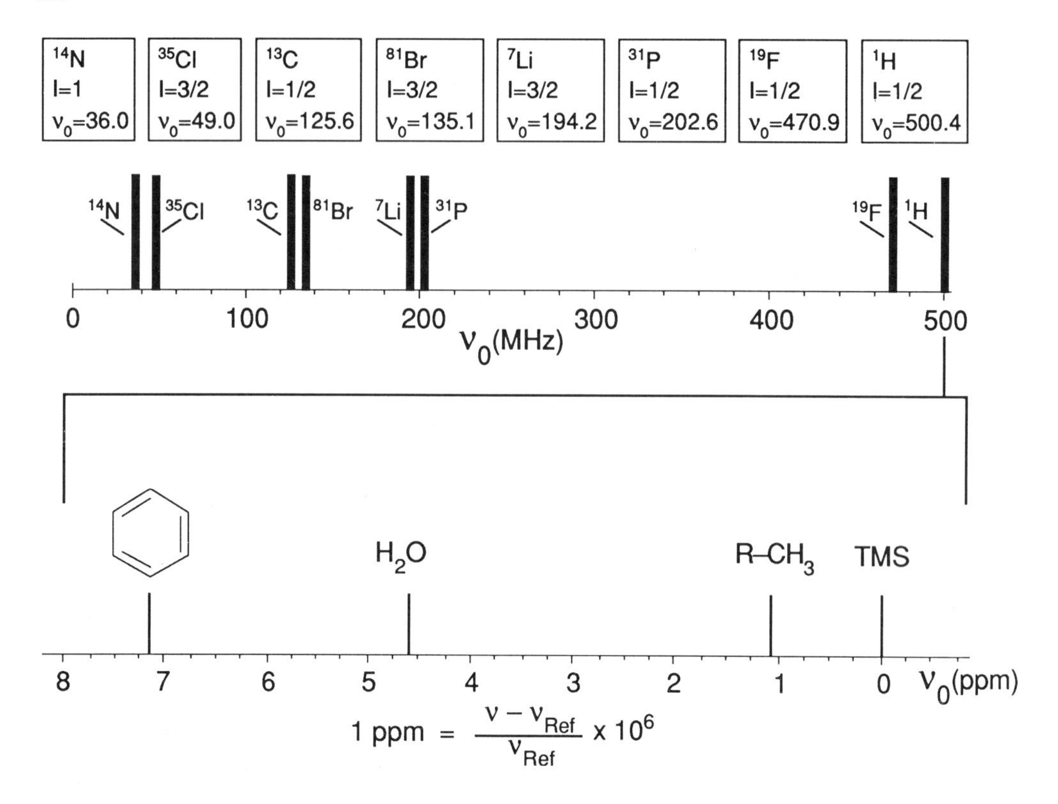

Figure 8.4 NMR Larmor frequencies (MHz) at 11.75 tesla for some representative chemically interesting magnetic nuclei. Top: schematic wide-range NMR spectrum, showing the huge separation ($>10^9$ spectral line widths!) between chemically different nuclei (e.g., ^1H and ^{13}C). Bottom: Schematic narrow-range NMR spectrum, showing the "chemical shift" variation in Larmor frequency for chemically identical nuclei (in this case, ^1H) in different chemical environments.

The Larmor frequency is proportional to magnetic field strength (Eq. 8.5). Because the inherent spectral resolution of NMR is so high [$\nu/\Delta\nu \cong$ (500,000,000 Hz/0.2 Hz) = 2.5 x 10^9 at 11.75 tesla], the NMR experiment can resolve tiny "chemical shift" differences (<1 ppb) in local magnetic field strength (and thus Larmor frequency) at different atoms in the *same* molecule.

The tiny intramolecular variations in magnetic field strength in turn arise from electric currents resulting from the electron circulation induced by the applied magnetic field, $\boldsymbol{B}_0$. Figure 8.5 shows and explains the benzene proton Larmor "chemical" frequency shift induced by an electric "ring current" effect. Although aromatic ring currents typically produce the largest ^1H and ^{13}C chemical shifts, *all* bonded atoms contain electrons whose motion is perturbed by $\boldsymbol{B}_0$ and which thus contribute to a net Larmor frequency shift which is characteristic of the chemical bonding at or near the atom in question. The relative scale of chemical shifts is arbitrary, and is generally referred to a standard compound [e.g., tetramethylsilane, $(CH_3)_4Si$, for ^1H NMR in non-aqueous solvents].

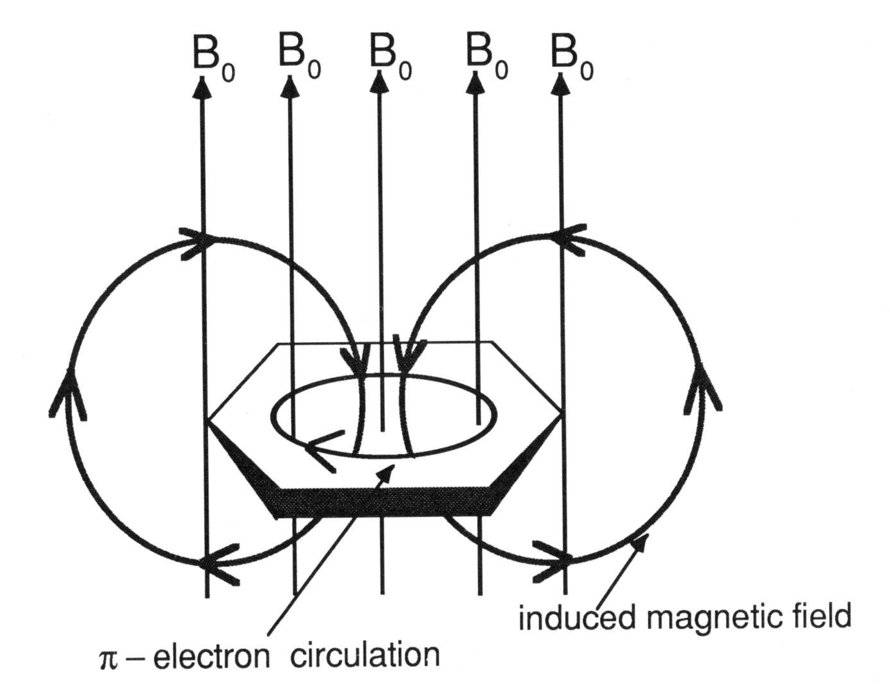

Figure 8.5 Ring-current induced shift in Larmor frequency for aromatic benzene protons. The spatially homogeneous applied magnetic field, B_0 **k** (shown by the solid lines) induces π- electrons to circulate in the plane of the benzene ring so as to produce a small counter-field whose magnetic flux lines are shown by the dashed lines. At a given ring proton, H, the applied and induced magnetic fields add to give a resultant magnetic *z* -field slightly larger than B_0, so that the ^{1}H NMR frequency is "chemically shifted" to a higher Larmor frequency (or "downfield"— see text) relative to that for a non-aromatic proton.

Modern FT/NMR instruments operate at fixed magnetic field strength. Therefore, an aromatic proton Larmor frequency is shifted upward (i.e., larger magnetogyric ratio) from a non-aromatic proton (see Figure 8.5). However, when ring-current shifts were first discovered in the 1950's, NMR spectra were acquired at fixed transmitter/detector frequency, by slowly scanning the applied magnetic field strength to bring nuclei of different magnetogyric ratio into resonance. Thus, a *smaller* magnetic field strength was required to bring an aromatic proton into resonance according to Eq. 8.5. To this day, aromatic protons are still said to resonate "downfield" of other protons, even though the experiment is no longer performed by field-sweep.

8.2.2 Scalar (J-) coupling through chemical bonds: dihedral angles

A magnetic nucleus, *A*, in a molecule induces a small electric current (and thus a small magnetic field) in the bonding electrons of its atom. That electric current can propagate through a few (1-4 or sometimes more) chemical bonds, and eventually produce a very small magnetic field at another magnetic nucleus, *B*, connected through a series of chemical bonds to the original nucleus, A. Since a given A nucleus can exhibit $2I+1$ possible observable magnetic moment values (e.g., 2 for a spin 1/2 particle, such as ^1H or ^{13}C), the magnetic field (and thus the Larmor frequency) for nucleus B will be split into $2I+1$ values, as shown in Figure 8.6.

The spectral "splitting" patterns produced by such through-bond ("scalar", or "J") coupling is formally analogous to two mechanical masses-on-springs coupled mechanically through a third spring, as described in Chapter 4.4. The number of multiplet peaks (but not their relative magnitudes) produced by n identical spin I nuclei is the same as that from a single nucleus of spin, nI (see Problems).

J-splittings in NMR spectra are important for two very practical reasons. First, the multiplet pattern helps to identify and distinguish different chemical functional groups: e.g., $-CH$ (^{13}C 1:1 doublet; ^1H 1:1 doublet) vs. $-CH_2$ (^{13}C 1:2:1 triplet; ^1H 1:1 doublet) vs. $-CH_3$ (^{13}C 1:3:3:1 quartet; ^1H 1:1 doublet), etc. Second, since the through-bond coupling depends on the overlap (and thus on the relative orientation) of the adjacent atomic orbitals, the strength of the J-coupling (manifested as the separation between the peaks of a J-split spectral multiplet) has a well-defined dependence on *dihedral angle* between two C–R groups at opposite ends of a C—C bond (see Figure 8.6). The *Karplus relation* (namely, the dependence of J on dihedral angle) shown in Figure 8.6 is one of the most highly cited papers in all of chemistry.

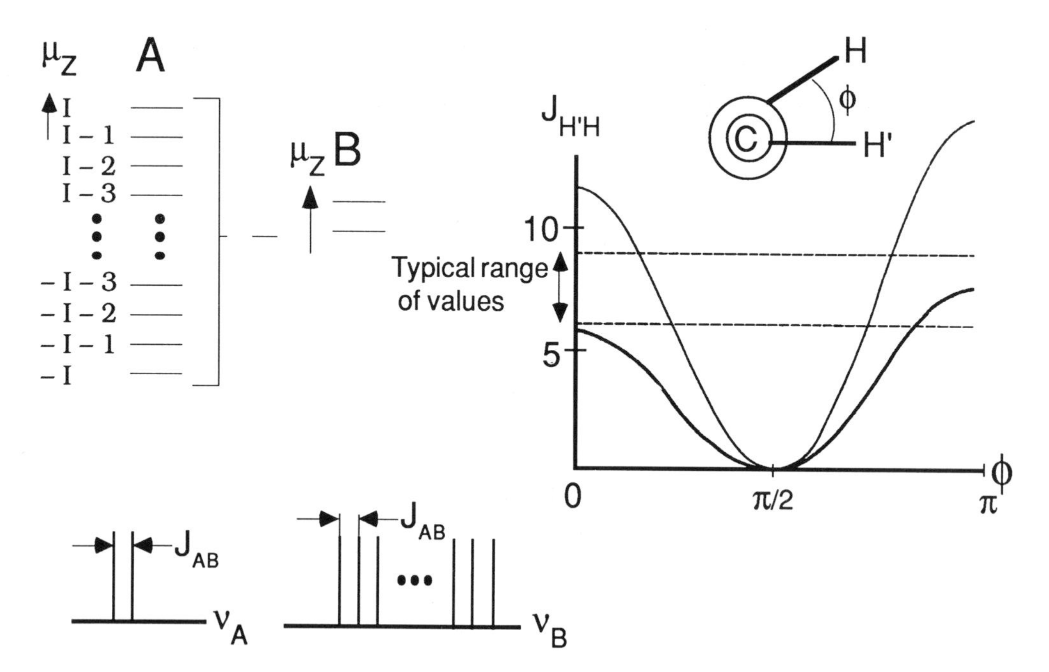

Figure 8.6 NMR scalar ("J", through-bonds) coupling. Left: Splitting of the nuclear magnetic resonance of nucleus, B, of spin 1/2 into $2I+1$ frequencies, by J-coupling from another nucleus, A, of spin, I. Right: Plots of J-coupling *vs.* dihedral angle for H–C–C–H (top curve) and C=C(H)–C-H (bottom curve).

Finally, it should be noted that NMR spectra show the simple behavior shown in Figure 8.6 only when the difference in Larmor frequency between the two nuclei is much larger than the J-coupling constant between them: $\nu_B - \nu_A >> J_{AB}$ (i.e., first-order time-independent Hamiltonian perturbation theory). That is one reason that NMR spectra are acquired at the highest possible magnetic field strength, since $(\nu_B - \nu_A)$ is proportional to B_0 and J_{AB} is independent of B_0.

8.2.3 Dipolar coupling through space: intra- and intermolecular distances

The NMR determination of molecular geometry in solution is based largely on chemical shifts and J-coupling, which in turn depend upon the chemical-bond linkages of a molecule. However, even in the absence of directly linked chemical bonds, the magnetic moment, μ_1, of one nucleus at the origin generates (through space) a small magnetic field, $\boldsymbol{B}_{\text{dipole}}$, at a second nucleus of magnetic moment, μ_2,

$$\boldsymbol{B}_{\text{dipole}} = -\frac{\mu}{r^3} + 3\left(\frac{\mu \cdot \boldsymbol{r}}{r^5}\right)\boldsymbol{r} \tag{8.13}$$

in which $\boldsymbol{r}$ is the (distance) vector between the two magnetic nuclei (see Figure 8.7). Since we will ultimately be interested only in the *energy* of interaction between the two dipole moments, and since that energy is independent of azimuthal angle, ϕ, we can average over ϕ in Eq. 8.13. Furthermore, since the dipole-dipole interaction energy is typically small compared to the energy of interaction of either dipole with the applied static magnetic field, it turns out that to first order, we may neglect terms involving μ_x and μ_y. The resulting dipolar field magnitude (along the z-direction) then becomes (see Problems)

$$B_{\text{dipole}}(z) = \frac{\mu_z}{r^3}(3\cos^2\theta - 1) \tag{8.14}$$

in which θ is the polar angle between $\boldsymbol{r}$ and the z-axis, so that $\cos\theta = z/r$.

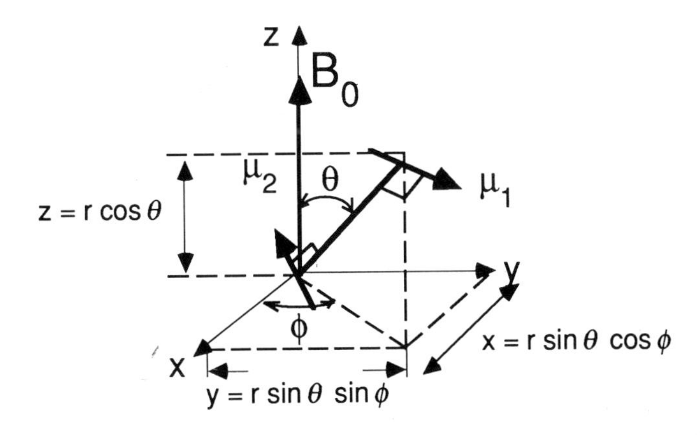

Figure 8.7 Coordinate frame for discussion of magnetic field induced at magnetic moment, μ_2, separated by (vector) distance, $\boldsymbol{r}$, from magnetic moment, μ_1 (see text).

Since the first magnetic moment, μ_1, takes on only $2I+1$ possible observable z-magnitudes (along or opposed to the $\boldsymbol{B}_0$-axis), the Larmor frequency, ν_0, of the second "dipolar-coupled" nucleus, μ_2, in a single crystal (in which all molecules have a common orientation with respect to $\boldsymbol{B}_0$) will be split into $2I+1$ separate frequencies distributed symmetrically about ν_0. For example, each proton in a single crystal will split the nuclear magnetic resonance of each other proton (see Eq. 8.14) into two resonances [2(1/2) + 1 = 2], separated by a factor proportional to $(3 \cos^2 \theta - 1)/r^3$.

Although dipolar splittings can be used to obtain molecular geometry from relative $(3 \cos^2 \theta - 1)/r^3$ values for various pairs of nuclei in a molecular single crystal, the geometrical factor in Eq. 8.14 is more commonly extracted from measurements of NMR relaxation times (see Chapter 8.4.2). Since dipole-dipole coupling occurs through space rather than through chemical bonds, it is suitable for measurement of relative distances *between* as well as *within* molecules.

In liquids, the rapid rotational reorientation of molecules effectively averages over all θ-values. Since the average of $\cos^2 \theta = 1/3$,

$$< 3 \cos^2 \theta - 1 > = 0 \tag{8.15}$$

so that no dipolar splittings are observed in liquids. In liquid crystals, molecules are partially ordered, and dipolar couplings (and thus molecular geometry) can be extracted from partially averaged splittings.

It should be noted that the magnetic moment of an (unpaired) electron is ~660 times larger than that of a proton. Thus, dipolar coupling from unpaired electrons can dominate NMR behavior when paramagnetic species are present.

A nucleus of spin ≥ 1 (e.g., ^{2}H, ^{14}N, ^{17}O, ^{23}Na, ^{35}Cl) is non-spherical, and therefore has an electric quadrupole moment which can interact with an electric field gradient to shift or split the NMR Larmor frequency. The geometric dependence of the splitting is similar (and, for nuclear spin, $I = 1$, identical) to that for the dipole-dipole interaction of Eq. 8.14. Since the electric field gradient direction is determined by the chemical bond(s) around a given atom, quadrupolar splittings are especially useful for determining molecular orientation in bilayer membranes and other partially or wholly ordered media.

8.3 Spin manipulations

8.3.1 90° pulse: FT/NMR excitation, T_2 relaxation, and detection

Detection in NMR, as in ICR, poses the immediate problem that a signal is not automatically generated by the presence of magnetic nuclei in a static magnetic field, $\boldsymbol{B}_0$. Although the nuclei all precess about $\boldsymbol{B}_0$, the vector sum of their magnetic moments, $\boldsymbol{M}_0$, lies along the z-axis (i.e., along $\boldsymbol{B}_0$) and is unobservable. It is necessary to rotate $\boldsymbol{M}_0$ (say, about the x-axis, as in Figure 8.8a) to give an observable component, $M_y \mathbf{j}$ along the y-axis (see Figure 8.8b). As $\boldsymbol{M}_0$ rotates about the z-axis, the electric current induced in a coil of wire wrapped around the y-axis oscillates from positive to zero to negative as the x-y component of $\boldsymbol{M}_0$ points along, perpendicular, or opposed to the positive y-axis (Figure 8.8c). That oscillating time-domain "free induction decay" (FID) signal can then be sampled, stored, and Fourier transformed to give a frequency-domain spectrum.

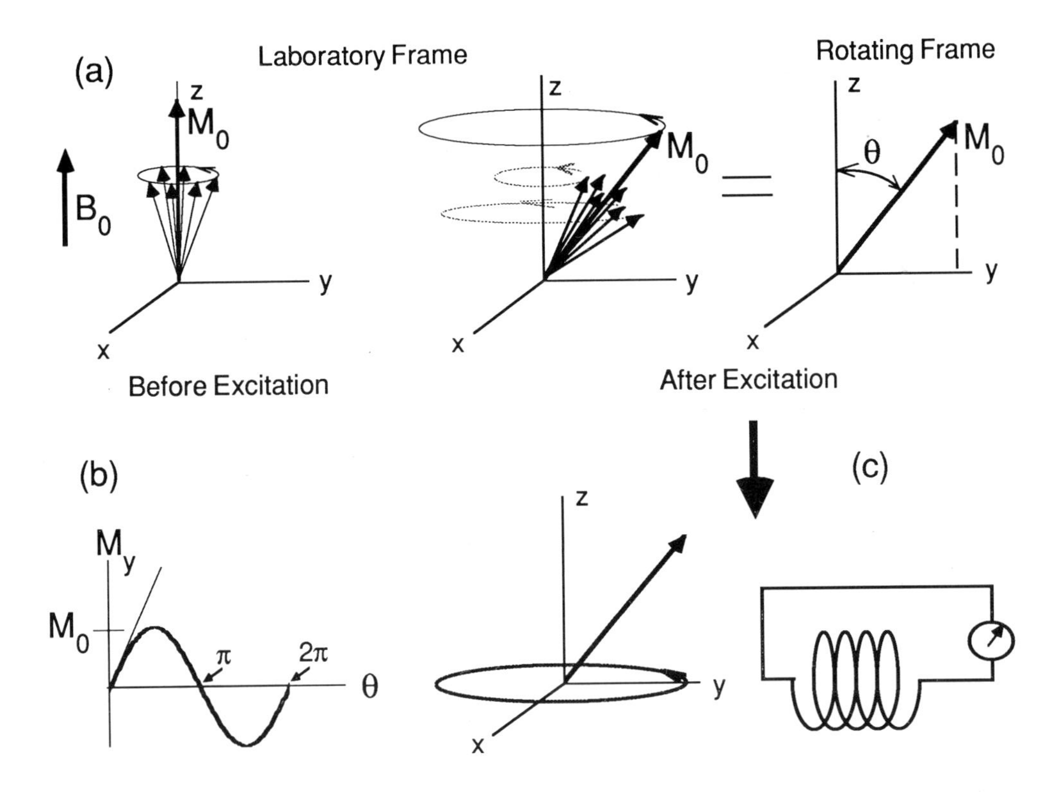

Figure 8.8 Generation and detection of an NMR time-domain "free induction decay" signal. (a) The equilibrium macroscopic magnetization, $\boldsymbol{M}_0$, is caused (see below) to rotate about the x-axis. (b) The resulting y-magnetization component, M_y, varies non-linearly with rotation angle, θ. (c) After $\boldsymbol{M}_0$ has rotated, precession of its x-y component about the z-axis produces an oscillating y-magnetization which induces an oscillating electric current (FID signal) in a detector coil wound around the y-axis (see next Figure).

The rotation required to accomplish the detection scheme of Figure 8.8 is achieved by applying an oscillating single-frequency rf magnetic field, $2B_1 \cos \omega_0 t$, along the (laboratory-frame) x-axis (Figure 8.9a). As noted in Chapter 4.3, a *linearly* oscillating field can be analyzed into two counter-rotating fields (Figure 8.9b), only one of which rotates in the same sense as the nuclear magnetization. [We can neglect the other rotating field component.]

The effect of the rf magnetic field is more readily understood in a coordinate frame (x', y', z') rotating at the Larmor frequency, ω_0, of the magnetic moment, $\boldsymbol{M}_0$ (Figure 8.9c). Since the effective magnetic field in the rotating frame is simply $B_1 \mathbf{i'}$, $\boldsymbol{M}_0$ rotates about $\mathbf{i'}$ at (angular) frequency,

$$\boxed{\omega_1 = \gamma B_1} \tag{8.16}$$

Thus, a 90° rotation of $\boldsymbol{M}_0$ about the rotating x'-axis requires that the rf field be applied for a period, $\tau_{90°}$,

$$\omega_1 \tau_{90°} = \frac{\pi}{2} \tag{8.17}$$

At the end of the excitation period, the $\boldsymbol{B}_1$ field is removed. The x-y magnetization, $M_x \mathbf{i} + M_y \mathbf{j}$, precesses at its Larmor frequency, and its y-component, M_y, is detected from the oscillating electric current (Figure 8.9d) it induces in the laboratory-frame receiver coil (Figure 8.8c) a free induction decay

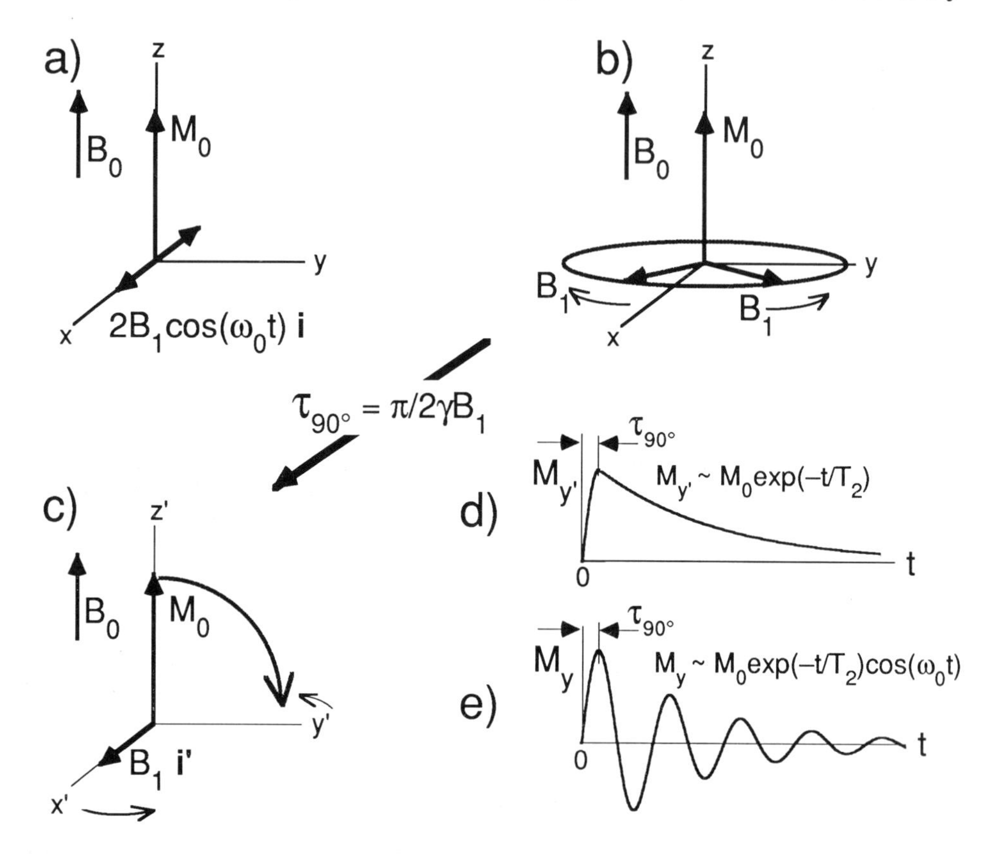

Figure 8.9 Laboratory and rotating-frame representations of an NMR single-frequency 90° excitation pulse. (a) Oscillating magnetic field, $2B_1 \cos \omega_0 t\ \mathbf{i}$ in the laboratory frame, where ω_0 is the Larmor frequency of $\boldsymbol{M}_0$. (b) Laboratory-frame analysis of (a) into two counter-rotating components, of which only the component rotating in the same sense as the magnetization, $\boldsymbol{M}_0$, need be considered further. (c) Rotating-frame representation of (b), in which $\boldsymbol{M}_0$ precesses at (angular) frequency, $\omega_1 = \gamma B_1$, about the rotating frame magnetic field, $B_1 \mathbf{i}'$, to reach a final orientation along the positive y-axis after excitation period, $\tau_{90°} = (\pi/2\omega_1)$. (d) and (e) show the detector coil free induction decay time-domain signal induced by the y-magnetization in rotating and laboratory frames (see Figure 8.8c and text).

signal shown in Figure 8.9d,e and expressed (in the lab frame) by Eq. 8.18:

$$\lim_{\tau_{90°} \to 0} M_{y'} = M_0 \exp(-t/T_2) \qquad \text{(Rotating frame)} \qquad (8.18a)$$

$$\lim_{\tau_{90°} \to 0} M_y = M_0 \exp(-t/T_2) \cos \omega_0 t \qquad \text{(Laboratory frame)} \qquad (8.18b)$$

in which T_2 is called the *transverse* or *spin-spin relaxation time*, which we shall discuss in the next section. It is convenient to describe the above experiment by the event sequence,

$$90°_x - \text{Observe} \quad \text{(Normal FID; } T_2 \text{ measurement)} \qquad (8.19)$$

in which $90°_x$ denotes a pulsed oscillating rf magnetic field applied along the x-axis for a period, $\tau_{90°} = (\pi/2\omega_1)$.

8.3.2 180° pulse: spin-lattice (T_1) relaxation; relation between T_1 and T_2

Similarly, the effect of a single-frequency rf magnetic field pulse, $2B_1 \cos \omega_0 \tau_{180°}\, \mathbf{i}$, in which

$$\omega_1 \tau_{180°} = \pi \qquad (8.20)$$

is to rotate $\boldsymbol{M}_0$ by 180° about the x-axis, to leave the magnetization pointing in the negative z-direction (see Figure 8.10a). In classical mechanical terms, the macroscopic magnetization has been "flipped over". In quantum mechanical terms, the rf pulse has produced a population "inversion", in which the original equilibrium populations in upper and lower energy levels have been interchanged.

By as yet unspecified means, the spin system must "relax" back to thermal equilibrium. However, z-magnetization is not directly observable. Therefore, in order to monitor the return of the system to equilibrium, it is necessary to introduce a second rf pulse, say, $2B_1 \cos \omega_0 \tau_{90°}\, \mathbf{i}$, to rotate the z-magnetization back into the x-y plane (Figure 8.10b) to generate a detectable FID signal, which may be Fourier transformed to yield a frequency-domain spectrum. A plot (Figure 8.10c) of frequency-domain peak area for spectra obtained from various "evolution periods", τ, between the $180°_x$ and $90°_x$ pulses,

$$180°_x - \tau - 90°_x - \text{Observe} \qquad \text{(Inversion/recovery } T_1 \text{ experiment)} \qquad (8.21)$$

then gives a direct measure of the y-magnetization immediately after each $90°_x$ "sampling" pulse:

$$\lim_{\tau_{180°} \to 0} M_{y'} = (1 - 2M_0) \exp(-t/T_1) \qquad \text{(Rotating frame)} \qquad (8.22a)$$

$$\lim_{\tau_{180°} \to 0} M_y = (1 - 2M_0) \exp(-t/T_1) \cos \omega_0 t \qquad \text{(Laboratory frame)} \qquad (8.22b)$$

in which T_1 is called the *longitudinal* or *spin-lattice relaxation time.*

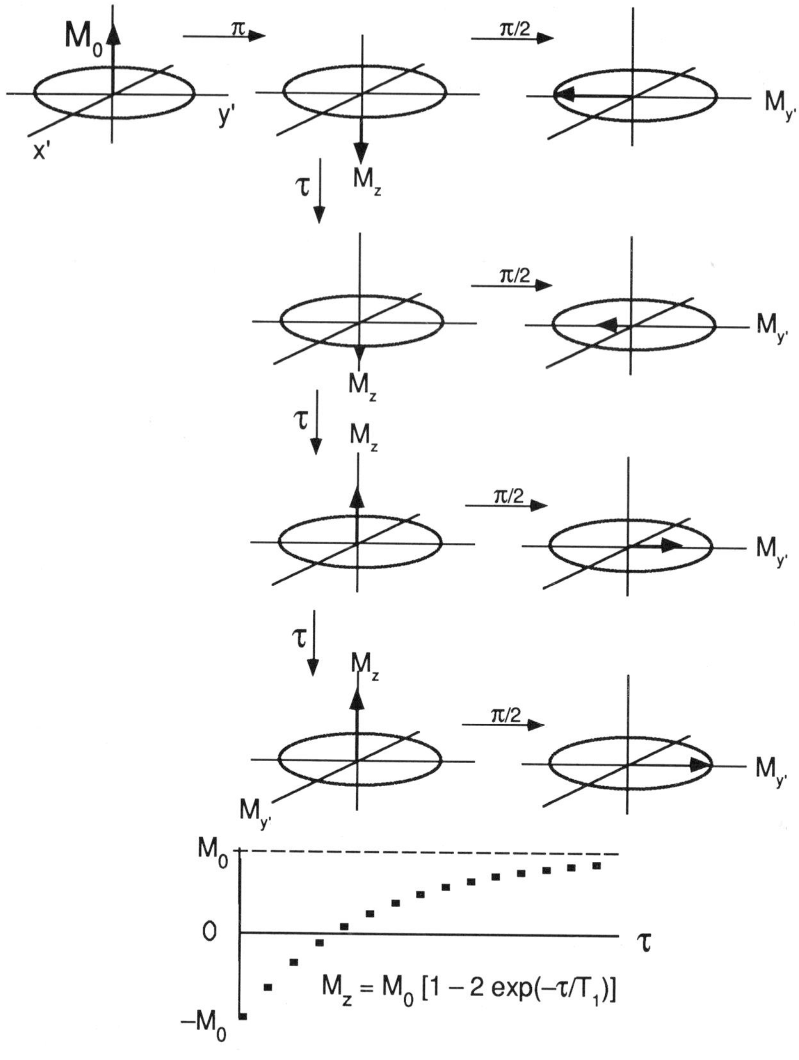

Figure 8.10 Inversion-recovery measurement of longitudinal (spin-lattice) relaxation, displayed in a coordinate frame rotating at angular frequency, ω_0, about the $\boldsymbol{B}_0$ axis (z -axis). Equilibrium magnetization, $\boldsymbol{M}_0$ (upper left) rotates by 180° about the (rotating) x'-axis under the influence of a pulsed B_1 (rotating-frame) magnetic field applied for a period, $(\pi/\gamma B_1)$ seconds. After a delay period, 0, τ, 2τ, or 3τ in the Figure, the z -magnetization is brought back into the x-y plane by a second (rotating-frame) B_1 pulse applied for a period, $(\pi/2\gamma B_1)$ seconds. A plot of rotating-frame y'-magnetization immediately after a $90°_x$ pulse, as a function of delay period, τ, is shown in the bottom trace.

Let us now consider the relationship between T_1 and T_2. T_1-relaxation can occur classically *only* by changing the *polar angle* between the magnetic moments and the applied magnetic field. In quantum mechanical language, "spin flips" must occur—i.e., change in z-component of angular momentum from one value to another, corresponding to a transition from one energy level to another.

In contrast (see Figure 8.11), T_2-relaxation can occur in two ways: *non-secular* relaxation, resulting from the T_1-process just described, and *secular* relaxation in which individual magnetic moments lose phase coherence (i.e., spread out in the x-y plane). De-phasing results from variation in Larmor frequency among different nuclei as a result of different local magnetic field strength at various nuclei. The magnetic field strength variation can in turn arise from chemical shift differences or from macroscopic spatial inhomogeneity in the applied magnetic field. In either case, since de-phasing results in an additional loss of (rotating-frame) $M_{y'}$ magnitude beyond that due to spin-flips (energy-level transitions), it is clear that $T_2 \leq T_1$. *Non-secular* and *secular* relaxation correspond to *homogeneous* and *inhomogeneous* spectral line-broadening, as discussed in Chapter 7.3.1 for ion cyclotron resonance spectroscopy.

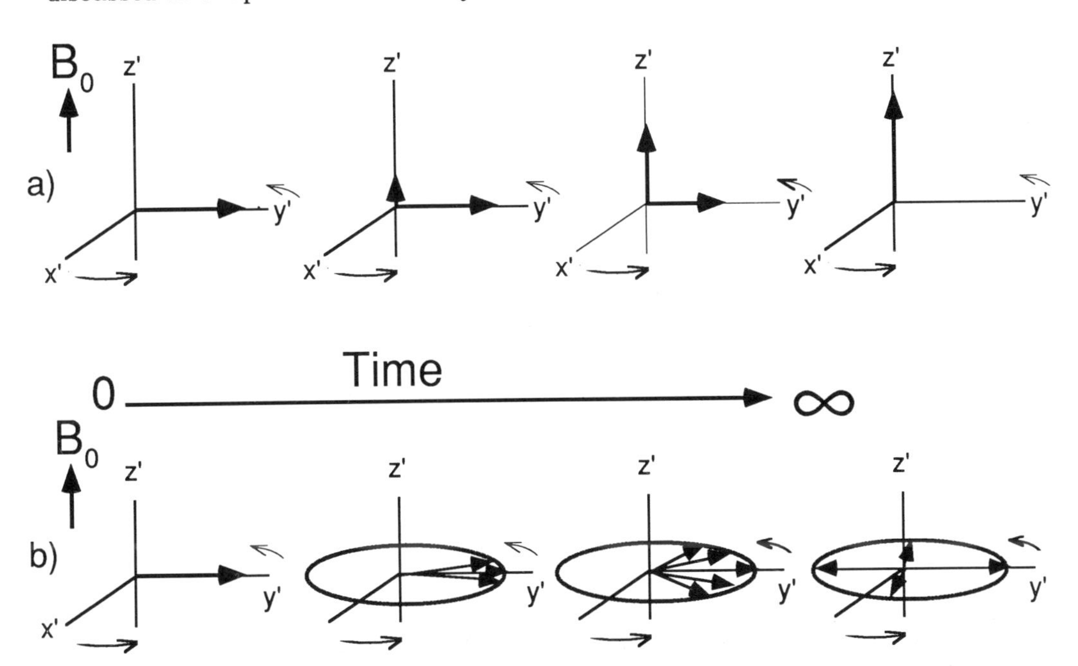

Figure 8.11 Mechanisms for secular and non-secular magnetic relaxation. The macroscopic (rotating-frame) y'-magnetization, $M_{y'}$, consists of a vector sum of all of the individual magnetic moments of the sample. $M_{y'}$ is therefore a vector whose length can decrease either by (a) *non-secular* relaxation, in which individual magnetic moments flip from "down" to "up", so that z-magnetization grows at the expense of y'-magnetization, or (b) *secular* relaxation, in which the individual magnetic moments "de-phase" ("fan out") in the x-y plane due to a distribution in their individual Larmor frequencies. In *non-secular* relaxation, the macroscopic $M_{y'}$ vector grows shorter as M_z increases with time. Compare to analogous ICR case (Figure 7.11).

$(1/T_1)$ is the rate constant for transitions between magnetic energy levels in the absence of an applied rf magnetic field. Thus, it should seem reasonable to learn that the rate of such induced transitions is proportional to the magnetic spectral density, $P(\omega_0)$, at the Larmor frequency, resulting from random fluctuations in magnetic field (e.g., molecular translational and rotational diffusion in liquids—see Chapter 5.2.1.).

Similarly, since "transverse" T_2–relaxation in the x-y plane can occur either as the result of energy-changing transitions or from de-phasing (which is a zero-energy process, since the individual magnetic moments do not change their z-component), it should seem reasonable that the transverse relaxation rate, $(1/T_2)$, should depend on the spectral density at *zero* frequency, $P(0)$, as well as at the Larmor frequency. See Chapter 5 Problems for a sample calculation of $P(0)$ and $P(\omega)$ for chemical exchange between two sites of different magnetic field strength.

The effects of static magnetic field and T_1 and T_2 relaxation on the magnitude and length of a macroscopic magnetization (vector), $\boldsymbol{M}$, can be combined in the "phenomenological" Bloch equations,

$$\frac{d\boldsymbol{M}}{dt} = \gamma \boldsymbol{M} \times \boldsymbol{B} - \frac{M_z - M_0}{T_1}\mathbf{k} - \frac{M_x \mathbf{i} + M_y \mathbf{j}}{T_2} \quad \text{(Laboratory frame)} \quad (8.23a)$$

in which

$$\boldsymbol{B} = 2B_1 \cos \omega_1 t \ \mathbf{i} + B_0 \mathbf{k} \quad (8.23b)$$

In a coordinate frame rotating at angular frequency, ω_1, about $\boldsymbol{B}_0 = B_0 \mathbf{k}$, the Bloch equations have the same form, except that the effective field (see Figure 8.12) is the vector sum of $\boldsymbol{B}_1 = B_1 \mathbf{i}$ and $\boldsymbol{B}_z = (B_0 - \omega_1/\gamma) \mathbf{k}$. Many NMR experiments can be interpreted from the Bloch equations alone, without the need for quantum mechanical derivations (see Problems).

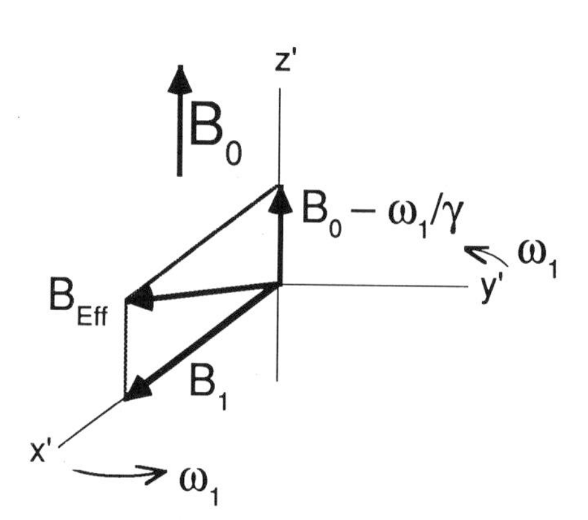

Figure 8.12 Effective magnetic field, $\boldsymbol{B}_{\text{eff}}$, in a coordinate frame rotating at angular frequency, ω_1, about the z-axis (i.e., the applied static magnetic field direction), in the presence of a magnetic field, $2B_1 \cos \omega_0 t$ oscillating along the lab-frame x-axis. A macroscopic magnetization (vector), $\boldsymbol{M}$, will precess about $\boldsymbol{B}_{\text{eff}}$ in the rotating frame coordinate system.

8.3.3 Spin-echo: elimination of chemical shift differences; double-sided FT

Static magnetic field inhomogeneity can significantly broaden NMR spectral peaks [remember that the width at half-maximum height of a Lorentzian absorption-mode peak is $\Delta\nu = (1/\pi T_2)$]. The overall transverse relaxation rate constant, [$1/T_2$(observed)], is the sum of the "intrinsic" relaxation rate constant (in the absence of static magnetic field inhomogeneity), $(1/T_2)$, and the "inhomogeneous" relaxation rate constant, $(1/T_2')$, due to the magnetic field inhomogeneity.

$$\frac{1}{T_2(\text{observed})} = \frac{1}{T_2} + \frac{1}{T_2'} \tag{8.24}$$

In practice, the NMR signal detector must be blanked off for a short time after the rf excitation pulse in order to allow the excitation pulse to "ring down" to zero. If $T_2' << T_2$ (as in NMR imaging experiments), the y-magnetization may thus decrease significantly before the detector can be turned on. An ingenious way to recapture much of the signal in such cases is the "spin-echo" event sequence

$$90°_{x'} - \tau - 180°_{x' \text{or } y'} - \tau - \text{Observe} \tag{8.25}$$

illustrated and explained in the schematic diagram of Figure 8.13. The re-focusing spin-echo principle is common to most multiple–pulse and/or two-dimensional NMR experiments, because it effectively eliminates chemical shift differences, thereby greatly simplifying an NMR spectrum (see Chapter 8.5).

It is worth noting that the FID obtained by detecting both halves of the spin-echo is "even" rather than "causal" and its FT is therefore a spectrum consisting of pure absorption-mode with no dispersion-mode. Unfortunately, the first half of the echo is generally of poorer quality [since it must be acquired sooner after the excitation (or, in the case of NMR imaging, after a magnetic field gradient pulse)], so that double-sided detection is seldom used in practice. Double-sided detection is however more feasible in FT/IR experiments (see Chapter 9).

8.3.4 Multiple-pulse excitation and non-linear phenomena: solvent suppression

As noted in Chapter 2.1.2., a single-frequency rectangular pulse of duration, T seconds produces frequency-domain excitation power that is flat to within ~10% over a frequency range of ~$1/8T$ Hz. Thus, an NMR spectrum spanning a frequency range of ~10 kHz can be excited with approximately flat power by a rectangular rf pulse of duration, $T \cong 12$ μs. If, in addition, we require that the same pulse accomplish a 90° rotation of $\boldsymbol{M}_0$ in ≤12 μs (for Larmor frequencies anywhere within the specified 10 kHz spectral range), then the time-domain rf amplitude of the rf pulse (see Problems) must be ≥ 5×10^{-4} tesla = 5 Gauss, an experimentally feasible level. [The earth's (static) magnetic field is ~0.5 Gauss.]

The single-frequency rf pulse is suitable for ordinary broadband FT/NMR excitation. However, some important NMR applications require highly selective excitation. For example, simple broadband excitation/detection of proton NMR signals from dilute (< 1 mM) biological molecules in H_2O (~100 M in protons) would require a dynamic range of ~10^5:1. Although the corresponding 16-bit analog-to-digital converter is available at the necessary speed (≤50 kHz), no currently available detector circuit is sufficiently linear over such a high dynamic

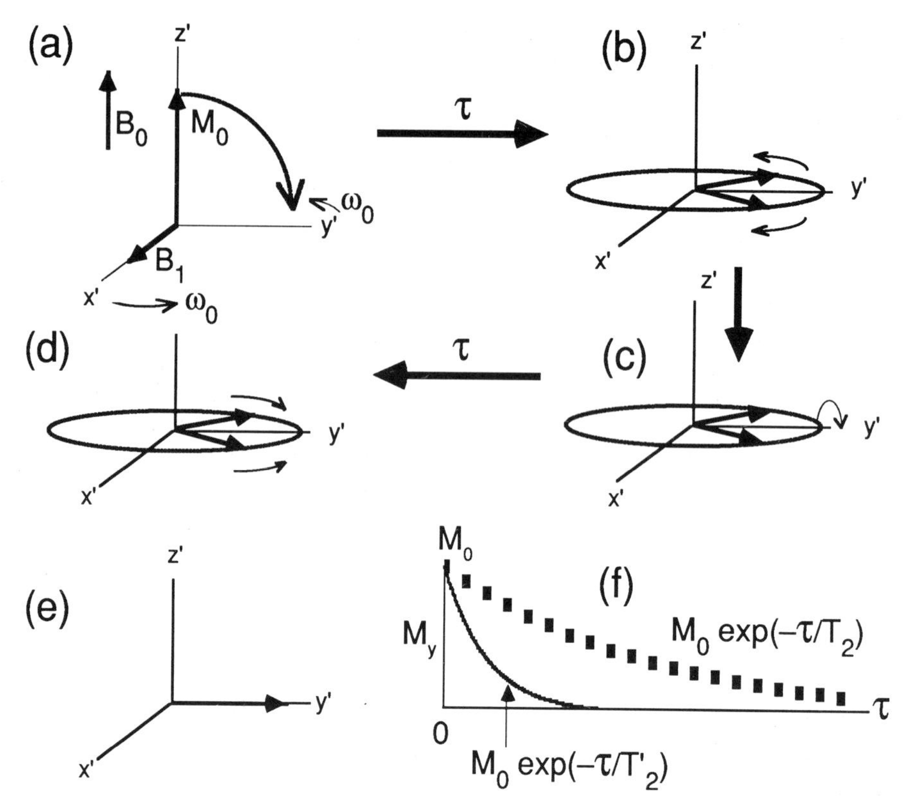

Figure 8.13 The spin-echo principle, shown schematically in a coordinate frame rotating at the Larmor frequency. (a) A 90° rf pulse applied along the (rotating) x'-axis rotates the equilibrium magnetization, $\boldsymbol{M}_0$, to the y'-direction. (b) During a delay period, τ, individual magnetic moments fan out according to their (slightly) different Larmor frequencies. (c) A 180° pulse applied along y' effectively interchanges the faster- and slower-precessing magnetic moments. (d) During a second delay period, τ, the faster-precessing components catch up and the slower-precessing components lag the Larmor rotation speed, ω_0. (e) Thus, at the end of the second delay period, the individual magnetic moments refocus to give an "echo" y'-magnetization at the detector. (f) Plot of the results of several such experiments, performed at various delay periods, τ. The (longer) "intrinsic" T_2 can be observed even in the presence of the (shorter) T_2'-relaxation due to magnetic field inhomogeneity.

range. [Also, the huge H_2O magnetic moment induces so much current in the receiver coil that I^2R power loss becomes significant; that power loss must in turn be drawn from the H_2O signal, which is thereby damped significantly (a phenomenon sometimes incorrectly described as "radiation damping"). The H_2O line width thus increases to ~10 Hz at ~12 tesla, and the "wings" of the H_2O resonance spread out to obscure small peaks up to several kHz away.] For example, Figure 8.14a shows a proton FT/NMR spectrum at 11.75 tesla (500 MHz proton Larmor frequency) for a dilute aqueous biological RNA macromolecule—the RNA base-pair protons are unobservable, even after vertical expansion of the spectrum.

More than 100 journal publications (see Meier review in Further Reading) have offered various excitation waveforms designed to excite the NMR signals of interest without exciting the solvent resonance (usually H_2O). The simplest idea is shown in Figure 8.14b: namely, to choose the frequency of a rectangular single-frequency rf pulse ~$(1/T)$ Hz away from the Larmor frequency of the solvent (e.g., H_2O), to give essentially zero on-resonance excitation of the H_2O signal. However, because the H_2O resonance is not infinitely narrow, even the relatively small "leakage" of excitation power *near* the center of the H_2O peak is sufficient to produce a large H_2O signal in the FT spectrum.

The Redfield $2\underline{1}4\underline{1}2$ sequence (in which the numbers denote the duration of each component pulse, and the underline denotes a 180° phase shift) of Figure 8.14c (see Problems) offered a major improvement, because its excitation power was much flatter (and near-zero) near the H_2O resonance. However, the Redfield pulse sequence introduces significant additional *phase variation* across the spectrum (Figure 8.14d), and has largely been replaced by pulse sequences such as $1\underline{3}3\underline{1}$ (see Problems). Finally, an additional trick with any excitation pulse sequence is to alternate successive acquisitions in which the delay period between excitation and detection differs by one-half of one cycle of the H_2O spectral frequency. Thus, the H_2O signal is preferentially canceled after every 2 acquisitions. For example, $1\underline{3}3\underline{1}$ excitation combined with alternating-delay-acquisition offers >50,000–fold reduction in the H_2O signal, as shown by the spectrum of ~0.75 mM ribosomal 5S RNA in Figure 8.14e).

At this stage, one might wonder why NMR spectroscopists do not employ stored-waveform inverse Fourier transform excitation (compare Chapter 7.4.2), which was in fact demonstrated for FT/NMR as early as 1973. First, most NMR spectroscopists have preferred to employ simpler (non-stored) pulse sequences, which are experimentally easier to generate. Second, as the excited NMR bandwidth increases, the direction of the *effective* magnetic field in the rotating frame begins to move significantly away from the (rotating) x'-axis (Figure 8.12). Thus, even if the rf magnitude is perfectly equal at different frequencies, $\boldsymbol{M}_0$ will rotate by a non-90° angle to a final position that does not even lie in the y'-z plane! (Many of the multiple-pulse excitation sequences of modern FT/NMR are designed to correct for the systematic errors resulting from imperfect tip angles.) Third, although ICR systems are highly *linear* (up to an experimentally useful ICR orbital radius), NMR systems become highly *non-linear* beyond tip angles of >10° or so (see Figure 8.8b). Finally, T_1– and T_2–relaxation can occur *during* the excitation, particularly when a relatively long excitation period is needed to achieve highly frequency-selective excitation.

The combined effect of the factors listed in the preceding paragraph is that the frequency-domain response to a given excitation can differ markedly from what one might predict from the Fourier transform magnitude-mode spectrum of the excitation itself. For example (see Figure 8.15b-e, solid curves), the simulated NMR frequency-domain responses to simple single-frequency rectangular pulse excitations of different time-domain amplitude begin to deviate significantly from the expected "sinc" spectral shape, as the excitation magnitude increases. Conversely, excitation time-domain pulse sequences with *near-identical* Fourier transform spectra can exhibit markedly *different* NMR response. Thus, we can no longer produce an optimal time-domain excitation waveform by a simple inverse FT of the desired frequency-domain excitation spectrum.

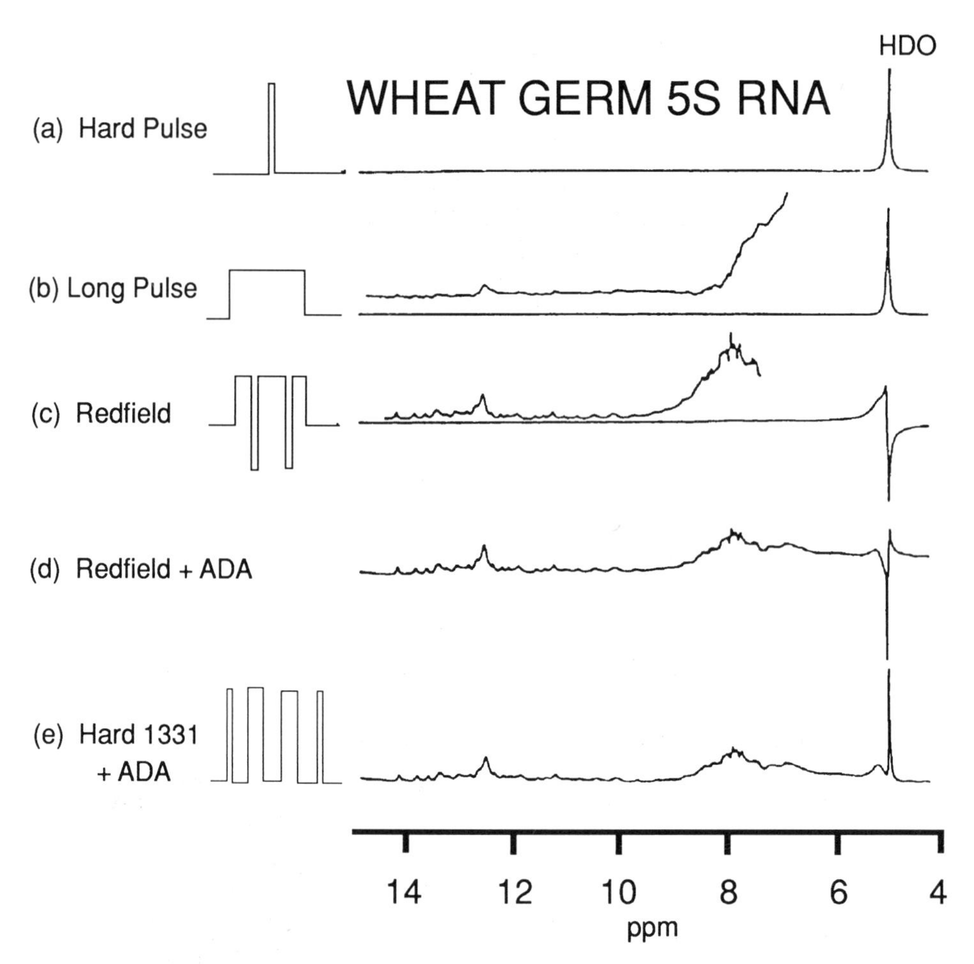

Figure 8.14 Time-domain excitation waveforms (left) and corresponding FT/NMR 500 MHz ^{1}H spectra (right) for an aqueous (95%/5% H_2O/D_2O) sample of ~0.75 mM ribosomal 5S RNA. [D_2O was added to provide a frequency reference for a field/frequency "lock" circuit which regulates the magnitude of the applied static magnetic field.] See text and Problems for discussion. (Taken, with permission, from A. G. Marshall and J. Wu, in *Biological Magnetic Resonance*, Vol. 9, Ed. L. J. Berliner (Plenum, NY, 1989), Chapter 2.

Interestingly, the complex time-domain (stored) waveform, (sech αT)$^{1+5i}$, shown in Figure 8.15a, produces a much more uniform and selective NMR response even for large excitation magnitude, as shown by the dotted curves in Figure 8.15b-e. Unfortunately, that waveform is almost the only one for which analytic NMR responses have to date been calculated under general conditions (e.g., frequency offset from resonance, time-domain amplitude, and initial condition of the spin system). As a result, the design of optimal "shaped" (stored-waveform, but not generated by inverse FT methods) excitation pulses continues as an active area of modern NMR technique development (see Further Reading).

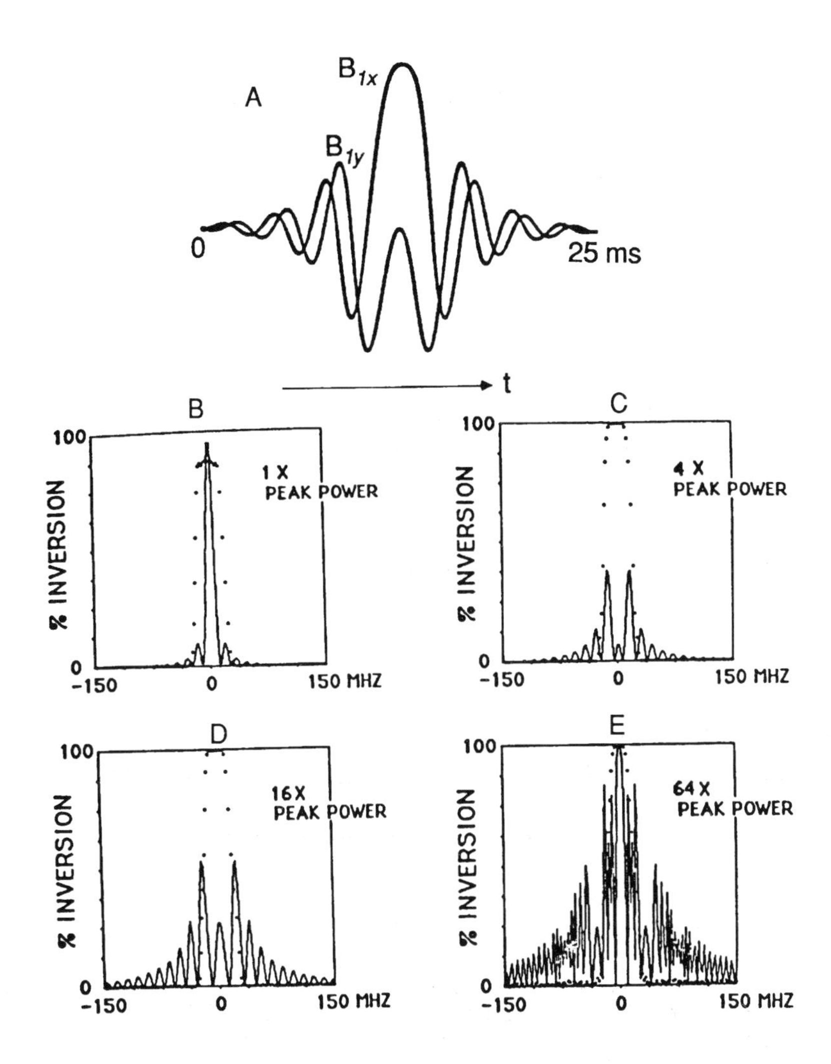

Figure 8.15 Simulated NMR frequency-domain magnitude response to two excitation waveforms. (A) Real and imaginary components of the complex time-domain waveform, (sech αT $)^{1+5i}$. (B)-(E) Response to single-frequency pulse of rectangular envelope and duration, T (solid curves) and response to the waveform of (A) (dotted curves) for time-domain amplitudes, 1X, 4X, 16X, and 64X peak power. A rectangular pulse of amplitude, 1X peak power, is scaled to produce complete inversion (i.e., ~180° pulse) at the Larmor frequency, labeled as 0 Hz in the Figure. (Because the original 180° rectangular pulse did not produce exactly 180° inversion on-resonance, the rectangular pulses of 4X, 16X, and 64X amplitude produce finite magnitude on-resonance.) (Adapted with permission from W. S. Warren, *Science* **1988**, *242*, 878-884)

8.4 NMR time scales

One commonly hears about "the NMR time scale". In fact, there are as many NMR time scales as there are NMR parameters. In principle, an NMR spectrum can be used to determine the rate constant, k, of any process for which k is comparable to the difference in Larmor frequency (or linewidth or energy level transition rate constant) between initial and final states of the process. A particular "time scale" depends on the NMR parameter which is changed by the motion: e.g., ω_0, T_1, T_2, J-coupling constant, dipole-dipole coupling constant, quadrupole coupling constant, etc. A change in the local magnetic field at a nucleus may occur either by discontinuous jumps between chemically or spatially different sites, or by diffusional translational or rotational motion. In this section, we briefly summarize the effect of chemical exchange and molecular motions on the appearance of NMR spectra.

8.4.1 Chemical exchange

The effect of chemical exchange and molecular motion on NMR spectra can be understood quantitatively and very generally from the simple case of a nucleus which jumps randomly back and forth between two sites, A and B, of different Larmor frequency:

$$A \underset{k_{-1}}{\overset{k_1}{\rightleftarrows}} B \tag{8.26}$$

Just as we ordinarily write expressions for the rates of change, $d[A]/dt$ and $d[B]/dt$, of *concentration* of A or B, we may write analogous equations for the rate of change of *magnetization* at the (two) Larmor frequencies, as if magnetization is transferred from one site to the other by the chemical exchange process:

$$\frac{d\,\mathbf{M}_A}{dt} = -k_1\,\mathbf{M}_A + k_{-1}\,\mathbf{M}_B \tag{8.27a}$$

$$\frac{d\,\mathbf{M}_B}{dt} = -k_{-1}\,\mathbf{M}_B + k_1\,\mathbf{M}_A \tag{8.27b}$$

Eqs. 8.27 can be combined with the Bloch Eqs. 8.23, from which we may solve for the final magnetization at either site for any given set of initial conditions. For example (see Problems), Figure 8.16 shows the absorption-mode spectrum obtained by Fourier transformation of the time-domain $M_{y'}$ signal following a 90° excitation pulse, for a spin system consisting of two uncoupled nuclei of different Larmor frequency, as a function of chemical exchange lifetime, τ,

$$\frac{1}{\tau} = k_1 + k_{-1} \tag{8.28}$$

for the chemical equilibrium of Eq. 8.26.

In the "slow-exchange" limit, $(1/\tau) << |\,\omega_A - \omega_B\,|$, the two resonances are clearly resolved at approximately ω_A and ω_B, and each spectral peak is broadened by the exchange process according to,

$$\Delta\nu_{\text{ave}} \cong \frac{1}{\pi}\left(\frac{1}{T_2} + \frac{1}{\tau}\right) \qquad \text{(Slow-exchange)} \quad (8.29)$$

In the "fast-exchange" limit, $(1/\tau) >> |\omega_A - \omega_B|$, the two spectral peaks coalesce to give a single resonance at an average Larmor frequency,

$$\omega_{\text{ave}} \cong \left(\frac{k_{-1}}{k_1 + k_{-1}}\right)\omega_A + \left(\frac{k_1}{k_1 + k_{-1}}\right)\omega_B = f_A\,\omega_A + f_B\,\omega_B \quad \text{(Fast-exchange)} \quad (8.30a)$$

and an average linewidth,

$$\Delta\nu_{\text{ave}} \cong \frac{1}{\pi}\left(\frac{1}{T_{2_A}} + \frac{1}{T_{2_B}}\right) \qquad (8.30b)$$

in which f_A and f_B represent the fraction of nuclei at sites A and B at chemical equilibrium.

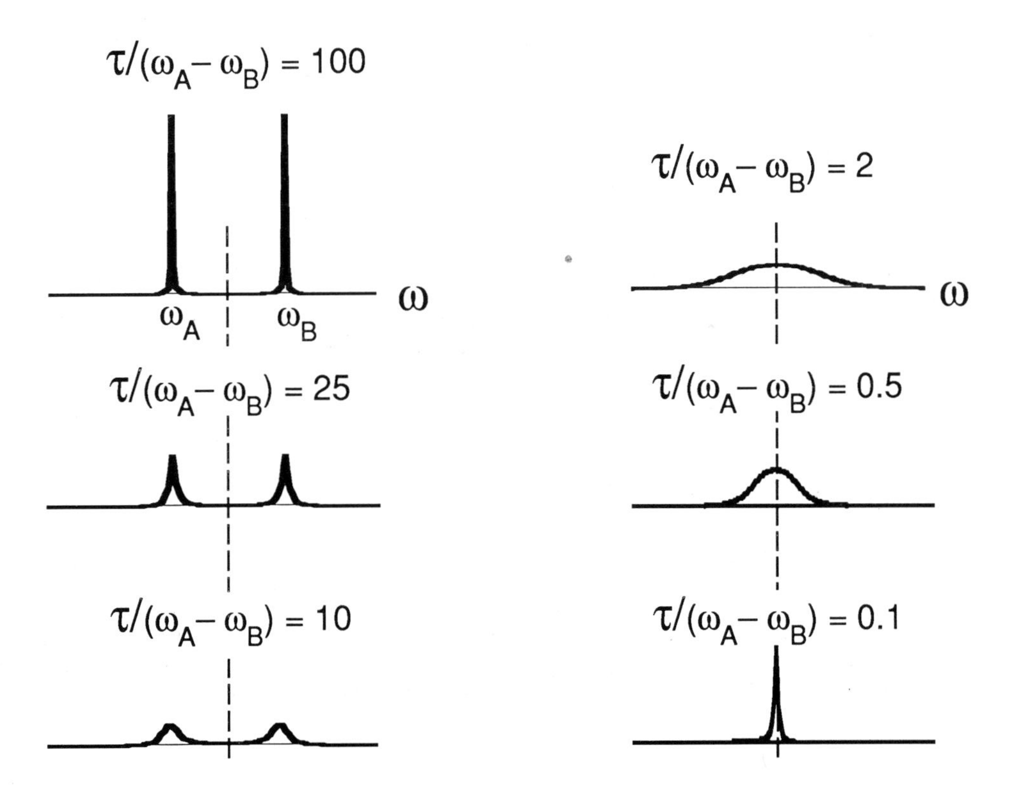

Figure 8.16 NMR spectra for a two-site system (Eq. 8.26), for several values of the lifetime, τ (see Eq. 8.28), for chemical exchange between two equally populated sites of different Larmor frequency, ω_A and ω_B .

The result shown in Figure 8.16 is extremely general. Essentially the same behavior is observed, no matter what the source of the frequency difference between the two sites: e.g., difference in J-coupling, dipolar coupling, quadrupolar coupling, T_1, T_2, etc. In each case, the NMR parameters for the two sites change from two values to a single average value when

$$\frac{1}{\tau} \geq |\omega_A - \omega_B| \tag{8.31}$$

where ω_A and ω_B in Eq. 8.31 now denote the appropriate NMR spectral peak frequencies or reciprocal relaxation times at sites A and B. Because the various NMR parameters span a huge range of frequencies (e.g., 0-200 Hz for J-couplings, 0-50,000 Hz for chemical shifts or dipolar couplings, etc.), a huge range of chemical and motional rates may be measured from analysis of NMR spectral peak positions and line widths (see Further Reading).

8.4.2 Decoupling: elimination of J-coupling

The preceding analysis can be extended to transitions between the two energy levels of a system of spin 1/2 nuclei of a common Larmor frequency, as shown in Figure 8.17. In this case, "sites" A and B of Eq. 8.26 correspond to the two spin states of different energy (i.e., $\gamma h B_0/4\pi$ or $-\gamma h B_0/4\pi$), and the "rate constants" of Eq. 8.26 correspond to the transition probabilities for upward and downward transitions, $p\uparrow$ and $p\downarrow$,

$$\frac{1}{T_1} = p\uparrow + p\downarrow \tag{8.32}$$

$$\frac{p\downarrow}{p\uparrow} = \exp(-\Delta E/kT) \tag{8.33}$$

In other words, randomly fluctuating magnetic fields from the surroundings (at temperature, T) induce transitions whose relative rates establish the equilibrium Boltzmann ratio of populations in the two energy levels. However, if this system is irradiated on-resonance with sufficient rf magnetic field amplitude and duration, then the *induced* ("stimulated") transition rate constants, which have equal probability upward and downward, become faster than the *spin-lattice* relaxation rate constant, $1/T_1$, and the spin system is "saturated"—i.e., *equal* populations in the upper and lower energy levels. (A 90° pulse has the same effect as saturation on energy level populations, but also preserves phase-coherence of the macroscopic magnetization.)

If each of the above A nuclei is J-coupled to a second nucleus, B, of different Larmor frequency (see Figure 8.17), then we can think of transitions between the two energy levels of nucleus A as a "chemical exchange" between the J-split resonances of nucleus B. Thus, if the irradiation at ω_A is sufficiently strong to induce transitions between the A energy levels at a rate constant, $(1/\tau) >> J$, then the J-splitting at ω_B will collapse by the same mechanism as the "fast-exchange" limit of Eq. 8.30. Therefore, the irradiation at A is known as "decoupling", since it effectively removes the J-coupling from nucleus A to nucleus B.

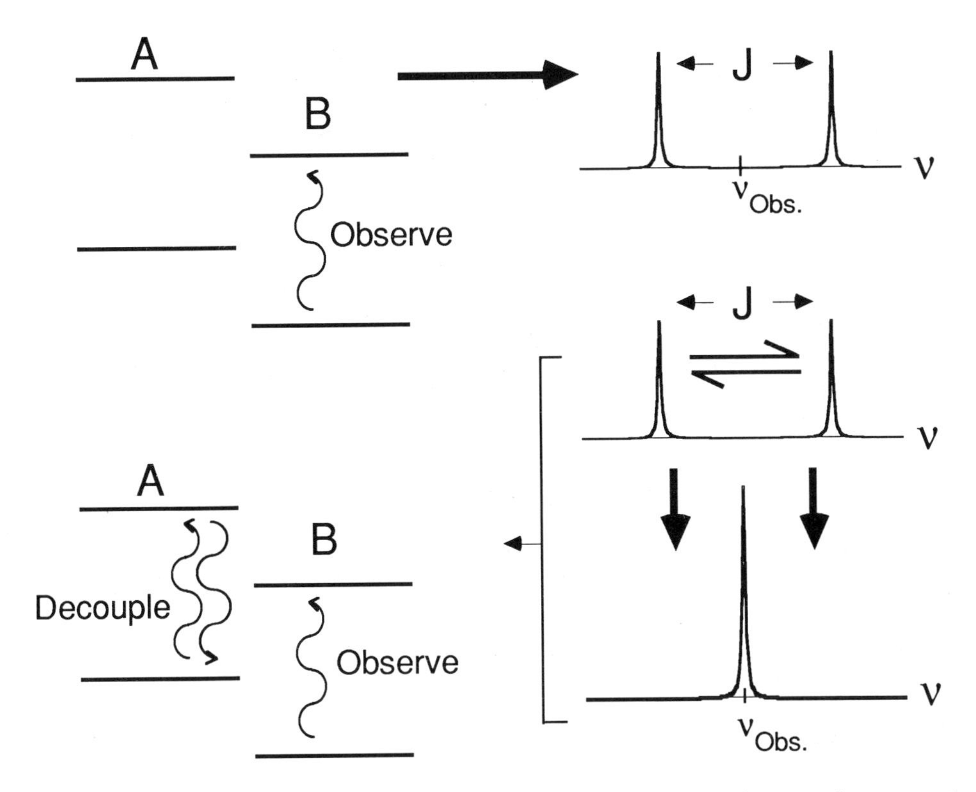

Figure 8.17 Effect of "decoupling" on a system of two spins 1/2. Irradiation at the Larmor frequency of nucleus A induces rapid transitions between its two energy levels. Since the two J-split resonances of J-coupled nucleus B correspond to A-nuclei in the upper and lower energy levels, transitions between those two A energy levels have the effect of rapid chemical exchange between the two J-split B resonances. For sufficiently high irradiation magnitude and duration, the "fast-exchange" limit effectively removes the J-splitting of the B resonance; hence, "decoupling" B from A.

8.4.3 Molecular motion: solids vs. liquids

As noted in Chapter 8.2.3, dipole-dipole through-space coupling in a *single crystal* produces an NMR spectral peak splitting in which the separation between the peaks is proportional to $(3 \cos^2 \theta - 1)$, where θ is the (polar) angle between the internuclear vector and the applied magnetic field direction.

Moreover, as argued in Chapter 8.2.1, the "chemical shift" in Larmor frequency results from currents induced in the electron distribution around a nucleus. Since a chemical bond polarizes the distribution of those electrons, the "chemical shift" depends on the direction of the chemical bond relative to the applied magnetic field, and it turns out that the magnitude of that shift also varies with angle as $(3 \cos^2 \theta' - 1)$, except that θ' now denotes the angle between the principal axis of the chemical shift tensor (which is in turn related to the orientation of the chemical bond) and the applied magnetic field direction.

Therefore, the NMR spectrum of a *polycrystalline powder* consists of a superposition of many dipolar doublets, each having different peak separation and average position. However, the distribution of possible θ-angles is non-uniform: e.g., there is only *one* way that the internuclear vector can be parallel to $\boldsymbol{B}_0$, whereas there are *many* ways (i.e., many ϕ-angles) in which the internuclear vector can be perpendicular to $\boldsymbol{B}_0$. Thus, the "powder pattern" NMR spectrum of a polycrystalline solid is inhomogeneously broadened by dipolar splittings and anisotropic chemical shifts, and exhibits characteristic cusps (see Fyfe reference in Further Reading). Typically, the polycrystalline spectral cusp-to-cusp separation produced by chemical shift anisotropy is a few kHz, whereas dipolar broadening can be much larger ($\geq$50 kHz). Unfortunately, because the "powder pattern" superposition of widely spaced doublets is spread over such a large frequency range that the spectral signal-to-noise ratio is very low (top spectrum in Figure 8.18). Moreover, T_1 values in solids can be quite long (minutes in some cases), so that it is no longer possible to accumulate thousands of time-domain transient signals in a reasonable length of time.

In a liquid, the internuclear vector reorients rapidly (e.g., typically by ~1 radian in $<10^{-10}$ s) and diffusionally (i.e., by many small angular jumps). Since the dipolar doublet peak separation varies with θ, and since θ changes with a rate constant much faster than the dipolar splitting, the "fast-exchange" model predicts that dipolar splittings should average to zero in liquids. Thus, except for J-couplings, the liquid-phase spectrum yields a single spectral peak for each chemically distinguishable nucleus in a molecule (bottom spectrum in Figure 8.18).

Although a powder pattern can be analyzed to yield the dipolar and chemical shift anisotropy components for very simple spin systems, the pattern becomes unmanageably complex for molecules with more than a few magnetic nuclei. In such cases, two further tricks are used to simplify the spectrum. First, in ^{13}C NMR, the dominant dipolar coupling for a given ^{13}C arises from its directly bonded proton(s). Thus, sufficiently high-power broadband decoupling irradiation at the ^1H NMR frequencies removes the dipolar coupling, just as lower-power decoupling removes J-coupling (Chapter 8.4.2).

Second, there is a "magic angle",

$$\theta = \cos^{-1}\left(\frac{1}{\sqrt{3}}\right) \tag{8.34}$$

at which the factor, $(3\cos^2\theta - 1) \rightarrow 0$. Thus, if a (polycrystalline) solid sample is mechanically spun very rapidly (~5 kHz!) about an axis tilted by the "magic angle" away from the applied magnetic field direction, then the chemical shift anisotropy is averaged to its component along the rotation axis. However, since the rotation axis happens to be oriented at the magic angle, any residual chemical shift anisotropy vanishes!

The combined "cross-polarization" (a result of high-power decoupling) "magic-angle spinning" (CP/MAS) experiment for a polycrystalline solid can produce a highly resolved FT/NMR spectrum that looks very much like that for a liquid (see Figure 8.18). However, the local environment in a solid is different than that in a liquid. For example, there may be two or more chemical shifts for each carbon of of a polycrystalline molecule if the unit cell of the crystal contains more than one crystallographically non-identical molecule (see Figure 8.18).

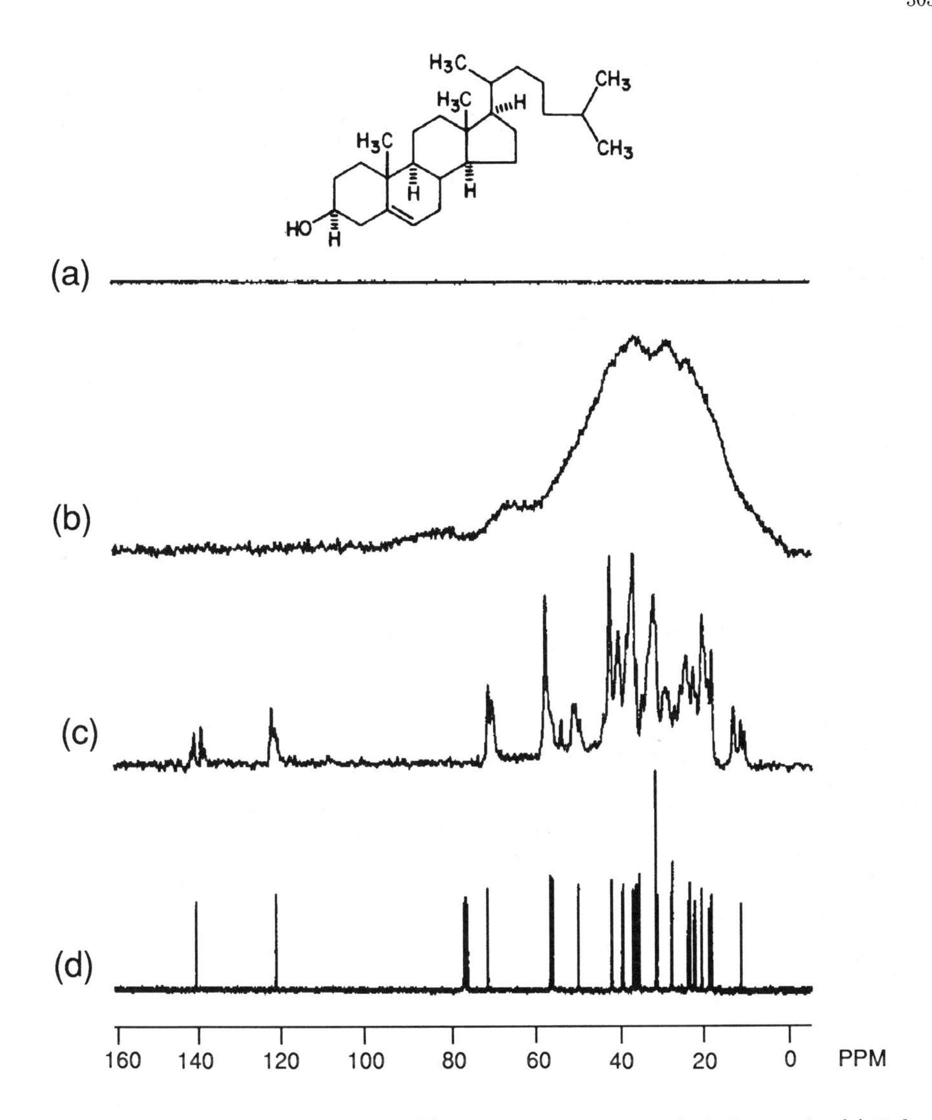

Figure 8.18 Solid- and liquid-phase ^{13}C FT/NMR spectra of cholesterol. (a) Polycrystalline solid, for which the spectrum is a superposition of peaks split by dipolar and chemical shift anisotropy effects (see text), to give an inhomogeneously broad spectrum too broad to be visible under these conditions. (b) Same sample subjected to high-power ^1H decoupling to remove dipolar splittings: the unresolved splittings from chemical shift anisotropy remain. (c) Same as (b), but with additional rapid spinning about an axis which is tilted by the "magic" angle, $\theta = \cos^{-1}(1/\sqrt{3})$, with respect to the applied magnetic field direction. (d) Spectrum of cholesterol dissolved in deuteriochloroform: rapid molecular tumbling averages the $(3 \cos^2 \theta - 1)$ factors to zero to give a sharp peak for each carbon. Note that the CP/MAS and liquid spectra are not identical, because not all cholesterol molecules in the solid phase have identical environments. (Spectra provided by C. E. Cottrell.)

8.5 Two-dimensional NMR

The experimental event sequence for a generalized two-dimensional Fourier transform (2D/FT) NMR experiment is shown in Figure 8.19. During the *preparation* period, the system usually relaxes to an equilibrium starting point. During the *evolution* period, the system is perturbed from equilibrium (e.g., by application of one or more rf pulses) and allowed to evolve for a period, t_1, whose N_1 equally incremented values form the time scale for the second Fourier transformation (see below). During the *mixing* period, the system is prepared (e.g., by application of one or more rf pulses) for detection. Finally, during the *detection* period, t_2, the time-domain FID is acquired in "real time" (i.e., N_2 regularly spaced samples of a single time-domain signal as that signal passes through the detector) and stored for the subsequent first Fourier transformation. N_1 need not be equal to N_2.

Preparation → Evolution → Mixing → Detection

Figure 8.19 Experimental event sequence for a generalized two-dimensional FT/NMR experiment (see text).

A typical procedure for processing the above two-dimensional t_1/t_2 array is shown in Figure 8.20. First, the FID acquired (as a function of t_2) for *each* of the t_1 values (Figure 8.20a) is Fourier transformed to give the ν_2/t_1 array shown in Figure 8.20b. The frequency/time data array is then reshuffled as shown to yield real and imaginary t_1-domain data sets (Figure 8.20c). A second FT (with respect to t_1) then yields the two-dimensional ν_2/ν_1 array shown as Figure 8.20d). In other words, the *first* FT is with respect to the *second* time-dimension (t_2), whereas the *second* FT is with respect to the *first* time-dimension (t_1).

The second FT of Figure 8.20 will produce spectral "peaks" at non-zero frequencies only if the NMR signal is somehow *modulated* as a function of the delay period, t_1. We therefore introduce 2D/FT/NMR with a discussion of the modulation resulting from scalar (J) coupling. The principle will then be extended (with decreasing detail) to three of the most useful classes of 2D/FT NMR experiments: COSY (COrrelated SpectroscopY), NOESY (Nuclear Overhauser Enhancement SpectroscopY), and INADEQUATE (Incredible Natural Abundance DoublE QUAntum Transfer Experiment). Since accurate description of the 2D/FT/NMR spectral peak frequencies, magnitudes, phases, and artifacts requires a full density-matrix quantum mechanical analysis, the interested reader is referred to any of several modern texts (see Further Reading) for details.

8.5.1 Spin-echo plus decoupling: J-Modulation; DEPT

Most two-dimensional FT/NMR experiments involve combinations of 90° and 180° pulses, delay periods, and decoupling. A one-dimensional example which illustrates all of these elements is the J-modulation experiment, which we shall now describe.

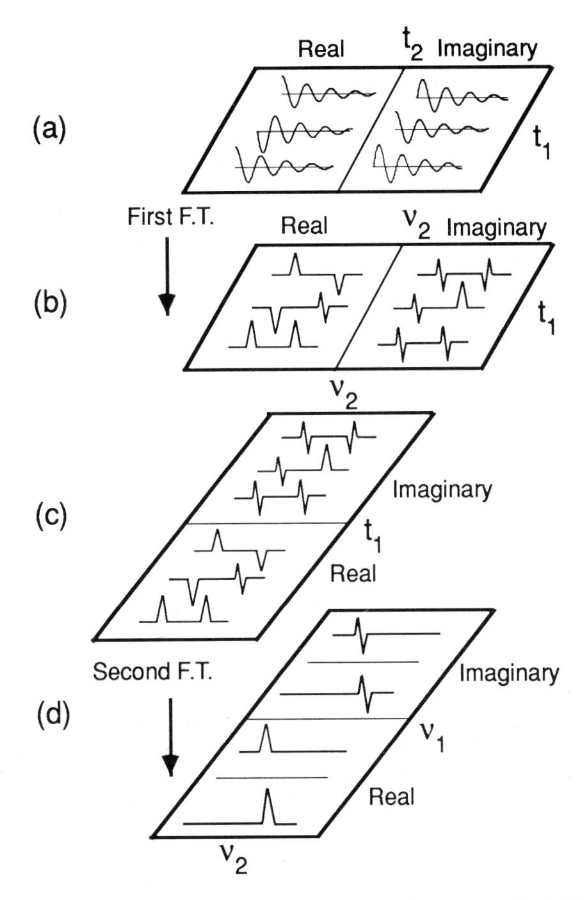

Figure 8.20 Schematic diagram of data manipulation in 2D/FT/NMR. Each t_2-row of the raw two-dimensional time/time (t_1/t_2) data set (a) is Fourier transformed to yield the frequency/time (ν_2/t_1) array shown in (b). After reshuffling (c), each t_1-column of that array is subjected to a second Fourier transform to yield the final two-dimensional frequency/frequency (ν_2/ν_1) spectral display shown in (d).

Consider a single carbon with a directly bonded proton. Approximately 1% of such carbons will be ^{13}C, and the the ^{13}C NMR spectrum of the ^{13}C-^{1}H system will consist of two peaks separated by $J_{^{13}C^{1}H} \cong$ 125-200 Hz. The reader should convince him/herself that a spin-echo event sequence (Eq. 8.25) of rf pulses at the ^{13}C Larmor frequency will refocus the magnetizations from both J-split components. In other words, the spin-echo experiment refocuses signals at all frequencies, whether the frequency differences arise from different chemical shifts or J-couplings.

However, a rather different pattern is observed for the spin-echo pulse sequence of Figure 8.21, in which proton decoupling irradiation is present *throughout* the refocusing and detection periods.

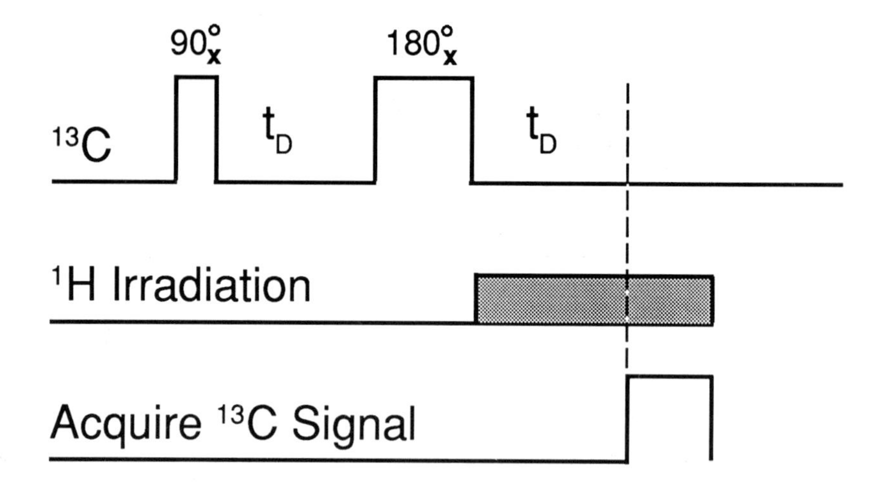

Figure 8.21 J-modulation spin-echo event sequence for a $^{13}C^1H$ system.

The effect of the event sequence of Figure 8.21 is shown in Figure 8.22. In the presence of proton decoupling irradiation, both J-components precess at the same (chemically shifted) Larmor frequency (i.e., the J-coupling is effectively removed—see Chapter 8.4.2). The Larmor precession is slightly faster than the rotating frame, so that the precession due to ω_0 moves away from the (positive) y'-axis during the first delay period, and back toward the (negative) y'-axis during the second delay period to give a final observed signal that always lies along the y'-axis. However, the two J-coupled components diverge away from the ω_0 direction during the *first* delay period, but precess at the *same* (ω_0) frequency during the second delay period. Thus, the final observed signal magnitude will be modulated at a frequency, $J/2$. For example, the observed signal goes to zero whenever the two J-components have precessed by 90°, 270°, 450°, $\cdots$ (corresponding to t_{delay} = 1/4, 3/4, 5/4, $\cdots$ of a J-period), so that their directions are opposed on the $\pm x'$-axis. Thus, only the chemical shift differences are completely refocussed during the second half of the event sequence. (Since decoupling is applied during ^{13}C signal detection, J-splittings will be absent in the final FT spectrum.)

Figure 8.23 shows experimental ^{13}C FT/NMR spectra of ethylbenzene, acquired at varying delay period, t_{delay}, of the event sequence of Figure 8.21. In each spectrum, there is a peak at the ^{13}C Larmor frequency, but the amplitude of that peak is modulated by ^{13}C–^1H J-coupling during the first delay (evolution) period of the experimental event sequence. Thus, a Fourier transform of the data set consisting of time-domain data as a function of t_{delay} at the Larmor frequency, ω_0, of each of the spectra in Figure 8.23 yields slices with a $^{13}CH_2$ triplet and $^{13}CH_3$ quartet, each with frequency separation of $J/2$ Hz between adjacent peaks.

Since decoupling effectively eliminates the J-splitting, the signal-to-noise ratio in the spectra of Figure 8.23 is approximately double that of a conventional ^{13}C FT/NMR spectrum. In addition, because the decoupling process perturbs the populations of all four energy levels of the two-spin system, there is a further change known as a *nuclear Overhauser enhancement*, or NOE (in this case, positive) in the ^{13}C signal-to-noise ratio (see Further Reading).

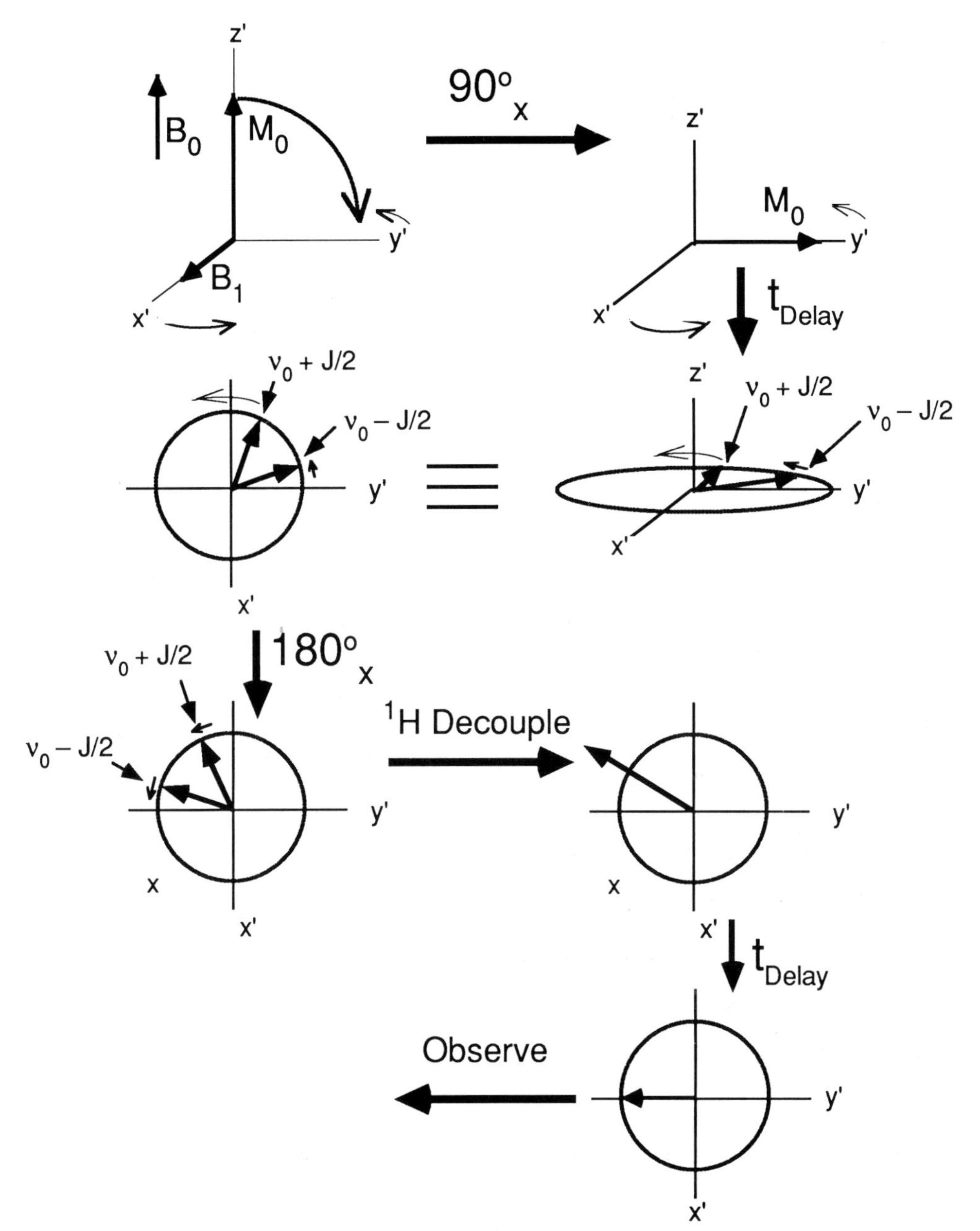

Figure 8.22 Rotating-frame diagram of the two J-coupled magnetization components of the ^{13}C signal of $CHCl_3$ at various stages of the event sequence of Figure 8.21. The coordinate frame rotation frequency, $\omega_1 = \gamma B_1$, is taken to be slightly different from the Larmor frequency, in order to show the refocussing of chemical shift differences (see text).

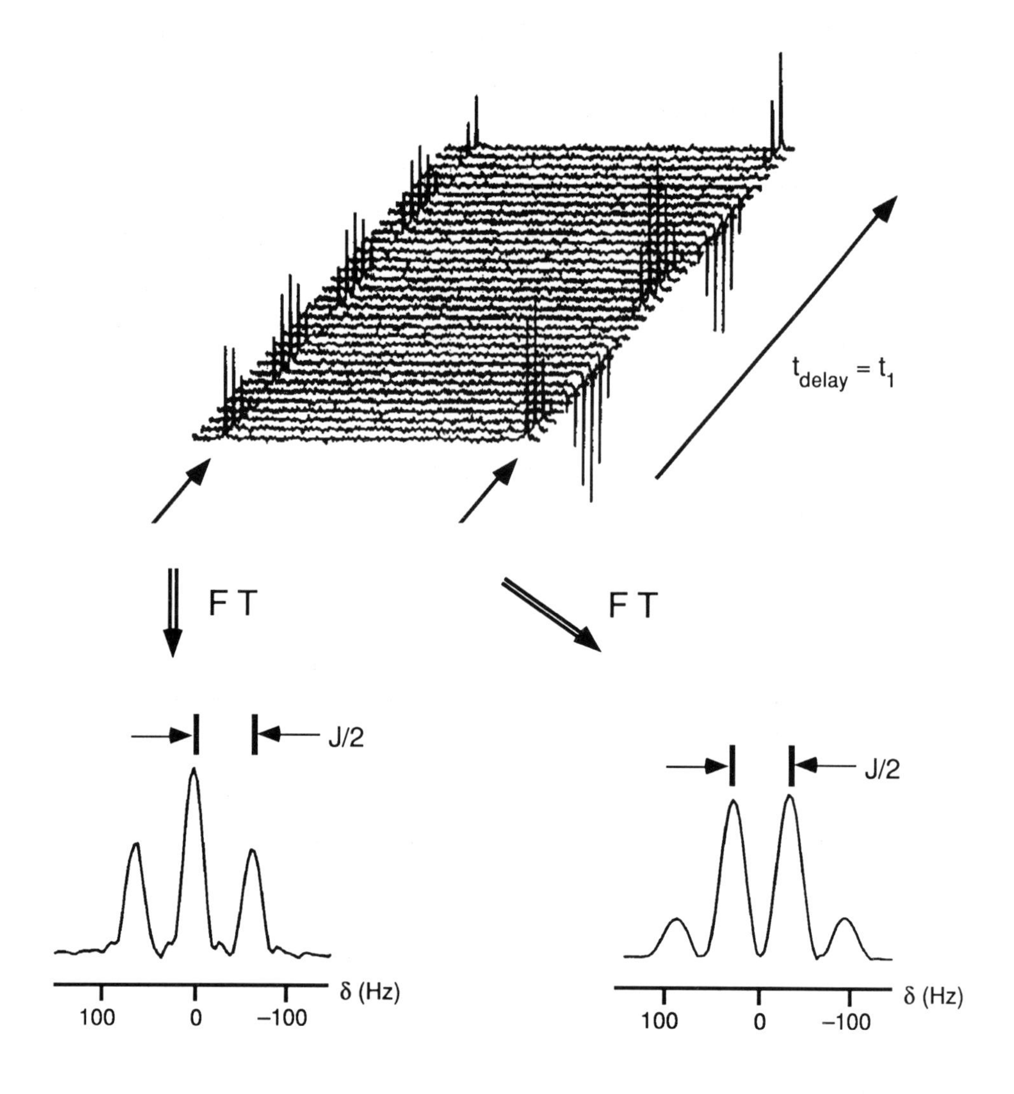

Figure 8.23 ^{13}C FT/NMR spectra at 11.75 tesla (top) from the CH_2–CH_3 portion of ethylbenzene (10%, in $CDCl_3$, at ~125 MHz), acquired at equally increments of the delay period, t_{delay}, of the event sequence of Figure 8.21. The amplitude of either ^{13}C spectral peak amplitude is modulated (at frequencies simply related to J) as a function of t_{delay}. Thus, a second FT of each of the two data "slices" designated by an arrow with respect to t_1 generates spectral multiplets (bottom) whose component peaks are separated by $J/2$ Hz. (Spectra provided by C. E. Cottrell.)

As noted above, the ^{13}C signal for a carbon J-coupled to its directly bonded proton can be made to disappear completely by appropriate choice of t_{delay} (namely, 1/4, 3/4, 5/4, $\cdots$ of a J-period). Similar, but different modulation patterns are observed for CH_2 and CH_3 groups. In fact, it is possible to design an event sequence ("Distortionless Enhancement by Polarization Transfer, or DEPT) whose results may be edited (i.e., linear combinations of spectra) to yield *subspectra* (Figure 8.24) showing *only* those carbons with one, or two, or three directly bonded protons. The conventional proton-decoupled ^{13}C spectrum (Figure 8.24d) consists of one peak for each chemically different carbon in the molecule. DEPT greatly facilitates the analysis of such a spectrum, by separating and identifying the carbons which have one or two or three carbons (Figures 8.24a-c) and (by comparison to the conventional spectrum) the carbons (e.g., quaternary) with no directly-bonded protons.

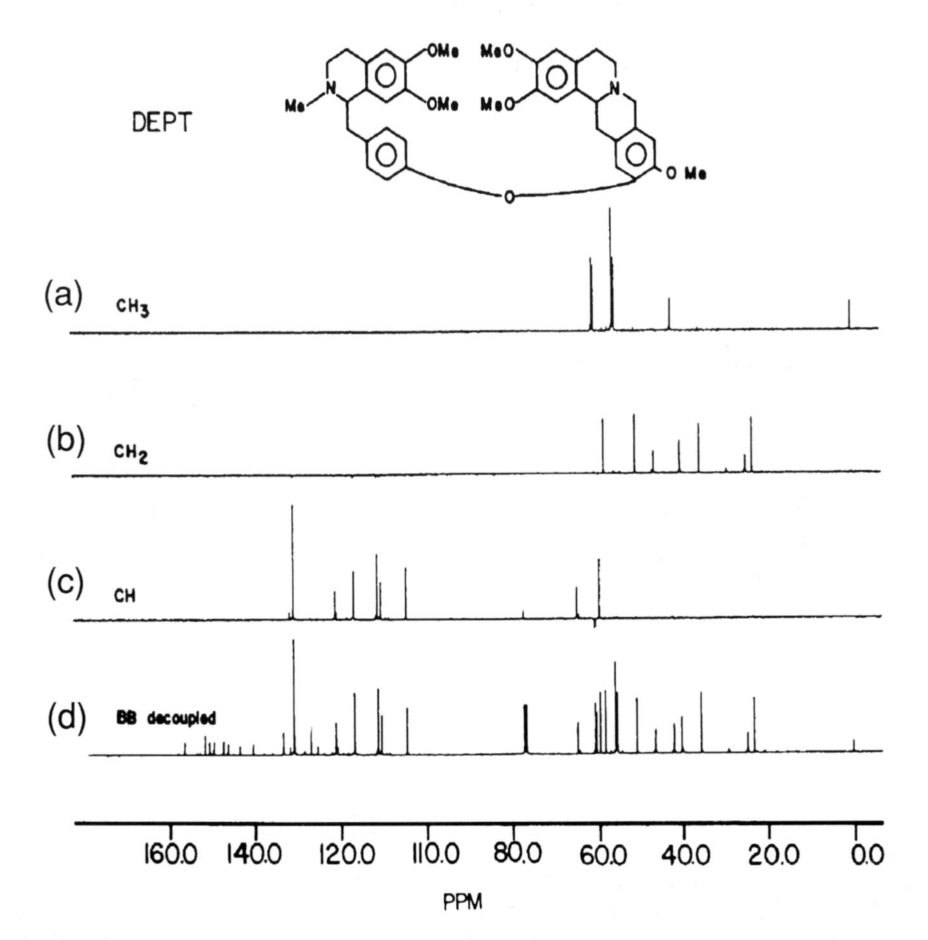

Figure 8.24 Edited DEPT ^{13}C FT/NMR subspectra (a)-(c) and the conventional proton-decoupled ^{13}C spectrum (d) of a natural product provided by R. Doskotch. The carbons with no directly-bonded protons can be identified by the difference between [(a) + (b) + (c)] and (d). (Spectrum provided by C. E. Cottrell)

8.5.2 Other 2D-FT/NMR Experiments

Although the J-modulation scheme discussed in Figures 8.21-23 can be extended to a full two-dimensional equivalent, the most useful 2D-FT/NMR experiments are considerably more complex in both event sequence (including phase cycling) and spin manipulation mechanism. Nevertheless, the principle of modulation as a function of incremented delay time exemplified in Figure 8.23 is common to all of them.

8.5.2.1 COSY

As noted in the preceding section, J-coupling between two nuclei of different Larmor frequency (chemical shift) can produce a frequency-domain peak in the spectrum produced by FT of a one-dimensional "slice" through a two-dimensional J-modulated $\nu_2 t_{delay}$ array (see Figure 8.23). The somewhat more complex COrrelated SpectroscopY (COSY) 2D-FT/NMR experiment produces a two-dimensional spectrum (Figure 8.25) in which the conventional one-dimensional spectrum appears on the diagonal and the horizontal and vertical projections of the off-diagonal peaks represent the chemical shifts of each pair of J-coupled nuclei. The data is displayed as a contour ("topographical") map consisting of lines of constant altitude above the baseline. It is useful to display the silhouette obtained by looking "south" or "east" across the "plain" of peaks (considered as "mountains") to give the conventional spectra shown at the left and top of the two-dimensional array. COSY (see, e.g., Sanders reference for details) is the most reliable and most frequently used of all of the two-dimensional FT/NMR experiments. If either the ^{13}C (or ^{1}H) NMR spectrum of a molecule has been "assigned" (i.e., each peak can be associated with a particular atom in the molecule), then the other can be assigned immediately from a COSY spectrum. For example, extending horizontal and vertical lines from each off-diagonal peak until they intersect the diagonal in Figure 8.25 shows that the proton at ~4.5 ppm is J-coupled to the proton at 5.5 ppm. In this way, one can establish connections between peaks which are unresolved due to overlap in the normal proton spectrum (e.g., the regions at 4.0 and 4.2 ppm in this example).

8.5.2.2 NOESY

As noted above, irradiation at the Larmor frequency of one magnetic nucleus will perturb the populations of the energy levels of any other magnetic nucleus to which the first nucleus is coupled. The coupling may occur either through-bonds (J-coupling) or through-space (dipole-dipole). The most useful 2D-FT/NMR experiment based on through-space coupling is the so-called Nuclear Overhauser Enhancement SpectroscopY (NOESY) experiment. As shown in Figure 8.26, the off-diagonal peaks in the NOESY spectrum identify the chemical shifts of pairs of nuclei which are spatially close to each other. In this case, the NOESY 2D-spectrum (only part of which is shown in Figure 8.26) serves to establish the base sequence of an enzymatic cleavage fragment of a eukaryotic ribosomal 5S RNA. NOESY experiments have proved especially useful in establishing the secondary and tertiary structures (see Wüthrich reference) of nucleic acids (both DNA and RNA), proteins, oligosaccharides, and other complex organic molecules.

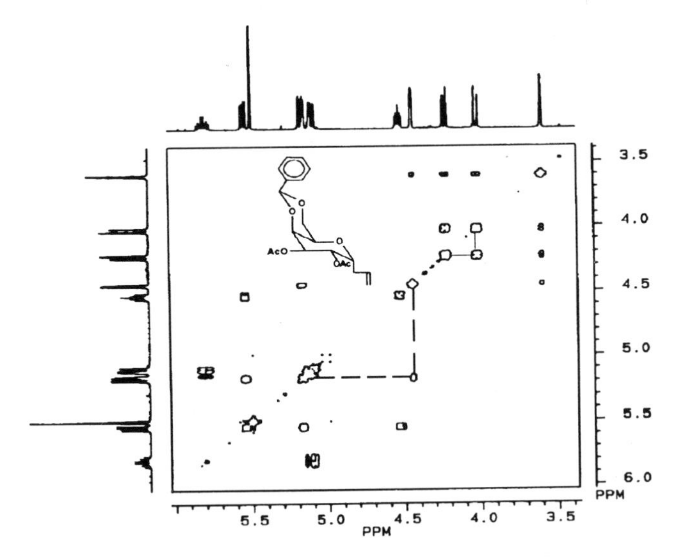

Figure 8.25 Homonuclear ^1H-^1H COSY 2D-FT/NMR spectrum of a carbohydrate provided by D. Horton. Off-diagonal peaks indicate which proton is J-coupled (through chemical bonds) to which other proton(s) (see text). (Spectra provided by C. E. Cottrell)

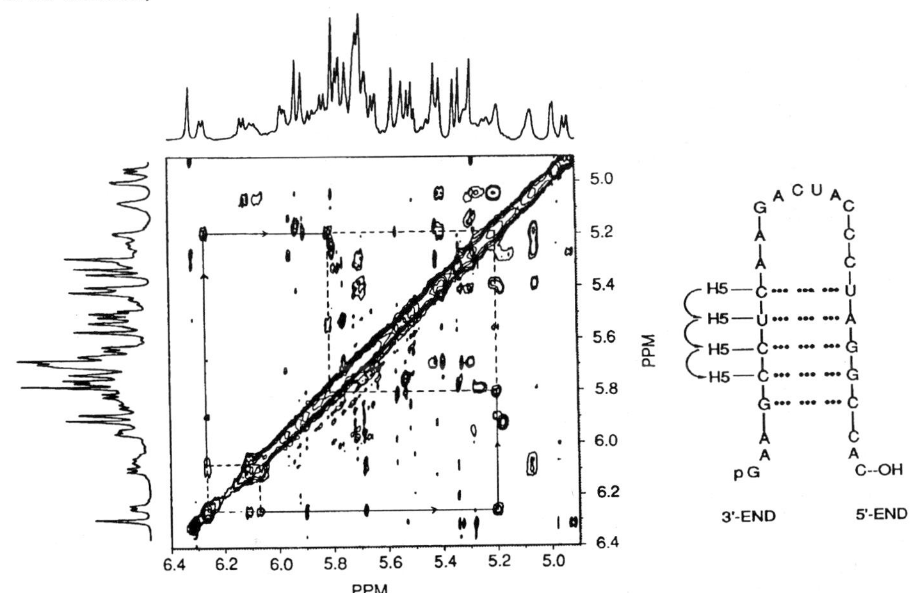

Figure 8.26 2D-FT/NMR ^1H NOESY spectrum of an RNA fragment. In this contour map, each off-diagonal peak identifies a through-space dipole-dipole coupling between a pair of protons whose chemical shifts are found by projecting horizontally or vertically from the peak to the conventional spectrum on the diagonal. The nucleotide sequence connections inferred from the data are shown at the right. (Compound and data provided by J. Wu)

8.5.2.3 Coherences: double-quantum NMR (INADEQUATE)

Although 2D-FT/NMR experiments can be intricately difficult to construct and perform, their interpretation can be extremely simple. Perhaps the most spectacular example is the 2D-FT/NMR experiment based on double-quantum transitions, namely "Incredible Natural Abundance DoublE QUAntum Transfer Experiment" (INADEQUATE). A meaningful explanation of this experiment requires quantum mechanical density matrix analysis, because double-quantum transitions are not directly observable. At this level, all we can say is that an event sequence consisting of as few as three consecutive 90° pulses can produce a "coherence" signal that precesses at the Larmor frequency but which arises from double-quantum transitions rendered detectable by J-coupling between two directly bonded carbons.

In the absence of artifacts (see below), interpretation of an INADEQUATE spectrum is particularly simple. Each pair of off-diagonal ^{13}C doublets which is equidistant (horizontally) from the diagonal identifies two directly-bonded carbons. Thus, we begin by labeling each ^{13}C resonance of the conventional proton-decoupled ^{13}C FT/NMR spectrum (Figure 8.27, bottom) according to its chemical shift. From the off-diagonal INADEQUATE spectral connections shown by horizontal lines, we can then infer that (e.g.,) carbon-1 is bonded to carbon-6; carbon-6 is bonded to carbon-13; carbon-13 is bonded to carbon-5; etc. From a series of such connections, it is possible (in many cases) to *trace out the entire carbon-bonded skeleton of a molecule from a single INADEQUATE experiment, without any further information*, as in the example of Figure 8.27.

Since carbon-13 is ~1% naturally abundant, only ~1 in 10^4 molecules will contain the required two directly-bonded ^{13}C atoms; thus, INADEQUATE spectra exhibit very low signal-to-noise ratio, and typically require extended data acquisition periods (e.g., several hours).

Multiple-quantum transitions ($n \geq 2$) have proved especially useful in unscrambling the very complex NMR single-quantum spectra of solid or partially oriented (as in membranes or liquid crystals) molecules (see Pines reference). For example, if a solid contains isolated clusters of (say) 8 like nuclei, then multiple-quantum spectra up to $n = 8$ may be observed.

In general, because 2D-FT/NMR experiments ideally require multiple pulses of precisely specified amplitude, phase, duration, and shape, as well as complete and instant "on/off" transmitter and receiver gating, the final 2D spectra can be plagued with spurious peaks and artifacts. Because the two-dimensional display already contains so many peaks, foldover can be more serious than for one-dimensional experiments. Phasing can present a major problem: e.g., some 2D event sequences produce a "twist" lineshape which approaches dispersion-mode far from resonance but absorption-mode on-resonance. A strong solvent signal is especially problematic. Even the choice of contour levels can be non-trivial—if contours are too high, small peaks may be overlooked, whereas if contours are too low, noise spikes may be interpreted as signal peaks. For all of the above reasons, most 2D-FT/NMR experiments involve elaborate multiple-acquisition phase-cycling designed to cancel out errors which cannot be avoided.

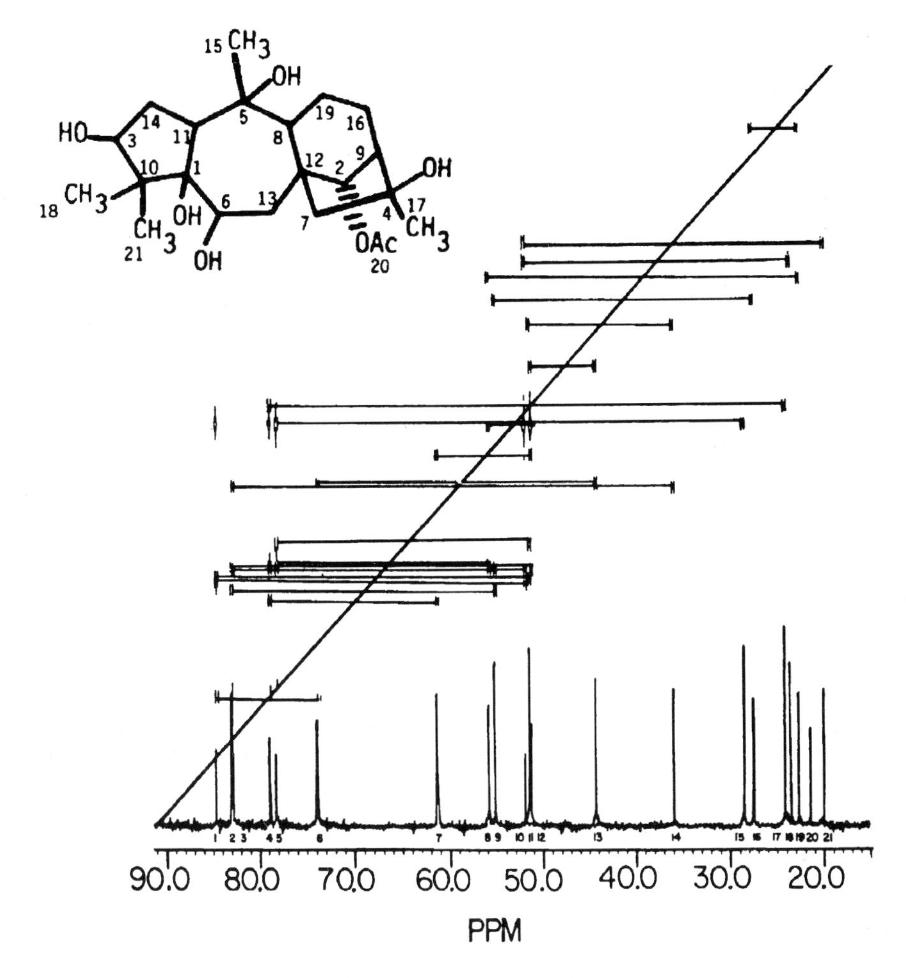

Figure 8.27 2D-FT/NMR INADEQUATE spectrum of a natural product provided by R. Doskotch. Virtually all of the carbon-bonded skeleton of the molecule can be inferred from the two-carbon fragments identified from this single spectrum (see text). (Data provided by C. E. Cottrell.)

8.6 NMR Imaging

The basic principle of magnetic resonance imaging (MRI) is that the proton NMR Larmor frequency (usually of H_2O) is directly proportional to applied magnetic field strength (Eq. 8.5). Thus, if the applied magnetic field strength varies monotonically with *position* across the object, then the corresponding NMR *spectral magnitude vs. frequency* will provide a direct measure of the *number of nuclei vs. spatial position* along the gradient.

Figure 8.28 shows the (one-dimensional) "projection" image of a simple object, based on continuous application of a magnetic field linear z-gradient during FT/NMR excitation and detection. One can think of such an image as a "shadow" of the object. Figure 8.28 is the simplest example of "frequency-encoding", in which the field gradient is left on during detection—see below. By varying the direction of the field gradient, one can produce "shadows" or "projections" from several viewing directions, and then reconstruct the object by "back-projection" from those "shadows", as in X-ray "CAT" scanning. Some of the earliest MR images were in fact produced in this way.

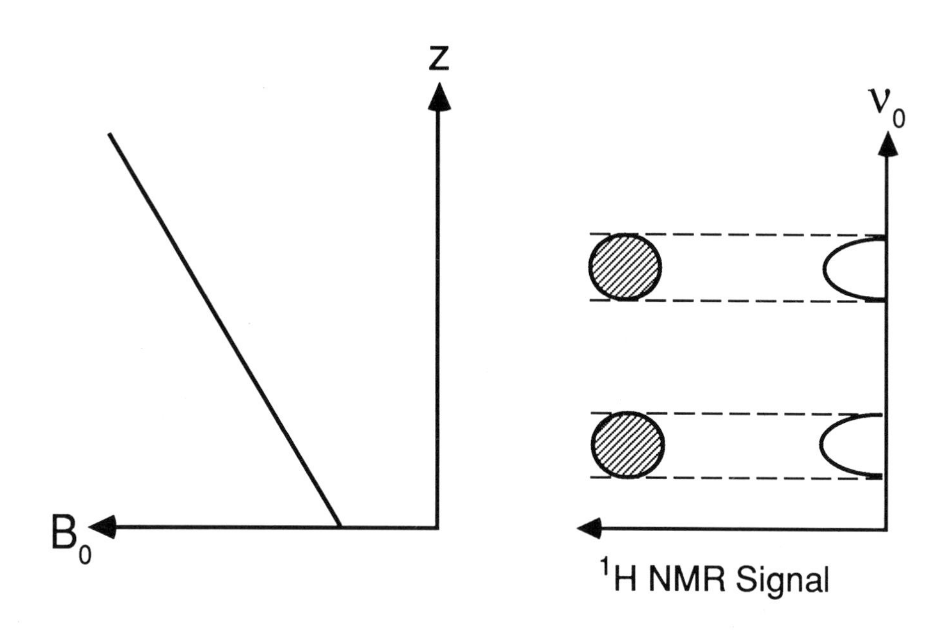

Figure 8.28 Schematic one-dimensional "projection" image of an object consisting of two water-filled test tubes of circular cross-section. The "image" is simply the ^{1}H NMR spectrum, whose magnitude reflects the relative number of protons and whose frequency is determined by the magnetic field strength at a given z-position.

However, Fourier transform methods provide a much more direct means for obtaining a direct MR image without the use of back-projection. By adding suitably timed magnetic field *gradient* pulses to the general 2D-FT/NMR event sequence of Figure 8.19, we can use 2D-FT techniques to generate two- and three-dimensional spatial images. Typically, a "slice" ~1 cm thick is selected by applying a z-gradient *excitation* by a sinc-shaped rf pulse. The remaining two dimensions are achieved by *phase-encoding* (e.g., y-gradient during the *evolution* period) and *frequency-encoding* (e.g., x-gradient during *detection*, as in Figure 8.28), as described in Chapter 8.6.1.

Since MR images can be weighted by "tuning" the experiment according to *number* of nuclei (ρ-image), or T_1 or T_2 (Chapter 8.6.2) parameters, the same object may appear light on a dark background or dark on a light background, depending on the experimental parameter weighting, as discussed in Chapter 8.6.2. *Flow* is detected by exciting spins in one location and detecting in another.

Finally, the magnetic field gradients used in MR imaging are so strong that differences in Larmor frequency ("chemical shifts") between chemically different protons are generally negligible. However, by combining frequency-selective excitation in the *absence* of gradients with MR imaging event sequences, one can generate MR images which are *chemically* as well as *spatially* selective (e.g., water-only vs. fat-only proton images). Images based on chemical shift differences are discussed in Chapter 8.6.3.

8.6.1 Spatial encoding methods: slice selection, phase- and frequency-encoding

The three basic techniques of spatial encoding are shown in Figure 8.29. First, a magnetic z-gradient disperses the nuclear magnetic Larmor frequencies along the z-direction. In the presence of that gradient, a single-frequency rf time-domain waveform with rectangular frequency-domain spectral magnitude (Figure 8.29, top) is applied for a period long enough to produce a 90° rotation of magnetization into the x-y plane. The net result is that only those nuclei whose Larmor frequencies are within the bandwidth of the excitation pulse will be observed. Since that band of frequencies corresponds to a particular range of z-positions, the effect of the excitation is to select a "slice" including all nuclei within a narrow (~1 cm) z-range. If the system were linear, the appropriate time-domain pulse shape would be the sinc function (Chapter 2.1.2). Although an NMR system is *not* linear, sinc excitation nevertheless is commonly employed.

Once a transverse magnetization corresponding to a given z-slice has been created, a magnetic y-gradient (i.e., B_z varies as a function of y-position) is applied. In the presence of the y-gradient, lab-frame y-magnetization will precess by a (phase) angle which varies directly with the y-distance away from the center of the magnet (Figure 8.29, middle). In high-resolution 2D-FT/NMR, the extent of such encoding is varied by increasing the evolution (delay) period in equal increments, and performing an FT with respect to t_{delay}. However, MRI experiments are more conveniently conducted with fixed delay period. Therefore, Bottomley proposed a "spin-warp" method, in which the y-gradient *slope* is incremented in successive event sequences, as shown in Figure 8.30 (see below).

Finally, a magnetic x-gradient (i.e., B_z varies as a function of x-position) is applied during real-time signal acquisition (Figure 8.29, bottom). Since the Larmor frequency now varies as a function of x-position, the frequency axis of the first FT of a two-dimensional "spin-warp" data set will correspond to the x-axis of the imaged object.

The event sequence described above will indeed generate a two-dimensional image of NMR signal as a function of x- and y-position for a given z-slice, but offers relatively low signal-to-noise ratio, because the NMR signal decays significantly between the initial 90° pulse and the final detection period due to T_2 relaxation and inhomogeneous broadening induced by the various gradients. Therefore, an actual "spin-warp" event sequence (Figure 8.30) contains additional (negative) gradients designed to refocus the inhomogeneously dispersed magnetization.

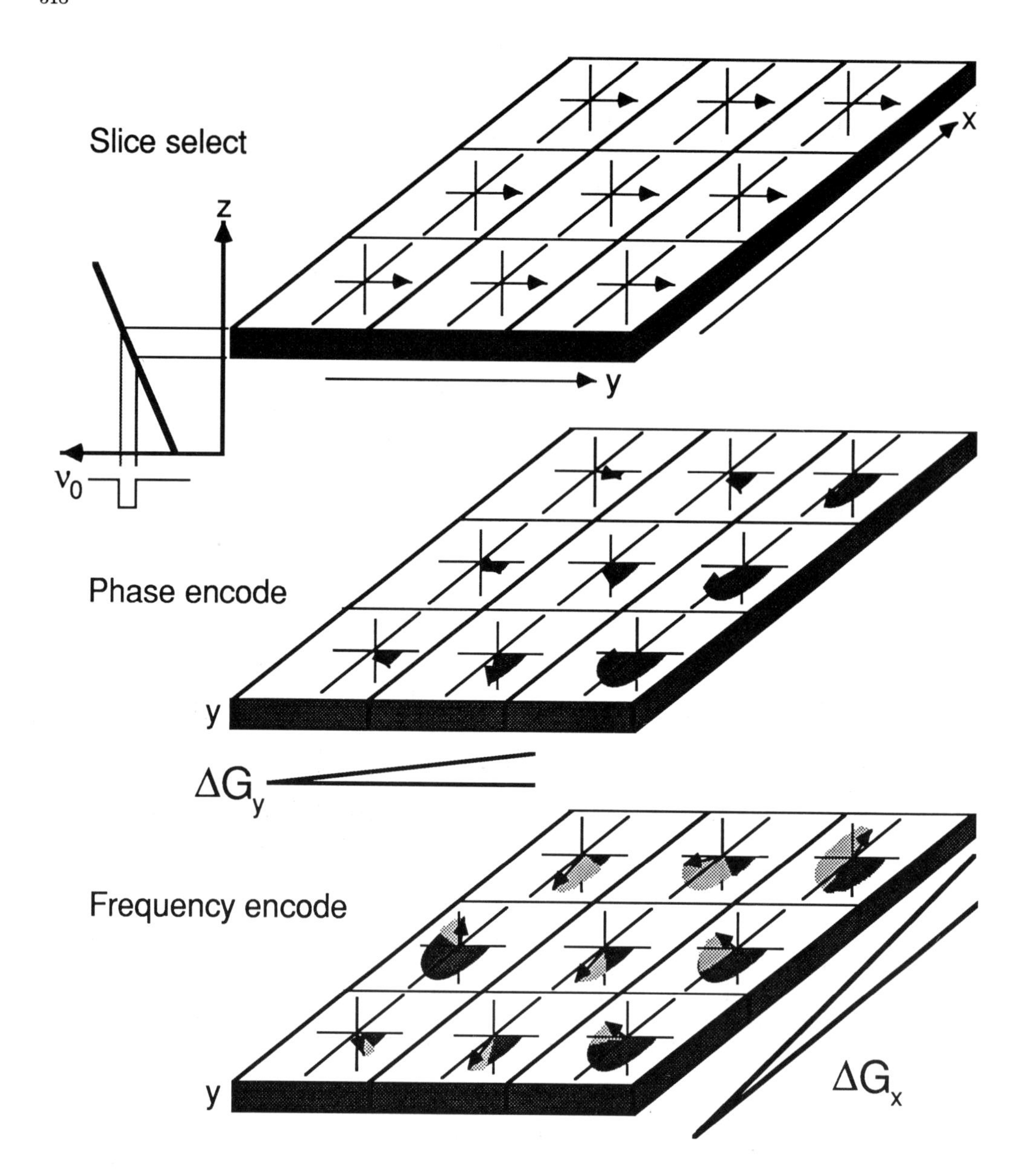

Figure 8.29 Use of magnetic field *z*-, *y*-, and *x*- gradients to encode the spatial locations of magnetic nuclei in the generation of a magnetic resonance image. Top: A *z*- slice is selected by exciting a narrow band of frequencies in the presence of a *z*- gradient. Middle: During an evolution period, the *z*- field is varied as a function of *y*- position, so that nuclei in different columns of the slice precess by different *phase* angles. Bottom: During the detection period, the *z*- field is varied as a function of *x*- position, so that nuclei in different rows of the slice precess at different Larmor *frequency* .

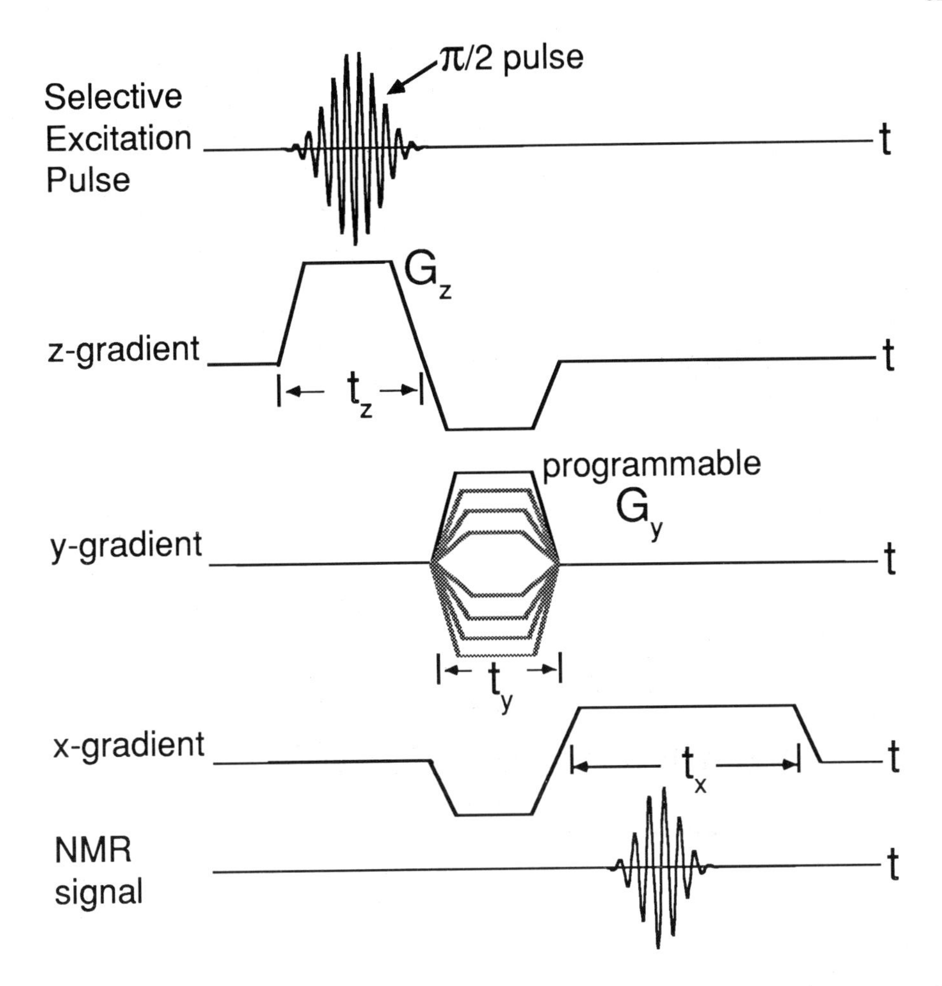

Figure 8.30 Spin-warp event sequence for two-dimensional (x and y) imaging of a single z-slice. G_i are magnetic field gradients applied along x-, y-, and z-axes. The detected time-domain signal appears as a spin-echo because of the refocusing from the negative gradients. The two-dimensional image consists of a 2D-FT/NMR spectrum obtained by FT with respect to real-time acquisition and incremented y-gradient (see text).

8.6.2 Image contrast: T_1-, T_2-, and ρ-weighting

In imaging, as in photography, it is not enough to obtain a *signal* from each part of the object—there must be a *difference* in signal (i.e., *contrast*) in order that the image be able to resolve one part of the object from another. X-ray image contrast is based on differences in a *single* parameter, namely X-ray absorbance, in different parts of an object. MR image contrast, on the other hand, arises from interplay of *three* parameters: number of magnetic nuclei in a given volume element ("voxel"), T_1, and T_2. Although all three parameters contribute to all MR images, it is possible to weight the result in favor of any one of the three by suitable choice of event sequence.

The appearance of an MR image is conveniently classified by the values of two experimental event sequence parameters, T_R, and T_E. T_R is the time-delay between successive event sequences. If $T_R >> T_1$, then signal-to-noise ratio is maximal per acquisition, because the magnetization has recovered to its full equilibrium value by the time the next event sequence begins. Unfortunately, one of the most useful sources of contrast in MR imaging is *variation* in T_1 between different tissues—e.g., the H_2O in tumor cells typically (but not always) has a longer proton T_1 than in normal cells. Thus, a "T_1-weighted" image is performed at $T_R \leq T_1$, in order to emphasize differences in T_1 between different voxels. In fact, because differences in T_1 between different tissues are largest at low applied field strength, it is possible to obtain images of significantly higher T_1-weighted contrast (although much poorer signal-to-noise ratio) at a B_0-field of only a few hundred Gauss (i.e., ~0.05 tesla)

T_E is the "echo" period between the center of the refocussing pulse and the spin-echo maximum (see Figure 8.30), as in Chapter 8.3.3. For $T_E << T_2$, signal-to-noise ratio will be maximal in the first echo, because T_2-relaxation is negligible, but the image will necessarily be insensitive to differences in T_2 between different voxels. Thus, a "T_2-weighted" image is performed at $T_E \geq T_2$, in order to display differences in T_2 between different voxels.

The most obvious source of MR image contrast is the *number* of magnetic nuclei in a given voxel. For example, in whole-body imaging, the lungs appear black because they are essentially empty of protons. Less obviously, blood vessel contents are also usually black, because the H_2O molecules in them move out of the range of the detector coil between the excitation and detection events (see below). The so-called "ρ-weighted" image ($T_E < T_2$; $T_R >> T_1$) is designed to suppress differences in both T_1 and T_2, to give an image that reflects mainly differences in *number of protons* between different voxels. Of the various MR images, the MR "ρ-weighted" image is most similar in principle to X-ray (computer assisted tomography, "CAT-scanning) images. Although signal-to-noise ratio is highest for ρ-weighted images, image *contrast* is typically higher for T_1- or T_2-weighted images, as shown in Figure 8.31.

8.6.3 Flow imaging

As noted above, blood flow may produce a dark MR image, partly because the nuclei initially excited while in a given voxel move to a different voxel before detection. However, that effect may be used to advantage, if the event sequence is tuned (see Wehrli *et al.* in Further Reading) to detect *only* those protons which have moved spatially during the time between creation of the initial transverse magnetization and the final detection period. A striking example is shown in Figure 8.32, in which the arterial (and therefore rapidly flowing) blood of the carotid arteries and their branches shows up in high contrast to the stationary background. As explained in Further Reading, the event sequence which produced this (negative) image was designed to enhance flowing blood, independent of the direction of flow.

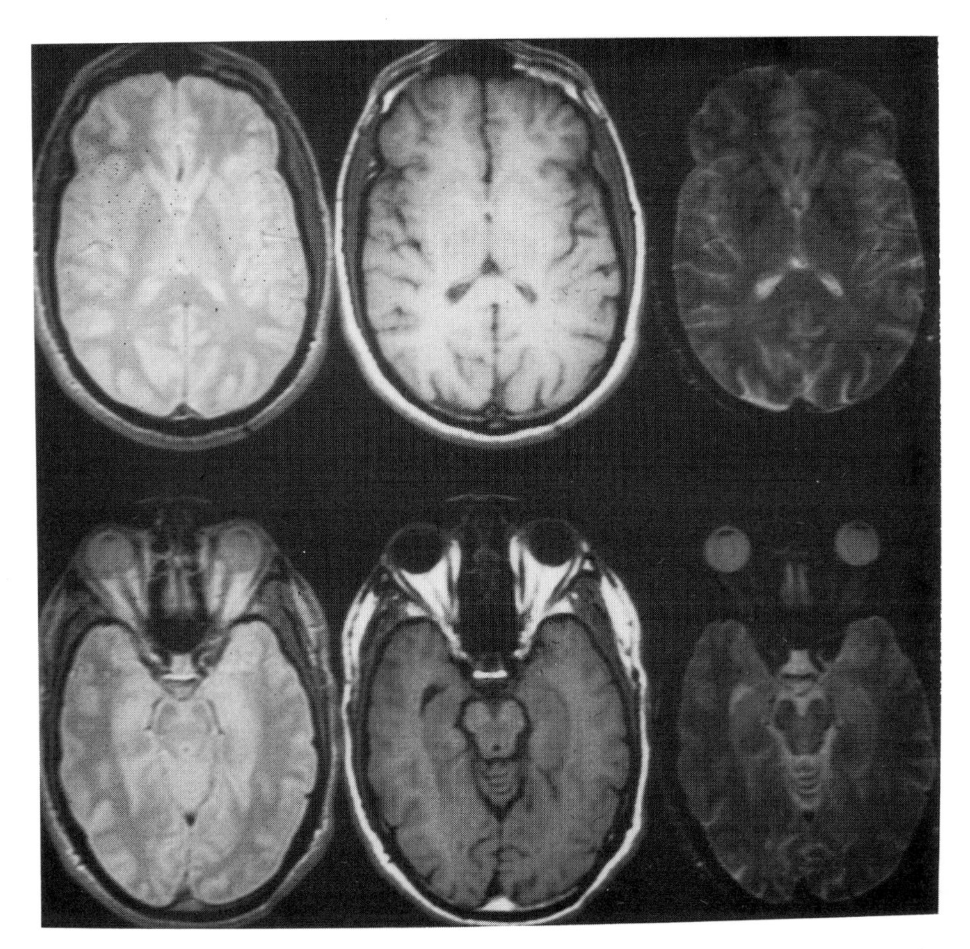

Figure 8.31 Two-dimensional MR cross-sectional "axial" images of the head of one of the authors (AGM), obtained by ρ - (left), T_1- (middle), and T_2-weighting (right). The upper row of images is at the level of the lateral ventricles and the lower row is through the orbits of the eye. MRI is especially useful for head imaging, since ultrasound is reflected by the skull, and X-ray contrast between white and gray matter is poor. (Images provided by P. Schmalbrock from Ohio State University's General Electric Signa 1.5 tesla whole-body imaging spectrometer)

8.6.4 Chemical shift imaging

Ordinarily, the magnetic field gradients in MR imaging are so large that the variation in Larmor frequency between voxels is greater than any chemical shift differences between different kinds of protons (e.g., water vs. fat) in a given voxel. Moreover, H_2O comprises ~90% of human body weight, and the H_2O signal therefore dominates most MR images.

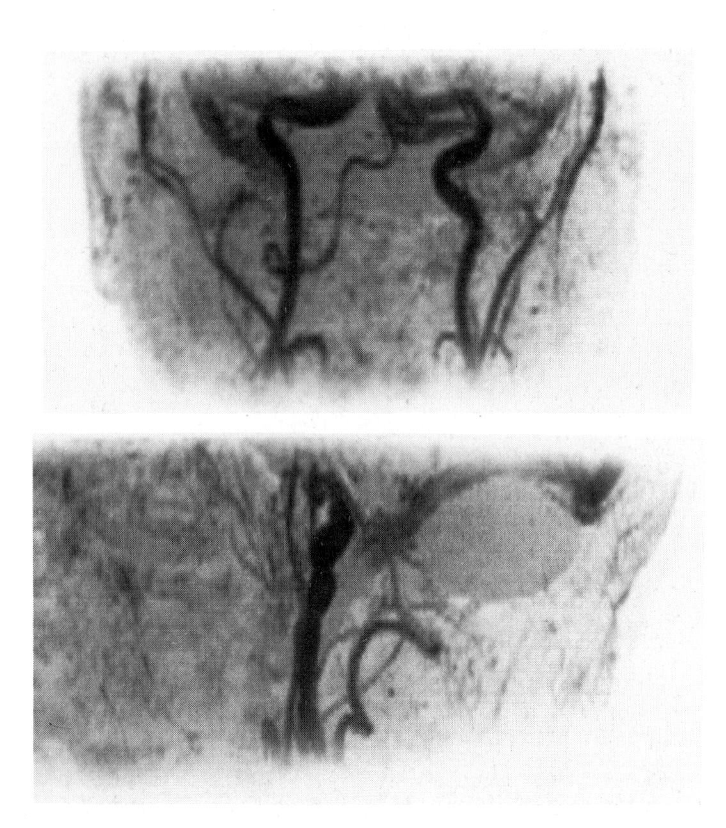

Figure 8.32 Magnetic resonance angiogram flow images (negative-printed) of arterial blood at the head/neck junction of one of the authors (AGM). Unlike most MR images, these pictures represent a superposition (rather than a cross-section) of all flow-containing voxels from a given viewing angle. Top: Anterior-to-posterior image (i.e., front view). The main arteries to the brain, including both carotids (the two large vertical vessels) and vertebral arteries are well visualized, as well as other more superficial branches supplying the face and scalp. Bottom: Lateral image (side view), in which a portion of the venous transverse sinus (upper right) is seen. These images offer good visualization of arterial flow, without the need for any injected contrast agents. (Images provided by P. Schmalbrock from Ohio State University's General Electric Signa 1.5 tesla whole-body imaging spectrometer.)

However, by use of frequency-selective excitation in the *absence* of magnetic field gradients, it is possible to discriminate between MR signals arising from H_2O (~4.7 ppm) and (e.g.) methylene ($-CH_2$) protons from fat (~1-2 ppm). Chemical shift thus adds a "fourth dimension" to MR imaging. Figure 8.33 shows MR images of the same anatomical region, but with selective excitation of water or fat Larmor frequencies. The images are clearly different, and offer yet another direction for future MR imaging applications.

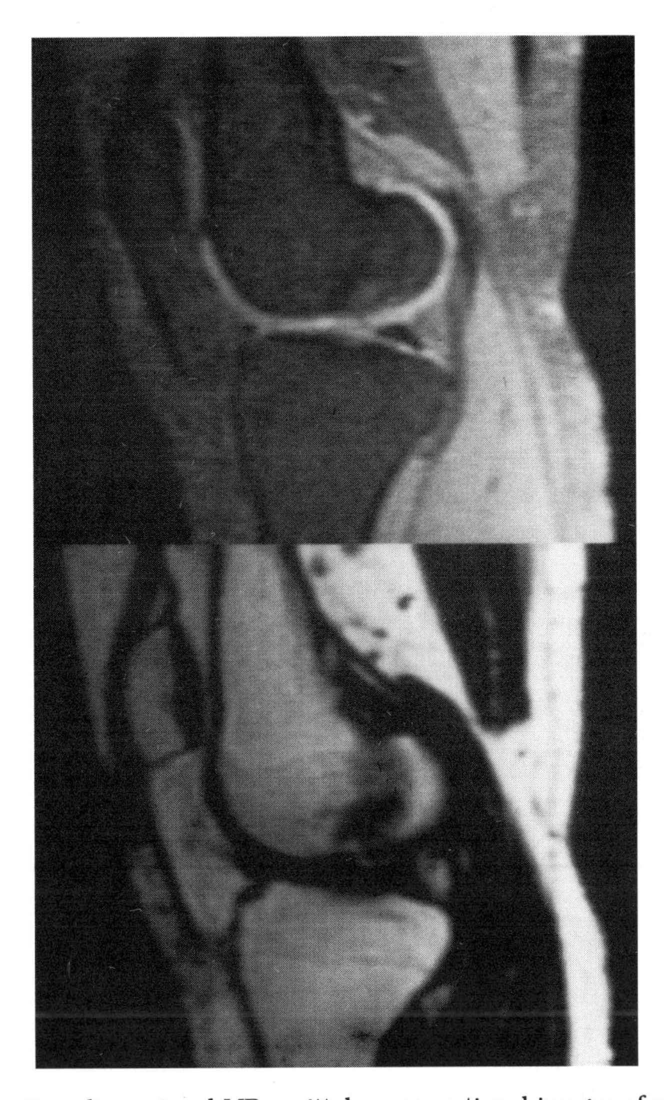

Figure 8.33 Two-dimensional MR sagittal cross-sectional images of a human knee, based on selective excitation at the Larmor frequency of either water (top image) or fat (bottom image) protons. Note the somewhat complementary appearance of the two images: fat is especially prominent in the muscle sheaths at the rear of the joint (i.e., right-hand side of the image), whereas the water image is especially bright along the articulating cartilage boundary of the femur (i.e, the thigh bone). (Images provided by P. Schmalbrock from Ohio State University's General Electric Signa 1.5 tesla whole-body imaging spectrometer.)

Further Reading

R. J. Abraham, J. Fisher, & P. Loftus, *Introduction to NMR Spectroscopy*, Wiley, Chichester, 1988. (Good qualitative introduction, with emphasis on chemical applications)

J. W. Akitt, *NMR and Chemistry: An Introduction to the Fourier Transform-Multinuclear Era*, 2nd ed., Chapman and Hall, London, 1983. [Very broad coverage, for such a short (250 pp.) treatment. Also includes applications for nuclei other than ^{1}H and ^{13}C]

R. Benn & H. Günther, *Angew. Chem. Int. Ed. Engl.* **1983**, *22*, 350-380. (Still one of the best short introductions to 2D-FT/NMR)

N. Chandrakumar & S. Subramanian, *Modern Techniques in High-Resolution FT-NMR*, Springer-Verlag,, New York, 1987. (Intermediate between Sanders & Hunter/Derome level and Ernst *et al.* level)

A. E. Derome *Modern NMR Techniques for Chemistry Research*, Pergamon Press, Oxford, 1987. (Good qualitative treatment)

R. R. Ernst, G. Bodenhausen, & A. Wokaun, *Principles of Nuclear Magnetic Resonance in One and Two Dimensions*, Clarendon Press, Oxford, 1987. (The most general and complete treatment of NMR at full quantum mechanical depth)

T. Farrar, *Pulsed Nuclear Magnetic Resonance Spectroscopy*, The Farragut Press, Chicago, 1987. (Good short entry-level description of basic NMR techniques)

R. Freeman, *A Handbook of Nuclear Magnetic Resonance*, Longman Scientific and Technical (Wiley in U.S.A.), 1988. (Unusual format--an extended glossary, with several-page explanations of ~50 NMR terms and techniques)

C. A. Fyfe, *Solid State NMR for Chemists*, C.F.C. Press, Guelph, Canada, 1983. (Excellent coverage of NMR of solids)

J. E. Meier & A. G. Marshall, in *Biological Magnetic Resonance*, Vol. 9, Ed. L. J. Berliner (Plenum, New York, 1989), Chapter 6. (Review of solvent suppression techniques in FT/NMR)

M. Munowitz & A. Pines, *Science* **1986**, *233*, 525-531. (Excellent review of multiple-quantum NMR, especially as applied to solids and partially ordered media)

J. K. M. Sanders & B. K. Hunter, *Modern NMR Spectroscopy: A Guide for Chemists*, Oxford University Press, Oxford, 1987. (Excellent qualitative discussion, with many chemical examples)

D. Shaw, *Fourier Transform N.M.R. Spectroscopy*, 2nd ed., Elsevier, Amsterdam, 1984. (Excellent coverage with minimal quantum mechanical jargon)

C. P. Slichter, *Principles of Magnetic Resonance*, 2nd ed., Springer-Verlag, Berlin, 1978. (Good introduction to density matrix and time-dependent NMR phenomena, especially the Nuclear Overhauser effect)

S. L. Smith, *Anal. Chem.* **1985**, *57*, 595A-608A. [The best short (9-page) introduction to MR imaging]

W. S. Warren, *Science* **1988**, *242*, 878-884. (Review of "shaped" rf pulses designed to overcome the non-linearities of the NMR response)

F. Wehrli, D. Shaw, & J. B. Kneeland, *Biological Magnetic Resonance Imaging: Principles, Methodology, Applications*, VCH, NY, 1988. (Newest and best treatment of MR imaging)

F. Wehrli & T. Wirthlin, *Interpretation of Carbon-13 NMR Spectra*, Heyden, London, 1976. (Comprehensive discussion of ^{13}C NMR parameters)

K. Wüthrich, *NMR of Proteins and Nucleic Acids*, Wiley, NY, 1986. (Excellent discussion of extraction of three-dimensional structure of large molecules by two-dimensional NMR techniques)

Problems

8.1 Obtain Eq. 8.12 from Eqs. 8.9 to 8.11. This exercise shows how the macroscopic magnetic moment (and thus the NMR signal magnitude) depends on temperature and magnetic field strength.

8.2 Show that the number of multiplet peaks (but not their relative magnitudes) produced in the NMR spectrum of a given magnetic nucleus by J-coupling to n identical spin I nuclei is the same as that produced by J-coupling to a single nucleus of spin, nI.

8.3 Obtain Eq. 8.14 from Eq. 8.13. This exercise establishes the magnetic field strength resulting from dipole-dipole through-space interaction, which in turn forms the basis for NMR determination of internuclear distances from T_1 and T_2 relaxation time constants.

8.4 As noted in Chapter 2.1.2, a single-frequency rectangular pulse of duration, T seconds produces frequency-domain excitation power that is flat to within ~10% over a frequency range of ~$1/8T$ Hz.

(a) Compute the longest excitation period, T, for a rectangular rf pulse which will excite NMR signals over a bandwidth of 10 kHz.

(b) Compute the magnetic rf field magnitude required to accomplish a 90° rotation of proton magnetization $\boldsymbol{M}_0$ during the period, T, from part (a). Repeat the calculation for carbon-13 magnetization. Finally, compare your result to the strength of the earth's magnetic field (~0.00005 tesla). This exercise should show that NMR excitation over a wide bandwidth can be performed with rf pulses of modest amplitude.

8.5 Chemical exchange is a good example of the value of the Bloch equations for quantitative representation of NMR spectral line shapes.

(a) Combine the basic Bloch Eqs. 8.23 with the chemical exchange Eqs. 8.27, to obtain equations which describe the time-behavior of the z-magnetization in the absence of an rf magnetic field for two-site chemical exchange (Eq. 8.26).

(b) For the special case that $T_{1A} = T_{1B} = T_1$, solve your equation for the initial condition that $\boldsymbol{M}_{zA} = \boldsymbol{M}_{zB} = 0$ at time zero. You should be able to show that the system perturbed from equilibrium follows a single-exponential relaxation with time constant, T_1. In other words, chemical exchange between two sites whose NMR parameters (in this case, T_1) are identical does not affect NMR behavior.

(c) The algebra for more general cases (e.g., $T_{1A} \neq T_{1B}$; $T_{2A} \neq T_{2B}$; or $\omega_{0A} \neq \omega_{0B}$ is considerably more complex and requires some experience at solving pairs of coupled differential equations (see Further Reading). However, the two-site *fast-exchange limit* [i.e., $1/\tau_{ex} << (a_A - a_B)$; $a = 1/T_1$ or $1/T_2$ or ω_0, etc.] reduces to the simple form,

$$a_{\text{observed}} = f_A\, a_A - f_B\, a_B \tag{8.5.1}$$

in which

$$f_A = \frac{k_{-1}}{k_1 + k_{-1}} \quad \text{and} \quad f_B = \frac{k_1}{k_1 + k_{-1}} \tag{8.5.2}$$

are the fractions of nuclei at sites A and B, respectively. This result (and the corresponding (algebraically more complex) result for transverse relaxation form the basis for determination of chemical exchange rate constants from NMR spectra.

8.6 Some of the most popular rf pulse sequences for selective excitation are based on trigonometric functions.

(a) Plot the frequency-domain functions, $\sin(\omega - \omega_0)^n$ and $\cos(\omega - \omega_0)^n$, for $n = 1$, 2, and 3. Note that the $n = 2$ and 3 cases produce frequency-domain spectra with relative flat-magnitude segments separated by near-null flat segments. Thus, such functions form the basis for selective excitation of NMR signals of a given range of chemical shifts, without exciting neighboring resonances of different chemical shift.

(b) Therefore, find the time-domain excitation signals corresponding to each of the above frequency-domain functions. Your answer should produce the famous binomial sequences, $1\underline{1}$, $1\underline{2}1$, $1\underline{3}3\underline{1}$, etc., which are widely used for solvent suppression in high-resolution NMR and for spatially selective excitation in magnetic resonance imaging. (The Redfield $2\underline{1}4\underline{1}2$ sequence was originally conceived as the difference between a single constant-magnitude rf pulse and a $1\underline{1}$ pulse sequence.)

8.7 Signal-to-noise ratio, *SNR*, in one-dimensional FT spectroscopy is proportional to $\sqrt{N}$, in which N is the number of time-domain accumulations. What is the relation between *SNR*, *N*, and voxel volume for two-dimensional or three-dimensional NMR imaging? What does your result suggest about the feasibility of magnetic resonance images based on natural-abundance carbon-13 NMR?

Solutions to Problems

8.1 The macroscopic equilibrium z-magnetization for an ensemble of N_0 nuclei of spin one-half is given by the sum of the z-component magnetic moments, $\mu_{upper} = (-\gamma h/4\pi)$ and $\mu_{lower} = (\gamma h/4\pi)$, for spins in the upper and lower energy levels whose respective energies are $-\mu_{upper} B_0$ and $-\mu_{lower} B_0$

$$M_0 = N_0 f_{upper} \mu_{upper} + N_0 f_{lower} \mu_{lower} \tag{8.1.1}$$

in which

$$f_{upper} = \frac{N_{upper}}{N_{upper} + N_{lower}} \text{ and } f_{upper} = \frac{N_{lower}}{N_{upper} + N_{lower}} \tag{8.1.2}$$

are the fractions of spins in the upper and lower energy levels. From the Boltzmann distribution (Eq. 8.10) and the high-temperature limit (Eq. 8.11),

$$f_{upper} = \frac{\exp\left(-\frac{\gamma h B_0}{4\pi k T}\right)}{\exp\left(\frac{-\gamma h B_0}{4\pi k T}\right) + \exp\left(\frac{+\gamma h B_0}{4\pi k T}\right)}$$

$$\cong \frac{1 - \frac{\gamma h B_0}{4\pi k T}}{1 - \frac{\gamma h B_0}{4\pi k T} + 1 + \frac{\gamma h B_0}{4\pi k T}} = \frac{1}{2}\left(1 - \frac{\gamma h B_0}{4\pi k T}\right) \tag{8.1.3a}$$

$$f_{lower} = 1 - f_{upper} = \frac{1}{2}\left(1 + \frac{\gamma h B_0}{4\pi k T}\right) \tag{8.1.3b}$$

Substituting Eqs. 8.1.3 and 8.1.2 into Eq. 8.1.1, we obtain

$$M_0 = \frac{N_0}{2}\left(\left(1 - \frac{\gamma h B_0}{4\pi k T}\right)\frac{-\gamma h}{4\pi} + \left(1 + \frac{\gamma h B_0}{4\pi k T}\right)\frac{\gamma h}{4\pi}\right) = N_0 \gamma^2 \left(\frac{h}{2\pi}\right)^2 \frac{B_0}{4 k T}$$

8.2 For spin, I, there will be $(2I + 1)$ equally probable and equally spaced values of the z-component of angular momentum. Therefore, the NMR spectrum of a magnetic nucleus, S, coupled to a nucleus of spin, I, will be split into $(2I + 1)$ peaks of equal magnitude, spaced evenly and symmetrically about the unshifted Larmor frequency of S.

As noted above Figure 8.6, the J-coupling multiplet magnitude patterns for 1, 2, or 3 spins one-half are: two peaks (1:1), three peaks (1:2:1), and four peaks (1:3:3:1). The binomial pattern should be clear: N spin one-half nuclei will produce a multiplet with $(2N + 1)$ peaks whose relative magnitudes are the binomial coefficients: $n!/(N!\,(N-n)!\,]$.

8.3 Beginning from Eq. 8.13, expand μ and $\mathbf{r}$ in terms of unit vectors,

$$\mathbf{B}_{\text{dipole}} = -\frac{\mu}{r^3} + 3\left(\frac{\mu \cdot \mathbf{r}}{r^5}\right)\mathbf{r} \tag{8.13}$$

$$= -\frac{1}{r^3}\left(\mu_x \mathbf{i} + \mu_y \mathbf{j} + \mu_z \mathbf{k}\right)$$

$$+ \frac{3}{r^5}\left(\mu_x\, x + \mu_y\, y + \mu_z\, z\right)\left(x\mathbf{i} + y\mathbf{j} + z\mathbf{k}\right) \tag{8.3.1}$$

$$= B_x \mathbf{i} + B_y \mathbf{j} + B_z \mathbf{k}$$

We are asked to find B_z:

$$B_z = -\frac{\mu_z}{r^3} + \frac{3}{r^5}\left(\mu_x\, xz + \mu_y\, yz + \mu_z\, z^2\right) \tag{8.3.2}$$

Next, substitute for

$$x = r \sin\theta \cos\phi \tag{8.3.3a}$$

$$y = r \sin\theta \sin\phi \tag{8.3.3b}$$

$$z = r \cos\theta \tag{8.3.3c}$$

in Eq. 8.3.2 to obtain

$$B_z = -\frac{\mu_z}{r^3} + \frac{3}{r^5}(\mu_x r^2 \sin\theta \cos\phi \cos\theta + \mu_y r^2 \sin\theta \sin\phi \cos\theta$$

$$+ \mu_z r^2 \cos^2\theta)$$

$$= \frac{\mu_z}{r^3}(3\cos^2\theta - 1) + \frac{3\cos\theta}{r^3}(\mu_x \sin\theta \cos\phi + \mu_y \sin\theta \sin\phi)$$

If we now average this result over all ϕ angles (since we are interested only in the z-component), only the first term (Eq. 8.14) remains, since

$$\int_0^{2\pi} \cos\phi \; d\phi = \int_0^{2\pi} \sin\phi \; d\phi = 0$$

8.4 (a) $\quad \dfrac{1}{8T} = 10^4$ Hz; $\quad T = 12.5\ \mu s$

(b) We know that the Larmor frequencies for ^1H and ^{13}C are approximately 500.4 MHz and 125.6 MHz at 11.75 tesla (Figure 8.4). Thus, we can compute their magnetogyric ratios:

$$\nu_0 = \frac{|\omega_0|}{2\pi} = \frac{\gamma |\mathbf{B}_0|}{2\pi} \quad \text{(Hz)} \tag{8.5b}$$

$$\gamma = 2\pi \nu_0 / B_0 \tag{8.4.1}$$

We require a 90° = $\pi/2$ radians pulse in 12.5 μs. Therefore,

$$\omega_1 T = \gamma B_1 T = \pi/2 \tag{8.4.2}$$

Solve Eq. 8.4.2 for B_1 and substitute for γ from Eq. 8.4.1 to yield

$$B_1 = \frac{\pi}{2\gamma T} = \frac{B_0}{4 \nu_0 T} \tag{8.4.3}$$

For ^1H,

$$B_1 = \frac{11.75}{(4)(500.4 \times 10^6)(12.5 \times 10^{-6})} = 4.7 \times 10^{-4} \text{ tesla} = 4.7 \text{ Gauss}$$

Similarly, for ^{13}C, B_1 = 18.7 x 10^{-4} tesla = 18.7 Gauss. Note that for a given 90° pulse duration, a stronger B_1 field is needed for ^{13}C than for ^1H, because of the smaller magnetogyric ratio for ^{13}C.

8.5 (a) There are two sets of Bloch equations, one for nuclei at each site:

$$\frac{d M_{zA}}{dt} = -k_1 M_{zA} + k_{-1} M_{zB} - \frac{M_{zA} - M_{A0}}{T_{1A}} \tag{8.5.3a}$$

$$\frac{d M_{zB}}{dt} = -k_{-1} M_{zB} + k_1 M_{zA} - \frac{M_{zB} - M_{B0}}{T_{1B}} \tag{8.5.3b}$$

(b) In general, one would need to solve Eqs. 8.53 as two coupled differential equations. However, if $T_{1A} = T_{1B} = T_1$ we may simply add the two equations:

$$\frac{d M_{zA}}{dt} + \frac{d M_{zB}}{dt} = \frac{d (M_{zA} + M_{zB})}{dt}$$

$$= -k_1 M_{zA} + k_{-1} M_{zB} - k_{-1} M_{zB} + k_1 M_{zA}$$

$$- \frac{M_{zA} - M_{A0}}{T_1} - \frac{M_{zB} - M_{B0}}{T_1} \text{, or equivalently}$$

$$\frac{d M_z}{dt} = -\frac{M_z - M_0}{T_1} \tag{8.5.4a}$$

in which $M_z = M_{zA} + M_{zB}$, and $M_0 = M_{A0} + M_{B0}$ (8.5.4b)

Eq. 8.54a is simply the usual one-site Bloch equation for time-evolution of M_z. Thus, for the initial condition, $M_z(0) = M_{z\,A}(0) + M_{z\,B}(0) = 0$, at time zero, Eq. 8.5.4 may easily be solved to yield

$$M_z(t) = M_0\left(1 - \exp(-t/T_1)\right)$$

8.6 Answers stated in problem.

8.7 Think of each voxel as a cube. If we reduce the length of each side of the cube by a factor of 2, then we reduce the volume of that voxel by a factor of $2^3 = 8$, and the number of acquisitions must be increased by a factor of $8^2 = 64 = 2^6$ to regain the original signal-to-noise ratio. (In two dimensions, the number of acquisitions increases "only" by a factor of $2^4 = 16$). This so-called sixth-power dependence of imaging time on spatial resolution severely limits the feasibility of MRI for nuclei whose natural abundance and/or magnetogyric ratio are lower than those of protons.

CHAPTER 9

FT/interferometry

9.1 Natural motions of vibrating molecules

Although a small (but increasing) number of interferometry measurements is conducted in the visible spectral range, by far the most experiments involve vibrational spectroscopy at infrared frequencies. Therefore, we begin this chapter with a short introduction to molecular vibrations as observed directly at infrared frequencies (Chapter 9.1.1) or indirectly by Raman spectroscopy at visible frequencies (Chapter 9.1.2). After a brief discussion of selection rules and spectral display modes (Chapter 9.1.3, 9.1.4), we discuss the generation of an interferogram and its multichannel advantages (Chapter 9.2). Finally, we present some general classes of applications for Hadamard and FT/IR, FT/Raman, and FT/visible spectrometry (Chapter 9.3). Generally speaking, vibrational spectra of molecules containing more than a few atoms are too complex to analyze completely, but provide a "fingerprint" and "group vibrations" for identification, whereas vibrational spectra of small molecules can be fully analyzed to yield force constants (an approximate measure of bond strength) for various normal coordinates (see below).

As for FT/NMR, there are numerous available texts and reviews of IR, Raman, and visible spectroscopy and interferometry. The most comprehensive treatment of FT/IR is the Griffiths/de Haseth book, to which the interested reader is referred (see Further Reading).

9.1.1 Normal modes: infrared spectra

The forces binding atoms together in molecules are complex in mechanism and in dependence on interatomic separation distance. Fortunately, if the atomic positions are perturbed only slightly (as by the oscillating electric field of coherent infrared radiation), then the interatomic force can (to a good approximation) be simplified to the simple linear restoring force of our mass-on-a-spring model (Chapter 1). (Stated another way, the potential energy varies approximately quadratically with interatomic distance: the "harmonic oscillator" model.) We need only extend that model from one mass connected by a spring to a wall to two masses connected by a spring to each other (e.g., a diatomic molecule, as shown in Figure 9.1). Analysis of the motions of a diatomic molecule establishes all of the principles needed to describe the vibrations of more complex molecules.

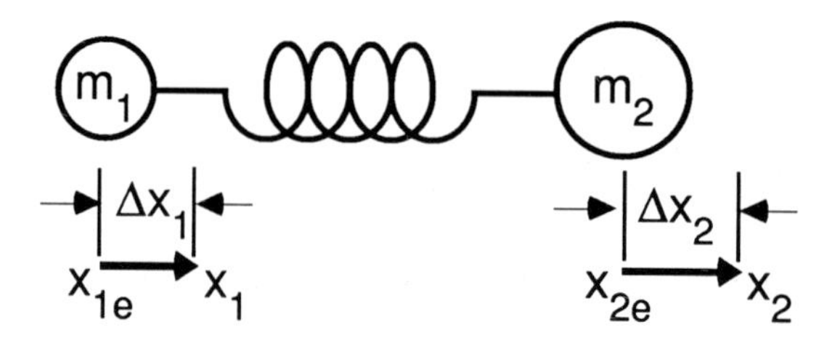

Figure 9.1 Heteronuclear diatomic molecule, modeled as two masses (m_1 and m_2) connected by a frictionless spring of force constant, k. The (restoring) force between the two masses is determined only by their net displacement, $\Delta x_1 - \Delta x_2$, from their equilibrium positions, x_{1e} and x_{2e}, in which $\Delta x_1 = x_1 - x_{1e}$ and $\Delta x_2 = x_2 - x_{2e}$.

We begin by treating just the x-axis motions (i.e., along the interatomic vector) of the diatomic molecule of Figure 9.1. The (restoring) x-force acting on each atom may then be written,

$$m_1 \frac{d^2 x_1}{dt^2} = -k\,(\Delta x_1 - \Delta x_2) \tag{9.1a}$$

$$m_2 \frac{d^2 x_2}{dt^2} = -k\,(\Delta x_2 - \Delta x_1) \tag{9.1b}$$

in which m_1 and m_2 are the masses, Δx_1 and Δx_2 are defined in Figure 9.1, and k is the force constant of the "spring" which represents the chemical bond between the two atoms. The trick for solving equations of this type is to change variables to so-called "normal coordinates",

$$q_1 = \sqrt{\frac{m_1 m_2}{m_1 + m_2}}\,(\Delta x_1 - \Delta x_2) \tag{9.2a}$$

$$q_2 = \frac{m_1}{\sqrt{m_1 + m_2}}\,\Delta x_1 + \frac{m_2}{\sqrt{m_1 + m_2}}\,\Delta x_2 \tag{9.2b}$$

We can now solve Eqs. 9.2a,b to obtain expressions for Δx_1 and Δx_2 in terms of q_1 and q_2 (see Problems),

$$\Delta x_1 = \frac{\sqrt{m_2}\, q_1 + \sqrt{m_1}\, q_2}{\sqrt{m_1(m_1 + m_2)}} \tag{9.3a}$$

$$\Delta x_2 = \frac{\sqrt{m_2}\, q_2 - \sqrt{m_1}\, q_1}{\sqrt{m_2(m_1 + m_2)}} \tag{9.3b}$$

which may then be substituted into Eqs. 9.1 to yield (see Problems),

$$m_1 m_2 \frac{d^2 q_1}{dt^2} + m_1 \sqrt{m_1 m_2} \frac{d^2 q_2}{dt^2} + k\,(m_1 + m_2)\, q_1 = 0 \tag{9.4a}$$

$$m_1 m_2 \frac{d^2 q_1}{dt^2} - m_2 \sqrt{m_1 m_2} \frac{d^2 q_2}{dt^2} + k\,(m_1 + m_2)\, q_1 = 0 \tag{9.4b}$$

Finally, we may add or subtract Eqs. 9.4a and 9.4b to obtain, respectively,

$$\frac{d^2 q_2}{dt^2} = 0 \tag{9.5a}$$

$$m_1 m_2 \frac{d^2 q_1}{dt^2} + k\,(m_1 + m_2)\, q_1 = 0 \tag{9.5b}$$

Eq. 9.5a states that the *acceleration* along normal coordinate, q_2, is zero, so that the *velocity* along q_2 must be constant

$$\frac{dq_2}{dt} = \text{constant} \tag{9.6}$$

If there is no motion along any other (normal) coordinates, then we may set $q_1 = 0$ (i.e., $\Delta x_1 = \Delta x_2$, from Eq. 9.2a) and infer from Eq. 9.2b and Eq. 9.6 that

$$\frac{dx_1}{dt} = \frac{dx_2}{dt} = \text{constant} \quad \text{(translational motion)} \tag{9.7}$$

Thus, motion along normal coordinate, q_2, represents simple linear *translation* of the diatomic molecule as a whole.

Eq. 9.5b is our familiar mass-on-a-spring equation of Chapter 1, except that the mass of one particle bound to a wall has been replaced by the *reduced mass*, μ,

$$\boxed{\mu = \frac{m_1 m_2}{m_1 + m_2}} \tag{9.8}$$

to give a "natural" vibrational frequency,

$$\boxed{\omega_0 = 2\pi\nu_0 = \sqrt{\frac{k}{\mu}}} \tag{9.9}$$

If there is no motion along any other (normal) coordinates, then we may set $q_2 = 0$ and infer from Eq. 9.2b that the relative displacements of atoms 1 and 2 are

$$\sqrt{\frac{m_2}{m_1(m_1 + m_2)}}\; x_0 \quad = \text{relative displacement of atom 1} \qquad (9.10a)$$

$$-\sqrt{\frac{m_1}{m_2(m_1 + m_2)}}\; x_0 \quad = \text{relative displacement of atom 2} \qquad (9.10b)$$

In other words, the relative vibrational displacements of the two atoms are determined by their masses. The two normal mode motions are shown in Figure 9.2.

In general, the kinematic description of a molecule composed of N atoms requires $3N$ variables: e.g., 3 Cartesian position coordinates for each of N atoms. However, as we begin to see from the preceding example, the motions of the atoms are more conveniently represented in terms of "normal" coordinates. For a diatomic (or any other) molecule, three of the normal coordinates correspond to *translations* along x- or y- or z-axes (e.g., q_2, q_4, and q_6 in Eqs. 9.2 and Figure 9.2). For a diatomic molecule, one of the normal modes (q_1 in Figure 9.1) is a *vibration*. The remaining two normal modes (q_3 and q_5 in Eqs. 9.2 and Figure 9.2) of a diatomic molecule are *rotations*. Just as all of the atoms move together in any one of the three *translations*, the atoms all *rotate* or *vibrate* at a common frequency along a given rotational or vibrational normal coordinate.

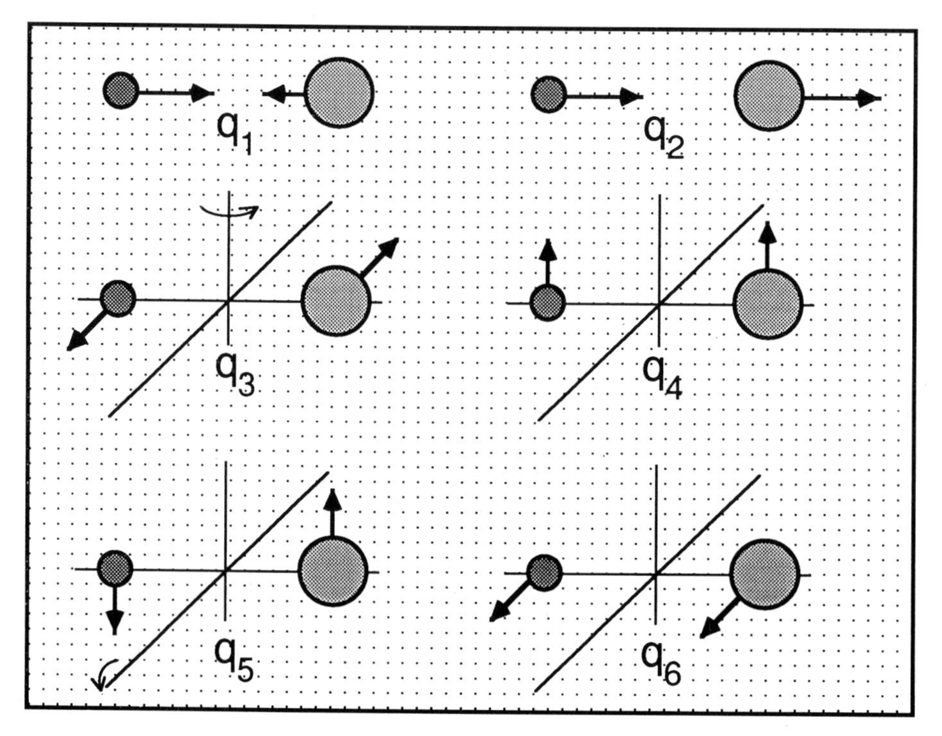

Figure 9.2 Motions along each of the six "normal" coordinates for a diatomic molecule. q_1 is the vibrational coordinate; q_2, q_4, and q_6 represent translations; and q_3 and q_5 are rotations.

In general, the number of possible vibrational normal modes is given by

$$\text{Number of vibrational normal modes} = 3N - 5 \text{ (linear molecule)} \qquad (9.11a)$$

$$= 3N - 6 \text{ (non-linear molecule)} \qquad (9.11b)$$

In the preceding treatment based on analysis of *forces*, normal coordinates were introduced as a *ad hoc* mathematical trick, and it was not obvious how to generalize the analysis to more complex molecules. The key to generalizing the method is to consider kinetic and potential *energy*. For example, the classical potential energy, V, for the x-vibration of the diatomic molecule of Figure 9.1 can be written:

$$V = \frac{1}{2} k \, (\Delta x_1 - \Delta x_2)^2 = \frac{1}{2} k \, (\Delta x_1^2 - 2 \, \Delta x_1 \, \Delta x_2 + \Delta x_2^2)$$

$$\text{or } 2V = k \, (\Delta x_1^2 - 2 \, \Delta x_1 \, \Delta x_2 + \Delta x_2^2) \qquad (9.12)$$

However, suppose that we did not happen to choose (or if it were not possible to choose) the coordinate frame such that the vibration of interest happens to produce a displacement exactly along just one of the coordinate axes. In that case, Eq. 9.12 would take the more general form (still for a diatomic molecule),

$$2V = a_{11} \Delta x_1^2 + a_{22} \Delta x_2^2 + a_{33} \Delta x_3^2 + a_{44} \Delta x_4^2 + a_{55} \Delta x_5^2 + a_{66} \Delta x_6^2$$

$$+ 2a_{12} \Delta x_1 \Delta y_1 + 2a_{13} \Delta x_1 \Delta z_1 + 2a_{14} \Delta x_1 \Delta x_2 + 2a_{15} \Delta x_1 \Delta y_2 + 2a_{16} \Delta x_1 \Delta z_2$$

$$+ 2a_{23} \Delta y_1 \Delta z_1 + 2a_{24} \Delta y_1 \Delta x_2 + 2a_{25} \Delta y_1 \Delta y_2 + 2a_{26} \Delta y_1 \Delta z_2$$

$$+ 2a_{34} \Delta z_1 \Delta x_2 + 2a_{35} \Delta z_1 \Delta y_2 + 2a_{36} \Delta z_1 \Delta z_2$$

$$+ 2a_{45} \Delta x_2 \, \Delta y_2 + 2a_{46} \Delta x_2 \, \Delta z_2$$

$$+ 2a_{56} \Delta y_2 \, \Delta z_2 \qquad (9.13)$$

Thus, the choice of normal coordinates reduces to the systematic combination of terms in Eq. 9.13 to produce a potential energy, V, with only quadratic terms:

$$V = \frac{1}{2} \left(\lambda_1 q_1^2 + \lambda_2 q_2^2 + \lambda_3 q_3^2 + \lambda_4 q_4^2 + \lambda_5 q_5^2 + \lambda_6 q_6^2 \right) \qquad (9.14a)$$

and (see Problems) just quadratic terms in kinetic energy, T, as well:

$$T = \frac{1}{2} \left(\frac{dq_1^2}{dt^2} + \frac{dq_2^2}{dt^2} + \frac{dq_3^2}{dt^2} + \frac{dq_4^2}{dt^2} + \frac{dq_5^2}{dt^2} + \frac{dq_6^2}{dt^2} \right) \qquad (9.14b)$$

For example, if we let

$$\lambda_1 = \frac{k\,(m_1 + m_2)}{m_1 m_2}; \quad \lambda_2 = \lambda_3 = \lambda_4 = \lambda_5 = \lambda_6 = 0 \qquad (9.15)$$

then we obtain Eq. 9.5b, corresponding to Eq. 9.12, and we have again found the harmonic vibrational mode for the diatomic molecule.

For the harmonic potential (i.e., quadratic distance-dependence) of Eq. 9.14a, we already know that the solution will be a sinusoidal oscillation. With that assumption, the values of λ_i in Eq. 9.14a may be obtained in a systematic classical mechanical "FG-matrix" method, to yield the various "normal mode" vibrational frequencies (see Further Reading). However, quantum mechanics is needed to determine the relative *amplitudes* of the various normal mode vibrations, and the *selection rules* (namely, which vibrations will be observed in a given experiment).

Some representative results are shown in Figure 9.3 for a linear triatomic molecule (CO_2, with 2 rotational and $3N - 5 = 4$ vibrational modes) and a non-linear triatomic molecule (SO_2, with 3 rotational and $3N - 6 = 3$ vibrational modes). Vibrations with displacements *along* the axes of chemical bonds are known as "stretches", and vibrations with displacements in other directions are variously known as "bending", "twisting", "rocking", etc.

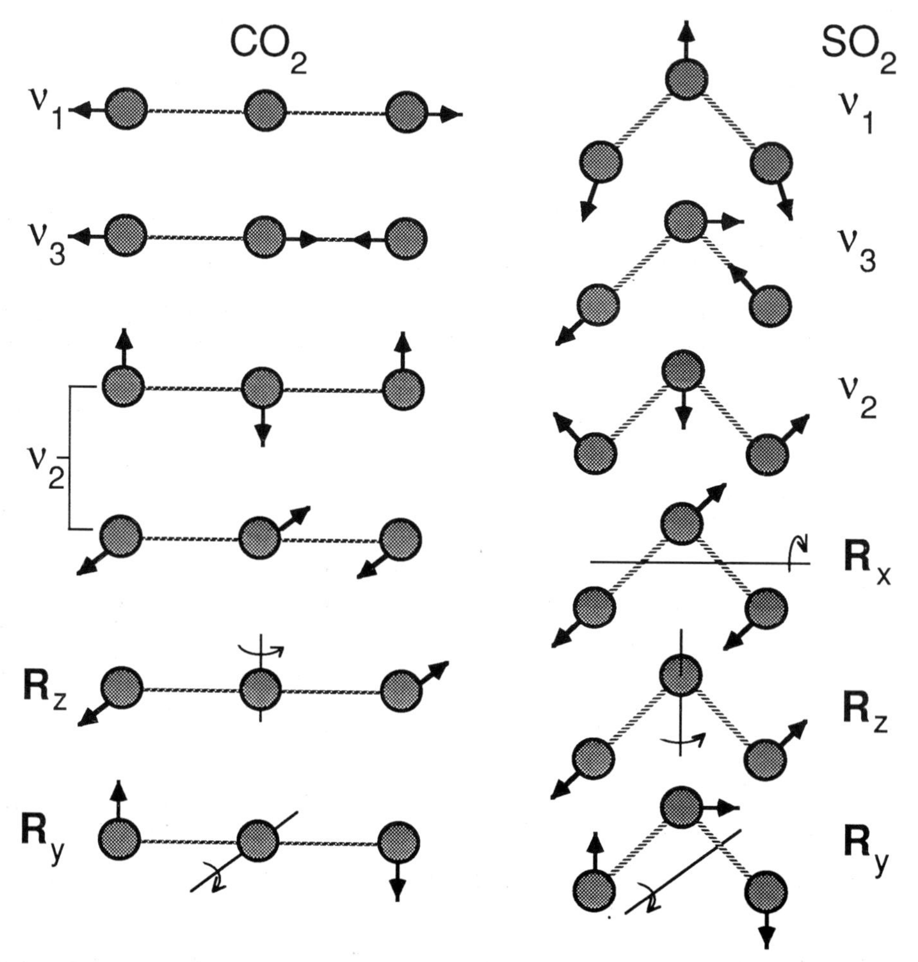

Figure 9.3 Normal modes of vibration and rotation of linear (CO_2) and non-linear (SO_2) triatomic molecules. The three translational modes are not shown.

All of the atoms move together at the same *frequency* (albeit with different *amplitudes*) in *each* normal mode vibration, but the relative *amplitudes* of their displacements depend on the relative masses of the atoms. For example, in a diatomic molecule, the lighter atom has the larger vibrational amplitude (see Eqs. 9.10). Thus, to the extent that we can treat a given atom as if it were bound by one "spring" to the rest of the (much more massive) molecule, then the amplitude of the "group vibrational frequency" of that atom (say, the oxygen in a carbonyl group) will be large and will be more characteristic of the functional group than of the rest of the molecule.

Molecular *rotational* frequencies typically fall in the *microwave* region of the electromagnetic spectrum, whereas *vibrational* frequencies typically fall in the *infrared* region. Interestingly, not all of the normal mode vibrations are experimentally observable by direct absorption of infrared electromagnetic radiation at the normal mode frequency. In Raman spectroscopy (see next section), infrared-frequency vibrations are observed indirectly by their modulation of visible-frequency electronic motions. We shall therefore digress briefly to introduce the Raman phenomenon, and then return to the issue of which vibrational normal modes are observable in each of the two experiments.

9.1.2 Coupling of nuclear and electronic motions: Raman spectra

Although FT interferometric methods have been applied to ultraviolet/visible electronic spectroscopy, by far the greatest impact of interferometry derives from its application to molecular vibrational spectroscopy. As noted in the previous section, vibrational frequencies may be observed directly in the infrared spectral range. In this section, we shall describe how vibrational frequencies may alternatively be observed from amplitude-modulation sidebands in the visible spectral range.

Infrared spectra arise from the natural vibrations of *atoms* bound to each other. *Ultraviolet/visible (uv/vis)* spectra arise from natural motions of *electrons* bound to atoms in a molecule. Electronic natural frequencies are much higher than those for vibrations of atoms, principally because the electron is so much lighter than a typical atom. Since the natural frequency of a mass-on-a-spring varies as $\mu^{-1/2}$ (see Eq. 9.9), and since $\mu \cong m_e$ for a (light) electron bound to a (heavy) molecule, we expect electronic "vibrational" frequencies to be larger than interatomic vibrational frequencies by a factor of ~100. (The remaining difference between electronic and interatomic vibrational frequency is due to the difference in "force constant" for the two kinds of motion.) Finally, from quantum mechanical analysis of hydrogen-like atoms, we know that an electron bound to an atom actually has several natural frequencies, corresponding to various electronic energy level spacings.

If a classical electron-on-a-spring is driven by a coherent oscillating off-resonance electric field of frequency, ω (Chapter 1.4.2), then the oscillating driven electron will radiate so-called "Rayleigh" light *at* the driving frequency, ω, as shown schematically in the top and middle diagrams of Figure 9.4. The origin of "Raman" scattering is shown schematically in Figure 9.4, and can be understood more quantitatively as follows.

Light scattering results from the *displacement*, x, of an electron from its equilibrium position. That displacement produces an *electric dipole moment*, ex, in which e represents electronic charge. An oscillating electric dipole moment

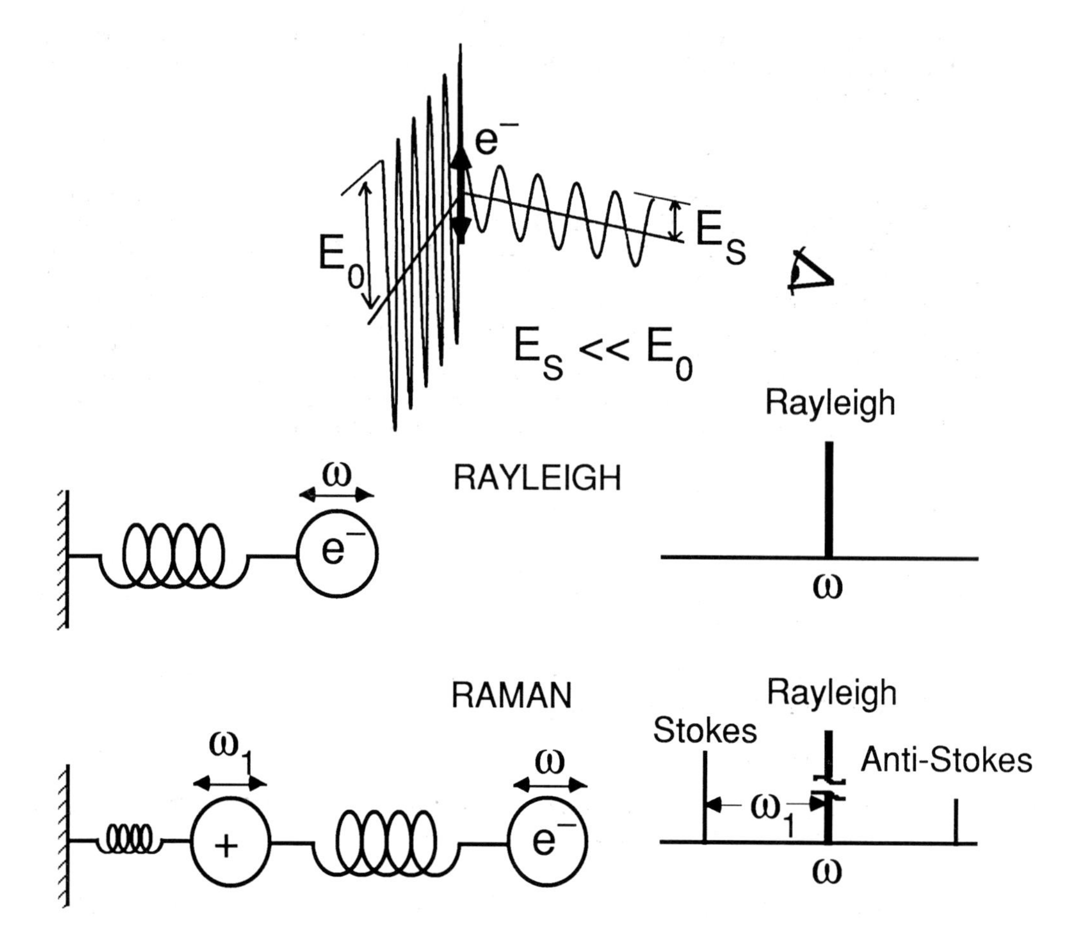

Figure 9.4 Classical description of Rayleigh (middle) and Raman (bottom) scattering. In both cases (top), an electron driven into oscillation by incident radiation, I_0, emits radiation whose component perpendicular to the incident beam may be detected as "scattered" light, I_s. The Rayleigh component (middle) corresponds to natural vibration of the electron bound to the molecule. The (small-amplitude) vibration at frequency, ω_1, of the atom to which the electron is bound (bottom) effectively modulates the amplitude of the electronic vibration, and therefore produces very small sidebands known as Raman scattering, located at $\omega - \omega_1$ (Stokes) and $\omega + \omega_1$ (Anti-Stokes) frequencies.

(Chapter 1.4.2) radiates (transverse) electromagnetic waves which we call "scattered" radiation. Thus, scattering may be understood by analyzing the amplitude of the electric dipole moment induced by application of an incident electric field, E,

$$E = E_0 \cos \omega t \tag{9.16}$$

To a first approximation, the induced electric dipole moment is proportional to the incident electric field magnitude, with a proportionality constant known as the "polarizability", α:

$$e\,x = \alpha E = \alpha\, E_0 \cos \omega t \tag{9.17}$$

However, electrons are not equally free to move in all directions. For example, the electron displacement (and thus the induced dipole moment) will depend on whether the induced displacement is along or perpendicular to a given chemical bond. Therefore, since the *atoms* of the molecule to which the electron is bound are continually vibrating (and thereby changing the way in which electrons are distributed in the molecule), it is reasonable to infer that the polarizability will change as the atoms vibrate to and from their equilibrium positions. So long as the change in polarizability induced by molecular vibrations is small, we may approximate the dependence of α on distance by just the first two terms of a Taylor series expansion about the equilibrium atomic positions, r_e :

$$\alpha = \alpha_{r = r_e} + \left(\frac{d\,\alpha}{d\,(r - r_e)}\right)_{r = re} (r - r_e) + \cdots \tag{9.18}$$

in which r is the (non-equilibrium) position of the vibrating atom to which the electron is bound, and $(r - r_e)$ represents the displacement of that atom away from its equilibrium position.

If we now recognize that the atom to which the electron is bound is itself oscillating at frequency, ω_1, then

$$r - r_e = A \cos \omega_1 t \tag{9.19}$$

in which A is the amplitude of the atomic oscillation. Substituting Eq. 9.19 into Eq. 9.18, and then Eq. 9.18 into Eq. 9.17, we obtain

$$e\,x = \alpha_{r = r_e} E_0 \cos \omega t + A\; E_0 \left(\frac{d\,\alpha}{d\,(r - r_e)}\right)_{r = re} \cos \omega t \, \cos \omega_1 t \tag{9.20}$$

The first term of Eq. 9.20 describes Rayleigh (centerband) scattering. The second term of Eq. 9.20 clearly introduces *amplitude modulation* at modulation frequency, ω_1. As shown in Chapter 4.4.3, amplitude modulation produces sidebands at $\omega \pm \omega_1$, resulting in "Raman" scattering (Figure 9.4, bottom diagram).

A quantum mechanical picture of Raman spectra is shown and described in Figure 9.5. The main difference between the classical and quantum mechanical results is that quantum mechanics explains why anti-Stokes scattering is weaker than Stokes scattering, namely because fewer molecules are in the first excited vibrational energy level of the ground state at ordinary temperatures (see Problems).

The net result of the above analysis is that molecular natural vibrational motions of *atoms* at *infrared* frequencies may be detected as amplitude-modulation sidebands at infrared frequency separation from the *uv/vis* centerband Rayleigh frequency originating from *electronic* natural motions. In the next section, we shall discuss why both IR and Raman observations may be needed to detect *all* of the molecular normal-mode vibrations.

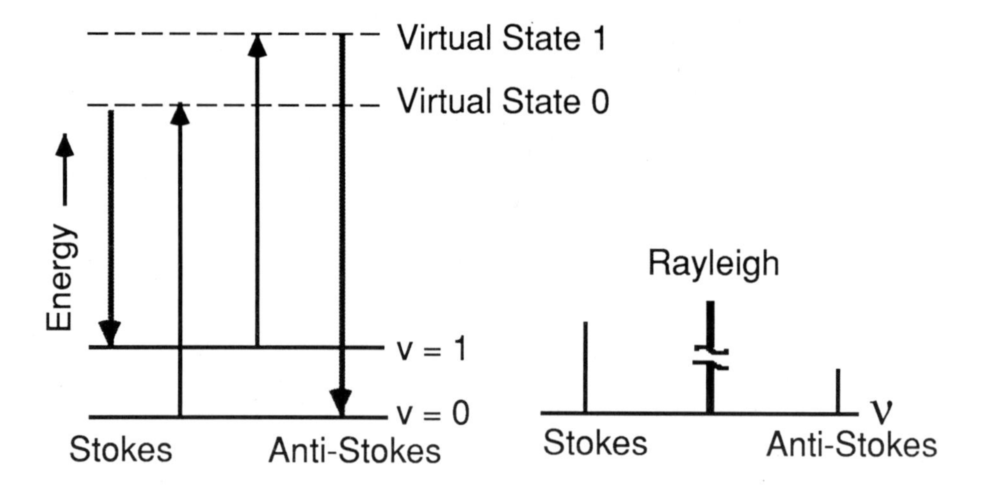

Figure 9.5 Quantum mechanical picture of Raman scattering. Intense monochromatic irradiation takes the molecule from either its zero'th or first vibrational energy level of the ground state (solid lines) to a higher-energy "virtual" state, from which the molecule immediately drops back to the zero'th or first ground state vibrational energy level, by emission of a photon whose frequency corresponds to either Stokes or anti-Stokes scattering. Because most molecules at ordinary temperature are in the ground vibrational level to start with, the Stokes path is much more probable, and anti-Stokes scattering is therefore much weaker than Stokes scattering (see Problems).

9.1.3 Infrared-active *vs.* Raman-active vibrations: selection rules

The classical mechanical mass-on-a-spring model suffices to compute all of the normal mode vibrational frequencies of a molecule for a purely harmonic potential (i.e., quadratic displacement terms only). The quantum mechanical harmonic oscillator yields an infinite number of energy levels; however, since the levels are equally spaced, there is still only one "natural" vibrational frequency, corresponding to a transition between any two adjacent energy levels. A more accurate representation of vibrational potential energy as a function of displacement from equilibrium includes "anharmonic" terms (e.g., terms which vary as $(\Delta x_1)^3$, $(\Delta y_2)^4$, etc.). In that case, the quantum mechanical energy level spacings are no longer equal, and there is an infinite number of different vibrational frequencies.

Even for a harmonic potential, quantum mechanics is necessary to decide which of the vibrational normal modes will be observable (and with what amplitude) in a given experiment. In classical mechanical terms, a *change in electric dipole moment during the vibration* is necessary for that normal mode vibration to be *infrared-active* (i.e., observable by direct absorption of infrared radiation at the frequency of that normal mode vibration), and a *change in polarizability during the vibration* is necessary for that normal mode vibration to be *Raman-active* (i.e., observable as an amplitude-modulation sideband of Rayleigh-scattered radiation).

For simple molecules, it is usually easy to decide whether or not a given normal mode vibration is IR-active, based upon the above (classical mechanical) criterion. For example, the reader should convince him/herself that all three normal mode vibrations of H_2O are IR-active, but that only three of the four CO_2 modes (all but ν_1 in Figure 9.3) are IR-active. Less obviously, it turns out that all of the totally symmetric stretch normal modes are Raman-active (e.g., the just-mentioned ν_1 mode of CO_2, or the symmetric stretch IR-inactive mode of molecular oxygen, O_2). For centrosymmetric molecules, such as CO_2 or O_2, a mode that is inactive in IR is active in Raman (and conversely), so that both IR and Raman experiments are needed to characterize all observable normal mode vibrations. In some cases, a given vibrational mode is unobservable by either IR or Raman.

Quantum mechanical calculations which predict which normal-mode vibrations will be IR- or Raman-observable ("allowed" *vs.* "forbidden) can be greatly simplified by use of *group theory*. Without going into detail, we simply note that normal modes can be classified according to their *symmetry*, and represented in a so-called *character table*, an example of which is shown in Table 9.1 for molecules with the same C_{3v} symmetry as (e.g.,) NH_3. Although non-trivial to derive, such character tables are easy to use (see Table legend), once a given vibration has been classified according to its symmetry (see Figure 9.6), as explained in (e.g.) the Cotton or Nakamoto references in Further Reading.

Table 9.1 Character table for the symmetry point group, $\mathbf{C}_{3v}$. Each normal-mode vibration may be symmetry-classified by its "Mulliken symbol" (leftmost column). A normal-mode vibration is IR-active if its Mulliken symbol is found in the same row as one of the x (or y or z) Cartesian coordinates. A normal-mode vibration is Raman-active if its Mulliken symbol is in the same row as one of polarizability tensor components, $x^2 + y^2$, z^2, $x^2 - y^2$, xy, yz, or xz. Parentheses denote "degenerate" transitions—i.e., two or more normal modes which happen to have the same vibrational frequency. The character sets (second box from left) are beyond the scope of this discussion.

C_{3v}	E	$2C_3$	$3\sigma_v$		
A_1	1	1	1	z	$x^2 + y^2, z^2$
A_2	1	1	-1		
E	2	-1	0	(x, y)	$(x^2 - y^2, xy)$ (xz, yz)

9.2 The Michelson interferometer

9.2.1 Generation of an interferogram: reflective vs. refractive optics

In FT/ICR and FT/NMR spectrometry, a frequency-domain spectrum is obtained by FT of a time-domain response following a brief excitation. Although such a procedure is practical even up to microwave frequencies (e.g., FT/electron paramagnetic resonance spectrometry), there are no analog-to-digital converters available at infrared or uv/vis frequencies, and a different approach is needed.

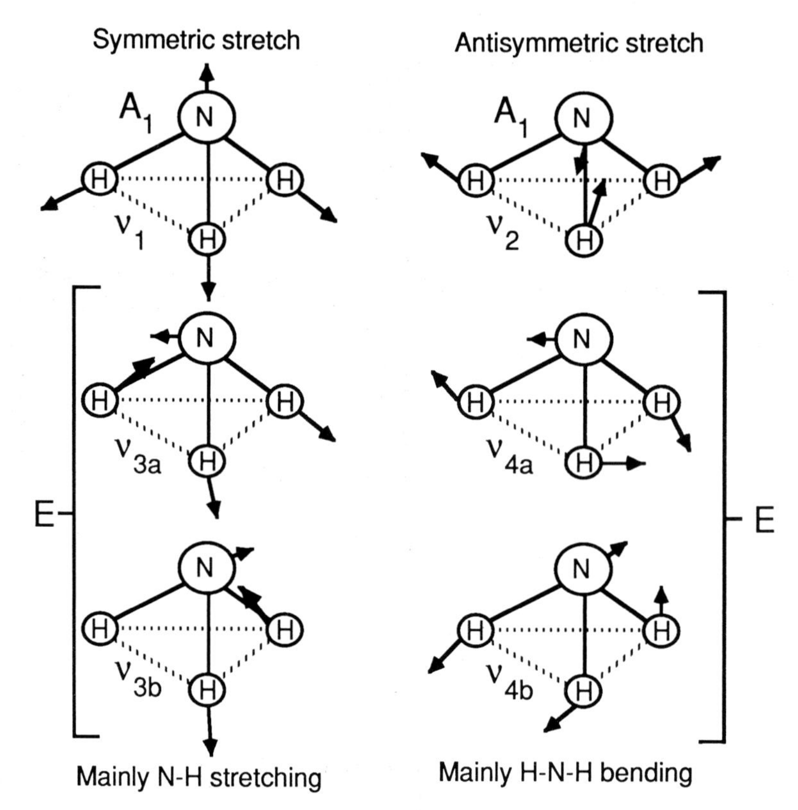

Figure 9.6 Normal-mode vibrations of ammonia, NH_3. Once the vibrational symmetry (A_1 or E in this case) has been determined for each normal mode, a group-theoretical "character table" can be used to determine which modes are IR-active or Raman-active (see Table 9.1 for the C_{3v} symmetry of this example).

Therefore, FT optical (IR or uv/vis) spectrometry is performed by dispersing the interference pattern in *space* rather than in *time*, by effectively heterodyning the signal against its time-delayed self. The general plan is to split the signal in half, and then arrange for the two half-signals to travel a different path length, so that subsequent recombination of the two half-signals will generate an interference pattern whose Fourier transform then yields a spectrum. The required path length difference may be generated *reflectively*, as by moving one mirror in a Michelson interferometer of Figure 9.7, or *refractively*, as by moving a wedge-shaped mirror in the refractively-scanned interferometer of Figure 9.8.

Because the wedge has a higher refractive index than air, one might expect that the wedge need not be displaced as far as the movable Michelson mirror to produce the same path length difference. However, in practice the wedge must move farther because of the narrow angle of the wedged optical element. Moreover, the refractive index and absorbance of the wedge may vary significantly with wavelength, so that it may be necessary to correct the final spectrum for those effects, or limit the spectral range accordingly. Both designs (and others based on the same principles) are in commercial use.

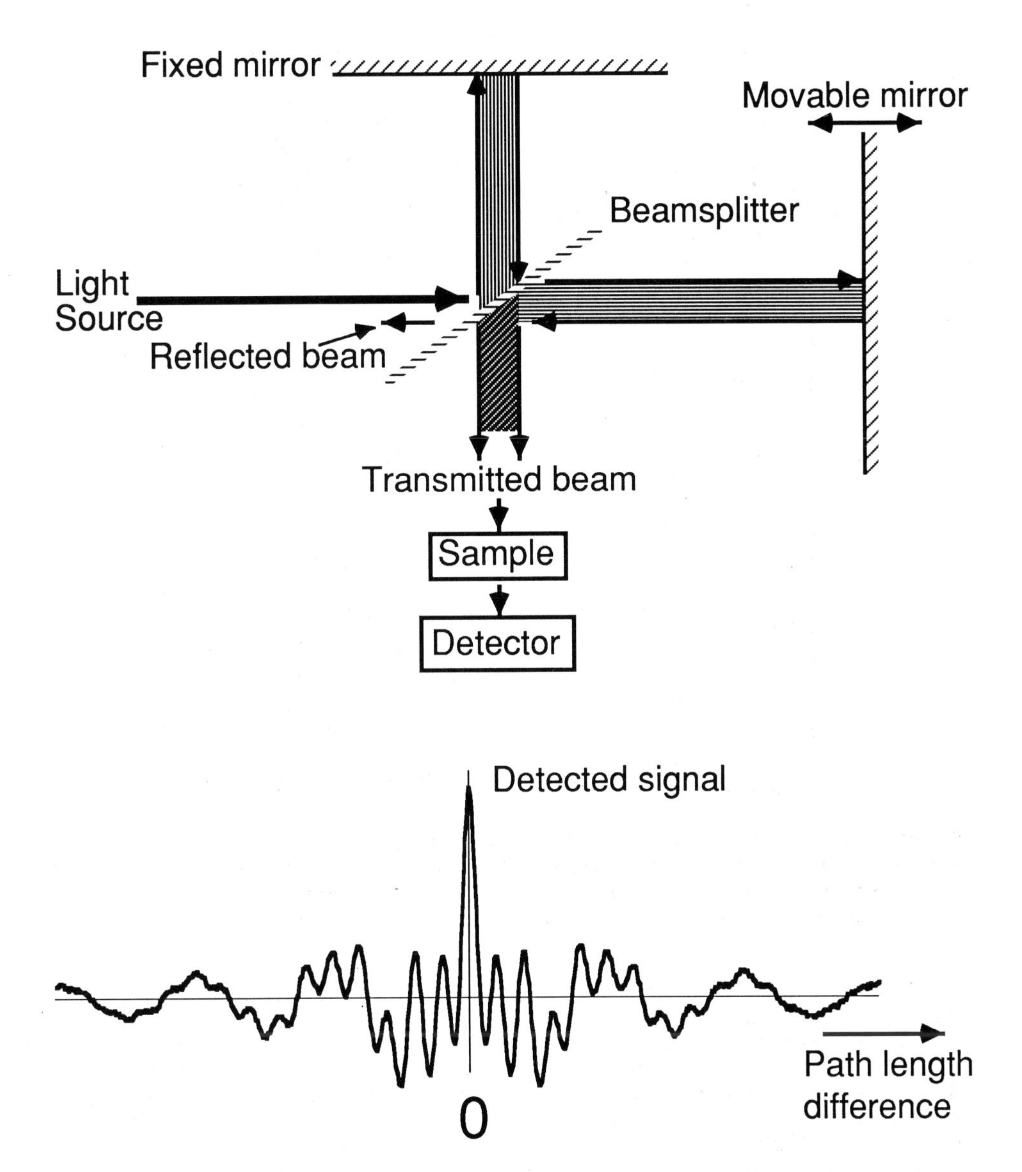

Figure 9.7 Schematic *reflectively* scanned (Michelson) interferometer. An incoming incoherent white light beam is split in half (e.g., by a half-silvered mirror). One component travels to and from a fixed-position mirror, and the other travels to and from a mirror whose position is movable. On reaching the beamsplitter, half of the recombined beam is reflected to pass through the sample to a detector, and the other half is reflected back toward the light source and is lost. An "interferogram", consisting of the detected beam intensity as a function of movable mirror position, has maximum magnitude when the two path lengths are equal (see Figure 9.9).

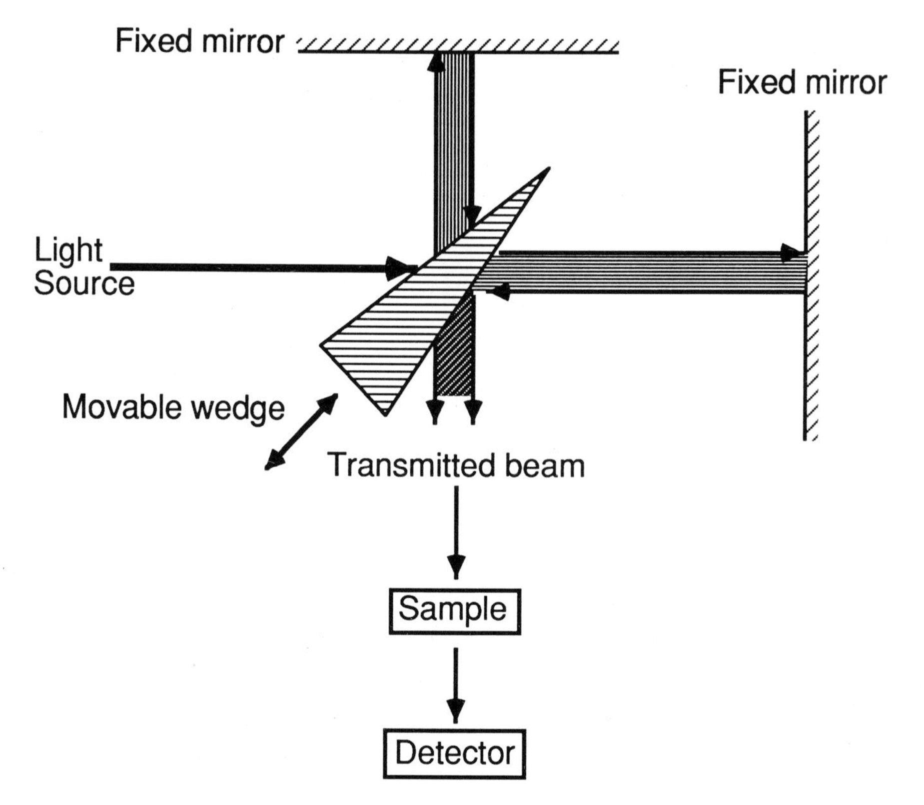

Figure 9.8 Schematic *refractively* scanned interferometer. The interferogram at the detector is based on linear variation in path length difference between two beams, as in Figure 9.7, except that the path difference is generated by moving a "wedge"-shaped beamsplitter as shown. (Since the refractive index of the wedge is larger than that of the surrounding air, the refracted beam is slowed, by passing twice through the wedge thickness, compared to the fixed-path beam.) The difference in path length for the beam reflected from the half-reflective side of the wedge and the beam passing (twice) through the other side of the wedge varies linearly with wedge position. (In practice, this type of interferometer employs corner retroreflectors rather than plane mirrors, and also includes a compensating wedge.)

In order to understand the nature of the detected interferometric signal, we need first to understand the difference between *coherent* and *incoherent* addition of electromagnetic waves. When two undamped monochromatic sinusoidal waves of equal amplitude and frequency but different phase are added (see Figure 9.9), the result (see Problems) is always a monochromatic sinusoidal wave of the same frequency but generally different phase and amplitude. If the two waves have the *same phase and same frequency*, then addition of the two waves is said to be *coherent*, and the resultant sinusoidal wave has maximal amplitude. If the two waves have different phase (Figure 9.9, middle), or frequency (Figure 9.9, bottom), then their additions is *incoherent* and the resulting wave is *periodic*.

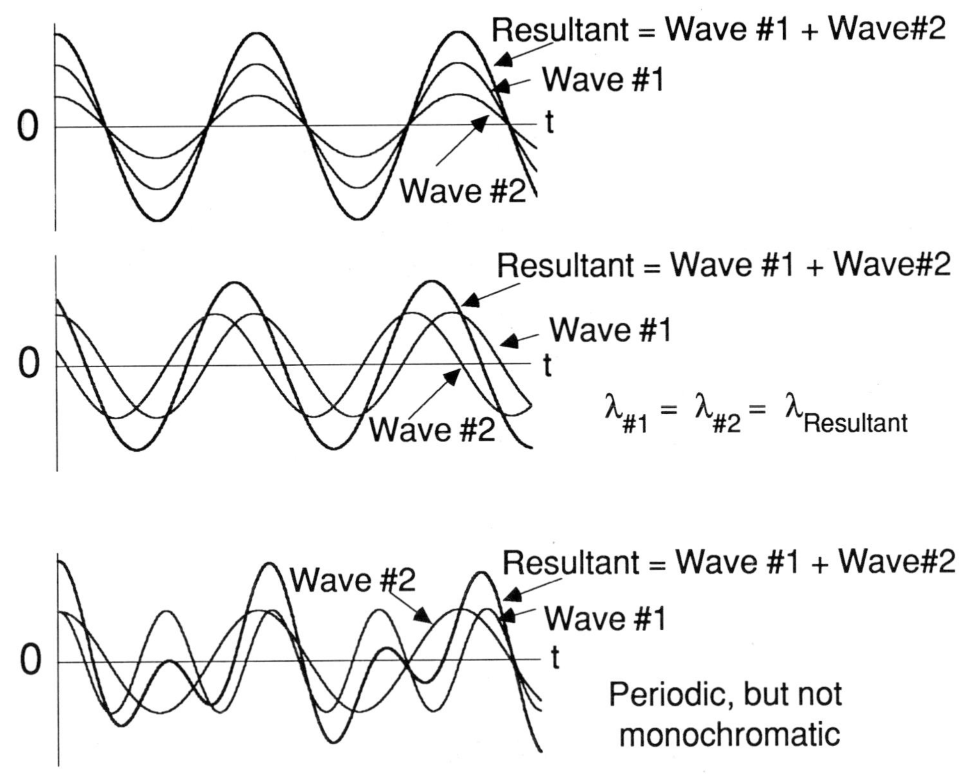

Figure 9.9 Coherent (top) and incoherent (middle and bottom) addition of sinusoidal waves. Addition is said to be *coherent* when the two waves have equal frequency (or wavelength) and phase, but not necessarily equal amplitude. Addition is *incoherent* when the two waves differ in either *phase* (middle), to yield a resultant sinusoidal wave which has the same frequency but smaller amplitude than for coherent addition, or *frequency* (bottom), to yield a resultant periodic (but no longer monochromatic) wave.

Next, consider what happens when a coherent monochromatic laser beam passes through the Michelson interferometer of Figure 9.7. If the movable mirror moves at constant velocity along the beam path, then the detected signal intensity, $I(\delta)$, will be maximal whenever the two path lengths are equal ($\delta = 0$) or differ by an integral number of wavelengths (as in Figure 9.9, top), and will be modulated sinusoidally (see Problems) as a function of movable mirror position, as shown in Figure 9.10 (top):

$$I(\delta) = \frac{1}{2} I_0 \left(1 + \cos \frac{2\pi\delta}{\lambda}\right) \tag{9.21}$$

in which I_0 is the intensity of the "source" beam, λ is the wavelength of the light, the "retardation", δ, is the path length difference (i.e., the movable mirror position measured from the position at which both optical beam component path lengths are equal) and there is no absorption or refraction along either beam path.

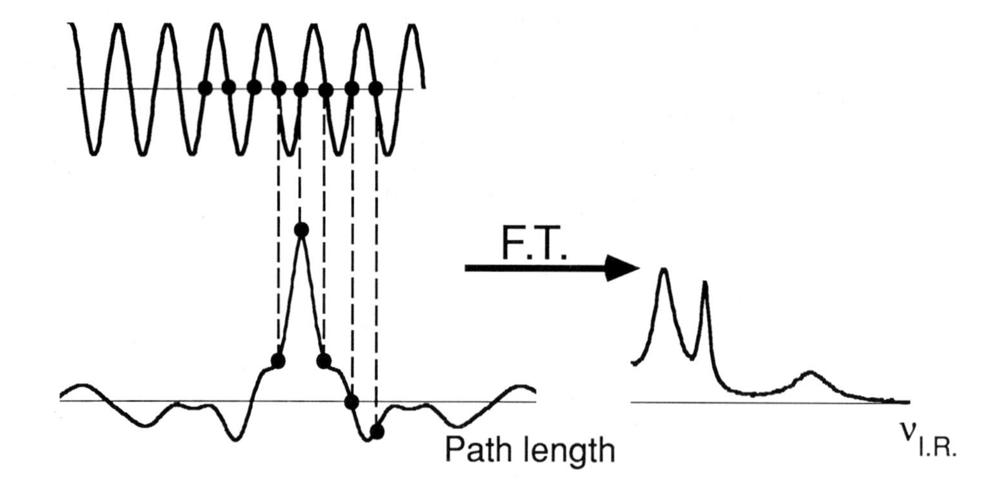

Figure 9.10 Interferograms (left) and corresponding Fourier transform (transmittance) spectra (right) produced by passing coherent monochromatic (e.g., laser) light (top) or white light (bottom) through a sample characterized by several "natural" absorption bands in the spectral range of interest. The regular "fringes" of a He/Ne *visible* laser interferogram pattern offer a simple way to calibrate the position of the moving mirror for *infrared* spectroscopy (see text).

The factor of 1/2 in Eq. 9.21 arises from the loss of half of the source beam intensity back to the source by reflection at the half-reflective mirror. The d.c. (zero-frequency) component arises because the signal is detected by its *energy* ; thus, the minimum signal is zero, and sinusoidal modulation must therefore occur with equal displacement about a non-zero average value. It is therefore customary to measure interferogram intensity as the deviation from its average value, so as to eliminate an otherwise huge d.c. component in the resulting FT spectrum. Finally, if signals of several frequencies are present, they will modulate the interferogram as shown in Figure 9.9 (bottom) to yield a two-sided interferogram as shown in Figure 9.10 (bottom).

The interferogram is formed by the *addition* of waves of different phase. However, photon detection is based on the *energy* of the light wave, which is proportional to the *square* of its amplitude (Eq. 1.8). In other words (see Problems), the *power* spectrum obtained by FT of the detected (i.e., squared) interferogram is the same as the FT of the true autocorrelation function of the of the incident light waveform (i.e., as if the interferometer acted to *multiply* together the undelayed and delayed components which actually recombine by *addition* at the beamsplitter). Finally, since sample *concentration* is related to spectral *intensity* rather than *amplitude* in optical absorbance spectra (see below), it is in fact the *power* spectrum (rather than magnitude-mode or absorption-mode) that is of interest in optical interferometry.

9.2.2 Sampling calibration by counting laser fringes

A standard (and ingenious) technique for calibrating the position of the moving mirror in a Michelson interferometer is to fix a visible-wavelength mirror to the back side of the moveable infrared mirror of Figure 9.7, and use that visible-wavelength mirror as the moving mirror of a second Michelson interferometer for which the source is (e.g.) a helium-neon laser (λ = 632.8 nm). The *infrared-wavelength* interferometer detector may then be triggered to sample the interferogram signal whenever the *visible-wavelength* interferometer detector signal crosses zero (see Figure 9.10, bottom).

For example, for a helium-neon laser, zero-crossings in the visible interferogram occur at intervals of 632.8/2 nm = 0.3164 μm. Since the Nyquist theorem requires at least two samples per cycle, the highest (infrared) frequency which could satisfy the Nyquist criterion (see Problems) would be 15,804 cm^{-1} (suitable for *near-infrared* sampling). For *mid-infrared* operation, one could sample at every other zero-crossing, to give a maximum Nyquist frequency of 7902 cm^{-1}, and thereby obtain a spectrum with twice the frequency-domain digital resolution for a given number of interferogram data points. Alternatively, one could electronically double the sampling frequency of the experimental helium-neon zero-crossings, to provide an interferogram whose FT spectrum extends to a Nyquist limit corresponding to a (minimum) wavelength of 316.4 nm.

Digital frequency-domain resolution (in cm^{-1} per data point) is simply $(1/\Delta)$, namely, the reciprocal of the maximum optical path difference, just as digital frequency-domain resolution (in Hz) in FT/ICR and FT/NMR is $(1/T)$, where T is the time-domain data acquisition period.

9.2.3 Transmittance and absorbance FT spectra from single-beam measurements

The detector in an interferometer measures *transmitted* light intensity. If I and I_0 are the respective transmitted spectral intensities (obtained by FT of the interferograms) of light passing through the sample and reference, then the *transmittance*, T, is obtained from their ratio:

$$\text{Transmittance} = T = \frac{I}{I_0} \tag{9.22}$$

T is usually reported as a percentage, $100(I/I_0)$. However, the (molar) *concentration*, c, of a dilute sample is proportional to *absorbance*, A,

$$A = \log_{10}\frac{I_0}{I} = -\log_{10} T \tag{9.23}$$

according to Beer's law:

$$A = \varepsilon\, l\, c \tag{9.24}$$

in which ε is molar absorptivity (i.e., absorbance per unit path length per unit molar concentration) and l is the path length of the light beam through the absorbing medium. When IR spectra were obtained directly from single-channel scanning instruments, it was more convenient to plot transmittance directly. However, now that FT/IR spectra can be computed with equal effort as transmittance or absorbance, the absorbance display is becoming more common.

The absorbance spectrum is generated simply from the log of the ratio of the transmitted spectra from reference and sample (Eq. 9.23). The reference spectrum may either be acquired in a separate experiment, or (in the newer dual-beam instruments) by flipping mirrors back and forth (say, every 20 scans or so) between reference and sample in the same compartment (see Griffiths & de Haseth in Further Reading). For very dilute samples, the background spectrum from CO_2 and water vapor can be significant.

9.2.4 Phasing

Although the interferogram detection process is phase-independent, the intensity of the light reaching the detector is modulated sinusoidally as a function of optical path difference (Eq. 9.21). Therefore, the final FT spectrum is represented by a "magnitude" (which is the intensity at that frequency) and a "phase" which is the phase of that frequency component at zero optical path difference (see below). Phase-correction is therefore necessary for a spectrum obtained by FT of an interferogram, even though the detection process is phase-independent.

Unlike ICR or NMR time-domain signals, which are causally generated, and therefore zero-valued until time zero, an optical interferogram is inherently *two-sided* (see Figure 9.7, bottom). Therefore, if both sides of the interferogram were acquired, then the interferogram would be centrosymmetric (i.e., an "even" function with respect to the "centerburst" position at which the two component beam paths are equal and all detected waves are in phase). The FT of such an even function yields a frequency-domain spectrum with no imaginary part—i.e., a perfectly "phased" absorption-mode spectrum. Thus, at first glance, it would appear that there is no need to phase-correct an FT/interferometer spectrum.

Unfortunately, phase variation across an FT/optical spectrum can arise from optical, electronic, and sampling effects. For example, the introduction of an electronic low-pass filter to remove high-frequency noise from the interferogram signal also introduces frequency-dependent phase shifts in the spectrum. Moreover, a phase shift will result if (as is usually the case) the midpoint of the sampled interferogram does not happen to fall at zero path-length difference. In addition, the interferometer optical components (e.g., beamsplitter, mirrors) can contribute frequency-dependent phase shifts. Finally, if phase varies non-linearly with frequency, then the interferogram can even be asymmetrical (see below).

In practice, most optical interferometers acquire only one half of the interferogram, in order to reduce the moving mirror travel distance by a factor of two and thereby make the instrument smaller and mechanically simpler. For all of the above reasons, it is necessary to phase-correct FT/optical spectra.

Fortunately, FT/optical spectral phase typically varies smoothly and relatively slowly with frequency. Therefore, the phase spectrum, $\phi(\omega)$, can be determined from the real and imaginary components of the frequency-domain spectrum, $\boldsymbol{F}(\omega)$, obtained by FT of a relatively small (say, 512 data points) two-sided interferogram

$$\phi(\omega) = \arctan\left(\frac{\mathrm{Im}[\boldsymbol{F}(\omega)]}{\mathrm{Re}[\boldsymbol{F}(\omega)]}\right) \tag{9.25}$$

as shown in Figure 9.11. Phase correction may then be performed on the *frequency-domain* spectrum as outlined in Chapter 2.4.2. Alternatively (see Problems), phase correction may be performed on the *interferogram* itself.

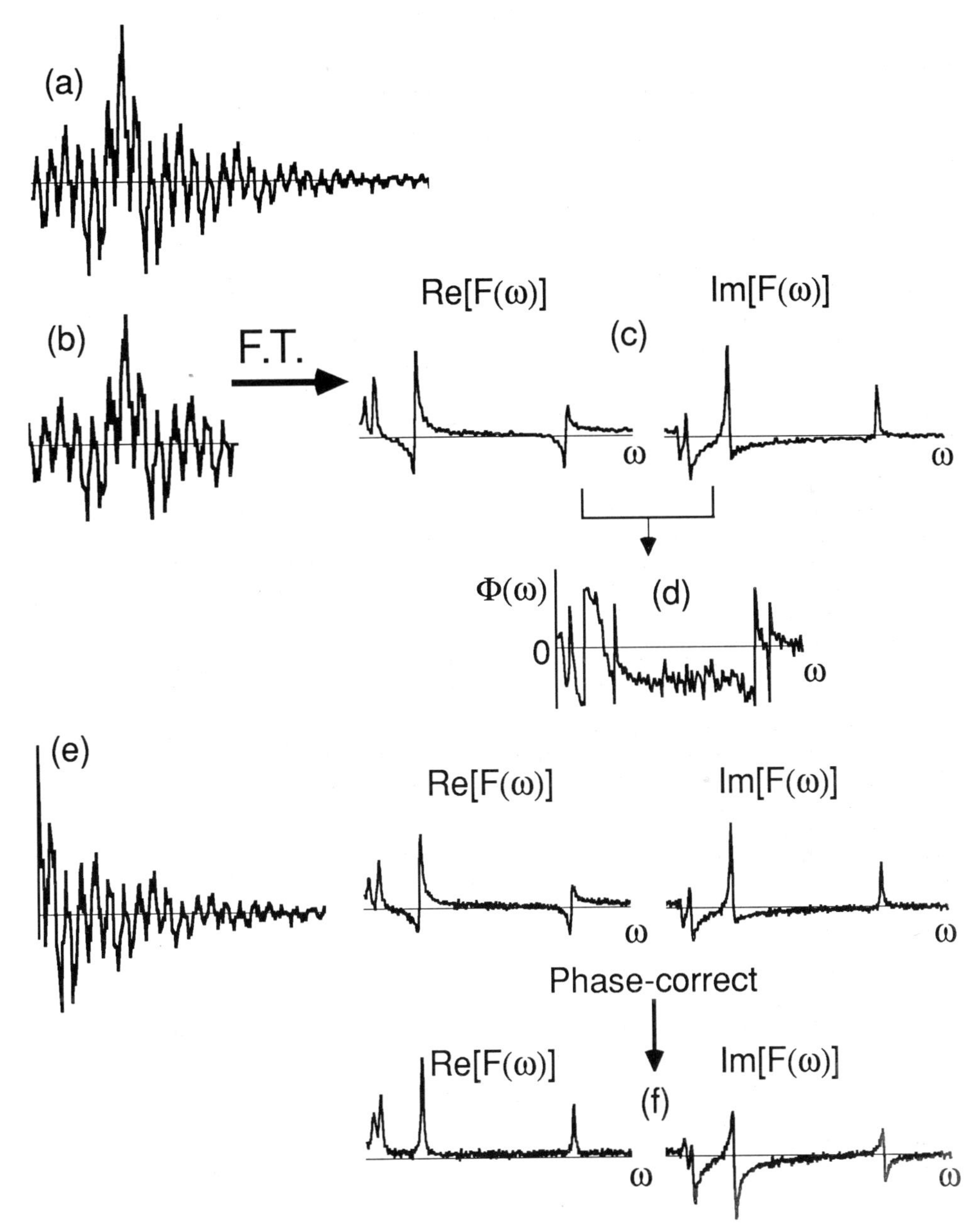

Figure 9.11 Phase correction of a simulated FT/IR spectrum. The original interferogram (a) extends slightly (256 points) beyond the centerburst. A two-sided 512-point interferogram (b) is Fourier transformed to give real and imaginary spectra (c) from which a phase spectrum (d) is computed from Eq. 9.25. Frequency-domain phase correction (Eqs. 2.37) of the (one-sided) interferogram (e) then gives the desired spectrum (f).

9.2.5 Dynamic range: gain-ranging; chirping

The "centerburst" of an interferogram results when the path lengths for the two recombined beams are equal, so that the signals at *all* frequencies are in-phase. The centerburst peak is similar to the maximum in the free-ion (FT/ICR) or free-induction (FT/NMR) time-domain signal immediately following an impulse excitation—each excited motion begins as a cosine wave at time zero (i.e., at the beginning of the detection period). As the interferometer path difference increases (or as an ICR or NMR response evolves with time) for a signal consisting of the sum of many oscillations of different frequency, the various oscillations get out-of-phase with each other, and the composite signal magnitude decreases (see Figure 9.7, and SWIFT phase modulation in Chapters 4.2.3 and 7.4.2).

Unfortunately, the detected interferogram represents *transmitted* radiation; therefore, the detected spectrum from a dilute sample results mainly from the (broadband) radiation *source*. Thus, the centerburst can be very large relative to the small change in the interferogram due to absorption by the sample. FT/IR dynamic range (ratio of largest to smallest signal) can easily exceed 10^5—beyond the range of even a 16-bit analog-to-digital converter.

One solution to the dynamic range problem is to use a *dual-beam (optical subtraction)* interferometer. The basic principle is that the two beams differing in phase by 180° are recombined. If the sample is inserted in one beam path and the reference in the other, then the net signal will represent only their difference. For the special case of vibrational circular dichroism, the detected beam is passed through a *phase modulator* which switches rapidly back and forth between right- and left-circularly polarized light. Phase-sensitive detection (i.e., mixer, followed by low-pass filter, as in Chapter 4.4.2) at the modulation frequency then yields a signal only when there is a difference in absorption of right- and left-circularly polarized components (see Griffiths & de Haseth in Further Reading). An ingenious related application is Guelachvili's use of magnetic field on-off modulation to detect dilute paramagnetic species by virtue of the Zeeman effect.

Another approach is to vary the *gain* of the detector, so that the centerburst is detected at relatively low gain and the data points measured at large retardation are detected at high gain. The resulting interferogram must of course be *un*-weighted before FT (e.g., by dropping the appropriate number of least significant bits from the data points acquired at increased gain).

However, since the cause of the dynamic range problem is that all of the signal components have a common *phase* at zero retardation (i.e., equal path length for the two recombined beams), a more fundamental solution is to introduce (as for SWIFT, Chapter 4.2.3) a *phase-variation* for signal components of different frequency. Such a "chirp" process may be achieved by inserting into the beam path a flat optical element whose refractive index varies with frequency. The result will be to disperse the zero-phase positions of signals of different frequency to different path-difference positions in the interferogram, thereby reducing the dynamic range by "spreading out" the centerburst, as shown in Figure 9.12. The center of the "chirped" interferogram is also shifted, since even if the refractive index were constant with frequency (so that each component wave is slowed by the same time-delay), the phase of a higher-frequency wave would be shifted proportionately more (because more oscillations occur during a given fixed-time period delay) than that of a low-frequency wave (see Chapters 2.4.1 and 4.2.3).

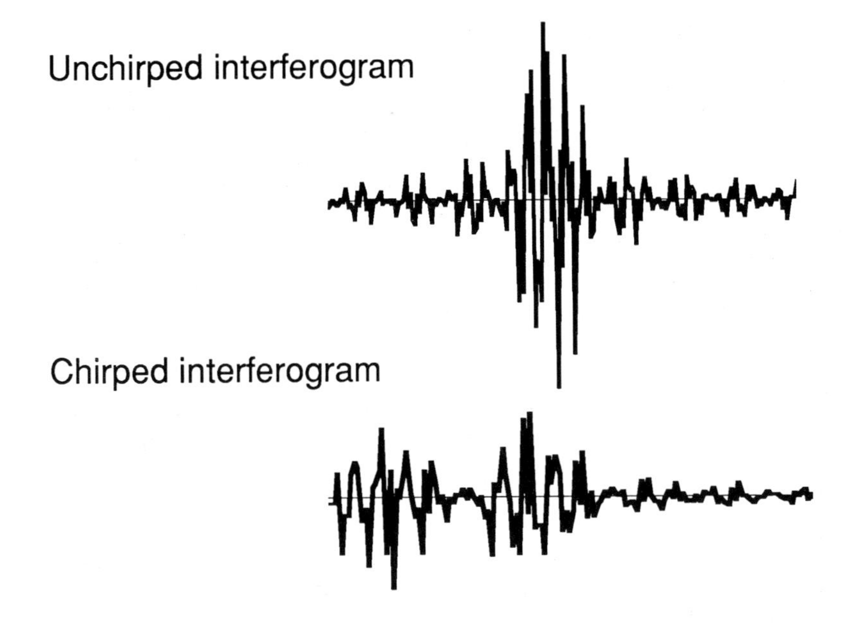

Figure 9.12 Effect of "chirping" on simulated interferograms. Top: Unchirped interferogram. Bottom: Chirped interferogram obtained by imposing a frequency-dependent phase shift on the original interferogram. Note the decrease in dynamic range in the chirped interferogram, resulting from different zero-path-difference positions for signals of different frequency.

In all of the preceding discussion, the sample has been located between the interferometer and the detector. Thus, both component beams from the interferometer are equally absorbed and slowed on passage through the sample. However, if the sample is placed in *one arm* of the interferometer (so-called "dispersive" FT interferometry), then the resulting interferogram will be *shifted* (if the refractive index of the sample is non-unity and independent of frequency) and *chirped* (if the refractive index varies with frequency).

9.2.6 Opening the exit and entrance slits: Fellgett (multichannel) and Jacquinot (throughput, étendue) advantages

There are two kinds of multichannel advantage in FT interferometry, compared to an instrument in which only a narrow band of frequencies is observed at a time. The familiar Fellgett advantage is essentially identical to that in FT/ICR and FT/NMR spectroscopy, and corresponds to opening the *exit* slit—i.e., *detection* of the whole spectrum at once. Although a factor of two in signal strength is generally lost by ignoring the beam component which is reflected back to the source, the multichannel advantage is nevertheless 10^4 or greater.

The second advantage is variously known as the "throughput", "étendue", or "Jacquinot" advantage. Throughput is the product of the area of an aperture along the optical path and its acceptable solid angle, and is a measure of the relative amount of incident light which is transmitted through the aperture. For example, for two succeeding apertures whose separation is larger than the aperture diameters (e.g., "entrance" and "exit" slits in a dispersive spectrometer), the throughput is the product of the areas of the apertures, divided by the square of the distance between them. For a grating spectrometer, throughput is severely limited by the area of the *entrance* slit. Although an interferometer also has an entrance aperture, its étendue advantage ranges from 10-250 over the IR frequency range. In fact, for typical FT/IR spectra at medium resolution (a few cm^{-1}), the étendue advantage often exceeds the Fellgett advantage. The throughput advantage is especially important for optical spectroscopy of astronomical weak emission sources [since a telescopic image (especially of an extended object) may be too large to put through a spectroscope slit] and was one of the reasons that FT/IR spectra of astronomical sources were produced even before the FFT algorithm was available. [The "throughput" advantage in ICR or NMR is achieved by use of a broadband excitation source (e.g., a short pulse or other stored-waveform brief excitation).]

9.3 Applications of FT/interferometry

In this section, we will briefly mention some of the principal applications of FT-interferometry in the optical (infrared/visible/ultraviolet) spectral range. Many more practical details and applications may be found in Further Reading.

9.3.1 FT/Infrared spectrometry

In FT/IR/interferometry, noise is generally detector-limited, as in FT/ICR or FT/NMR spectroscopy (see Chapter 9.3.2.1). Therefore, most applications are based on use of the étendue and Fellgett advantages to produce a spectrum with better signal-to-noise ratio for a given data acquisition period (Chapter 9.3.3.1) or a spectrum of a given signal-to-noise ratio in a shorter data acquisition period (Chapter 9.3.1.2), compared to a dispersive single-channel spectrometer.

9.3.1.1 *Molecular structure from vibrational group frequencies*

For small molecules, the methods outlined in the first part of this chapter may be used to determine all of the normal mode vibrations, from infrared and/or Raman spectra of the molecule and all of its available isotopic variations (e.g., $^{1}H^{35}Cl$, $^{1}H^{37}Cl$, $^{2}H^{35}Cl$, $^{2}H^{37}Cl$). However, such effort becomes prohibitive for large molecules. For example, the complete normal-mode analysis for the retinal molecule ($C_{20}H_{30}O$) took more than 20 man-years of combined effort in organic synthesis (to make the isotopically substituted ^{2}H and ^{13}C species) and vibrational spectral analysis.

Therefore, most practical FT/IR applications are based on the correlation between IR frequency and chemical functional group, as illustrated in Table 9.2. As for NMR chemical shifts, IR group frequencies may vary considerably for a given functional group (e.g., CO stretch frequency can vary from ~1100 cm^{-1} to ~2300 cm^{-1}). Thus, group frequencies can be useful aids in identifying an

Table 9.2 Typical stretching and bending vibrational frequencies (ν, in cm^{-1}) associated with particular chemical functional groups. (Adapted from J. M. Hollas, *Modern Spectroscopy*, Wiley, NY, 1987, p. 131)

Bond-stretching		Bond-stretching	
Group	ν (in cm^{-1})	Group	ν (in cm^{-1})
≡C—H	3,300	—O—H	3,600†
=C<(H)	3,020	>N—H	3,350
Except O=C<(H)	2,800	—P=O	1,295
⋝C—H	2,960	>S=O	1,310
—C≡C—	2,050	**Angle-bending**	
>C=C<	1,650	≡C—H	700
⋝C—C⋜	900	=C<(H)(H)	1,100
⋝Si—Si⋜	430	—C(H)(H)(H)	1,000
>C=O	1,700		
—C≡N	2,100		
⋝C—F	1,100	>C<(H)(H)	1,450
⋝C—Cl	650		
⋝C—Br	560	—C≡C—C	300
⋝C—I	500		

†may be reduced in condensed phases by hydrogen-bonding

unknown compound (by comparison to the same group frequency in a molecule of closely related structure). Alternatively, because of the huge number of possible normal modes for a molecule of even modest size, the IR spectrum provides a useful fingerprint which is nearly unique for a given molecule. Thus, if the IR spectrum of a molecule is known, its presence (even in a mixture) can be established from an IR spectrum. Moreover, even if the IR spectra of the components of a mixture are not known in advance, any of several "chemometric" methods can yield best-fit spectra for each of an arbitrarily specified number of components—such schemes of course work best when signal-to-noise ratio is large and the number of components is known in advance.

Because of the étendue and Fellgett advantages for such applications, FT/IR has effectively completely displaced dispersive single-channel IR spectroscopy to the extent that at least one scientific journal has proposed that the "FT" designation is no longer necessary to denote an FT/IR experiment!

9.3.1.2 *GC/FT/IR analysis of mixtures*

The Fellgett *speed* advantage is obviously most important for experiments in which the detected spectrum changes rapidly with time. By far the most important such application is chromatography (gas chromatography or high-performance liquid chromatography) with FT/IR detection. For example, a typical capillary GC peak might elute over a period of ~1 second. At one spectrum per second for the ≥20 min of a typical capillary GC run, more than 1,200 digitized interferograms (say, 4K each) must be stored, preferably on a high-capacity magnetic disk. The sensitivity of FT/IR detection clearly depends on the IR molar absorbance of the sample—detection limits of <100 pg have been demonstrated for highly-absorbant samples.

Once the FT/IR spectrum for each of the interferogram samples of a GC or HPLC eluent has been computed, the data may be reduced in various ways. First, for any chromatograph peak which contains only one chemical component, the FT/IR spectrum of the peak provides an identification fingerprint, and the relative concentration of that component may be estimated from the relative intensity of the absorption and the (previously known) molar absorbance.

Alternatively (Figure 9.13), one can ask for the time-variation of the integrated spectral intensity over any of several specified frequency ranges chosen to correspond to particular functional groups (e.g., carbonyl stretches). Figure 9.13 shows chromatograms reconstructed from the integrated absorption over each of five different IR frequency bands. Analysis of eluate components according to chemical functional class is a major aid in identification of unknowns. Note that the effective chromatographic resolution is improved, since two peaks eluting close together in time may have different functional groups. The lowermost trace in Figure 9.13 is a "Gram-Schmidt" chromatogram (GSC) which represents a plot of IR absorbance (integrated over the full spectral range) versus scan time. The GSC is actually computed directly from the interferogram (see Further Reading).

9.3.2 Ultraviolet/visible interferometry

FT/interferometry in the uv/visible spectral range differs in several important ways from FT/IR interferometry. First, because the inherent spectral frequencies are much larger, a given retardation corresponds to more cycles of uv/vis than of IR oscillation. For example, a spacing of 8 cm^{-1} between successive frequency-domain data points corresponds to a resolving power of ≥500 across the mid-IR spectrum (say, up to 4,000 cm^{-1}), but a uv/vis FT spectrum acquired for the same maximum displacement of the moveable mirror (and thus the same digital resolution) would have an order of magnitude higher resolving power (5,000) at 250 nm (40,000 cm^{-1}). Second, because the uv/vis frequencies are an order of magnitude larger than IR frequencies, the interferogram must be sampled about 10 times as often per unit mirror displacement in order to avoid foldover (see below). Third, because uv/vis photons can be detected with nearly unit efficiency, the noise in FT/uv/vis interferometry is no longer independent of the signal, and the Fellgett multichannel advantage becomes peak-dependent (see Chapter 9.3.2.2).

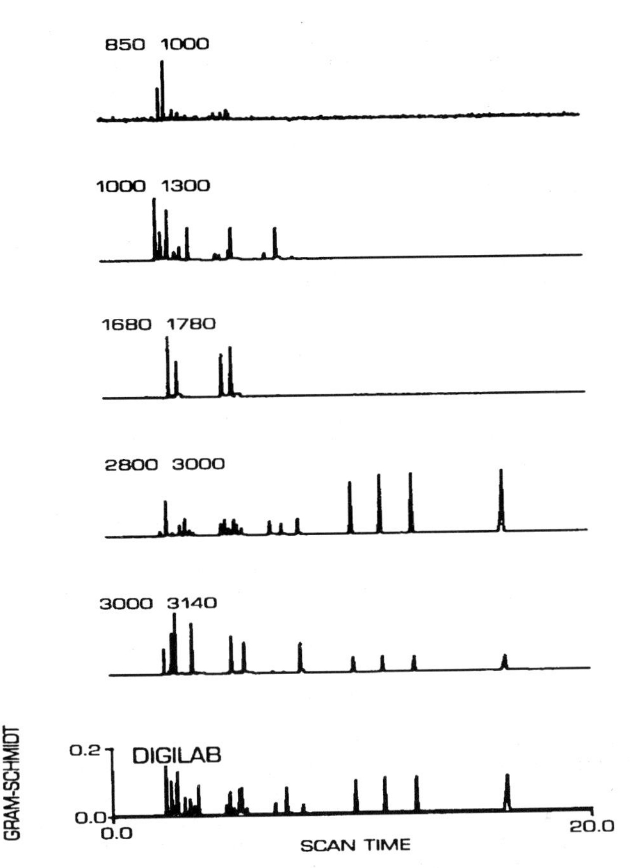

Figure 9.13 Gas chromatograms reconstructed from the integrated FT/IR absorbance over each of five spectral regions. Proceeding from top to bottom: 850-1000 cm^{-1} (primary amine, aromatics, unsaturates); 1000-1300 cm^{-1} (C—O stretch: ethers, esters, alcohols, etc.); 1680-1780 cm^{-1} (C$=$O stretch); 2800-3000 (aliphatic C—H); 3000-3140 (aromatic C—H, unsaturates). Note the presence of different functional groups (as evidenced by different IR group vibration frequencies) in different GC peaks. The lowermost trace is a Gram-Schmidt reconstructed chromatograph (see text) representing integrated IR absorbance over the full detected spectral range, ~700-4400 cm^{-1} . (Taken, with permission, from: S. V. Compton, BioRad (Digilab Division), FTS®/IR Notes No. 50, June, 1987)

9.3.2.1 *Resolution, bandwidth, and foldover*

A practical advantage of FT/interferometry for uv/vis spectroscopy is that measurement of peak location is not only very precise (<0.1 cm^{-1}, or ~1 part in 10^6 to 10^7) by means of laser fringe-counting, but very reproducible. However, in order to take advantage of the very high available FT/uv/vis resolution (say, 0.05 cm^{-1}), the number of sampled interferogram data points must be very large (≥1,000,000) if the Nyquist limit is to be satisfied for a broadband spectrum.

Fortunately, the foldover in the spectrum obtained by FT of an intentionally undersampled interferogram can be used to advantage if the spectral peak *positions* are already known and we are asked only to determine their relative *magnitudes* (as in elemental analysis based on atomic emission in the uv/vis range). For example, Figure 9.14 shows FT/visible spectra produced by sampling at $2\nu_{Nyquist}$, $\nu_{Nyquist}$, $\nu_{Nyquist}/2$ and $\nu_{Nyquist}/4$, in which $\nu_{Nyquist}$ represents the largest spectral frequency which can be represented without foldover. In the

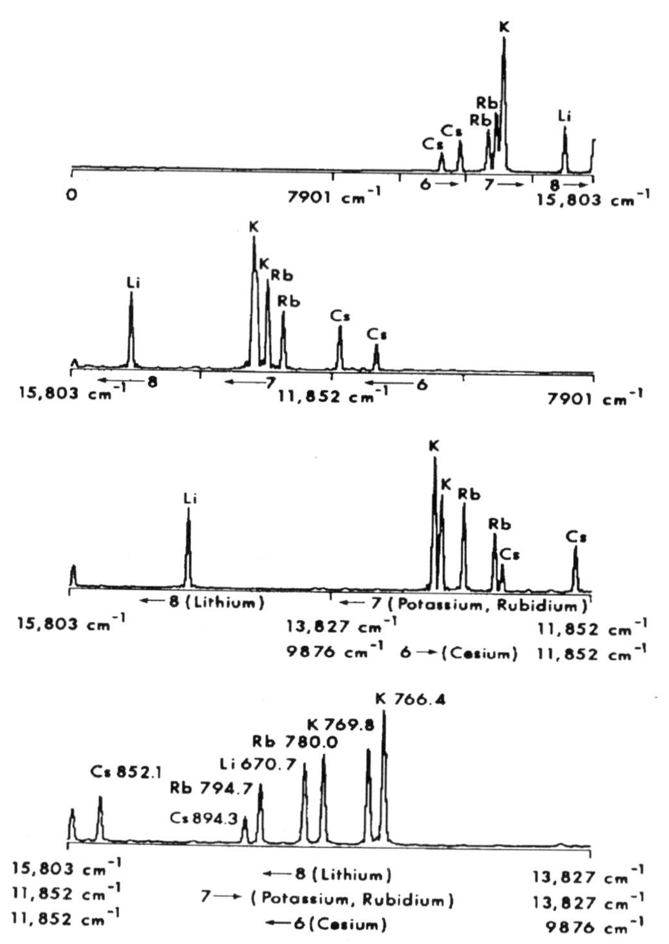

Figure 9.14 Flame emission FT/visible spectra of alkali metal atoms, based on interferogram sampling at every 0.3164 μm (top spectrum), 0.6328 μm, 1.2656 μm, and 2.5312 μm (bottom spectrum). Each successive undersampling by a factor of 2 effectively folds the spectrum in half at its center, thereby providing twice as many data points per unit frequency. The peak magnitudes are unchanged by the folding process (see text). [Reproduced, with permission, from G. Horlick, R. H. Hall, and W.K. Yuen, Chapter 2 in *Fourier Transform Infrared Spectroscopy: Applications to Chemical Systems*, Vol. 3 (J. R. Ferraro and L. J. Basile, Eds.), Academic Press, NY, 1982, pp. 37-81]

topmost spectrum, the Nyquist limit is satisfied and all of the emission peaks appear at their correct frequencies. However, because the number of data points is limited, the spectral resolution is insufficient to resolve (for example) the two potassium lines. Undersampling by a factor of 8 (bottom spectrum) folds the peaks into aliased *positions*, but with the same relative *magnitudes*, and with higher *resolution* (since the same number of data points is now distributed over only 1/8 the frequency range (compare to FT/ICR undersampling, Figure 7.14).

9.3.2.2 *Source-limited vs. detector-limited noise: multiplex disadvantage for sparse spectra*

As noted in Chapter 5.1.1, noise for IR detectors is typically "detector-limited" (i.e., independent of signal strength), whereas noise for uv/vis detectors is typically "shot-noise-limited" (i.e., proportional to the square root of signal strength). Thus (see Figure 5.1), the source-limited noise in a spectrum acquired with a single-channel scanning instrument will be largest at the frequencies of the largest spectral peaks, whereas source-limited noise in an FT spectrum should be distributed over *all* frequencies.

The Fellgett advantage (or disadvantage) in an FT/uv/vis spectrum therefore would appear to depend upon the number and magnitude of spectral peaks, and the frequency at which the signal-to-noise ratio is determined. For example, the signal-to-noise ratio for a large peak should be *enhanced* (by a factor of up to $N^{1/2}$, if it is the only peak in the spectrum) whereas the signal-to-noise ratio for a small peak will be *reduced* (by a factor of up to $N^{1/2}$, if there is large spectral magnitude at all other frequencies) by source-limited noise.

9.3.3 Raman spectroscopy: the centerband dynamic range problem

There are several features of Raman spectroscopy that affect the performance of an FT/interferometry experiment. First and most important, the central Rayleigh band is several orders of magnitude more intense than the Raman sidebands of interest, creating a severe dynamic range problem for interferometric detection. Second, the relative intensity of Raman scattering varies directly with frequency to the fourth power:

$$I_{\text{Raman}} \propto \nu^4 \qquad (9.26)$$

Thus, Raman scattering is strongest at frequencies (uv/vis) at which the Fellgett advantage of FT interferometry may be lost. Third, irradiation in the uv/vis range can also produce *fluorescence* emission at frequencies which may fall in the range of the desired Raman sidebands. In principle, one could hope to distinguish Raman scattered light (which is emitted instantly upon irradiation) from fluorescence (whose intensity decreases exponentially with time after an excitation pulse of incident light) by their time-dependence, but in practice fluorescence is nevertheless a severe problem when the incident irradiation falls at or near an absorption band. Finally, since Raman scattering is usually excited by a focused laser beam, the Jacquinot (throughput) advantage of an interferometer is unused in an FT/Raman experiment.

There are presently three general methods for achieving part or all of the multichannel advantage in Raman spectroscopy: multidetector arrays, Fourier transform, and Hadamard transform techniques. Although *multi-* detector arrays are much more feasible in the uv/vis than in the IR spectral range, they are expensive (if single-photon counting sensitivity is needed), and are currently limited to relatively low-resolution detection (~1,000 resolution elements). We shall therefore consider the two *multiplex* (Fourier and Hadamard transform) methods, each of which uses a *single* detector.

9.3.3.1 *FT/Raman spectroscopy*

The FT/Raman experiment is shown schematically in Figure 9.15. As noted above, except for wavelength calibration (which is very important for spectral subtraction experiments), the FT experiment holds no particular advantage over conventional single-channel scanned detection (and may even be disadvantageous in signal-to-noise ratio for small peaks) when conducted in the *uv/vis* range, due to source-limited noise.

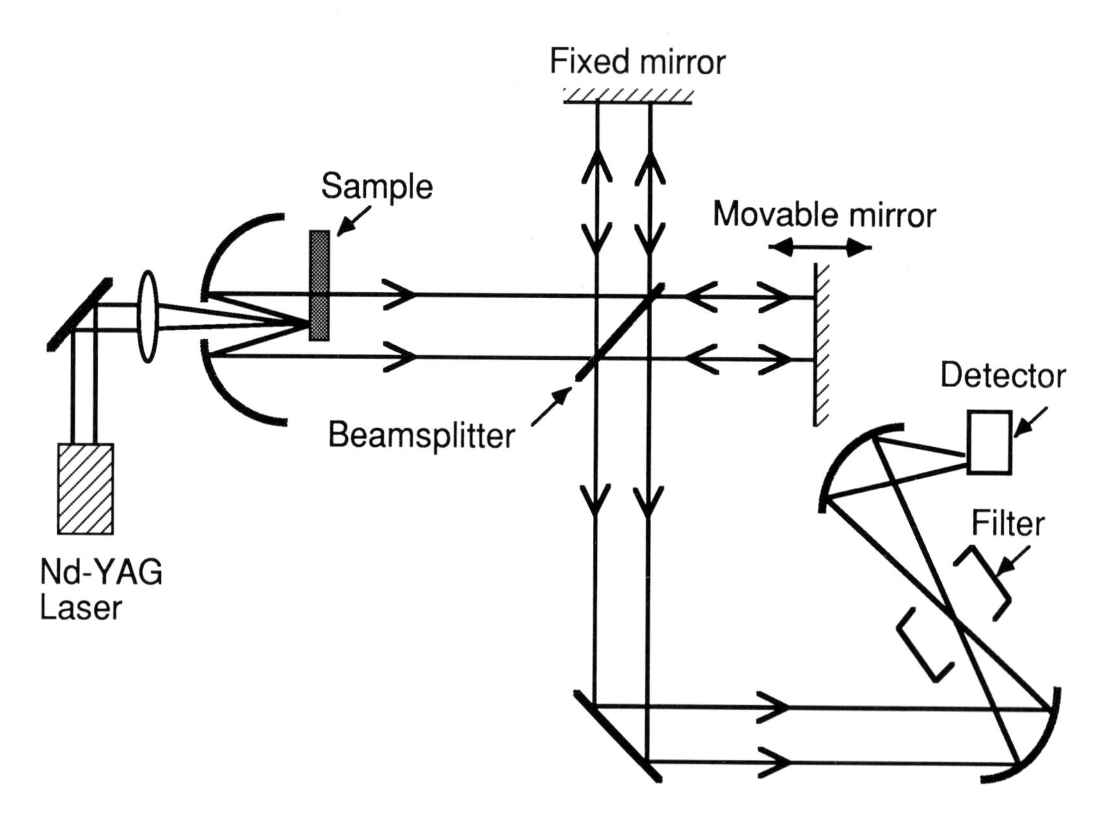

Figure 9.15 Optical diagram of an FT/Raman spectrometer. 1.06 μm laser irradiation elicits Raman scattered light which is collected by a parabolic mirror and then passed through a Michelson interferometer. An optical filter then selectively reduces the intensity of the Rayleigh centerband before detection of the interferogram. (Adapted, with permission, from B. Chase, *Anal. Chem.* **1987**, *59*, 881A-889A.)

If, on the other hand, the experiment is conducted in the *infrared* (say, by use of Nd/YAG laser excitation at 1.06 μm), then the noise becomes detector-limited, and the Fellgett advantage is regained (compared to a single-channel scanning Raman experiment at the same 1.06 μm wavelength). However, single-channel scanning Raman experiments are conventionally conducted in the uv or visible range, typically at 500-600 nm wavelength. Thus, shifting the experiment to the infrared results in an immediate loss of Raman scattered intensity by about an order of magnitude (Eq. 9.26). Fortunately, part of that loss can be made up by use of higher laser power at the Nd/YAG frequency.

The principal advantage of infrared FT/Raman over visible single-channel scanning Raman spectroscopy is the elimination of fluorescent background, as shown dramatically in Figure 9.16. The fluorescent background in the visible Raman spectrum poses two problems. First, the curved baseline makes quantitation of peak heights or areas more difficult. Second, because the visible noise is generally shot-noise-limited, the noise level is proportional to the square root of the *overall* signal strength. Therefore, a large fluorescent *signal* (even if monochromatic) produces additional baseline *noise* throughout the spectrum (see Figure 5.1). Conduction of the FT/Raman experiment in the infrared (i.e., at a frequency well below the electronic absorption frequencies for anthracene) virtually eliminates the fluorescent background.

In the absence of significant fluorescence, the remaining practical advantage of infrared FT/Raman over visible single-channel scanning Raman spectroscopy is the higher frequency precision of the FT/Raman data. Thus, spectral subtraction (e.g., to remove the spectral contribution from a known component of a mixture) is much improved in the FT experiment.

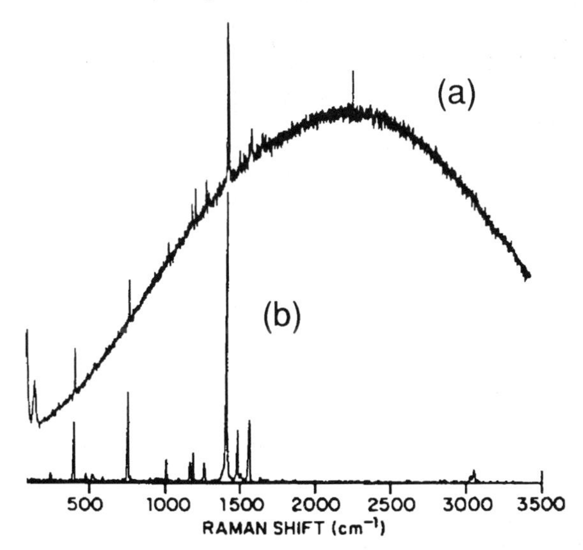

Figure 9.16 Raman spectra of anthracene. (a) Conventional single-channel scanning, with excitation at 514.5 nm (argon-ion laser). (b) FT of the interferogram produced with excitation at 1.06 μm (Nd/YAG laser). Note the elimination of the broad fluorescent background [and its high ("shot") noise level] in the infrared FT/Raman spectrum. (Taken, with permission, from: B. Chase, *J. Amer. Chem. Soc.*. ***1986***, *108*, 7485-7488.)

9.3.3.2 *Hadamard transform Raman spectroscopy*

The principles of Hadamard transform spectroscopy were discussed in Chapter 4.1.3.1. From a practical standpoint, the key feature is how to switch from one "mask" pattern to another (see Figure 4.3). It is clearly impractical to construct (say) 1,023 separate masks and mechanically switch from one to another to generate the corresponding 1,023 intensities recorded by a single broadband detector.

When Hadamard encoding was originally introduced to IR spectroscopy in 1969 by Decker and Harwit, the encoding was performed by translating a *single* $(2N+1)$-channel mask stepwise across a spectral window spanned by N channels. Each mask position then encodes one row of the Hadamard matrix whose transmitted light yields a single observed data point (see Problems). Although the encoding was thereby simplified from N masks to 1 mask, in practice the mechanical stepwise movement of the mask proved difficult to reproduce, and instruments operating on that principle have pretty much disappeared.

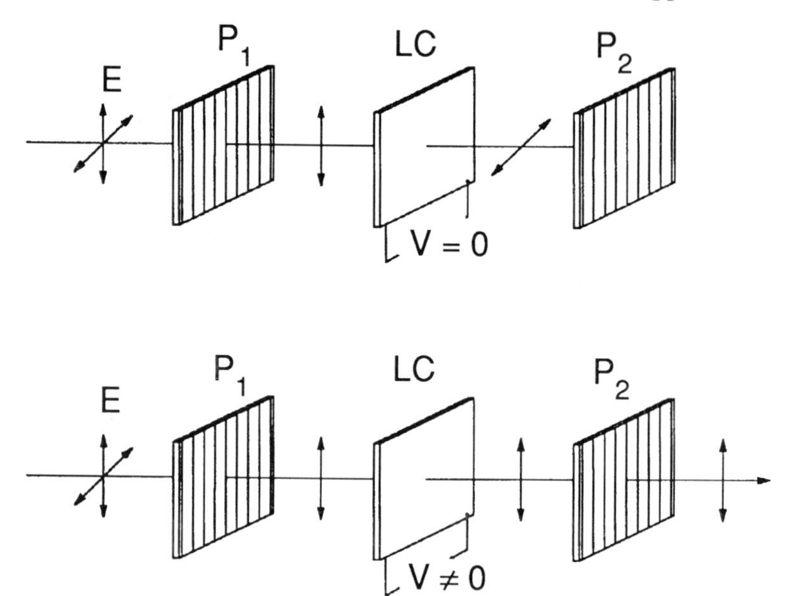

Figure 9.17 Schematic diagram of a liquid crystal electro-optic switch in "closed" (top) and "open" (bottom) position. On passing through the first polarizer, P_1, one of the two linearly-polarized components of unpolarized incident radiation, **E**, is removed. On passing through an unenergized ($V = 0$) liquid crystal cell (LC), the plane of polarization is rotated by 90°, so that none of that light is able to pass through the second polarizer (P_2) in the top diagram. However, if a small ac voltage ($V \neq 0$) is applied to the liquid crystal, then the liquid crystal no longer rotates the plane of polarization of the plane-polarized light from P_1, and the light can now pass through the second polarizer (P_2), as shown in the bottom diagram. The net effect is that incident light is either blocked (top diagram) or fully transmitted (bottom diagram), by turning the ac voltage to the liquid crystal cell off or on, respectively. (Adapted, with permission, from: D. C. Tilotta & W. G. Fateley, *Spectroscopy* **1987**, *3*, 14-25.)

In 1987, Fateley and co-workers devised an ingenious solution to the Hadamard encoding problem, consisting of a single N-channel mask in which the "open" and "shut" slits were achieved by appropriate electro-optical switching, as shown in Figure 9.17. A given slit is rendered either opaque (black) or transmissive (clear) depending on whether or not a voltage is applied to a liquid crystal sandwiched between two (parallel) linear polarizing filters—the same switching principle is used in liquid-crystal displays for wristwatches or computer terminals.)

A representative Hadamard "encodegram" and its corresponding Hadamard transform spectrum are shown in Figure 9.18. A given data point in the encodegram represents the total detected intensity reaching the detector through a given mask. Each of the 127 masks transmits a specified linear combination (with weight factors of 0 or 1 corresponding to open or shut "slits") of the spectral signals throughout the detected spectral range.

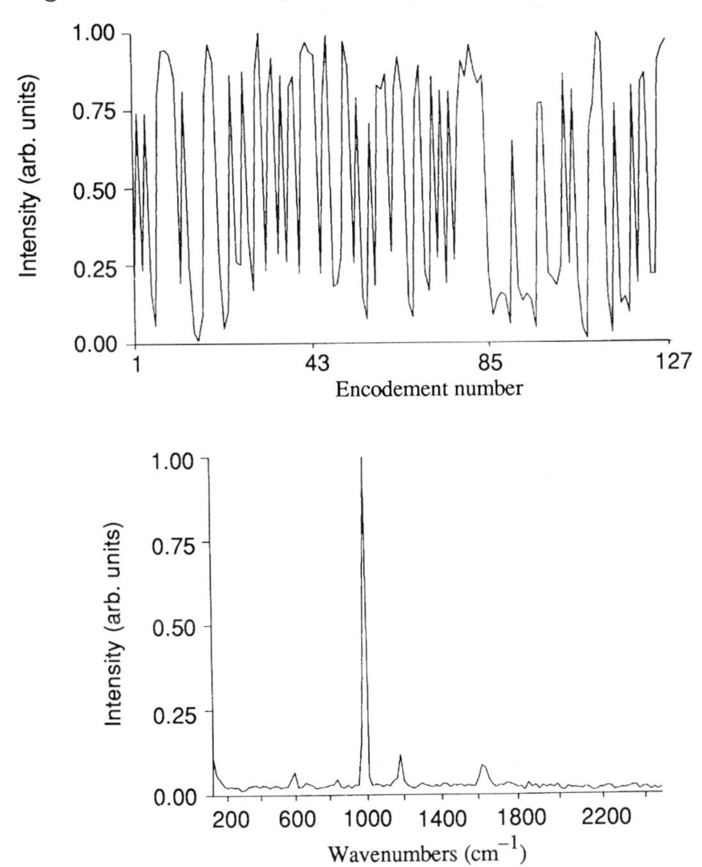

Figure 9.18 Raman spectrum of benzene (bottom) obtained by Hadamard transformation of the "encodegram" (top) measured by a single silicon photodiode detector. Note the effective suppression of the Rayleigh scattering at 514.5 nm (corresponding to 0 cm^{-1} on the abscissa). (Taken, with permission, from: D. C. Tilotta & W. G. Fateley, *Spectroscopy* **1987**, *3*, 14-25)

A Hadamard transform optical spectrometer based on the encodement scheme described above differs from other optical spectrometers in the following ways. First, the Hadamard instrument has no moving parts—a major practical feature for applications requiring a mechanically rugged instrument. Second, one- or two-dimensional Hadamard encoding can be used for *spatial imaging* of an object (see below). Third, when operated in the visible range for emission or scattering (for which the Fellgett advantage may not apply), the Hadamard instrument has the advantage over an FT/interferometer that the signal from one or more intense peaks can be permanently "blacked-out" (by switching that slit "closed" in every encodement mask), thereby eliminating the "multiplex disadvantage" arising from source-limited noise. Thus, the Hadamard technique is particularly well-suited to multiplex detection of dilute species mixed with more concentrated species in atomic emission spectra, or (more generally) for Raman spectroscopy.

Electro-optical encoding is imperfect in at least three ways. First, switching from one mask to another is relatively slow (~10 s^{-1}), so that data acquisition is slower than by FT/interferometry or multidetector methods. Second, actual masks are neither perfectly transmissive nor perfectly opaque. Nevertheless, even imperfectly opaque or transmissive channels can yield a significant multiplex advantage when the number of channels is large (say, 1023). Third, the slits themselves do not have infinitely thin boundaries; i.e., part of the spectrum is blacked out by the opaque region between adjacent "slits" in the mask.

The Hadamard method is perhaps most attractive for Raman spectroscopy, because the mask position(s) at or near the Rayleigh centerband can be blacked-out in each mask code, as shown in Figure 9.19. In this way, Raman spectra can be acquired in the visible range much more simply by Hadamard transform than by FT/interferometry.

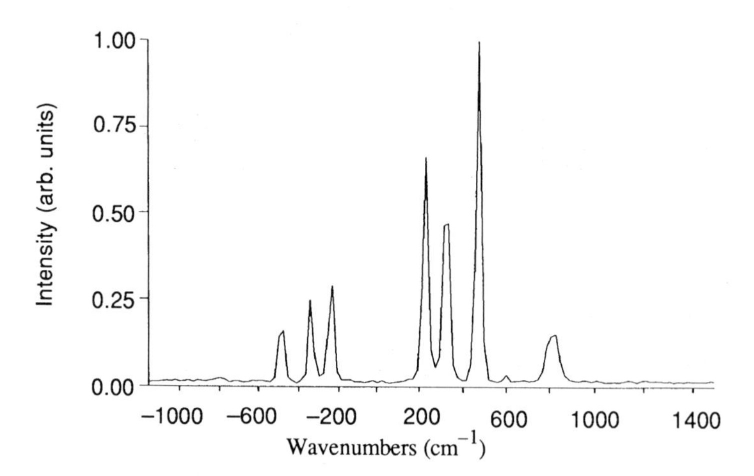

Figure 9.19 Stokes (right) and anti-Stokes (left) Hadamard transform Raman spectra of neat carbon tetrachloride (CCl_4), measured from the frequency of the argon-ion laser frequency at 514.5 nm. Note the effective suppression of the Rayleigh centerband by permanent blanking-off of the Hadamard mask at the position(s) of the Rayleigh peak (i.e., zero-frequency on this scale). (Taken, with permission, from: D. C. Tilotta, R. M. Hammaker, & W. G. Fateley, *Appl. Spectrosc.* **1987**, *41*, 1280-1287)

Although approximately half of the spectral intensity is removed by the Hadamard mask, half of the spectral intensity is also lost at the beamsplitter of an FT/interferometer, so that both techniques offer multiplex advantages of approximately $N^{1/2}/2$ compared to single-channel scanning detection under detector-limited noise conditions (e.g., IR).

9.3.3.3 *Hadamard encodement for spatial imaging*

Hadamard encodement is uniquely suited for spatial imaging, as shown in Figure 9.20. In the example shown here, the individual encodements were varied by physically translating the mask horizontally and/or vertically across the image plane.

A potentially attractive application for Hadamard-encoded spatial imaging, developed by M. D. Morris, is based on Raman scattering, as shown in Figure 9.21 for a crystalline mixture of two compounds. By selecting a wavelength at which Raman scattering for a particular compound is maximal, one can obtain a surface image which reflects the relative amount of that compound in each pixel.

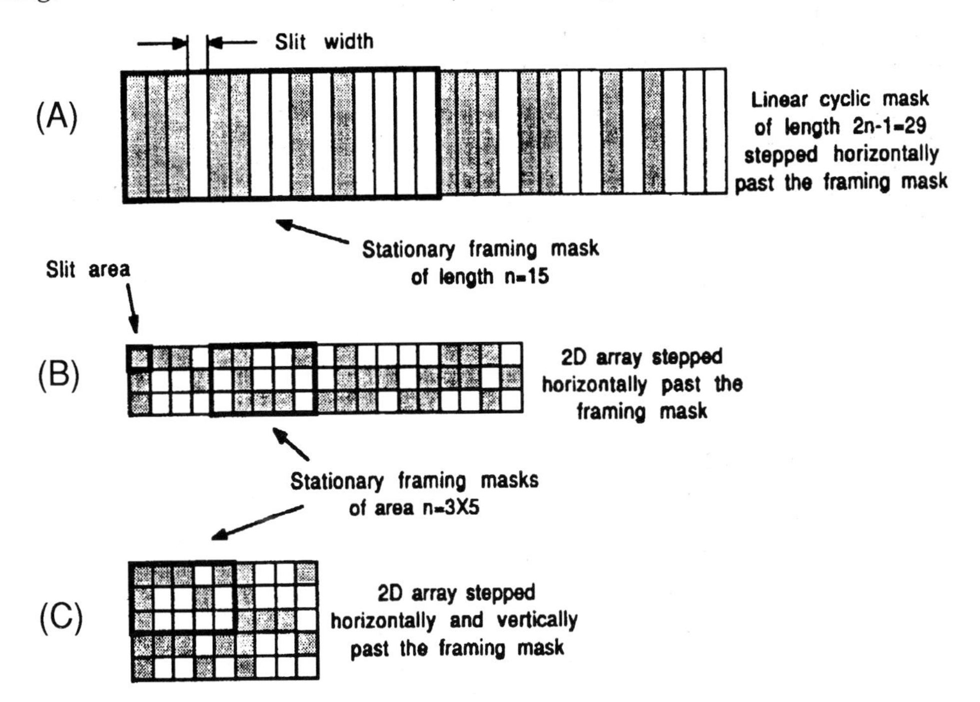

Figure 9.20 One- and two-dimensional Hadamard mask systems. (A) 15-element cyclic mask arranged for line-encodement. Each encodement is formed by shifting the mask one element to the left past the framing mask. (B) 3 x 5 array encodement in two dimensions. The entire array is shifted one element to the left to generate the next Hadamard sequence. (C) A more compact two-dimensional mask, in which both horizontal and vertical translation of the mask are used to generate the Hadamard encodements. [Taken, with permission, from P. J. Treado & M. D. Morris, *Proc. International Laser Symp. IV*, **1989** (in press)]

The images in Figure 9.21 were generated from a 255-element mask, folded into a 15 x 17 element array as shown schematically in Figure 9.20 (C). The two top diagrams in Figure 9.21 have different patterns, because the two kinds of microcrystals occupy different locations on the surface. However, after heating to their melting temperature (by irradiation with 200 mW laser excitation), the melted solid is now homogeneous, and the Hadamard Raman scattering images are the same for both compounds (Figure 9.21, bottom). Hadamard-encoded spatial imaging based on Raman scattering thus provides a simple means for performing spectroscopically-selective microscopy (see Further Reading).

N,N-dimethyl-*p*-nitroaniline (1310 cm^{-1})

Benzoic acid (992 cm^{-1})

5 mW imaging

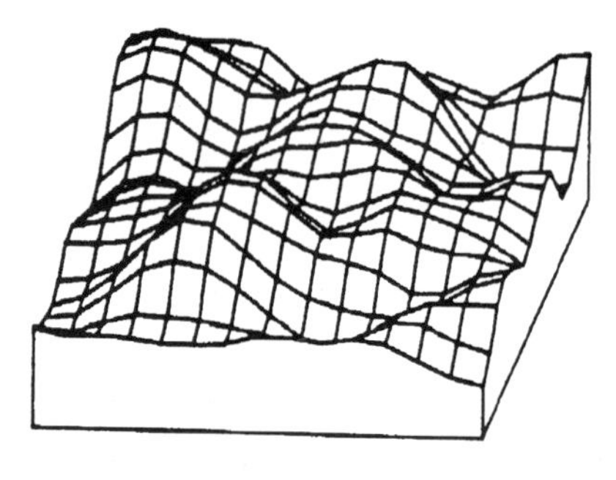

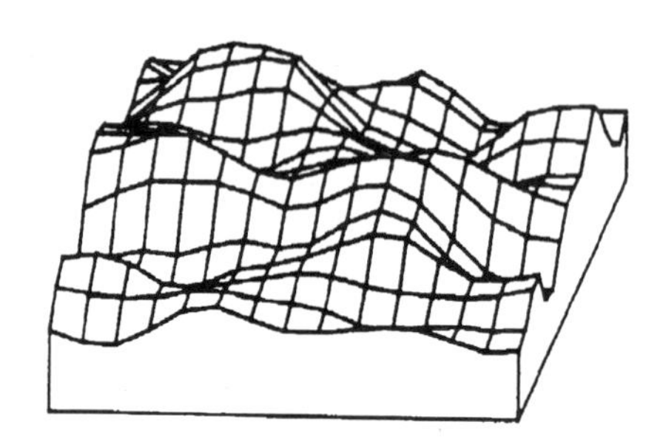

N,N-dimethyl-*p*-nitroaniline (1310 cm^{-1})

Benzoic acid (992 cm^{-1})

200 mW imaging

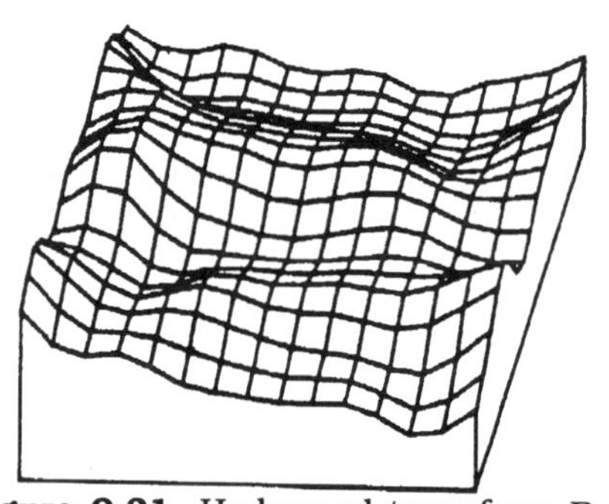

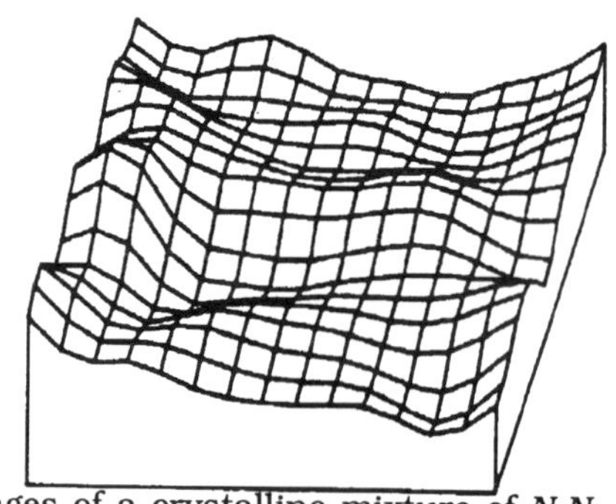

Figure 9.21 Hadamard transform Raman images of a crystalline mixture of *N,N*-dimethyl-*p*-nitroaniline (detected at 1310 cm^{-1}) and benzoic acid (detected at 992 cm^{-1}). Top: The two compounds are spectrally and spatially resolved. The image was obtained at 4.4 μm spatial resolution and 10 cm^{-1} spectral resolution, with an Ar^+ ion laser excitation at 5 mW. Bottom: Same mixture after exposure to 200 mW excitation power. The spatial profiles of both compounds are now essentially identical in the homogeneous crystalline melt. (Unpublished spectra provided by M. D. Morris)

Further Reading

R. J. Bell, *Introductory Fourier Transform Spectrometry*, Academic Press, NY, 1972. (Good treatment of FT interferometry history and fundamentals)

B. Chase, *Anal. Chem.* **1987**, *59*, 881A-889A. (Good introduction and survey of FT/Raman spectroscopy)

F. A. Cotton, *Chemical Applications of Group Theory*, Interscience, NY, 2nd ed., 1971. (Still the most readable introduction to group theory as used to classify molecular vibrational symmetry)

P. R. Griffiths & J. A. de Haseth, *Fourier Transform Infrared Spectrometry*, Wiley-Interscience, NY, 1986. (Comprehensive coverage of FT/IR, with emphasis on practical details and applications)

M. Harwit & N. J. A. Sloane, *Hadamard Transform Optics*, Academic Press, NY, 1979. (The fundamental reference for Hadamard transform spectroscopy)

K. Nakamoto, *Infrared and Raman Spectra of Inorganic and Coordination Compounds*, Wiley-Interscience, NY, 3rd Ed., 1978. (Good treatment of calculation of normal mode frequencies, with a wide range of chemical examples)

D. C. Tilotta & W. G. Fateley, *Spectroscopy* **1987**, *3*, 14-25. D. C. Tilotta, R. M. Hammaker, and W. G. Fateley, *Appl. Spectrosc.* **1987**, *41*, 727-734. D. C. Tilotta, R. M. Hammaker, and W. G. Fateley, *Appl. Optics* **1987**, *26*, 4285-4292. [The fundamental papers describing the development of Hadamard transform spectrometry for both absorption and emission (e.g., Raman)]

P. J. Treado & M. D. Morris, *Appl. Spectrosc.* **1989**, *43*, 190-193. (Good explanation of Hadamard transform Raman spatial imaging)

E. B. Wilson, J. C. Decius, & P. C. Cross, *Molecular Vibrations*, McGraw-Hill, 1955. (Classic text, written by the inventor of the FG-matrix method now used universally for computation of vibrational normal mode frequencies)

L. A. Woodward, *Introduction to the Theory of Molecular Vibrations and Vibrational Spectroscopy*, Oxford, Clarendon Press, 1972. (Good classical mechanical introduction, followed by quantum mechanical theory, with several worked-out examples of the mathematical methods involved)

Problems

9.1 Combine Eqs. 9.1 and 9.2 to yield Eqs. 9.4. Then show that Eqs. 9.14 result from Eqs. 9.2 and 9.13. These exercises complete the derivation of the normal mode motions for a diatomic molecule.

9.2 Predict the relative magnitudes of Stokes and anti-Stokes Raman emission, for vibrational energy level spacings of 100 and 1000 cm^{-1}, for a sample at a temperature of 300 K.

9.3 Show that the addition of the electric fields from the undelayed and delayed components of the light beam emerging from a Michelson interferometer produces a beam whose detected intensity is sinusoidally modulated (Eq. 9.21). Hint: remember that the IR detector records intensity, not amplitude.

9.4 Explain how one might phase-correct an FT/IR spectrum by suitable manipulation of the interferogram data prior to FT (rather than by frequency-domain phase correction, as in Chapter 2.4.2).

9.5 As noted in the text, "shot" noise is proportional to the square root of signal magnitude. Thus, a spectrum with approximately uniform signal frequency-domain signal magnitude will exhibit a "multiplex disadvantage" of a factor of $\sim 1/\sqrt{N}$ in signal-to-noise ratio compared to a single-channel scanning instrument. However, suppose that instead of spending an equal amount of time at each movable mirror position, we arrange to spend just enough time at each retardation to allow the accumulated signal magnitude to fill the full analog-to-digital converter word length. Then the signal at each retardation will be the same (and so will the noise), so that the interferogram noise is now effectively independent of signal magnitude! Before FT, one will of course have to weight each data value according to the time it took to fill its ADC word. Can such a "step-scan" interferometry scheme overcome the "multiplex" disadvantage of FT/uv/vis spectrometry?

9.6 Hadamard spectroscopy performed with a succession of different code masks would be impractical if one had to introduce a completely new mask for each row of the Hadamard code matrix.

(a) Therefore, for an $N = 3$ channel Hadamard encodement, devise a single 5-element linear mask, such that exposure of elements 1-3, 2-4, and 3-5 generate the three desired Hadamard codes for the $N = 3$ case.

(b) Now try to come up with a 15-element linear mask which will accomplish the same thing for the $N = 7$ case (refer to Figure 4.3 for the needed masks).

(c) From these two examples, see if you can see the pattern for the general case of a $(2N - 1)$-element linear mask that will produce all of the N rows of an N-channel Hadamard encodement. This form of Hadamard encodement is the basis for the Hadamard transform Raman spatial imaging applications discussed at the end of this Chapter. It was also the method employed by Harwit and Decker in the first Hadamard transform IR spectrometer.

Solutions to Problems

9.1 See Nakamoto, Woodward, and Wilson *et al.* references for diatomic and more complex cases.

9.2 Assume that the relative Stokes and anti-Stokes emission probabilities are determined wholly by the relative populations of the $v = 0$ and $v = 1$ energy levels in Figure 9.5. We need then simply compute the relative populations from the Boltzmann factor, in which $\Delta E = hc/\lambda$, and $(1/\lambda) = 100$ or 1000 cm^{-1}:

$$\frac{N_{(v=1)}}{N_{(v=0)}} = \exp(-\Delta E / kT)$$

$$= \exp\left(-\frac{(6.62 \times 10^{-34}\text{ J s})(3.00 \times 10^{8}\text{ m s}^{-1})(1000\text{ cm}^{-1})(100\text{ cm m}^{-1})}{(1.38 \times 10^{-23}\text{ J K}^{-1})(300\text{ K})}\right)$$

= 0.0082, or a population ratio of ~120:1 for 1000 cm^{-1} vibration energy,

= 0.62, or a population ratio of ~1.6:1 for 100 cm^{-1} vibration energy.

Thus, the anti-Stokes peaks become much less intense than the Stokes peaks as the vibrational frequency increases (see Figure 9.19).

9.3 The argument here is essentially the same as that for the non-linear detector in Chapter 4.4.2. The electric field of each of the two light beams recombining at the beamsplitter may be represented as two cosinusoids of different phase: say, $\cos \omega t$ and $\cos [\omega (t - \delta / c)]$, in which c is the speed of light and δ is the path length difference. The electric field of the recombined beam is the sum of the electric fields of the two component beams. However, since the detector reports the *square* of the recombined beam electric field, the detected signal will be of the form, $(\cos A + \cos B)^2$, which includes a term proportional to $\cos A \cos B$, which in turn may be represented as a sum of $\cos(A + B)$ and $\cos(A - B)$. Thus, the final detected beam intensity is modulated at the difference frequency between the two component waves, as in Eq. 9.21. Thus, although the interferometer output is detected as *intensity* (i.e., an FT *power* spectrum), the incident beam intensity is in fact modulated sinusoidally, and the *power* spectrum therefore is characterized by a phase which may vary with frequency (see text).

9.4 The trick is to notice that frequency-domain phase correction consists of *multiplication* of the complex FT data by a suitable complex phase factor, and that *multiplication* in one domain is equivalent to *convolution* in the FT domain. Thus, one determines the magnitude and phase spectrum for an interferogram extending somewhat beyond the centerburst on one side, and then convolves the FT of the phase spectrum with the measured interferogram. For additional details and limitations of the method, see Griffiths & de Haseth, pp. 94-97.

9.5 See R. Williams, *Appl. Spectrosc.* **1989**, *43*, 235-238.

9.6 For the $N = 3$ case, a suitable 5-element mask is: $\underline{1\ 1\ 0}\ 1\ 1$, in which the first row of the Hadamard encodement (i.e., the portion of the mask which is placed over the spectral window) is underlined.

For the $N = 7$ case, a suitable 13-element mask (refer to Problem 4.1) is:

$\underline{1\ 1\ 1\ 0\ 1\ 0\ 0}\ 1\ 1\ 1\ 0\ 1\ 0$

The pattern should now be clear: we obtain a suitable $(2N + 1)$-element mask by extending the N-element sequence of the first row of the Hadamard encodement matrix by repeating its first $(N - 1)$ elements.

CHAPTER 10

Epilog: Fourier transforms in other types of spectroscopy

This book has been devoted to three of the most popular forms of Fourier transform spectroscopy. In this very brief epilog, we simply note that FT methods have been extended into many other types of spectroscopy, including (but certainly not limited to) nuclear quadrupole resonance, a.c. dielectric response, electron-nuclear double resonance, electron spin resonance (see K. Holczer & D. Schmalbein reference), pure rotational (microwave) spectroscopy (see Dreizler), a.c. electrochemical response, mu spin resonance, and others (see Further Reading). Fourier transform methods have also been adapted to both time-of-flight (see Knorr *et al.*) and all-electric quadrupole ion trap (see Syka & Fies reference) forms of mass spectrometry.

Two-dimensional correlation spectroscopy (COSY) experiments have been extended to microwave pure rotational (see Vogelsanger *et al.*) and electron paramagnetic resonance (see Gorcester & Freed) applications.

Interferometry has been extended to the microwave range (see Ramsey & Whitten). FT methods with pseudorandom noise modulation have even been used for multiplex chromatography (Phillips *et al.*), by modulating the sample introduced onto the column rather than using the usual single injection.

Hadamard imaging, originally devised for infrared spectroscopy (see Harwit & Sloane reference) has been extended to Raman (Chapter 9), NMR (see Bolinger & Leigh reference), and photoelectron spectroscopy (see Hoffman *et al.* reference). More general spatial imaging codes have been applied to x-ray images (see Skinner reference). The interested reader is encouraged to seek out recent reviews of the above and other areas of application for Fourier and Hadamard transform spectroscopy and imaging.

Finally, any discussion of Fourier transforms in chemistry, physics, and engineering, must acknowledge their use in the generation of an electron density image from its corresponding x-ray diffraction image from a single crystal. As described in more detail elsewhere (e.g., Marshall, 1978), ray-tracing of the light from an object through a focusing lens system generates two kinds of images: a *diffraction* image and a *true* image, which are related by a Fourier transform. In the optical (infrared/visible/ultraviolet) range, lenses are available for conversion of the diffraction image into the true image. However, since ordinary lenses are not available at x-ray wavelengths (see Chapter 1.4.2), a mathematical (three-dimensional) Fourier transform is required to convert the diffraction image to a true image. The regular spacings between unit cells in a single crystal effectively produce equally spaced spatial sampling of the object, to give a three-dimensional diffraction pattern whose intensity "spot" spacings are inversely related to the unit cell dimensions of the object. The relative intensities of the spots in the diffraction image represent an "interferogram-type" code for the relative locations (and atomic numbers, since x-ray scattered intensity varies as the square of the number of electrons per atom) of the various atoms in the unit cell. In fact, the situation is much worse than in FT spectroscopy, because the x-ray diffraction pattern must be detected from its scattered radiation *energy*, so that there is no direct method for determining the *phase* of each FT component frequency. Thus, one is forced to analyze the results from a "Patterson function", which is the Fourier transform of the detected *magnitude* (rather than *magnitude* and *phase*) of the diffraction (three-dimensional "interferogram") signal. For all but the smallest molecules, it is usual to introduce a "heavy atom" (i.e., one with large atomic number) whose location is found from the Patterson map and then used to introduce phase information into the rest of the diffraction pattern. A related experiment (EXAFS, for x-ray absorption fine structure) provides less detailed images from polycrystalline samples. The interested reader is referred to any of several modern texts on x-ray diffraction for more details.

Further Reading

L. Bolinger & J. S. Leigh, *J. Magn. Reson.* **1988**, *80*, 162-167. (Hadamard encoding o magnetic resonance images)

P. Connes, *Mikrochim. Acta (Wien)* **1987**, *III*, 337-352. (Historical review of the development of FT/interferometry, FT/NMR, and FT/ICR spectroscopy)

H. Dreizler, *Mol. Phys.* **1986**, *59*, 1-28. (Review of FT microwave rotational spectroscopy of gases)

L. M. Faires, *Anal. Chem.* **1986**, *58*, 1023A-1034A. (Review of FT/uv-vis interferometry for atomic spectroscopy)

J. Gorcester and J. H. Freed, *J. Chem. Phys.* **1988**, *88*, 4678-4693. (Two-dimensional FT/ESR COSY)

M. Harwit & N. J. A. Sloane, *Hadamard Transform Optics*, Academic Press, NY, 1979, 249 pp. (Fundamentals of Hadamard codes for spectroscopy and spatial imaging)

D. G. Hoffmann, A. Proctor and D. M. Hercules, *Appl. Spectrosc.* **1989**, *43*, 899-908. (Spatially resolved ESCA by use of Hadamard masks)

K. Holczer & D. Schmalbein, *BRUKER Report*, **1987** (1), p. 22. (Pulsed FT/EPR spectrometer)

T. Kallard, *Laser Art and Optical Transforms*, Optosonic Press, NY, 1979, 170 pp. (Fascinating collection of two-dimensional Fourier transforms produced by optical methods, and introduction to holography)

F. J. Knorr, R. L. Eatherton, W. F. Siems, & H. H. Hill, Jr., *Anal. Chem.* **1985**, *57*, 402-406. (FT ion mobility spectrometry)

A. G. Marshall, *Biophysical Chemistry: Principles, Techniques, and Applications*, Wiley, NY, 1978, 812 pp. (See Chapter 22 for a qualitative introduction to x-ray diffraction analysis.)

A. G. Marshall, in *Physical Methods in Modern Chemical Analysis*, Vol. 3, Ed. T. Kuwana, Academic Press, 1983, 57-135. (A condensed version of the following more comprehensive treatment)

A. G. Marshall, Ed. *Fourier, Hadamard, and Hilbert Transforms in Chemistry*, Plenum, NY, 1982, 562 pp. (Chapters on FT/ICR, FT/NQR, FT/dielectric, FT microwave, FT/ENDOR, FT/μ SR, FT/IR, FT/uv-vis, and FT/Faradaic admittance spectroscopy, FT methods in spectroelectrochemistry, and Hilbert transforms)

J. B. Phillips, D. Luu, J. B. Pawliszyn, & G. C. Carle, *Anal. Chem.* **1985**, *57*, 2779-2787. (Multiplex gas chromatography)

J. M. Ramsey & W. B. Whitten, *Rev. Sci. Instrum.* **1986**, *57*, 1329-1337. (Microwave Michelson interferometer)

G. K. Skinner, *Scientific American* **1988** (8), 84-89. (x-ray spatial imaging with coded masks)

J. E. P. Syka & W. J. Fies, Jr., *35th Amer. Soc. Mass Spectrometry Ann. Conf. on Mass Spectrometry & Allied Topics*, Denver, CO, 1987, pp. 767-768. (FT quadrupole ion-trap mass spectrometry)

B. Vogelsanger, M. Andrist, & A. Bauder, *Chem. Phys. Lett.* **1988**, *144*, 180-186. (Microwave two-dimensional COSY pure rotational spectrometry)

Appendix A

Integrals and theorems for FT applications

Logarithms

Definitions

If $\qquad y = a^x$

then $\qquad x = \log_a y \qquad$ (A.1)

In particular, $\log_e x = \ln x \qquad$ (A.2)

Properties

$$\log (a \cdot b) = \log a + \log b \qquad (A.3)$$

$$\log\left(\frac{a}{b}\right) = \log a - \log b \qquad (A.4)$$

$$\log x^n = n \log x \qquad (A.5)$$

Conversion from one base to another

If $\qquad a^y = x$

and $\qquad a^z = w$

then $\qquad a = w^{1/z}$

Also, $\qquad x = w^{y/z}$

and $\qquad \log_w x = \frac{y}{z} = \frac{\log_a x}{\log_a w} \qquad$ (A.6)

For example, $\log_{10} x = \log_{10} e \cdot \log_e x = 0.434 \log_e x$

Series expansions and approximations

Taylor expansion [approximates $f(x)$ in the vicinity of $x = a$]

$$f(x) = f(a) + \frac{x-a}{1!} f'(a) + \frac{(x-a)^2}{2!} f''(a) + \cdots + \frac{(x-a)^{n-1}}{(n-1)!} f^{(n-1)\prime}(a) + R_n \qquad (A.7)$$

$$R_n = \frac{(x-a)^n}{n!} f^{n\prime}(r) \ , \quad a < r < x$$

Factorials

$$N! = N \cdot (N-1) \cdot (N-2) \cdots 3 \cdot 2 \cdot 1 \qquad (A.8)$$

$$0! = 1 \text{ (definition)} \qquad (A.9)$$

$$\ln N! = N \ln N - N, \text{ for large } N \text{ (Stirling approximation)} \qquad (A.10)$$

Power series representations of simple functions

$$\exp(\pm x) = 1 \pm \frac{x^2}{2!} \pm \frac{x^3}{3!} \pm \frac{x^4}{4!} \pm \cdots + \frac{(\pm x)^n}{n!}, \; x \text{ may be complex} \quad (A.11)$$

$$\cos x = 1 - \frac{x^2}{2!} + \frac{x^4}{4!} - \frac{x^6}{6!} \pm \cdots \quad (A.12)$$

$$\sin x = x - \frac{x^3}{3!} + \frac{x^5}{5!} - \frac{x^7}{7!} \pm \cdots \quad (A.13)$$

$$\ln(1 \pm x) = \pm x - \frac{x^2}{2} \pm \frac{x^3}{3} - \frac{x^4}{4} \pm \cdots ; \quad -1 < x < +1 \quad (A.14)$$

Binomial expansion

$$(a+b)^n = a^n + n\,a^{(n-1)}\,b + \cdots + \frac{n!}{m!(n-m)!}\,a^{(n-m)}\,b^m + \cdots + n\,ab^{(n-1)} + b^n \quad (A.15)$$

Trigonometric functions

Definition

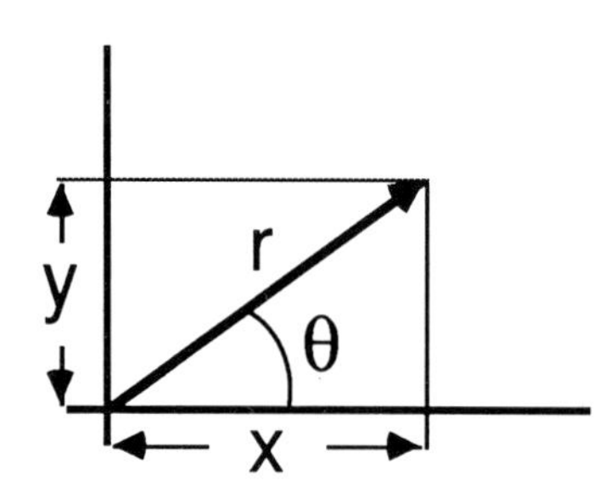

$$\sin\theta = \frac{y}{r} \quad (A.16)$$

$$\cos\theta = \frac{x}{r} \quad (A.17)$$

$$\tan\theta = \frac{y}{x} = \frac{\sin\theta}{\cos\theta} \quad (A.18)$$

$$\cos(-\theta) = \cos\theta \quad (A.19)$$

$$\sin(-\theta) = -\sin\theta \quad (A.20)$$

$$2\pi \text{ radians} = 360 \text{ degrees} \quad (A.21)$$

Relation between cartesian and polar coordinates

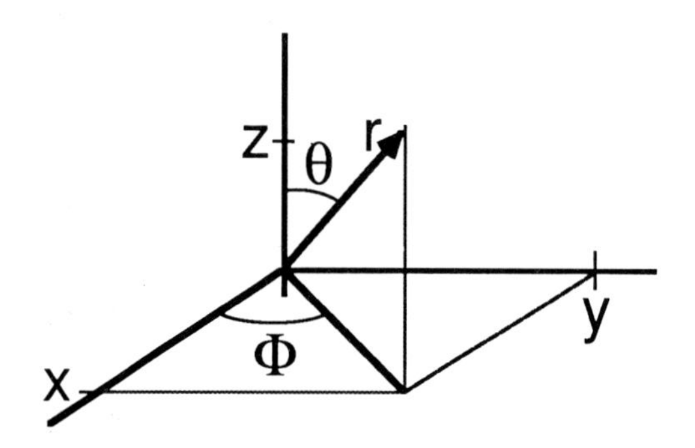

$$x = r\sin\theta\cos\phi \quad (A.22)$$

$$y = r\sin\theta\sin\phi \quad (A.23)$$

$$z = \cos\theta \quad (A.24)$$

Trigonometric identities

$$\sin(a \pm b) = \sin a \cos b \pm \cos a \sin b \tag{A.25}$$

$$\cos(a \pm b) = \cos a \cos b \mp \sin a \sin b \tag{A.26}$$

$$\sin a \cos b = \frac{1}{2}[\sin(a+b) + \sin(a-b)] \tag{A.25a}$$

$$\sin a \sin b = \frac{1}{2}[\cos(a-b) - \cos(a+b)] \tag{A.26a}$$

$$\cos a \cos b = \frac{1}{2}[\cos(a+b) + \cos(a-b)] \tag{A.26b}$$

$$\cos^2 a + \sin^2 a = 1 \tag{A.27}$$

$$\cos 2a = \cos^2 a - \sin^2 a \tag{A.28}$$

$$= 2\cos^2 a - 1 = 1 - 2\sin^2 a \tag{A.28a}$$

$$\sin 2a = 2 \sin a \cos a \tag{A.29}$$

$$\cos\left[\frac{\pi}{2} - a\right] = \sin a \tag{A.30}$$

$$\sin\left[\frac{\pi}{2} - a\right] = \cos a \tag{A.31}$$

Calculus : derivatives

Definition

$$\frac{d f(x)}{dx} = f'(x); \quad \frac{d^2 f(x)}{dx^2} = f''(x) \text{ ; and so on} \tag{A.32}$$

Derivatives of simple functions (a is a constant in Eqs. A.33 to A.37):

$$\frac{d x^a}{dx} = a x^{(a-1)} \tag{A.33}$$

$$\frac{d \exp(ax)}{dx} = a \exp(ax) \tag{A.34}$$

$$\frac{d \ln ax}{dx} = \frac{1}{x} \tag{A.35}$$

$$\frac{d \sin ax}{dx} = a \cos ax \tag{A.36}$$

$$\frac{d \cos ax}{dx} = -a \sin ax \tag{A.37}$$

Derivative of a sum (*u* and *v* are functions of *x* in Eqs. A.38 to A.42)

$$\frac{d(u+v)}{dx} = \frac{du}{dx} + \frac{dv}{dx} \tag{A.38}$$

Derivative of a product

$$\frac{d(u \cdot v)}{dx} = u\frac{dv}{dx} + v\frac{du}{dx} \tag{A.39}$$

Derivative of a quotient

$$\frac{d\left(\frac{u}{v}\right)}{dx} = \frac{v\frac{du}{dx} - u\frac{dv}{dx}}{v^2} \tag{A.40}$$

Chain rule

$$\frac{d\big(f(u)\big)}{dx} = \frac{df}{du} \cdot \frac{du}{dx} \tag{A.41}$$

$$\frac{d\big(f(u, v, \cdots)\big)}{dx} = \frac{\partial f}{\partial u} \cdot \frac{du}{dx} + \frac{\partial f}{\partial v} \cdot \frac{dv}{dx} + \cdots \tag{A.42}$$

Calculus : integrals

Basic properties

$$\int_a^b f(x)\, dx = -\int_b^a f(x)\, dx \tag{A.43}$$

$$\int_a^b \big(f_1(x) + f_2(x)\big)\, dx = \int_a^b f_1(x)\, dx + \int_a^b f_2(x)\, dx \tag{A.44}$$

$$\int_a^b c\, f(x)\, dx = c\int_a^b f(x)\, dx; \quad c \text{ is a constant} \tag{A.45}$$

$$\int_a^b f(x)\, dx = \int_a^c f(x)\, dx + \int_c^b f(x)\, dx \tag{A.46}$$

Even and odd functions

$$f(x) \text{ is even if } f(-x) = f(x) \tag{A.47}$$

Examples: x^2, $\cos x$, $\sin^2 x$

$$f(x) \text{ is odd if } f(-x) = -f(x) \tag{A.48}$$

Examples: x, $\sin x$

Let $E(x)$ be an even function and $O(x)$ be an odd function; then

$$E(x) \cdot E(x) = \text{even} \tag{A.49}$$

$$E(x) \cdot O(x) = \text{odd} \tag{A.50}$$

$$O(x) \cdot O(x) = \text{even} \tag{A.51}$$

Integrals of even and odd functions have the properties,

$$\int_{-a}^{a} E(x)\, dx = 2 \int_{0}^{a} E(x)\, dx \tag{A.52}$$

$$\int_{-a}^{a} O(x)\, dx = 0 \tag{A.53}$$

Change of variables

Given the integral, $\int_{x=a}^{x=b} f(x)\, dx$, suppose that we are asked to change variables from x to $u = u(x)$ and then to integrate over u, from $u = u(a)$ to $u = u(b)$. We first express x as a function of u, $x = x(u)$, and then use the relation:

$$\int_{x=a}^{x=b} f(x)\, dx = \int_{u=u(a)}^{u=u(b)} f\,[x(u)]\, \frac{dx}{du}\, du \tag{A.54}$$

Integration by parts (for generating integrals not listed in the table)

$$\int_{a}^{b} u(x) \cdot v'(x)\, dx = u(x) \cdot v(x) \Big|_{a}^{b} - \int_{a}^{b} v(x) \cdot u'(x)\, dx \tag{A.55}$$

Indefinite integrals of selected functions. Note: Each indefinite integral is defined only to within a constant which may be evaluated when definite integration limits are specified.

$$\int x^n\, dx = \frac{x^{(n+1)}}{n+1}, \qquad n \neq -1 \tag{A.56}$$

$$\int \frac{dx}{x} = \ln|x|, \quad x \neq 0 \tag{A.57}$$

$$\int \sin ax \, dx = -\frac{\cos ax}{a} \tag{A.58}$$

$$\int \cos ax \, dx = \frac{\sin ax}{a} \tag{A.59}$$

$$\int \exp(ax) \, dx = \frac{\exp(ax)}{a} \tag{A.60}$$

$$\int \ln ax \, dx = x \ln ax - x \tag{A.61}$$

$$\int x \exp(ax) \, dx = \frac{ax - 1}{a^2} \exp(ax) \tag{A.62}$$

$$\int \sin ax \cos bx \, dx = -\frac{\cos (a+b)x}{2(a+b)} - \frac{\cos (a-b)x}{2(a-b)}, \quad a^2 \neq b^2 \tag{A.63}$$

[For integrals involving $\sin ax \ \sin bx$ or $\cos ax \ \cos bx$, first use the trigonometric identities A.26a and A.26b; then use integrals A.58 and A.59.]

$$\int \exp(ax) \sin bx \, dx = \frac{\exp(ax)}{a^2 + b^2} (a \sin bx - b \cos bx) \tag{A.64}$$

$$\int \exp(ax) \cos bx \, dx = \frac{\exp(ax)}{a^2 + b^2} (a \cos bx + b \sin bx) \tag{A.65}$$

By applying Eqs. A.26a and A.26b to Eqs. A.64 and A.65, one can show that

$$\int \exp(ax) \sin bx \ \sin cx \, dx = \frac{\left((b-c)\sin(b-c)x + a\cos(b-c)x\right)}{2\left(a^2 + (b-c)^2\right)} \exp(ax) - \frac{\left((b+c)\sin(b+c)x + a\cos(b+c)x\right)}{2\left(a^2 + (b+c)^2\right)} \exp(ax) \tag{A.66}$$

$$\int \exp(ax) \cos bx \ \cos cx \, dx = \frac{\left((b-c)\sin(b-c)x + a\cos(b-c)x\right)}{2\left(a^2 + (b-c)^2\right)} \exp(ax) + \frac{\left((b+c)\sin(b+c)x + a\cos(b+c)x\right)}{2\left(a^2 + (b+c)^2\right)} \exp(ax) \tag{A.67}$$

Finally, by applying Eq. A.25a to Eq. A.64, one can show that

$$\int \exp(ax) \sin bx \ \cos cx \ dx = \frac{\left(a \ \sin(b-c)x \ - \ (b-c) \cos(b-c)x \right)}{2\left(a^2 + (b-c)^2\right)} \exp(ax)$$

$$+ \frac{\left(a \ \sin(b+c)x \ - \ (b+c) \cos(b+c)x \right)}{2\left(a^2 + (b+c)^2\right)} \exp(ax) \quad \text{(A.68)}$$

Definite integrals of selected functions

$$\int_0^\infty \exp(-ax)\, dx = \frac{1}{a} \quad \text{(A.69)}$$

$$\int_0^\pi \sin^2 mx\, dx = \int_0^\pi \cos^2 mx\, dx = \frac{\pi}{2} \quad \text{(A.70)}$$

$$\int_0^\infty \exp(-a^2x^2)\, dx = \frac{\sqrt{\pi}}{2a} \quad \text{(A.71)}$$

$$\int_0^\infty \exp(-a^2x^2) \cos bx\, dx = \frac{\sqrt{\pi}}{2a} \exp\left[\frac{-b^2}{4a^2}\right] \quad \text{(A.72)}$$

$$\int_0^\infty x \exp(-x^2)\, dx = \frac{1}{2} \quad \text{(A.73)}$$

$$\int_0^\infty x^2 \exp(-x^2)\, dx = \frac{\sqrt{\pi}}{4} \quad \text{(A.74)}$$

$$\int_0^\infty \exp(-ax) \cos bx\, dx = \frac{a}{a^2+b^2} \quad \text{(A.75)}$$

$$\int_0^\infty \exp(-ax) \sin bx\, dx = \frac{b}{a^2+b^2} \quad \text{(A.76)}$$

$$\operatorname{erf} x = \frac{2}{\sqrt{\pi}} \int_0^x \exp(-t^2)\, dt = \text{error function;} \quad x \text{ may be complex} \quad \text{(A.77)}$$

$$\operatorname{erfc} x = \frac{2}{\sqrt{\pi}} \int_x^\infty \exp(-t^2)\, dt = 1 - \operatorname{erf} x = \text{error function complement} \quad \text{(A.78)}$$

$$\int \exp\left(-\left(ax^2+2bx+c\right)\right)\,dx = \frac{\sqrt{\pi/a}}{2}\exp\left(\frac{b^2-ac}{a}\right)\,\mathrm{erf}\,[\,x\,\sqrt{a}\,] \tag{A.79}$$

$$\int_0^\infty \frac{\exp\,(-\,at^2)}{t^2+x^2}\,dt = \frac{\pi}{2x}\,\exp(ax^2)\;\mathrm{erfc}(x\sqrt{a})\;, \qquad a>0,\quad x>0 \tag{A.80}$$

Fresnel integrals

$$\mathrm{C}(x) = \int_0^x \cos\left(\frac{\pi}{2}t^2\right)dt \tag{A.81}$$

$$\mathrm{S}(x) = \int_0^x \sin\left(\frac{\pi}{2}t^2\right)dt \tag{A.82}$$

$$\mathrm{C}_1(x) = \sqrt{\frac{2}{\pi}}\int_0^x \cos t^2\,dt \tag{A.83}$$

$$\mathrm{S}_1(x) = \sqrt{\frac{2}{\pi}}\int_0^x \sin t^2\,dt \tag{A.84}$$

$$\mathrm{C}_2\,(x) = \frac{1}{\sqrt{2\pi}}\int_0^x \left(\frac{\cos t}{\sqrt{t}}\right)dt \tag{A.85}$$

$$\mathrm{S}_2\,(x) = \frac{1}{\sqrt{2\pi}}\int_0^x \left(\frac{\sin t}{\sqrt{t}}\right)dt \tag{A.86}$$

$$\mathrm{C}(x) + i\cdot\mathrm{S}(x) = \frac{1+i}{2}\,\mathrm{erf}\left(\frac{\sqrt{\pi}}{2}(1-i)\,x\right) \tag{A.87}$$

$$\int_0^\infty \exp(-ax)\cos x^2\,dx = \sqrt{\frac{\pi}{2}}\left\{\left[\frac{1}{2}-\mathrm{S}(\frac{a}{2}\sqrt{\frac{2}{\pi}})\right]\cos\frac{a^2}{4}-\left[\frac{1}{2}-\mathrm{C}(\frac{a}{2}\sqrt{\frac{2}{\pi}})\right]\sin\frac{a^2}{4}\right\} \tag{A.88}$$

$$\int_0^\infty \exp(-ax)\sin x^2\,dx = \sqrt{\frac{\pi}{2}}\left\{\left[\frac{1}{2}-\mathrm{C}(\frac{a}{2}\sqrt{\frac{2}{\pi}})\right]\cos\frac{a^2}{4}+\left[\frac{1}{2}-\mathrm{S}(\frac{a}{2}\sqrt{\frac{2}{\pi}})\right]\sin\frac{a^2}{4}\right\} \tag{A.89}$$

Struve Functions

$$H_n(x) = \frac{2\left[\frac{x}{2}\right]^n}{\sqrt{\pi}\,\Gamma(n+\frac{1}{2})} \int_0^1 \left[1-t^2\right]^{n-\frac{1}{2}} \sin xt \; dt \tag{A.90}$$

$$\Gamma(n+1) = n! \quad , \qquad \text{(gamma-function)} \tag{A.91}$$

Bessel functions of the first kind (of integer order)

$$J_n(x) = \frac{1}{\pi} \int_0^{\pi} \cos(x \sin\theta - n\theta)\, d\theta \tag{A.92a}$$

$$J_n(x) = \frac{i^{-n}}{\pi} \int_0^{\pi} (\cos n\theta) \cdot [\exp(i\, x \cos\theta)]\, d\theta \tag{A.92b}$$

$$J_n(x) = \frac{2\left[\frac{x}{2}\right]^n}{\sqrt{\pi}\,\Gamma(n+\frac{1}{2})} \int_0^1 \left[1-t^2\right]^{n-\frac{1}{2}} \cos xt \; dt \tag{A.92c}$$

$$\cos(x \sin\theta) = J_0(x) + 2 \sum_{k=1}^{\infty} J_{2k}(x) \cos(2k\theta) \tag{A.93}$$

$$\sin(x \sin\theta) = 2 \sum_{k=0}^{\infty} J_{2k+1}(x) \sin\big((2k+1)\theta\big) \tag{A.94}$$

$$\cos(x \cos\theta) = J_0(x) + 2 \sum_{k=1}^{\infty} (-1)^k J_{2k}(x) \cos(2k\theta) \tag{A.95}$$

$$\sin(x \cos\theta) = 2 \sum_{k=0}^{\infty} (-1)^k J_{2k+1}(x) \sin\big((2k+1)\theta\big) \tag{A.96}$$

Probability and Statistics

Measures of the *average* value of a series of measurements

Let $p_i \geq 0$ be the probability of observing x_i, namely the i'th possible result of a measurement of x. Various average values of x are defined as follows.

$$\text{Arithmetic mean} = \langle x \rangle = \sum_i p_i x_i \quad ; \text{ in which } \sum_i p_i = 1 \qquad \text{(A.97)}$$

$$\text{Root mean square (rms) average} = \sqrt{\langle [x - \langle x \rangle]^2 \rangle}$$

$$= \sqrt{\sum_i p_i x_i^2 - 2 \sum_i p_i x_i + \left(\sum_i p_i x_i \right)^2} \qquad \text{(A.98)}$$

$$\text{Geometric mean } (x_i > 0) = \sum_i p_i \ln x_i \qquad \text{(A.99)}$$

Measures of the *deviation* from the mean value

$$\mu_k = k^{\text{th}} \text{ moment about } \langle x \rangle = \sum_i p_i (x_i - \langle x \rangle)^k \qquad \text{(A.100)}$$

In particular,

$\mu_0 = 1$ if the probability distribution is normalized,

$\mu_1 = 0$ if the distribution is symmetrical about its mean value,

and μ_2 = Variance (A.101)

$$\sigma = \text{Standard Deviation} = \sqrt{\mu_2} \qquad \text{(A.102)}$$

$$\text{Mean absolute deviation} = \sum p_i \, | x_i - \langle x \rangle | \qquad \text{(A.103)}$$

Normal distribution

The random variable, x, is said to be normally distributed if its (continuous) probability distribution is given by:

$$p(x) = \frac{1}{\sqrt{2\pi}} \exp\left(\frac{-(x - \langle x \rangle)^2}{2\,\sigma^2} \right) \qquad \text{(A.104)}$$

in which the mean value of x is $\langle x \rangle$, the variance is σ^2, and the standard deviation is σ.

Chi-square distribution

If $y_1, y_2, \cdots, y_n$ are normally and independently distributed, with mean = 0 and variance = 1, then the distribution, χ^2,

$$\chi^2 = \sum_{i=1}^{n} y_i^2$$

is known as a Chi-square distribution with n degrees of freedom, with a probability distribution given by:

$$p(\chi^2) = \frac{(\chi^2)^{(1/2)(n-2)}}{2^{(n/2)}\Gamma\left(\frac{n}{2}\right)} \exp(-\chi^2/2) \qquad \text{(A.105)}$$

for which the mean value $<\chi^2> = n$, and the variance $= \sigma^2 = 2n$.

Matrices

A *vector*, **x**, may be represented as a *one*-dimensional row or column of numbers:

$$\mathbf{x} = \{x_i\} = \begin{pmatrix} x_1 & x_2 & x_3 & . & . & . \end{pmatrix}$$

or

$$\mathbf{x} = \{x_j\} = \begin{pmatrix} x_1 \\ x_2 \\ x_3 \\ \vdots \end{pmatrix} \qquad \text{(A.106)}$$

A *matrix*, **A**, is a *two*-dimensional array of numbers. The element in the i'th row and j'th column of the matrix is denoted, a_{ij}. The following rules apply to manipulations of vectors and matrices.

$$\mathbf{A} = \{a_{ij}\} = \begin{pmatrix} a_{11} & a_{12} & a_{13} & a_{14} & \cdots \\ a_{21} & a_{22} & a_{23} & a_{24} & \cdots \\ a_{31} & a_{32} & a_{33} & a_{34} & \cdots \\ a_{41} & a_{42} & a_{43} & a_{44} & \cdots \\ \vdots & \vdots & \vdots & \vdots & \vdots \end{pmatrix} \qquad \text{(A.107)}$$

Vector or Matrix addition or multiplication by a constant

$$\mathbf{z} = k(\mathbf{x} + \mathbf{y})$$

$$z_i = k(x_i + y_i) \qquad \text{(A.108a)}$$

$$\mathbf{C} = k(\mathbf{A} + \mathbf{B})$$

$$c_{ij} = k(a_{ij} + b_{ij}) \qquad \text{(A.108b)}$$

Product of a matrix and a (column) vector to give another (column) vector

$$\mathbf{y} = \mathbf{A}\,\mathbf{x}$$

$$y_i = \sum_j a_{ij}\, x_j \qquad \text{(A.109a)}$$

For example, $y_1 = a_{11}x_1 + a_{12}x_2 + a_{13}x_3 + \cdots$, as shown below.

$$\begin{pmatrix} \mathbf{a_{11}} & \mathbf{a_{12}} & \mathbf{a_{13}} & \cdots \\ a_{21} & a_{22} & a_{23} & \cdots \\ a_{31} & a_{32} & a_{33} & \cdots \\ \vdots & \vdots & \vdots & \vdots \end{pmatrix} \cdot \begin{pmatrix} \mathbf{x_1} \\ \mathbf{x_2} \\ \mathbf{x_3} \\ \vdots \end{pmatrix} = \begin{pmatrix} \boldsymbol{\sum_j a_{1j}\, x_j} \\ \sum_j a_{2j}\, x_j \\ \sum_j a_{3j}\, x_j \\ \vdots \end{pmatrix} \qquad \text{(A.109b)}$$

Product of two matrices

$$\mathbf{C} = \mathbf{A} \cdot \mathbf{B}$$

$$c_{ij} = \sum_k a_{ik}\, b_{kj} \qquad \text{(A.110a)}$$

In other words, the ij'th element of the product matrix, **C**, is obtained by multiplying the elements of the i'th *row* of the first matrix, **A**, by the elements of the j'th *column* of the second matrix, **B**. For example, $c_{23} = a_{21}b_{13} + a_{22}b_{23} + a_{23}b_{33} + \cdots$ as shown in Eq. A.110b.

$$\begin{pmatrix} a_{11} & a_{12} & a_{13} & \cdots \\ \mathbf{a_{21}} & \mathbf{a_{22}} & \mathbf{a_{23}} & \cdots \\ a_{31} & a_{32} & a_{33} & \cdots \\ \vdots & \vdots & \vdots & \vdots \end{pmatrix} \cdot \begin{pmatrix} b_{11} & b_{12} & \mathbf{b_{13}} & \cdots \\ b_{21} & b_{22} & \mathbf{b_{23}} & \cdots \\ b_{31} & b_{32} & \mathbf{b_{33}} & \cdots \\ \vdots & \vdots & \vdots & \vdots \end{pmatrix} = \begin{pmatrix} c_{11} & c_{12} & c_{13} & \cdots \\ c_{21} & c_{22} & \mathbf{c_{23}} & \cdots \\ c_{31} & c_{32} & c_{33} & \cdots \\ \vdots & \vdots & \vdots & \vdots \end{pmatrix} \qquad \text{(A.110b)}$$

Appendix B

The Dirac δ-functional

As noted at the outset of Chapter 2, the Fourier transform of a time-domain sinusoid of infinite duration leads directly to a strange kind of frequency-domain spectrum, consisting of "peaks" which appear to be infinitely narrow but must yet have finite area. If one were to attempt to represent such a peak as an ordinary function, $\delta(x - x_0)$, one would encounter the immediate paradox that

$$\delta(x - x_0) = \begin{cases} 0 \text{ if } x \neq x_0 \\ 1 \text{ if } x = x_0 \end{cases} \tag{B.1a}$$

and $$\int_{-\infty}^{\infty} \delta(x - x_0)\, dx = 1 \tag{B.1b}$$

It is hard to see how an infinitely narrow line can yet have finite area. Some people try to avoid the issue by defining the δ-function as a Gaussian curve in the limit of zero peak width:

$$\delta(x - x_0) = \lim_{b \to 0} \left(\frac{1}{|b|} \exp(-\pi x^2 / b) \right) \tag{B.2}$$

However, the fundamental problem is that the so-called delta-"function" is really not a function at all, but rather another mathematical object called a "functional", as shown in Table B.1.

The Dirac δ-function(al) converts a *function*, $f(x)$, into one of the *values* of that function, $f(x_0)$. Since a definite integral also converts a function into a number (see Table 1), it is common to represent the Dirac δ-functional as a definite integral:

$$\int_{x_1}^{x_2} f(x)\, \delta(x - x_0)\, dx = f(x_0) \tag{B.3}$$

in which it is understood that $x_1 < x_0 < x_2$. Eq. B.3 closely resembles the convolution integral defined by Eq. 2.20, and it is in fact permissible to regard the δ-functional as a convolution of the δ-"function" with $f(x)$. In other words, although the δ-"function" is *not* a true function, it behaves the same way when represented by the definite integral of Eq. B.3.

Table B.1. Recipes for converting one kind of mathematical object into another. Note that the so-called δ-"function" is actually a "functional", since it converts a *function* into a *number.*

x	y	Recipe, $x \to y$	Example
number	number	**function**	$y = \cos x$
			$y = \exp x$
			$y = x^3$
function	number	**functional**	$y = \int_0^1 \exp(-x)\ dx$
			$y = f(x_0) = \int_{-\infty}^{\infty} f(x)\ \delta(x - x_0)\ dx$
function	function	**operator**	d/dx
			multiply by x
operator	operator	**superoperator (Liouville operator)**	$L(A) = AB - BA$; A, B are operators

Next, from the definition of a complex Fourier transform,

$$\boldsymbol{F}(\omega) = \int_{-\infty}^{\infty} \boldsymbol{f}(t)\ \exp(-i\,\omega t)\ dt \tag{1.37}$$

we can proceed to evaluate the Fourier transform, $\mathbf{X}(\omega)$, of the complex sinusoid, $\exp(i\,\omega_0\,t)$.

$$\mathbf{X}(\omega) = \int_{-\infty}^{\infty} \exp(i\,\omega_0\,t)\exp(-i\,\omega t)\ dt \tag{B.4}$$

However, since the inverse Fourier transform of $\mathbf{X}(\omega)$ must give back the original function [in this case, $\exp(i\,\omega_0\,t)$],

$$\exp(i\,\omega_0 t) = \frac{1}{2\pi}\int_{-\infty}^{\infty} \mathbf{X}(\omega)\exp(-i\,\omega t)\ d\omega \tag{B.5}$$

comparison of Eqs. B.3 and B.5 shows that $\mathbf{X}(\omega) = 2\pi\,\delta(\omega - \omega_0)$. Thus, the Fourier transform of a complex sinusoid, $\exp(i\,\omega_0\,t)$, is simply a Dirac δ-function centered at $\omega = \omega_0$:

	F.T		
$\exp(i\,\omega_0\,t)$	$\rightarrow$ $\leftarrow$	$2\pi\,\delta(\omega - \omega_0) \;=\; \delta(\nu - \nu_0)$	(B.6)
	Inverse F.T		

Conversely, we may consider the time-domain "function", $\delta(t - t_0)$, to be an infinitely narrow time-domain pulse, with the property that its Fourier transform is

$$\int_{-\infty}^{\infty} \delta(t - t_0) \exp(-i\,\omega t)\; dt \;=\; 2\pi \exp(i\,\omega_0\,t) \qquad \text{(B.7)}$$

which, when inverse Fourier transformed, must give back $\delta(t - t_0)$:

$$\delta(t - t_0) = \frac{1}{2\pi} \int_{-\infty}^{\infty} 2\pi \exp(i\,\omega_0\,t)\; \exp(i\,\omega t)\; d\omega \qquad \text{(B.8)}$$

(Eq. B.9 is the same as Eq. B.4 with variables interchanged.) Thus,

	F.T		
$\delta(t - t_0)$	$\rightarrow$ $\leftarrow$	$2\pi \exp(i\,\omega_0\,t)$	(B9)
	Inverse F.T		

Eqs. B.6 and B.9 may be stated in more physical terms in two ways:

(a) A time-domain sinusoid of infinite duration has a frequency spectrum consisting of an infinitely narrow spike at the frequency of the sinusoid.

(b) An infinitely short time-domain pulse produces a frequency spectrum having perfectly flat magnitude at all frequencies: $|\exp(i\,\omega\,t_0\;| = 1$ for all ω.

Finally, by use of the by now familiar identities [remember that $(1/i) = -i$],

$$\cos a \;=\; \frac{1}{2}\;[\exp(i\,a) \;+\; \exp(-\,i\,a)] \qquad \text{(B.10a)}$$

$$\sin a \;=\; \frac{1}{2i}\;[\exp(i\,a) \;-\; \exp(-\,i\,a)] \qquad \text{(B.10b)}$$

the reader can readily extend Eq. B.6 to obtain the Fourier transforms of time-domain cosine and sine signals listed on the next page. Graphical displays of Eqs. B.6, B.11, and B.12 are shown in Figure 2.1.

$$\int_{-\infty}^{\infty} \cos \omega_0 t \;\; \cos \omega t \;\; dt = \pi \;\delta(\omega - \omega_0) \;+\; \pi \;\delta(\omega + \omega_0) \qquad \text{(B.11)}$$

$$= \frac{1}{2} [\delta(\nu - \nu_0) \;+\; \delta(\nu + \nu_0)] \qquad \text{(2.2a)}$$

$$\int_{-\infty}^{\infty} \sin \omega_0 t \;\; \sin \omega t \;\; dt = \pi \;\delta(\omega - \omega_0) \;-\; \pi \;\delta(\omega + \omega_0) \qquad \text{(B.12)}$$

$$= \frac{1}{2} [\delta(\nu - \nu_0) \;-\; \delta(\nu + \nu_0)] \qquad \text{(2.3b)}$$

Other convenient properties of the δ-functional follow from its definition in Eq. B.3:

$$\delta\left(\frac{x - x_0}{b}\right) = |b| \;\; \delta(x - x_0) \qquad \text{(B.13)}$$

$$\delta(ax - x_0) = |a|^{-1} \;\; \delta(x - \frac{x_0}{a}) \qquad \text{(B.14)}$$

$$\delta(-x) = \delta(x) \qquad \text{(B.15)}$$

Appendix C

The FFT algorithm: conceptual basis and program listings

Eq. 3.5a defines a procedure for generating a given frequency-domain (complex) spectral data point, $\boldsymbol{F}(\nu_m)$, from N (real or complex) time-domain data points [$\boldsymbol{f}(t_n)$, $n = 0,1,2,\cdots, N-1$].

$$\boldsymbol{F}(\nu_m) = \sum_{n=0}^{N-1} \boldsymbol{F}_{nm}\boldsymbol{f}(t_n) \qquad (3.5a)$$

in which $\boldsymbol{F}_{nm} = \exp\left(\frac{-i\,2\pi nm}{N}\right)$

When repeated N times to generate all N frequency-domain data points, the algorithm defined by Eq. 3.5a requires N^2 complex multiplications and $N(N-1)$ complex additions.

However, that computation is inefficient, because of redundancy arising from symmetry and periodicity in the FT code matrix. In 1965, Cooley and Tukey introduced a fast Fourier transform (FFT) algorithm designed to reduce drastically the number of arithmetic operations. In this section, we will describe their "Radix-2" algorithm, so named because it consists of successive two-fold divisions of an original N-point time-domain data set into smaller subsets until only a two-point FT remains. Thus, the Cooley-Tukey and related algorithms require that $N = 2^n$, $n = 2,3,4,\cdots$. Dozens of other algorithms have since been proposed (e.g., $N = a\cdot b\cdot c\cdots$, in which a, b, c, $\cdots$ are any prime numbers). In particular, the Hartley transform offers all of the advantages of the FFT without the need for mathematically complex variables. The reader is referred to Further Reading for a critical comparison of FFT algorithms.

The basic idea of the Cooley-Tukey algorithm is that if we could break the original N-point time-domain data set into two $(N/2)$-point sets, then perform a discrete FT on each of those two halves, and somehow combine the results to give the desired N-point frequency-domain data set, we would then require two discrete FT computations each involving $(N/2)\cdot(N/2)$ multiplications, or $(N^2/2)$ total multiplications in place of the original N^2 multiplications. [The same idea can then be applied again to break each of the $(N/2)$-point sets in half again, and so on successively, until only two data points are left in each subset.] The "trick" is to decide how to break the original data set into halves, and then how to recover the desired N-point frequency-domain data set from the discrete FT's of each of the two time-domain halves. In this section, we shall first describe a general FFT algorithm, and then use it to perform a discrete FT of an 8-point data set.

Definitions

For compact notation, we will henceforth denote the m'th frequency-domain data point, $\boldsymbol{F}(\nu_m)$, simply as $\boldsymbol{F}(m)$, and the n'th time-domain (complex) data point, $\boldsymbol{f}(t_n)$, simply as $\boldsymbol{t}(n)$. Eq. 3.5a then becomes

$$\boldsymbol{F}(m) = \sum_{n=0}^{N-1} {}^{N}\boldsymbol{F}_{nm}\, \boldsymbol{t}(n); \qquad n = 0, 1, 2, \cdots, N-1 \tag{C.1}$$

The index, N, has been introduced into the Fourier "code" element, ${}^{N}\boldsymbol{F}_{nm}$, in order to specify that the discrete FT is applied to an N-point data set.

Properties of the FT "code" matrix

$${}^{N}\boldsymbol{F}_{nm} \text{ is periodic with period, } N \tag{C.2}$$

In other words, we can increase either index of ${}^{N}\boldsymbol{F}_{nm}$ by an integral multiple of N, without changing the value of ${}^{N}\boldsymbol{F}_{nm}$. For example,

$${}^{N}\boldsymbol{F}_{(nm+rN)} = \exp\left(\frac{-i\,2\pi(nm+rN)}{N}\right); \qquad r = 0, 1, 2, \cdots$$

$$= \exp\left(\frac{-i\,2\pi nm}{N}\right) \exp\left(\frac{-i\,2\pi rN}{N}\right)$$

$$= \exp\left(\frac{-i\,2\pi nm}{N}\right) \exp(-i\,2\pi r)$$

But if r is an integer, $[\exp(-i\,2\pi r) = 1]$, so that

$${}^{N}\boldsymbol{F}_{(nm+rN)} = \exp\left(\frac{-i\,2\pi nm}{N}\right) = {}^{N}\boldsymbol{F}_{nm} \qquad \textbf{Q.E.D.} \tag{C.3}$$

A second property relates FT "code" elements for discrete FT of an $(N/2)$-point data set to FT code elements for discrete FT of an N-point data set:

$${}^{N}\boldsymbol{F}_{2nm} = \exp\left(\frac{2(-i\,2\pi nm)}{N}\right)$$

$$= \exp\left(\frac{-i\,2\pi nm}{(N/2)}\right)$$

$$= {}^{N/2}\boldsymbol{F}_{nm} \qquad \textbf{Q.E.D.} \tag{C.4}$$

A third property relates FT code matrix elements for any given row (or column) to elements differing by $(N/2)$ in row (or column) number. For example,

$${}^{N}\boldsymbol{F}_{[nm+(N/2)]} = \exp\left(\frac{-i\,2\pi[nm+(N/2)]}{N}\right)$$

$$= \exp\left(\frac{-i\,2\pi nm}{N}\right) \exp\left(\frac{-i\,2\pi N}{2N}\right)$$

$$= \exp\left(\frac{-i\,2\pi nm}{N}\right)\ \exp(-i\,\pi)$$

But $\exp(-i\,\pi) = -1 = \exp(i\,\pi)$

So $${}^{N}\boldsymbol{F}_{[nm+(N/2)]} = \exp\left(\frac{-i\,2\pi nm}{N}\right)\ \exp(i\,\pi)$$

$$= \exp\left(\frac{-i\,2\pi nm}{N}\right)\ \exp\left(\frac{i\,2\pi N}{2N}\right)$$

$$= \exp\left(\frac{-i\,2\pi[nm-(N/2)]}{N}\right)$$

$$= {}^{N}\boldsymbol{F}_{[nm-(N/2)]} \qquad \textbf{Q.E.D.} \qquad \text{(C.5)}$$

One cycle of the FFT algorithm

The first step is to separate the N time-domain data points, $\boldsymbol{t}(n)$, $n = 0, 1, 2, \cdots, N-1$, into two parts: $\boldsymbol{t}_1(n)$ and $\boldsymbol{t}_2(n)$ containing respectively the even and odd members of $\boldsymbol{t}(n)$:

$$\boldsymbol{t}_1(n) = \boldsymbol{t}(2n); \qquad n = 0, 1, 2, \cdots, (N/2-1)$$
$$\boldsymbol{t}_2(n) = \boldsymbol{t}(2n+1); \qquad n = 0, 1, 2, \cdots, (N/2-1) \qquad \text{(C.6)}$$

The Fourier transform (Eq. C.1) of $\boldsymbol{t}(n)$ can then be rewritten equivalently as

$$\boldsymbol{F}(m) = \sum_{n=0}^{(N/2)-1} {}^{N}\boldsymbol{F}_{2nm}\ \boldsymbol{t}(2n) + \sum_{n=0}^{(N/2)-1} {}^{N}\boldsymbol{F}_{(2n+1)m}\ \boldsymbol{t}(2n+1) \qquad \text{(C.7)}$$

Applying the property of Eq. C.4 to Eq. C.7, we obtain

$$\boldsymbol{F}(m) = \sum_{n=0}^{(N/2)-1} {}^{N/2}\boldsymbol{F}_{nm}\,\boldsymbol{t}_1(n) + {}^{N}\boldsymbol{F}_{1m} \sum_{n=0}^{(N/2)-1} {}^{N/2}\boldsymbol{F}_{nm}\,\boldsymbol{t}_2(n) \qquad \text{(C.8)}$$

$$= \boldsymbol{T}_1(m) + {}^{N}\boldsymbol{F}_{1m}\,\boldsymbol{T}_2(m); \qquad 0 \le m \le (N/2) - 1 \qquad \text{(C.9a)}$$

in which the notation, $\boldsymbol{T}_1(m)$ and $\boldsymbol{T}_2(m)$, represents the discrete FT's of the two halves of the original time-domain data set:

$$\boldsymbol{T}_1(m) = \sum_{n=0}^{(N/2)-1} {}^{N/2}\boldsymbol{F}_{nm}\,\boldsymbol{t}_1(n) \qquad \text{(C.10a)}$$

$$\boldsymbol{T}_2(m) = \sum_{n=0}^{(N/2)-1} {}^{N/2}\boldsymbol{F}_{nm}\,\boldsymbol{t}_2(n) \qquad \text{(C.10b)}$$

From Eq. C.9a, we can evaluate $\boldsymbol{F}(m)$ only for $0 \leq m \leq (N/2) - 1$. However, the remaining $\boldsymbol{F}(m)$ values for $N/2 \leq m \leq (N-1)$ can readily be generated by means of Eqs. C.3 and C.5. Specifically,

$$\boldsymbol{F}(m) = \boldsymbol{T}_1(m - N/2) - {}^{N}\boldsymbol{F}_{1(m-N/2)}\, \boldsymbol{T}_2(m - N/2); \quad N/2 \leq m \leq (N-1) \quad \text{(C.9b)}$$

Explicit FFT for an 8-point data set

For an 8-point time-domain data set, the first cycle of the FFT algorithm requires 2 discrete FT's of the 4-point data sets constructed from the even and odd elements of the original 8-point data set, as summarized in Figure C.1.

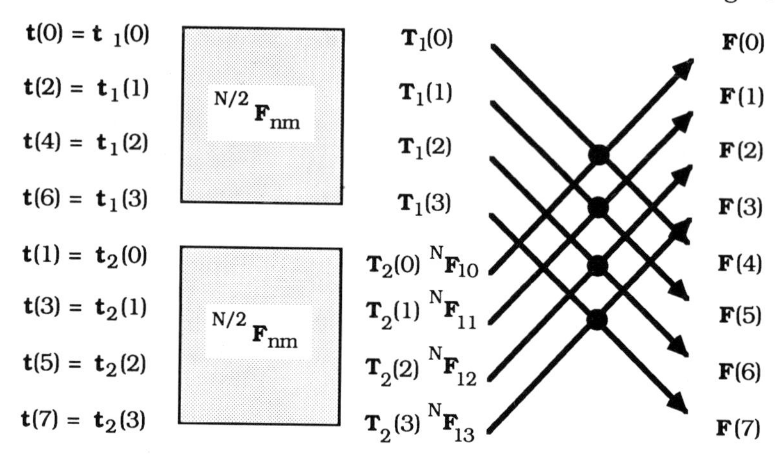

Figure C.1. Flow chart corresponding to Eqs. C.9. The even-numbered and odd-numbered original time-domain data points are separated and relabeled as shown at the far left. Each half is then subjected to discrete FT (with code matrix elements, ${}^{N/2}\boldsymbol{F}_{nm}$) to yield the frequency-domain spectral data points denoted as $\boldsymbol{T}_n(m)$. Each value shown at the right is the sum or difference of the values of the left side, according to the recipe shown below. For each intersecting pair of lines in the diagram, the upper right value is the sum of the two left-hand quantities, whereas the lower right value is the difference between the two left-hand quantities.

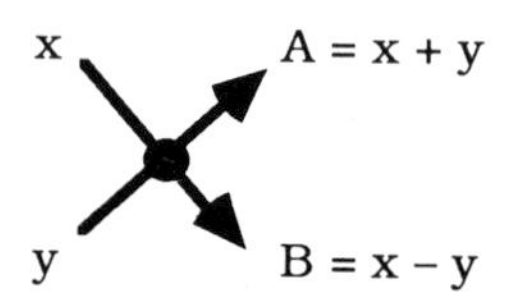

For example, $\boldsymbol{F}(0) = \boldsymbol{T}_1(0) + \boldsymbol{T}_2(0) \cdot {}^{N}\boldsymbol{F}_{10}$, whereas $\boldsymbol{F}(4) = \boldsymbol{T}_1(0) - \boldsymbol{T}_2(0) \cdot {}^{N}\boldsymbol{F}_{10}$.

For the next cycle of the FFT procedure, we again divide each of the two time-domain data sets in half. For example, let $\boldsymbol{t}_a(n)$ and $\boldsymbol{t}_b(n)$ denote respectively the even and odd members of $\boldsymbol{t}_1(n)$. Then Eqs. C.9 can be rewritten as

$$\boldsymbol{T}_1(m) = \boldsymbol{T}_a(m) + {}^{N/2}\boldsymbol{F}_{1m}\,\boldsymbol{T}_b(m) \qquad 0 \le m \le (N/4)-1$$

$$\boldsymbol{T}_1(m) = \boldsymbol{T}_a(m-N/4) - {}^{N/2}\boldsymbol{F}_{1(m-N/4)}\,\boldsymbol{T}_b(m-N/4) \quad (N/4) \le m \le (N/2)-1 \qquad \text{(C.11)}$$

and the flow chart of Figure C.1 can then be extended as shown in Figure C.2.

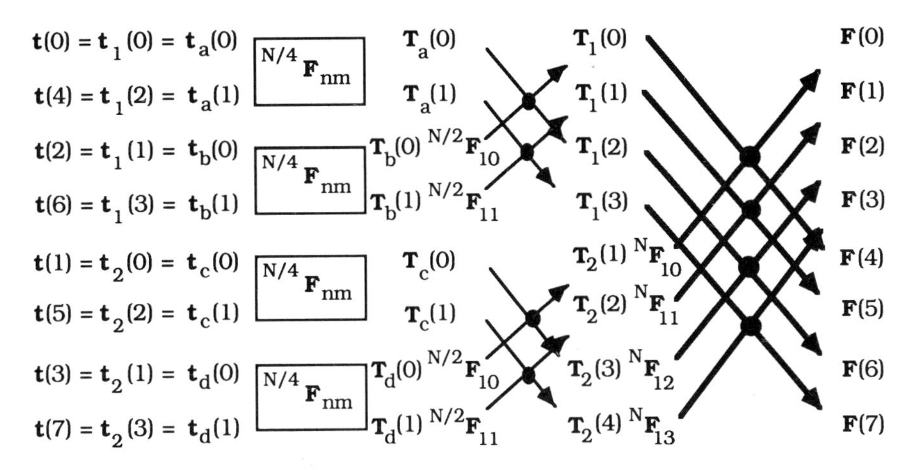

Figure C.2 Flow chart for construction of an 8-point discrete FT from four 2-point discrete FT's, which are in turn constructed from two 4-point discrete FT's (see Figure C.1). Notation is as in Figure C.1.

The net result of Figure C.2 is that we have replaced a single $N = 8 = 2^3$ point discrete FT by four 2-point discrete FT's, in a three-stage procedure (i.e., two reductions (by a factor of two) in the number of data points per FT, and a final step to reconstruct the desired spectrum from the 2-point FT's. Thus, for arbitrary N, we would need $\log_2 N$ stages to divide N in half, and again in half, and so on, until only 2-point subsets remain, followed by the final reconstruction step. From Figures C.1 and C.2, it is clear that at each stage of the FFT (i.e. each halving of the previous data set size per FT), $N/2$ complex multiplications are required. Since there are $\log_2 N$ stages, the total number of multiplications is therefore approximately $\sim(N/2)\log_2 N$; rather than N^2. The FFT number is approximate, since multiplication by factors such as ${}^{N}\boldsymbol{F}_{10}$, ${}^{N/2}\boldsymbol{F}_{1(N/2)}$, etc. actually amounts to a complex addition or subtraction.

As a result of the successive separation of the even and odd members of the preceding data (sub)set, the final time-domain data set (e.g., leftmost column of Figure C.2) must be re-ordered (shuffled) in preparation for the 2-point FT computations. Fortunately, it turns out that the shuffled sequence (0, 4, 2, 6, 1, 5, 3, 7 for the 8-point case) can be easily generated by bit-reversal, as illustrated for the $N = 8$ case in Table C.1.

Finally, there is an even simpler way to write out the final bit-reversed binary sequence. Note that the first column of the list of bit-reversed numbers has the sequence, 0,1,0,1,0,1,...; the second column has the sequence, 0,0,1,1,0,0,1,1,...; the third column has the sequence, 0,0,0,0, 1,1,1,1,...; and so on.

Table C.1 Generation of the shuffled sequence of time-domain data points required for the 2-point FT's of the FFT algorithm, for an 8-point data set. The original decimal sequence is converted to binary numbers, whose bit order is then reversed left-to-right. The desired sequence is then obtained from the decimal equivalent of the bit-reversed binary values.

Original sequence	Binary representation	Bit-reversed binary representation	Bit-reversed sequence
0	000	000	0
1	001	100	4
2	010	010	2
3	011	110	6
4	100	001	1
5	101	101	5
6	110	011	3
7	111	111	7

Inverse Fourier transform

The forward discrete Fourier transform (e.g., from time-domain to frequency-domain) is defined by Eq C.1:

$$\boldsymbol{F}(m) = \sum_{n=0}^{N-1} {}^{N}\boldsymbol{F}_{nm}\ \boldsymbol{t}(n), \qquad n = 0, 1, 2, \cdots, N-1 \qquad \text{(C.1)}$$

The inverse transform (e.g., from frequency-domain to time-domain) is given by (compare Eqs. 3.5b and 3.9b):

$$\boldsymbol{t}(m) = \sum_{n=0}^{N-1} {}^{N}\boldsymbol{F}_{nm}^{-1}\ \boldsymbol{F}(n), \qquad n = 0, 1, 2, \cdots, N-1 \qquad \text{(C.12)}$$

The forward and inverse discrete FT formulas differ only by a sign in one index of each term (and a factor of $1/N$). Thus, exactly the same FFT algorithm may be used to perform forward or inverse FT, by simply changing the sign of the sine term (equivalent to replacing a positive angle by a negative angle in the complex-plane representation—see Program P.2).

Fast Fourier transform of real data

The above FFT development is predicated on complex input data [i.e., $\boldsymbol{t}(n)$ are complex numbers]. Fortunately, a real discrete FT can be performed with the same algorithm as follows. First, the even- and odd-numbered members of the (real) input data set are treated as if the even-numbered points were real and the odd-numbered points were imaginary. Following complex FFT, a one-pass multiplication converts the FFT output into the true discrete FT of the original (real) data set. The one-pass multiplication is based on additional properties of the discrete FT and requires a few pages of derivation. The interested reader is referred to Further Reading. An FFT algorithm for complex data is listed in Program P.2 and for real data is listed in Program P.3, for both forward and inverse discrete FT. Alternatively, the *Hartley* transform (see Program P.4) offers the same speed advantages as the above algorithm, without ever involving mathematically complex notation. Programs P.1, and P.2 are written in Standard Fortran 77; Programs P.3 and P.4 are in Basic.

Program P.1 Direct discrete Fourier transform subroutine with lookup table, adapted from a program originally written by C. S. Burrus from Rice University. (Standard Fortran 77 version).

```
C  ***************************************************************************************************
C *                                                                                                  *
C *                    Subroutine Direct DFT with lookup table                                       *
C *                           Adapted by F. R. Verdun                                                *
C *                                                                                                  *
C  ***************************************************************************************************
C
C TIR, TIM: Real and imaginary parts of complex time-domain data
C FREAL, FIMAG: Real and imaginary parts of frequency-domain data
C  N: Number of real data points
C
      SUBROUTINE DFT(TIR, TIM, FREAL, FIMAG, N)
      REAL TIR(N),TIM(N), FREAL(N), FIMAG(N), C(N), S(N)
C
C -------------------- Calculation of lookup table of sines and cosines -------------------
C
            Q = 6.2831853071796/N
            DO 10 J = 1, N
            C(J) = COS(Q*(J – 1))
            S(J) = SIN(Q*(J – 1))
10          CONTINUE
C
C -------------------- Fourier transform loops -------------------
C
            DO 20 J = 1, N
             AT = TIR(1)
             BT = TIM(1)
             K = 1

                DO 30 I = 2, N
                  K = K + J – 1
                  IF (K.GT.N)  THEN  K = K – N
                  AT = AT + C(K)*TIR(I) + S(K)*TIM(I)
                  BT = BT + C(K)*TIM(I) – S(K)*TIR(I)
30              CONTINUE

            FREAL(J) = AT
            FIMAG(J) = BT
20          CONTINUE
C
      RETURN
      END
```

Program P.2 Cooley-Tukey Radix-2 FFT Subroutine, adapted from a program originally written by C. S. Burrus from Rice University (Standard Fortran 77 version).

```
C   ****************************************************************************
C *                                                                            *
C *                        Cooley-Tukey Radix-2 FFT                            *
C *                         Adapted by F. R. Verdun                            *
C *                                                                            *
C   ****************************************************************************
C
C  X, Y:  Real and imaginary parts of complex time- or frequency-domain data
C  N:  Number of real data points (must be an integral power of two);  M = log2(N)
C  ISIGN= + 1 for direct FFT;  - 1 for inverse FFT
C
       SUBROUTINE DFT(X, Y, N, M, ISIGN)
       REAL X(N),Y(N)
C
C- Note:  To avoid a discontinuity at the time-domain origin, one should
C  divide the first time-domain data point  by two.
C
C-------------------- FFT loops----------------------
C
      N2 = N
      DO 10 K = 1, M
      N1 = N2
      N2 = N2/2
      E = 6.2831853071796/N1
      A = 0.
            DO 20 J = 1, N2
            C = COS(A)
            S = ISIGN*SIN(A)
            A = J*E
                  DO 30 I = J, N, N1
                  L = I + N2
                  XT = X(I) – X(L)
                  X(I) = X(I) + X(L)
                  YT = Y(I) – Y(L)
                  Y(I) = Y(I) + Y(L)
                  X(L) = C*XT + S*YT
                  Y(L) = C*YT – S*XT
30                CONTINUE
20          CONTINUE
10    CONTINUE
```

```
C
C--------------------- Digit reverse counter -----------------------
C
50          J = 1
            N1 = N - 1
            DO 54 I = 1, N1
                IF (I.GE.J)  GOTO  51
                XT = X(J)
                X(J) = X(I)
                X(I) = XT
                XT = Y(J)
                Y(J) = Y(I)
                Y(I) = XT
51              K = N/2
52              IF (K.GE.J)  GOTO  53
                J = J - K
                K = K/2
                GOTO  52
53              J = J + K
54          CONTINUE
            RETURN
            END
```

Program P.3 Real FFT (Macintosh Microsoft "Quick-Basic" Version) program based on a routine originally written by N. Brenner of Lincoln Laboratories. The version given here is an adapted form of the program given in Press et al. in Further Reading.

```
' ***********************************************************************************************
' *                                                                                             *
' *                  Real FFT, with cosines and sines determined by recursion                   *
' *                                  Adapted by F. R. Verdun                                    *
' *                                                                                             *
' ***********************************************************************************************

REM: sdata contains the time-domain or frequency-domain data
REM: ndp% is the total number of real data points (must be a power of two).
REM: ntp% is the total number of complex data points.(ndp%/2)

            DIM sdata(ndp%), re(ntp%), im(ntp%)

REM The data points can be generated external to this program and then read into sdata.

      REM ********** Subroutine to read external data *********
      I% = 0
      OPEN "FILENAME" FOR INPUT AS #1
      WHILE NOT EOF (1)
      I% = I% + 1
      INPUT #1, SDATA(I%)
      WEND
      CLOSE 1
      ndp% = i%
      REM ********** Go to the FFT part of the program *********
```

```
REM Data points generated just before the FFT routine.

REM  ********** Forward FFT case*********

REM Generation of a time-domain sinusoid
INPUT "Input  number of real data points: ",ndp%
REM BWI: Bandwidth in Hz
REM XDW: dwell time in seconds.

pi = 4.*ATN(1.)
BWI = 1000.
XDW = 1./(2.*BWI.)
 INPUT " Enter the frequency value (f  <  1000):  ",  f
 f1 = 2.*pi*f

      FOR t% = 1 TO ndp%
      t1 = CSNG(t% – 1)*XDW
      omega = f1*t1
      sdata(t%) = 100.*cos(omega)
      NEXT t%

REM: To avoid a discontinuity, one should divide the first data point by two.

sdata(1) = sdata(1)/2.

REM then input 1 to generate the frequency-domain spectrum

                  REM  ********** FFT routine *********

                  ntp% = ndp%/2
INPUT "Enter 1 For forward FT or –1 For inverse FT :  ",  isign%

      REM Shuffling operation:

IF isign% = –1 THEN
      FOR i% = 1 TO ntp%
      re(i%) = sdata(i%)
      im(i%) = sdata(i% + ntp%)
      NEXT i%
              K% = 0
              FOR i% = 1 TO ntp%
              K% = K% + 1
              sdata(K%) = re(i%)
              K% = K% + 1
              sdata(K%) = im(i%)
              NEXT i%
END IF

GOSUB realft

      REM Shuffling operation:
```

```
IF isign% = 1 THEN
      FOR i% = 1 TO ntp%
      I2% = 2*i% – 1
      re(i%) = sdata(I2%)
      im(i%) = sdata(I2% + 1)
      NEXT i%
            FOR i% = 1 TO ntp%
            sdata(i%) = re(i%)
            sdata(i% + ntp%) = im(i%)
            NEXT i%
END IF
GOSUB Graph
END

realft:

      THETA# = – 3.141592653589793#/CDBL(ntp%)

      IF isign% = 1 THEN
      GOSUB FFT
      THETA# = – 3.141592653589793#/CDBL(ntp%)
      c1 = .5
      c2 = – .5
      END IF

            IF isign% = – 1 THEN
            c1 = .5
            c2 = .5
            THETA# = – THETA#
            END IF
WPR# = – 2#*(SIN(.5#*THETA#))^2
WPI# = SIN(THETA#)
WR# = 1# + WPR#
WI# = WPI#
N2P3% = 2*ntp% + 3
      FOR i% = 2 TO ntp%/2 + 1
      I1% = 2*i% – 1
      I2% = I1% + 1
      I3% = N2P3% – I2%
      I4% = I3% + 1
      WRS = CSNG(WR#)
      WIS = CSNG(WI#)
      H1R = c1*(sdata(I1%) + sdata(I3%))
      H1I = c1*(sdata(I2%) – sdata(I4%))
      H2R = – c2*(sdata(I2%) + sdata(I4%))
      H2I = c2*(sdata(I1%) – sdata(I3%))
      sdata(I1%) = H1R + WRS*H2R – WIS*H2I
      sdata(I2%) = H1I + WRS*H2I + WIS*H2R
      sdata(I3%) = H1R – WRS*H2R + WIS*H2I
      sdata(I4%) = –H1I + WRS*H2I + WIS*H2R
      WTEMP# = WR#
      WR# = WR#*WPR# – WI#*WPI# + WR#
      WI# = WI#*WPR# + WTEMP#*WPI# + WI#
      NEXT i%
```

```
IF isign% = 1 THEN
H1R = sdata(1)
sdata(1) = H1R + sdata(2)
sdata(2) = H1R – sdata(2)
ELSE

      H1R = sdata(1)
      sdata(1) = c1*(H1R + sdata(2))
      sdata(2) = c1*(H1R – sdata(2))
      GOSUB FFT
      END IF

              RETURN

FFT:

ndp% = 2*ntp%
J% = 1
      FOR i% = 1 TO ndp% STEP 2
              IF J% > i% THEN
              TEMPR = sdata(J%)
              TEMPI = sdata(J% + 1)
              sdata(J%) = sdata(i%)
              sdata(J% + 1) = sdata(i% + 1)
              sdata(i%) = TEMPR
              sdata(i% + 1) = TEMPI
              END IF

M% = ntp%

1                     IF M% >= 2  AND  J% > M%  THEN
                      J% = J% – M%
                      M% = M%/2
                      GOTO  1
                      END IF
                      J% = J% + M%
      NEXT i%

MMAX% = 2
2     IF ndp% > MMAX%   THEN
      ISTEP% = 2*MMAX%
      THETA# = – 6.28318530717959#/CDBL((isign%*MMAX%))
      WPR# = – 2#*(SIN(.5#*THETA#))^2
      WPI# = SIN(THETA#)
      WR# = 1#
      WI# = 0#
```

```
            FOR M% = 1 TO MMAX% STEP 2
                FOR i% = M% TO ndp% STEP ISTEP%
                J% = i% + MMAX%
                TEMPR = CSNG(WR#)*sdata(J%) – CSNG(WI#)*sdata(J% + 1)
                TEMPI = CSNG(WR#)*sdata(J% + 1) + CSNG(WI#)*sdata(J%)
                sdata(J%) = sdata(i%) – TEMPR
                sdata(J% + 1) = sdata(i% + 1) – TEMPI
                sdata(i%) = sdata(i%) + TEMPR
                sdata(i% + 1) = sdata(i% + 1) + TEMPI
                NEXT i%

            WTEMP# = WR#
            WR# = WR#*WPR# – WI#*WPI# + WR#
            WI# = WI#*WPR# + WTEMP#*WPI# + WI#
            NEXT M%

        MMAX% = ISTEP%
        GOTO 2
        END IF

RETURN

graph:

REM  ********** Graphic routine *********

REM This routine is written in "Quick Basic" for a Macintosh II computer.
REM Frequency spectrum scaled from zero frequency to highest frequency.
REM No vertical scaling factor has been included..
x = 600
y = 300

        FOR i = 1 TO ndp%
        IF ABS(sdata(i)) > ymax THEN ymax = ABS(sdata(i))
        NEXT i

y = y/2
xfac = x/ndp%
yfac = 0.9*y/ymax

PICTURE ON

        REM Draw x-axis
        CALL PENSIZE(1, 1)
        CALL MOVETO(0, y)
        CALL LINETO(ndp%*xfac, y)

            REM Draw y-axis through center of x-axis
            x0 = x/2
            ya = 10
            yb = 2*y – 10
            CALL PENSIZE(1,1)
            CALL MOVETO(x0, ya)
            CALL LINETO(x0, yb)
```

```
          REM Draw either time or frequency domain
          CALL PENSIZE(2, 2)
          CALL MOVETO(0, y – sdata(1)*yfac)
          FOR i = 1 TO ndp%
          CALL LINETO(i*xfac, y – sdata(i)*yfac)
          CALL MOVETO(i*xfac, y – sdata(i)*yfac)
          NEXT i

PICTURE OFF

CLS

PICTURE

    OPEN "clip:picture" FOR OUTPUT AS 2
    PRINT #2,PICTURE$
    CLOSE#2

WHILE ((MOUSE(0) = 0) AND (INKEY$=""))

WEND

RETURN
```

Program P.4 Fast Hartley Transform (FHT) (Macintosh Microsoft Basic Version) routine based on a routine originally written by R. N. Bracewell of Stanford University. The version given here is the program adapted by Williams & Marshall from Bracewell (see Further Reading). The graphic routine of Program P.3 can be used to display the data from this program.

```
' ****************************************************************************************
' *                                                                                      *
' *                              Fast Hartley Transform                                  *
' *                             Adapted by C. P. Williams                                *
' *                                                                                      *
' ****************************************************************************************

Pi = 4.*ATN(1.)
OPTION BASE 0
REM Generate Sample Data
ndp% = 1024
        REM Number of Data Points
DIM sdata(ndp%)
FOR I = 0 TO ndp% – 1
    sdata(I) = 1000.*COS(Pi/3 + 100.*I*Pi/CSNG(ndp%))
NEXT I
Power% = 0
                    REM  Calculate Power of 2 for ndp%
I% = ndp%
WHILE I% > 1
    Power% = Power% + 1
    I% = I%/2
WEND
```

```
N9 = 2^(Power% – 2)
N = 4*N9
C5 = N – 1
C6 = Power% – 1

DIM M(Power%)
            REM Get Powers of 2
FOR I% = 0 TO Power%
      M(I%) = 2^I%
NEXT I%

DIM S(ndp%/4)
            REM Get Sines
S(N9) = 1
FOR I = 1 TO 3
      S(I*N9/4) = SIN(I*Pi/8)
NEXT I
H = .5/COS(Pi/16)
      REM  Initial half secant
C4 = Power% – 4
            REM  Fill sine table
FOR I% = 1 TO Power% – 4
      C4 = C4 – 1
      V0 = 0
      FOR J% = M(C4) TO (N9 – M(C4)) STEP M(C4 + 1)
            V1 = J% + M(C4)
            S(J%) = H*(S(V1) + V0)
            V0 = S(V1)
      NEXT J%
      H = 1/SQR(2 + 1/H)  REM  half secant recursion
NEXT I%

DIM T(ndp%/4)
            REM  Get Tangents
C0 = N9 – 1
FOR I% = 1 TO N9 – 1
      T(I%) = (1 – S(C0))/S(I%)
      C0 = C0 – 1
NEXT I%
T(N9) = 1

Q% = Power%/2
            REM Fast Permute
C2 = M(Q%)
Q% = Q% + (Power% MOD 2)
DIM A(1024)
                        REM Cell ordinate
A(1) = 1
FOR I% = 2 TO Q%
      FOR J% = 0 TO M(I% – 1) – 1
            A(J%) = 2*A(J%)
            A(J% + M(I% –1)) = A(J%) + 1
      NEXT J%
NEXT I%
```

```
FOR I% = 1 TO C2 – 1
      REM Permute
      V4 = C2*A(I%)
      V5 = I%
      V6 = V4
      V7 = sdata(V5)
      sdata(V5) = sdata(V6)
      sdata(V6) = V7
      FOR J% = 1 TO A(I%) – 1
            V5 = V5 + C2
            V6 = V4 + A(J%)
            V7 = sdata(V5)
            sdata(V5) = sdata(V6)
            sdata(V6) = V7
      NEXT J%
NEXT I%

FOR I% = 0 TO n – 2 STEP 2
      REM First Stages (1 and 2)
      V6 = sdata(I%) + sdata(I% + 1)
            REM  2-Element DHT
      V7 = sdata(I%) – sdata(I% + 1)
      sdata(I%) = V6
      sdata(I% + 1) = V7
NEXT I%

FOR I% = 0 TO n – 4 STEP 4
            REM  4-Element DHT
      V6 = sdata(I%) + sdata(I% + 2)
      V7 = sdata(I% + 1) + sdata(I% + 3)
      V8 = sdata(I%) – sdata(I% + 2)
      V9 = sdata(I% + 1) – sdata(I% + 3)
      sdata(I%) = V6
      sdata(I% + 1) = V7
      sdata(I% + 2) = V8
      sdata(I% + 3) = V9
NEXT I%

U% = C6
                        REM Last Stages (3 and up)
S% = 4
FOR L% = 2 TO C6
      V2 = 2*S%
      U% = U% – 1
      V3 = M(U% – 1)
      FOR Q% = 0 TO C5 STEP V2
            I% = Q%
            D% = I% + S%
            V6 = sdata(I%) + sdata(D%)
            V7 = sdata(I%) – sdata(D%)
            sdata(I%) = V6
            sdata(D%) = V7
            K% = D% – 1
```

```
            FOR J% = V3 TO N9 STEP V3
                        I% = I% + 1
                        D% = I% + S%
                        E% = K% + S%
                        V9 = sdata(D%) + (sdata(E%)*T(J%))
                        X = sdata(E%) – V9*S(J%)
                        Y = X*T(J%) + V9
                        V6 = sdata(I%) + Y
                        V7 = sdata(I%) – Y
                        V8 = sdata(K%) – X
                        V9 = sdata(K%) + X
                        sdata(I%) = V6
                        sdata(D%) = V7
                        sdata(K%) = V8
                        sdata(E%) = V9
                        K% = K% – 1
            NEXT J%
            E% = K% + S%
      NEXT Q%
      S% = V2
NEXT L%

Norm.factor = 1/SQR(ndp%)
                        REM Normalize
FOR I% = 0 TO ndp% – 1
      sdata(I%) = Norm.factor*sdata(I%)
NEXT I%

END
```

Further Reading

G. D. Bergland, *I.E.E.E. Spectrum*, **1969**, *6*, 41-52. (Good description of various FFT algorithms)

R. N. Bracewell, *The Hartley Transform*, Oxford University Press, NY, 1986. (Detailed description and properties of the Hartley transform—see Williams & Marshall for corrected fast Hartley transform algorithm)

E. O. Brigham, *The Fast Fourier Transform*, Prentice-Hall, Englewood Cliffs, NJ, 1974. (Good presentation of fast Fourier transform algorithms)

C. S. Burrus & T. W. Parks, *DFT/FFT and Convolution Algorithms—Theory and Implementation*, J. Wiley & Sons, NY, 1985. (Listings of several FFT and convolution algorithms)

J. W. Cooley & J. W. Tukey, *Math. Comput.*, **1965**, *19*, 297-301. (The original FFT algorithm—now the most highly cited paper in all of mathematics).

J. W. Cooley, P. A. W. Lewis, & P. D. Welch, *J. Sound Vib.*, **1970**, *12*, 315-337. (Discusses real and complex FFT algorithms)

J. W. Cooper, in *Transform Techniques in Chemistry*, Ed. P. R. Griffiths, Plenum, NY, 1978, pp. 84-108. (Good discussion of FFT procedure)

P. Duhamel, *I.E.E.E. Trans. ASSP-34*, **1986**, *2*, 285-295. (Implementation of "Split-Radix" FFT algorithms for complex, real and real-symmetrical data; very good explanation of the "Split-Radix" procedure)

P. Duhamel & M. Vetterli, *I.E.E.E. Trans. ASSP-35*, **1987**, *6*, 818-824. (Improved Fourier and Hartley algorithms with the application to convolution of mathematically real data)

H. Hao & R. N. Bracewell, *Proc. I.E.E.E.*, **1987**, *75*, 264-266. (Description of a three dimensional DFT algorithm using the Fast Hartley Transform)

W. H. Press, B. P. Flannery, S. A. Teukolsky, & W. T. Vetterling, *Numerical Recipes —The Art of Scientific Computing*, Cambridge University Press, NY, 1986, pp 381-427. (Good presentation of the FFT algorithm)

L. R. Rabiner & B. Gold, *Theory and Applications of Digital Signal Processing* Prentice-Hall, Englewood Cliffs, NJ, 1975, pp. 356-433. (Good introduction to the "Radix-2" FFT).

C. P. Williams & A. G. Marshall, *Anal. Chem.*, **1989**, *61*, 428-431. (Application of Hartley transform to spectroscopy of linearly-polarized time-domain signals; includes corrected fast HT algorithm)

Appendix D

Fourier transform properties and pictorial atlas of Fourier transform pairs

Definitions

Cosine Fourier Transform:

$$\mathrm{CFT}[f(t)] = \left. \begin{array}{l} C(\omega) = \int_{-\infty}^{+\infty} f(t) \cos \omega t \; dt \\ \\ C(\nu) = \int_{-\infty}^{+\infty} f(t) \cos 2\pi \nu t \; dt \end{array} \right\} \quad \text{for real } f(t)$$

Sine Fourier Transform:

$$\mathrm{SFT}[f(t)] = \left. \begin{array}{l} S(\omega) = \int_{-\infty}^{+\infty} f(t) \sin \omega t \; dt \\ \\ S(\nu) = \int_{-\infty}^{+\infty} f(t) \sin 2\pi \nu t \; dt \end{array} \right\} \quad \text{for real } f(t)$$

Complex Fourier Transform:

$$\left. \begin{array}{l} \boldsymbol{F}(\omega) = \int_{-\infty}^{+\infty} \boldsymbol{f}(t) \exp(-i\,\omega t)\, dt \\ \\ \boldsymbol{F}(\nu) = \int_{-\infty}^{+\infty} \boldsymbol{f}(t) \exp(-i\,2\pi\nu t)\, dt \end{array} \right\} \quad \text{for complex } \boldsymbol{f}(t)$$

Inverse Fourier Transform:

$$\boldsymbol{f}(t) = \frac{1}{2\pi} \int_{-\infty}^{+\infty} \boldsymbol{F}(\omega) \exp(i\,\omega t)\, d\omega \qquad \text{for complex } \boldsymbol{F}(\omega)$$

$$\boldsymbol{f}(t) = \int_{-\infty}^{+\infty} \boldsymbol{F}(\nu) \exp(i\,2\pi\nu t)\, d\nu \qquad \text{for complex } \boldsymbol{F}(\nu)$$

Relationships

$\boldsymbol{F}(\omega) = \mathrm{Re}[\boldsymbol{F}(\omega)] + \mathrm{Im}[\boldsymbol{F}(\omega)]$

For real $f(t)$, positive and negative frequencies cannot be distinguished, and

$\mathrm{Re}[\boldsymbol{F}(\omega)] = A(\omega) =$ Absorption mode spectrum if $\boldsymbol{F}(\omega)$ is correctly phased.
$\mathrm{Im}[\boldsymbol{F}(\omega)] = D(\omega) =$ Dispersion mode spectrum if $\boldsymbol{F}(\omega)$ is correctly phased.

$\boldsymbol{F}(\omega) = C(\omega) - i\,S(\omega)$

For complex $f(t) = a(t) + i\,b(t)$, positive and negative frequencies can be distinguished, and

$\boldsymbol{F}(\omega) = \mathrm{Re}[\boldsymbol{F}(\omega)] + \mathrm{Im}[\boldsymbol{F}(\omega)]$

$\boldsymbol{F}(\omega) = \mathrm{CFT}[a(t)] + \mathrm{SFT}[b(t)] + i\{\mathrm{CFT}[b(t)] - \mathrm{SFT}[a(t)]\}$

General Properties

Existence:

$f(t)$ has a Fourier transform if $\int f(t)\,dt < \infty$.

Symmetry:

$f(t)$	$\boldsymbol{F}(\omega)$
Complex, no symmetry	Complex, no symmetry
Complex, even	Complex, even
Complex, odd	Complex, odd
Real, even	Real, even
Real, odd	Imaginary, odd
Imaginary, even	Imaginary, even
Imaginary, odd	Real, odd

Linearity:

F.T. of $\{a f(t) + b\,h(t)\} = a\boldsymbol{F}(\omega) + b\boldsymbol{H}(\omega)$

Central Ordinate:

$$\boldsymbol{F}(0) = \int_{-\infty}^{+\infty} \boldsymbol{f}(t)\,\exp(-i\,0t)\,dt = \int_{-\infty}^{+\infty} \boldsymbol{f}(t)\,dt$$

The area of a function is equal to the central ordinate of its Fourier transform.

Scaling:

F.T. of $f(t/a) = |a|\,\boldsymbol{F}(a\omega)$

For example, if $a = -1$, then F.T. of $f(-t) = \boldsymbol{F}(-\omega)$

Shifting

$$\text{F.T. of } f(t - t_0) = \exp(-i\, 2\pi\omega t_0).\boldsymbol{F}(\omega)$$

Transform of a conjugate:

$$\text{F.T. of } f^*(t) = \boldsymbol{F}^*(-\omega)$$

Transform of a convolution:

$$\text{F.T. of } [f(t) * h(t)] = \boldsymbol{F}(\omega)\cdot \boldsymbol{H}(\omega)$$

Transform of a product:

$$\text{F.T. of } [f(t)\cdot h(t)] = \boldsymbol{F}(\omega) * \boldsymbol{H}(\omega)$$

Transform of a derivative:

$$\text{F.T. of } f^{(k)}(t) = (i\, 2\pi\omega)^k\, \boldsymbol{F}(\omega)$$

Transform of an integral:

$$\text{F.T. of} \left(\int_{-\infty}^{t} \boldsymbol{f}(t)\, dt \right) = \frac{1}{i\, 2\pi\omega} \mathbf{F}(\omega) + \frac{\boldsymbol{F}(0)}{2}\, \delta(\omega)$$

Notation for special functions

Heaviside unit step function: $H(t) = 1$ for $t \geq 0$, $H(t) = 0$ elsewhere.

Rectangle (Dirichlet) function: $\Pi(t) = 0$ for $|t| \leq \frac{1}{2}$; $\Pi(t) = 0$ elsewhere.

Shah (comb) function: $\text{III}(t) = \sum_{n=-\infty}^{\infty} \delta(t - n\,\Delta t)$, n is an integer

Sign (signum) function: $\text{sgn}(t) = -1$ for $t < 0$, $\text{sgn}(0) = 0$, $\text{sgn}(t) = 1$ for $t > 0$.

Sinc function: $\text{sinc}(t) = \dfrac{\sin \pi t}{\pi t}$

Triangle function: $\Lambda(t) = 1 - |t|$ for $|t| \leq 1$; $\Lambda(t) = 0$ elsewhere.

Further Reading

H. Bateman, *Tables of Integral Transforms (Vol. II)*, Ed. A. Erdélyi, McGraw-Hill, New-York, **1954**.

D. C. Champeney, *Fourier Transforms and Their Physical Applications*, Academic Press, New York, **1973**.

Handbook of Mathematical Functions with Formulas, Graphs and Mathematical Tables, Ed. M. Abramowitz and I. A. Stegun, N.B.S. Applied Mathematics Series 55, Washington, **1964**.

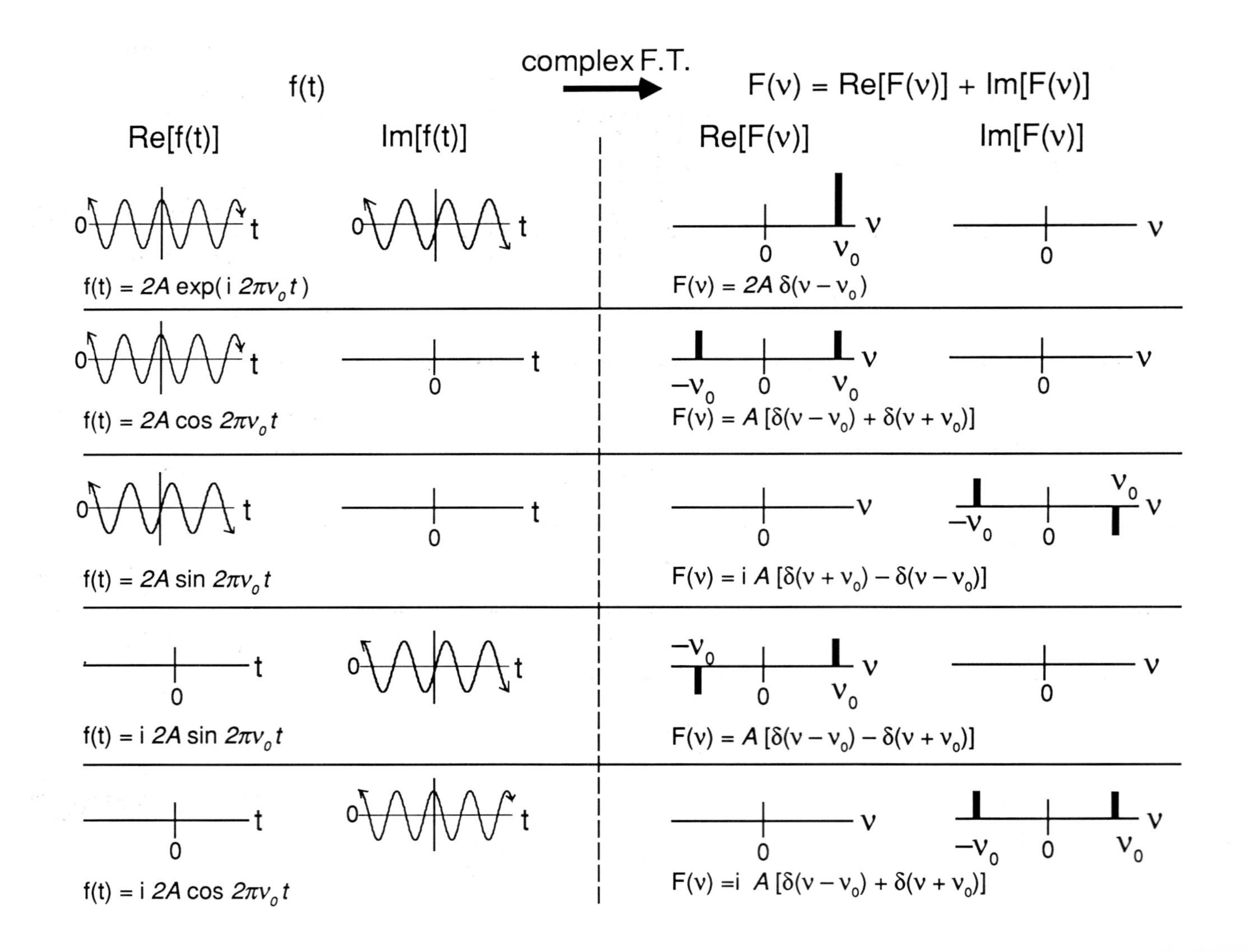
f(t)
complex F.T.
F(ν) = Re[F(ν)] + Im[F(ν)]
Re[f(t)]
Im[f(t)]
Re[F(ν)]
Im[F(ν)]
f(t) = 2A exp(i 2πν0 t)
F(ν) = 2A δ(ν − ν0)
f(t) = 2A cos 2πν0 t
F(ν) = A [δ(ν − ν0) + δ(ν + ν0)]
f(t) = 2A sin 2πν0 t
F(ν) = i A [δ(ν + ν0) − δ(ν − ν0)]
f(t) = i 2A sin 2πν0 t
F(ν) = A [δ(ν − ν0) − δ(ν + ν0)]
f(t) = i 2A cos 2πν0 t
F(ν) =i A [δ(ν − ν0) + δ(ν + ν0)]

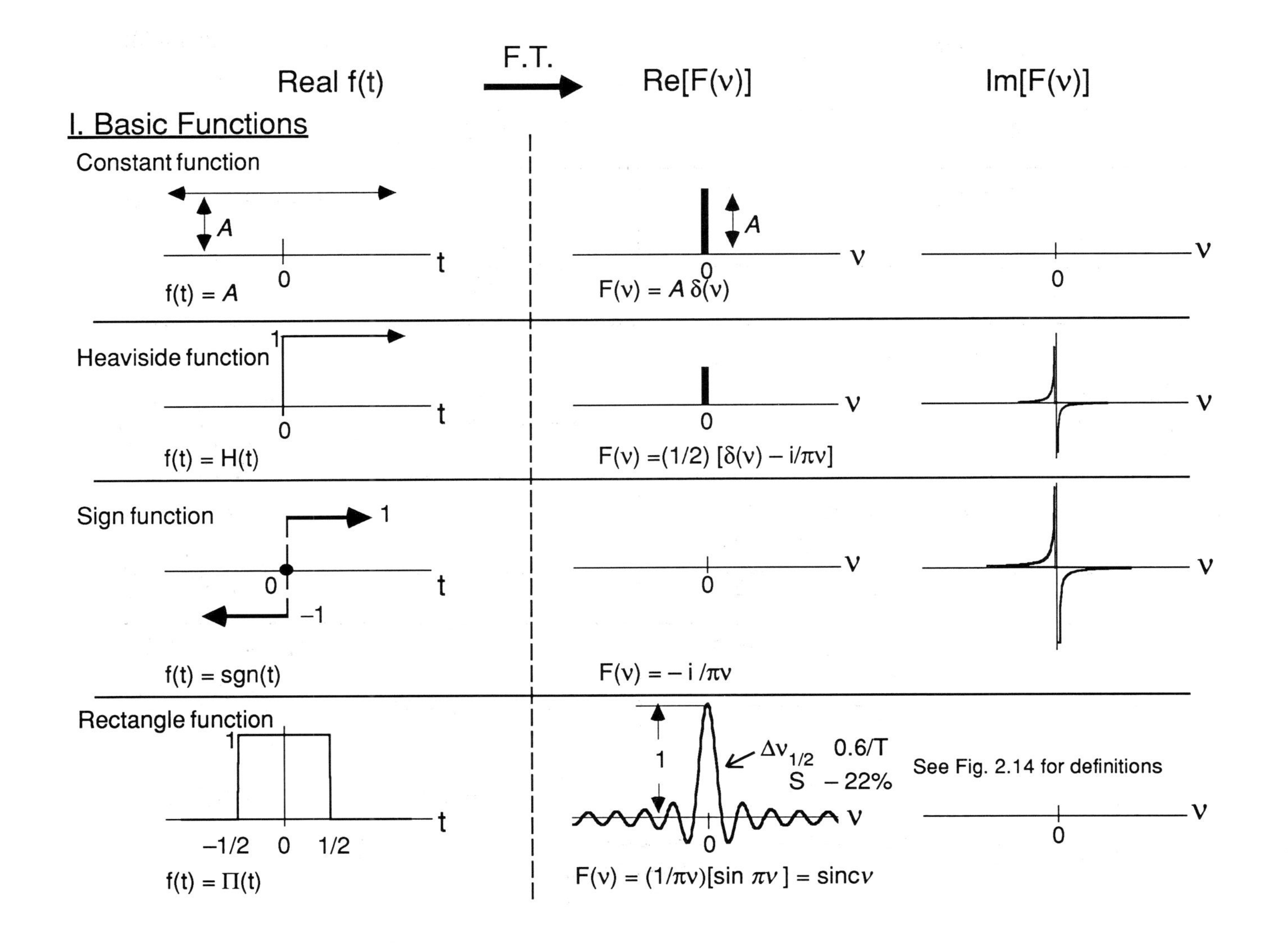

Real f(t)
F.T.
Re[F(ν)]
Im[F(ν)]
I. Basic Functions
Constant function
A
t
0
f(t) = A
ν
F(ν) = A δ(ν)
Heaviside function
1
f(t) = H(t)
F(ν) =(1/2) [δ(ν) – i/πν]
Sign function
1
–1
f(t) = sgn(t)
F(ν) = – i /πν
Rectangle function
–1/2
1/2
f(t) = Π(t)
Δν1/2 0.6/T
S – 22%
See Fig. 2.14 for definitions
F(ν) = (1/πν)[sin πν] = sincν

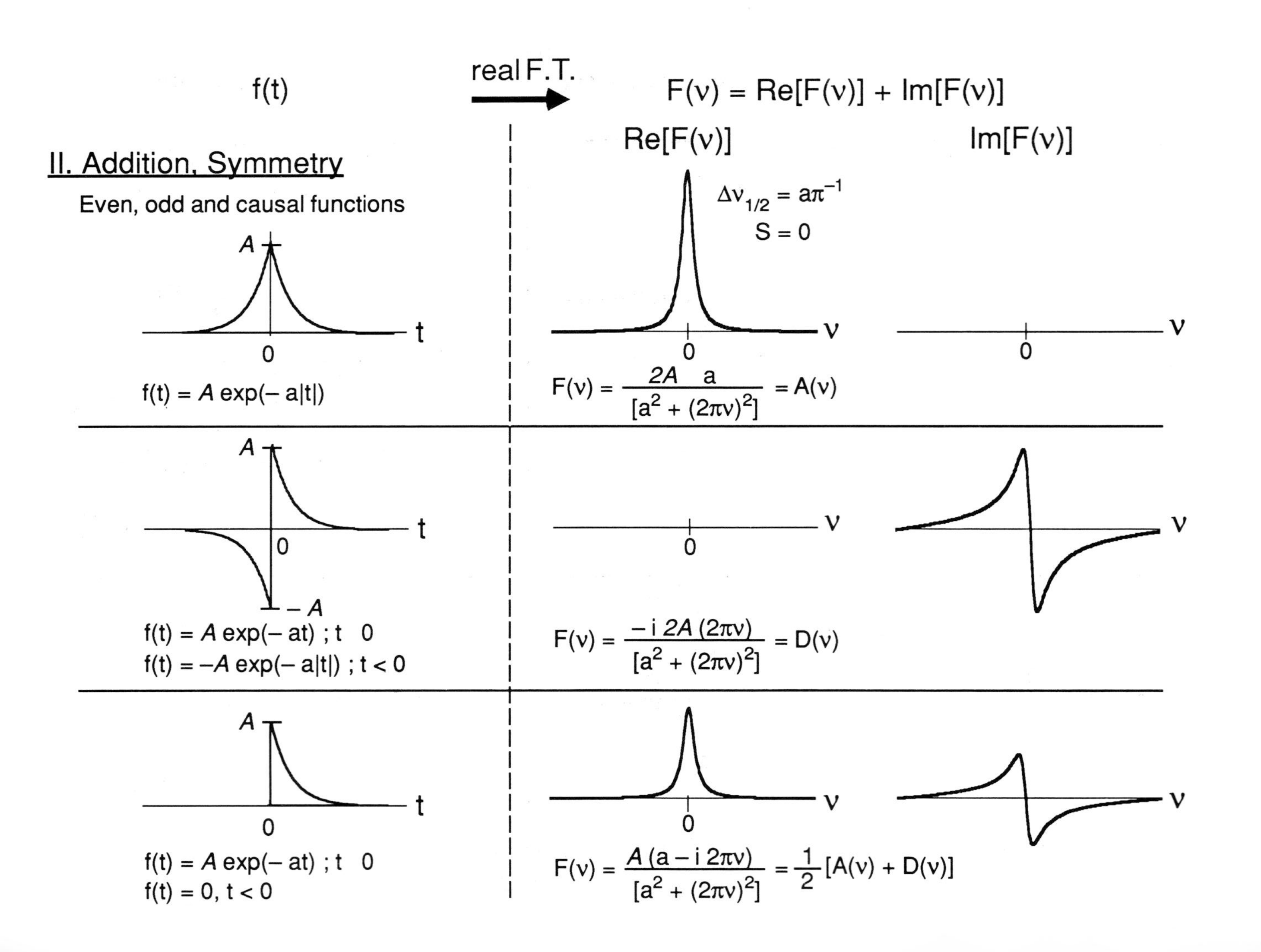
f(t)
real F.T.
F(ν) = Re[F(ν)] + Im[F(ν)]
Re[F(ν)]
Im[F(ν)]
II. Addition, Symmetry
Even, odd and causal functions
A
0
t
ν
Δν1/2 = aπ−1
S = 0
f(t) = A exp(− a|t|)
F(ν) = 2A a / [a2 + (2πν)2] = A(ν)
−A
f(t) = A exp(− at) ; t 0
f(t) = −A exp(− a|t|) ; t < 0
F(ν) = − i 2A (2πν) / [a2 + (2πν)2] = D(ν)
f(t) = A exp(− at) ; t 0
f(t) = 0, t < 0
F(ν) = A (a − i 2πν) / [a2 + (2πν)2] = 1/2 [A(ν) + D(ν)]

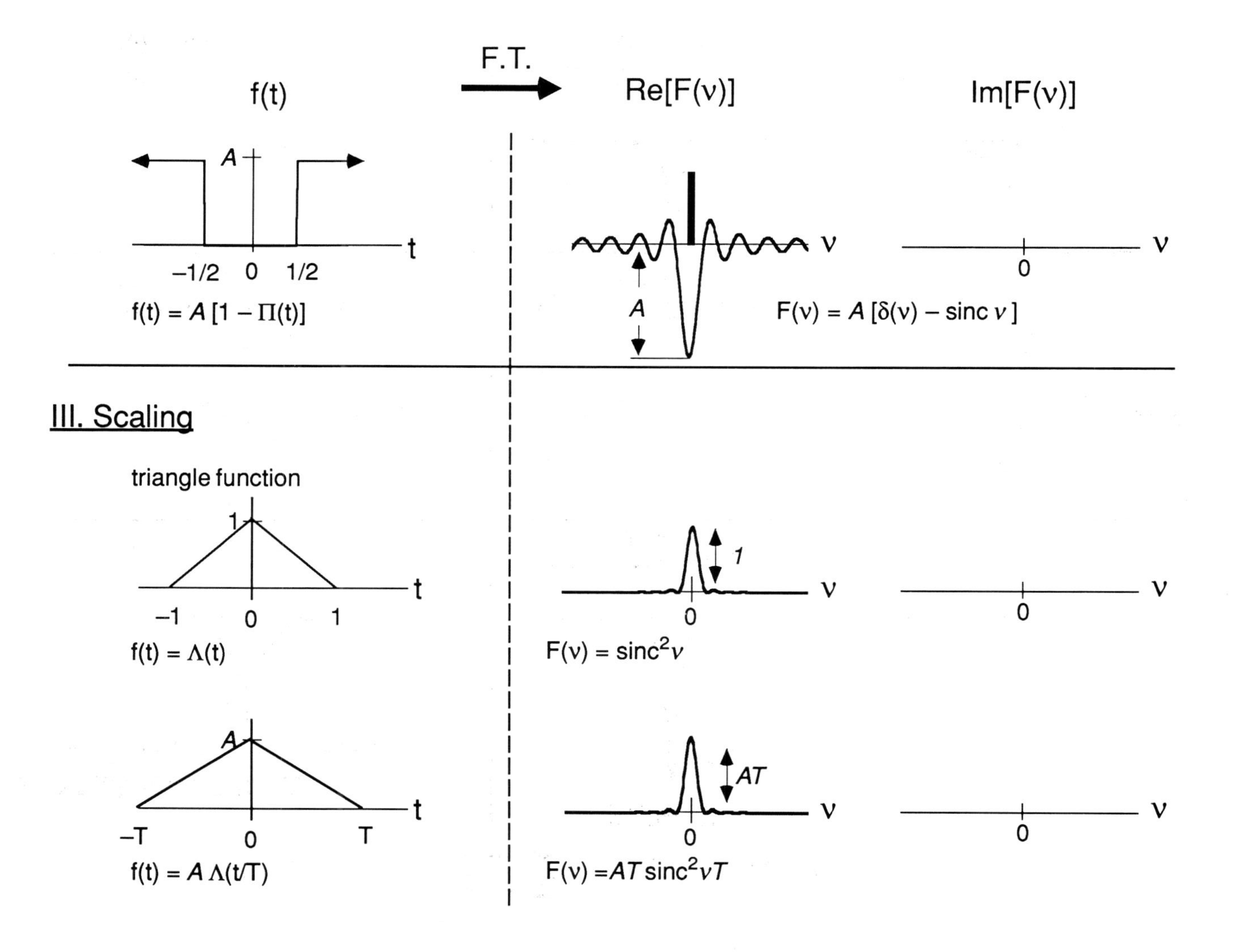

f(t)
F.T.
Re[F(ν)]
Im[F(ν)]
A
t
−1/2 0 1/2
f(t) = A [1 − Π(t)]
ν
A
F(ν) = A [δ(ν) − sinc ν]
ν
0
III. Scaling
triangle function
1
t
−1 0 1
f(t) = Λ(t)
1
ν
0
F(ν) = sinc²ν
ν
0
A
t
−T 0 T
f(t) = A Λ(t/T)
AT
ν
0
F(ν) = AT sinc²νT
ν
0

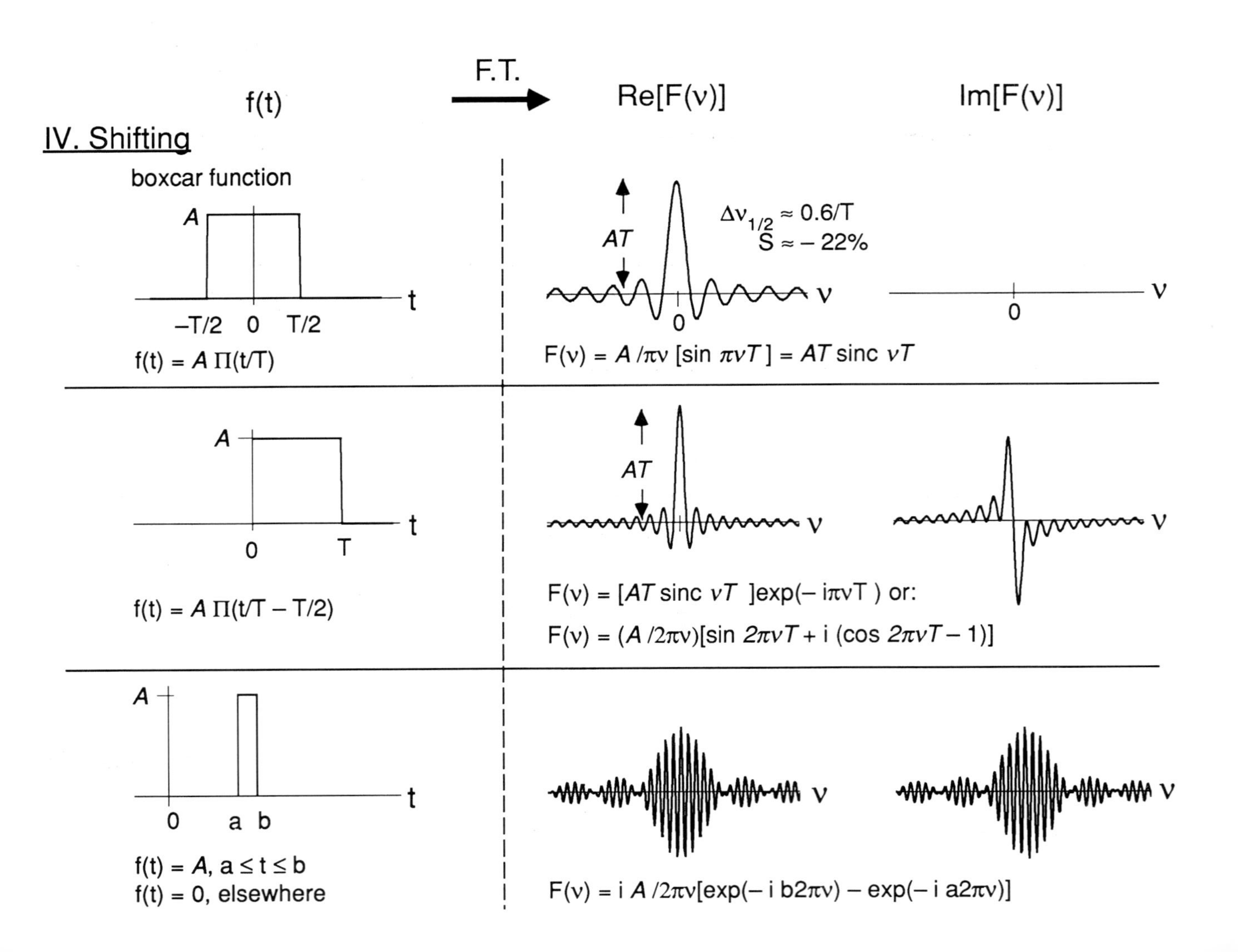

f(t)
F.T.
Re[F(ν)]
Im[F(ν)]
IV. Shifting
boxcar function
A
–T/2 0 T/2
t
f(t) = A Π(t/T)
AT
Δν1/2 ≈ 0.6/T
S ≈ – 22%
ν
0
F(ν) = A /πν [sin πνT] = AT sinc νT
A
0 T
t
f(t) = A Π(t/T – T/2)
AT
ν
F(ν) = [AT sinc νT]exp(– iπνT) or:
F(ν) = (A /2πν)[sin 2πνT + i (cos 2πνT – 1)]
A
0 a b
t
f(t) = A, a ≤ t ≤ b
f(t) = 0, elsewhere
ν
F(ν) = i A /2πν[exp(– i b2πν) – exp(– i a2πν)]

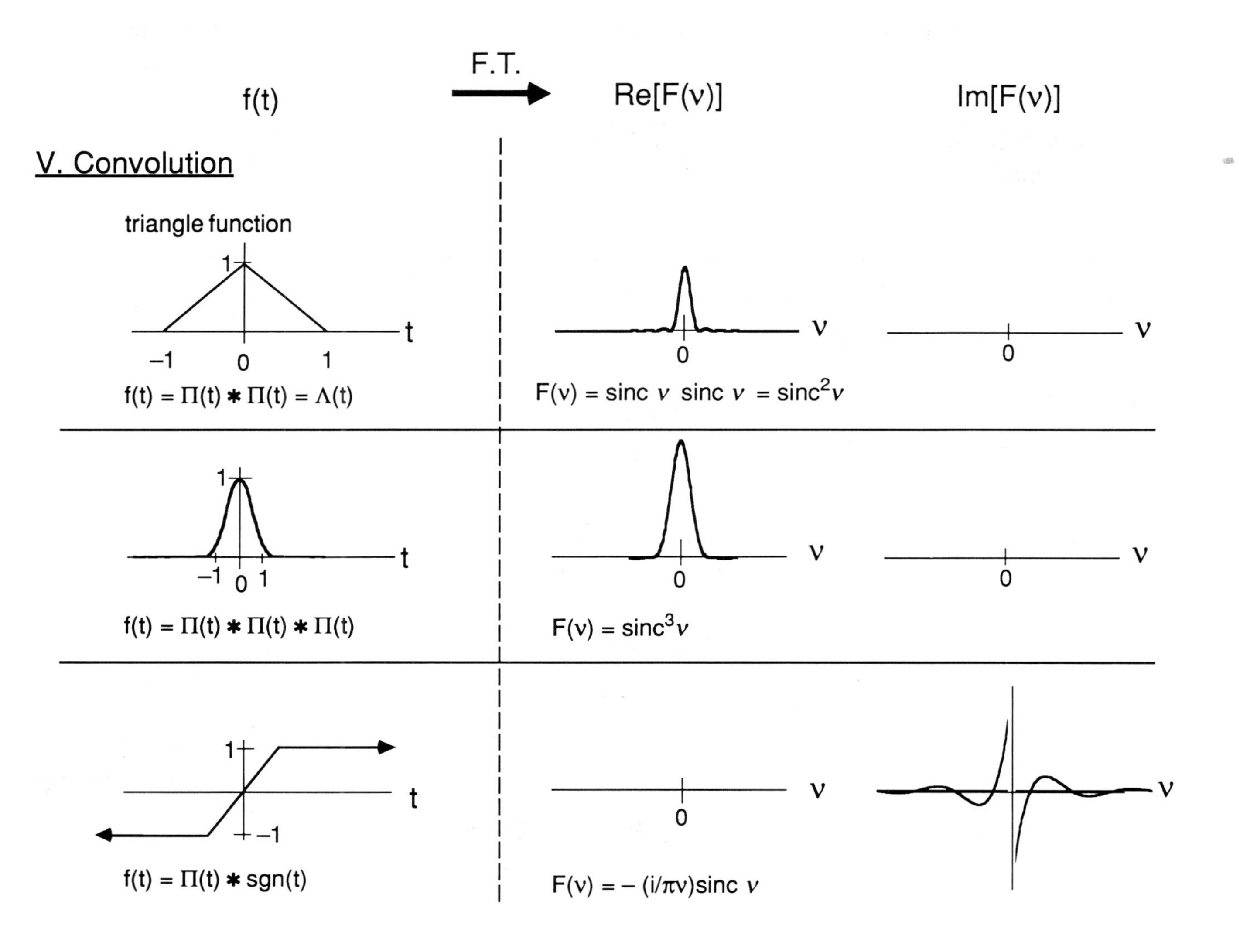

F.T.
f(t)
Re[F(ν)]
Im[F(ν)]
V. Convolution
triangle function
1
−1
0
1
t
ν
0
f(t) = Π(t) ∗ Π(t) = Λ(t)
F(ν) = sinc ν sinc ν = sinc²ν
−1
0
1
f(t) = Π(t) ∗ Π(t) ∗ Π(t)
F(ν) = sinc³ν
f(t) = Π(t) ∗ sgn(t)
F(ν) = − (i/πν)sinc ν

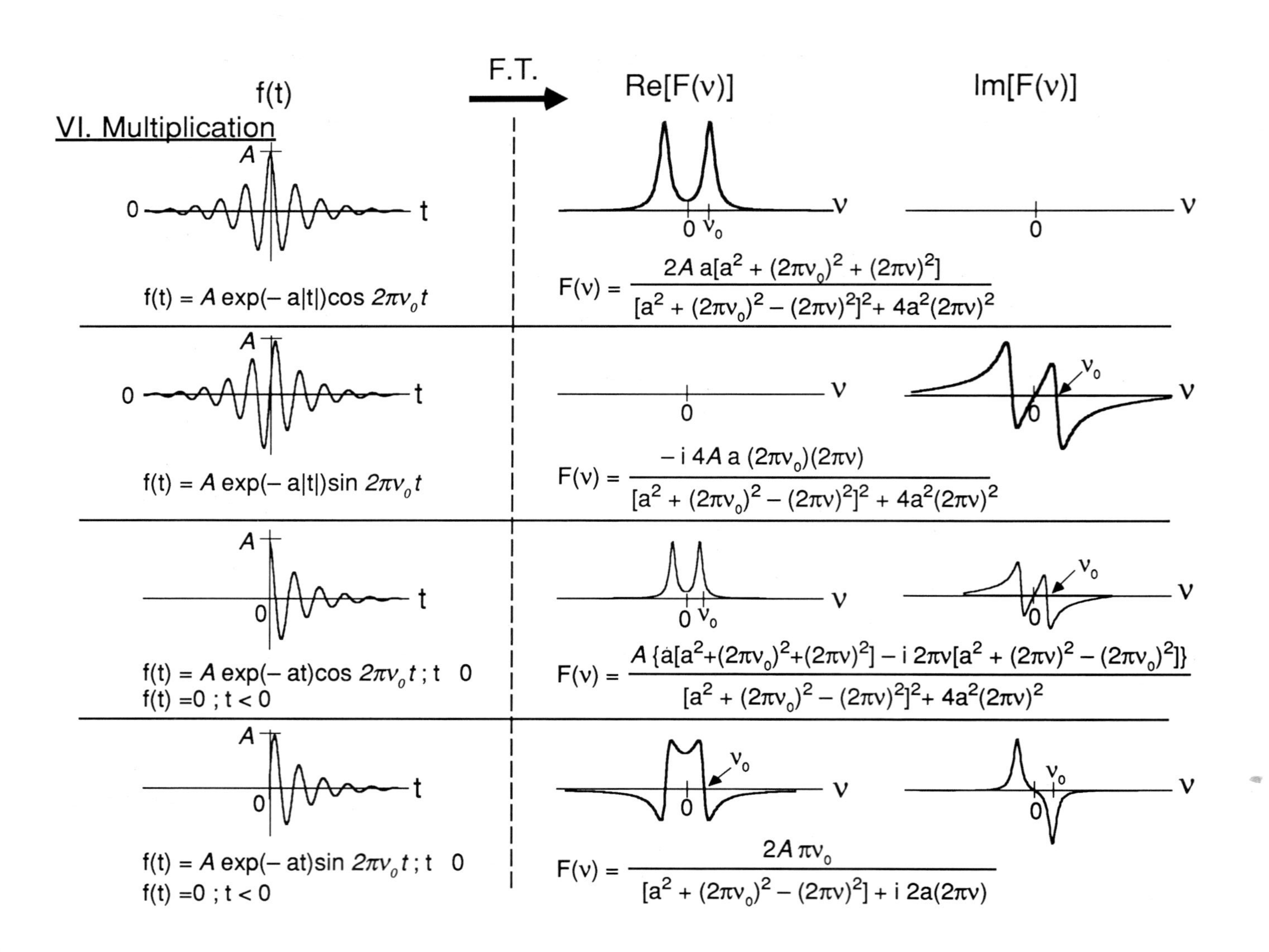
VI. Multiplication
f(t)
F.T.
Re[F(ν)]
Im[F(ν)]
A
0
t
ν
ν₀
f(t) = A exp(– a|t|)cos 2πν₀t
F(ν) = 2A a[a² + (2πν₀)² + (2πν)²] / [a² + (2πν₀)² – (2πν)²]² + 4a²(2πν)²
f(t) = A exp(– a|t|)sin 2πν₀t
F(ν) = – i 4A a (2πν₀)(2πν) / [a² + (2πν₀)² – (2πν)²]² + 4a²(2πν)²
f(t) = A exp(– at)cos 2πν₀t ; t 0
f(t) =0 ; t < 0
F(ν) = A {a[a²+(2πν₀)²+(2πν)²] – i 2πν[a² + (2πν)² – (2πν₀)²]} / [a² + (2πν₀)² – (2πν)²]²+ 4a²(2πν)²
f(t) = A exp(– at)sin 2πν₀t ; t 0
f(t) =0 ; t < 0
F(ν) = 2A πν₀ / [a² + (2πν₀)² – (2πν)²] + i 2a(2πν)

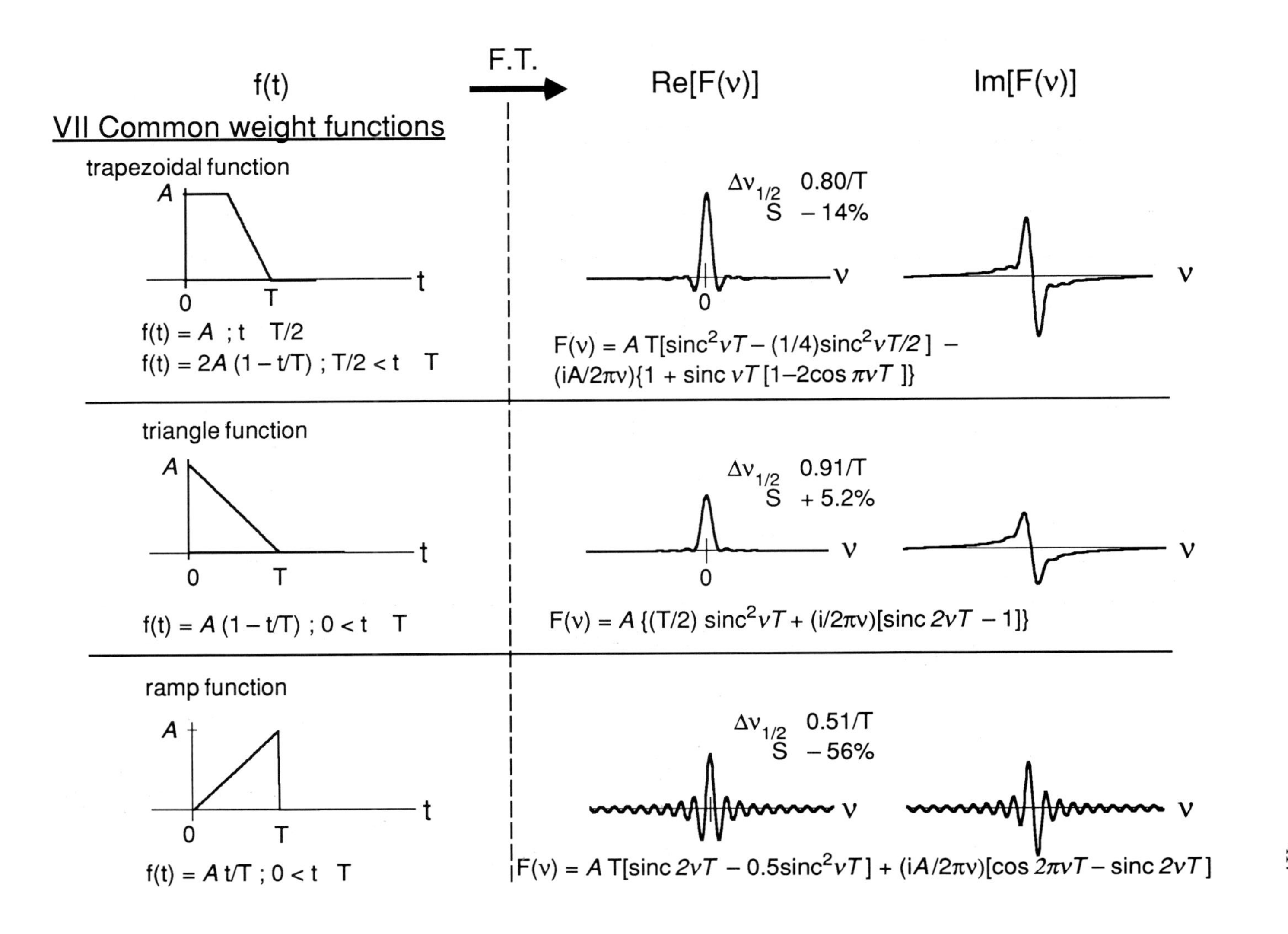
f(t)
F.T.
Re[F(ν)]
Im[F(ν)]
VII Common weight functions
trapezoidal function
A
0
T
t
Δν1/2 0.80/T
S − 14%
ν
0
ν
f(t) = A ; t T/2
f(t) = 2A (1 − t/T) ; T/2 < t T
F(ν) = A T[sinc²νT − (1/4)sinc²νT/2] −
(iA/2πν){1 + sinc νT [1−2cos πνT]}
triangle function
A
0
T
t
Δν1/2 0.91/T
S + 5.2%
ν
0
ν
f(t) = A (1 − t/T) ; 0 < t T
F(ν) = A {(T/2) sinc²νT + (i/2πν)[sinc 2νT − 1]}
ramp function
A
0
T
t
Δν1/2 0.51/T
S − 56%
ν
ν
f(t) = A t/T ; 0 < t T
F(ν) = A T[sinc 2νT − 0.5sinc²νT] + (iA/2πν)[cos 2πνT − sinc 2νT]

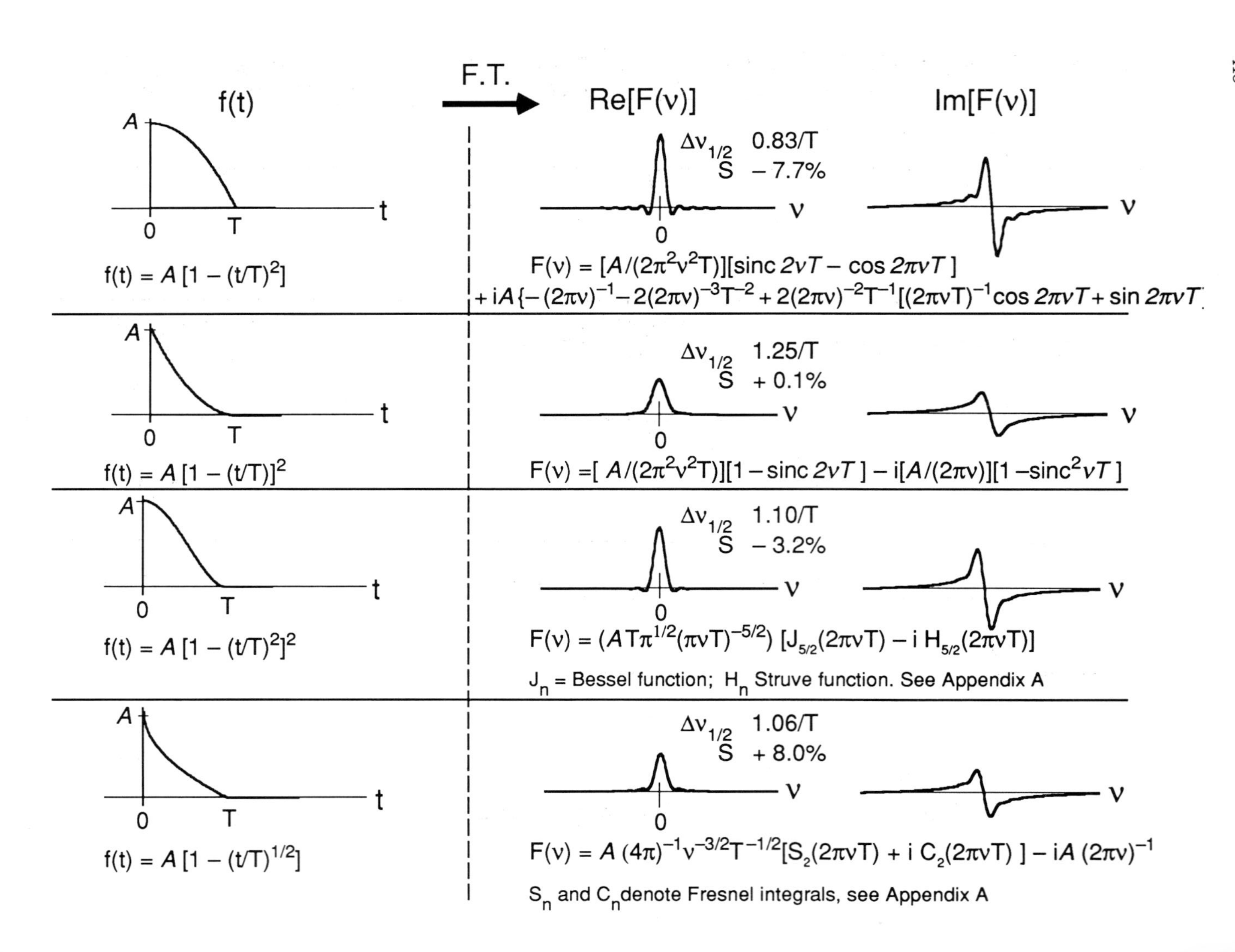
f(t)
F.T.
Re[F(ν)]
Im[F(ν)]
A
0
T
t
ν
f(t) = A [1 – (t/T)²]
Δν1/2 0.83/T
S – 7.7%
F(ν) = [A/(2π²ν²T)][sinc 2νT – cos 2πνT]
+ iA{– (2πν)⁻¹ – 2(2πν)⁻³T⁻² + 2(2πν)⁻²T⁻¹[(2πνT)⁻¹cos 2πνT + sin 2πνT
f(t) = A [1 – (t/T)]²
Δν1/2 1.25/T
S + 0.1%
F(ν) =[A/(2π²ν²T)][1 – sinc 2νT] – i[A/(2πν)][1 –sinc²νT]
f(t) = A [1 – (t/T)²]²
Δν1/2 1.10/T
S – 3.2%
F(ν) = (ATπ^1/2(πνT)^–5/2) [J5/2(2πνT) – i H5/2(2πνT)]
Jn = Bessel function; Hn Struve function. See Appendix A
f(t) = A [1 – (t/T)^1/2]
Δν1/2 1.06/T
S + 8.0%
F(ν) = A (4π)⁻¹ν^–3/2T^–1/2[S2(2πνT) + i C2(2πνT)] – iA (2πν)⁻¹
Sn and Cn denote Fresnel integrals, see Appendix A

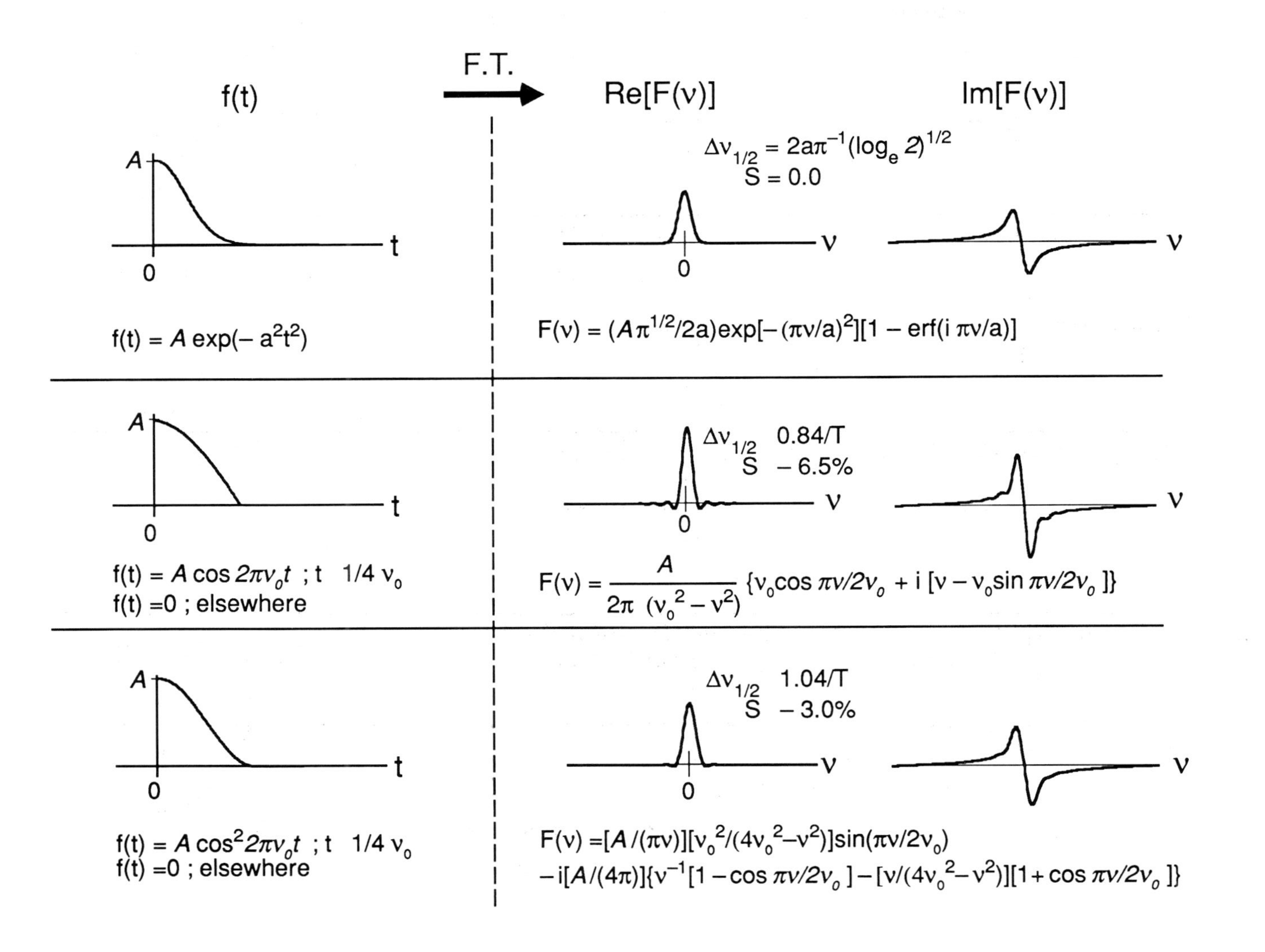

f(t)
F.T.
Re[F(ν)]
Im[F(ν)]
A
0
t
Δν1/2 = 2aπ−1(loge 2)1/2
S = 0.0
ν
f(t) = A exp(− a2t2)
F(ν) = (Aπ1/2/2a)exp[−(πν/a)2][1 − erf(i πν/a)]
Δν1/2 0.84/T
S − 6.5%
f(t) = A cos 2πν0t ; t 1/4 ν0
f(t) = 0 ; elsewhere
F(ν) = A / 2π (ν0² − ν²) {ν0cos πν/2ν0 + i [ν − ν0sin πν/2ν0]}
Δν1/2 1.04/T
S − 3.0%
f(t) = A cos²2πν0t ; t 1/4 ν0
f(t) = 0 ; elsewhere
F(ν) = [A/(πν)][ν0²/(4ν0² − ν²)]sin(πν/2ν0)
− i[A/(4π)]{ν−1[1 − cos πν/2ν0] − [ν/(4ν0² − ν²)][1 + cos πν/2ν0]}

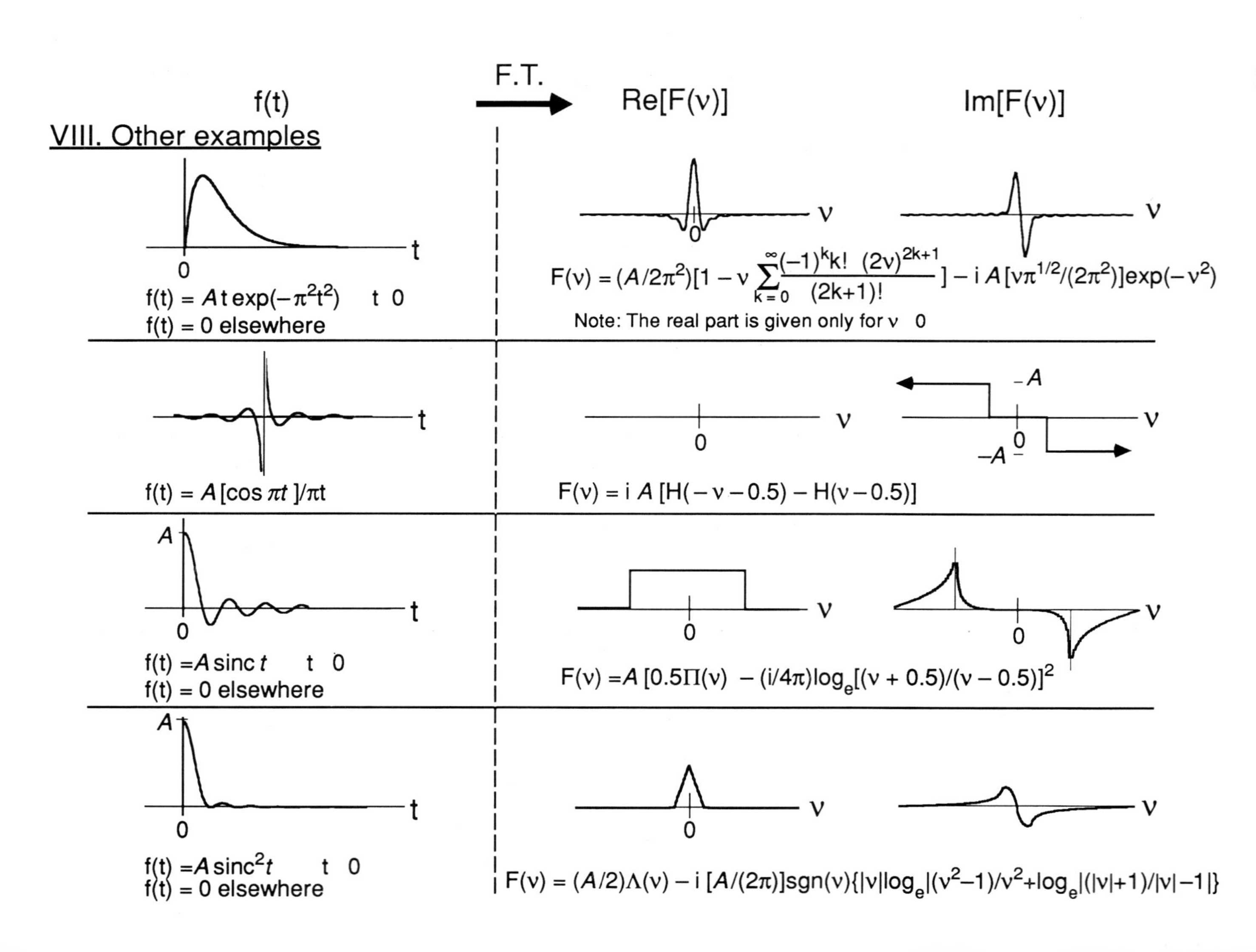

f(t)
F.T.
Re[F(ν)]
Im[F(ν)]
VIII. Other examples
f(t) = A t exp(−π²t²)   t 0
f(t) = 0 elsewhere
F(ν) = (A/2π²)[1 − ν Σ_{k=0}^{∞} (−1)^k k! (2ν)^{2k+1} / (2k+1)!] − i A[νπ^{1/2}/(2π²)]exp(−ν²)
Note: The real part is given only for ν 0
f(t) = A[cos πt]/πt
F(ν) = i A [H(−ν−0.5) − H(ν−0.5)]
−A
f(t) = A sinc t   t 0
f(t) = 0 elsewhere
F(ν) = A [0.5Π(ν) − (i/4π)log_e[(ν + 0.5)/(ν − 0.5)]²
f(t) = A sinc²t   t 0
f(t) = 0 elsewhere
F(ν) = (A/2)Λ(ν) − i [A/(2π)]sgn(ν){|ν|log_e|(ν²−1)/ν²+log_e|(|ν|+1)/|ν|−1|}

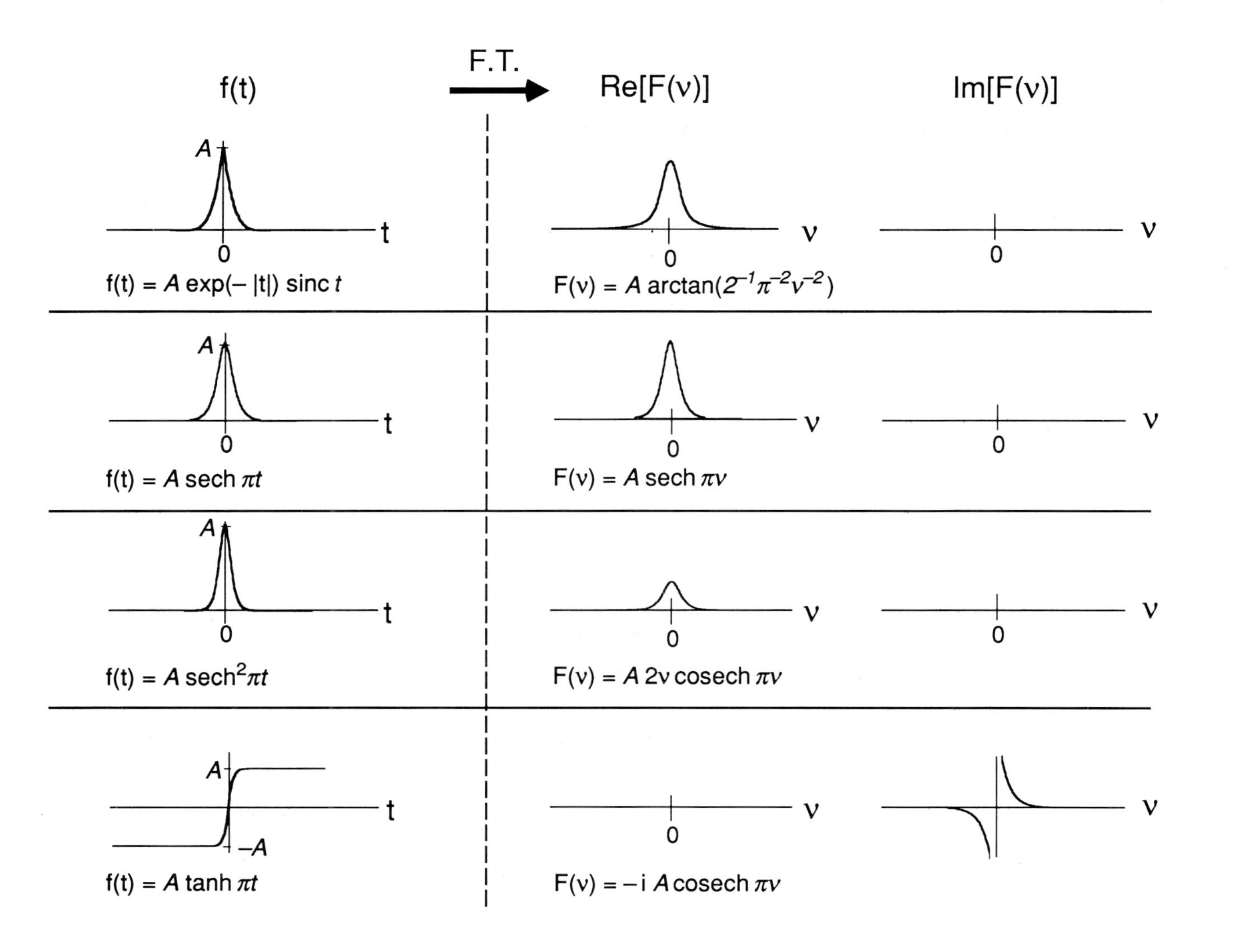
f(t)
F.T.
Re[F(ν)]
Im[F(ν)]
A
0
t
ν
f(t) = A exp(− |t|) sinc t
F(ν) = A arctan(2⁻¹π⁻²ν⁻²)
f(t) = A sech πt
F(ν) = A sech πν
f(t) = A sech²πt
F(ν) = A 2ν cosech πν
−A
f(t) = A tanh πt
F(ν) = − i A cosech πν

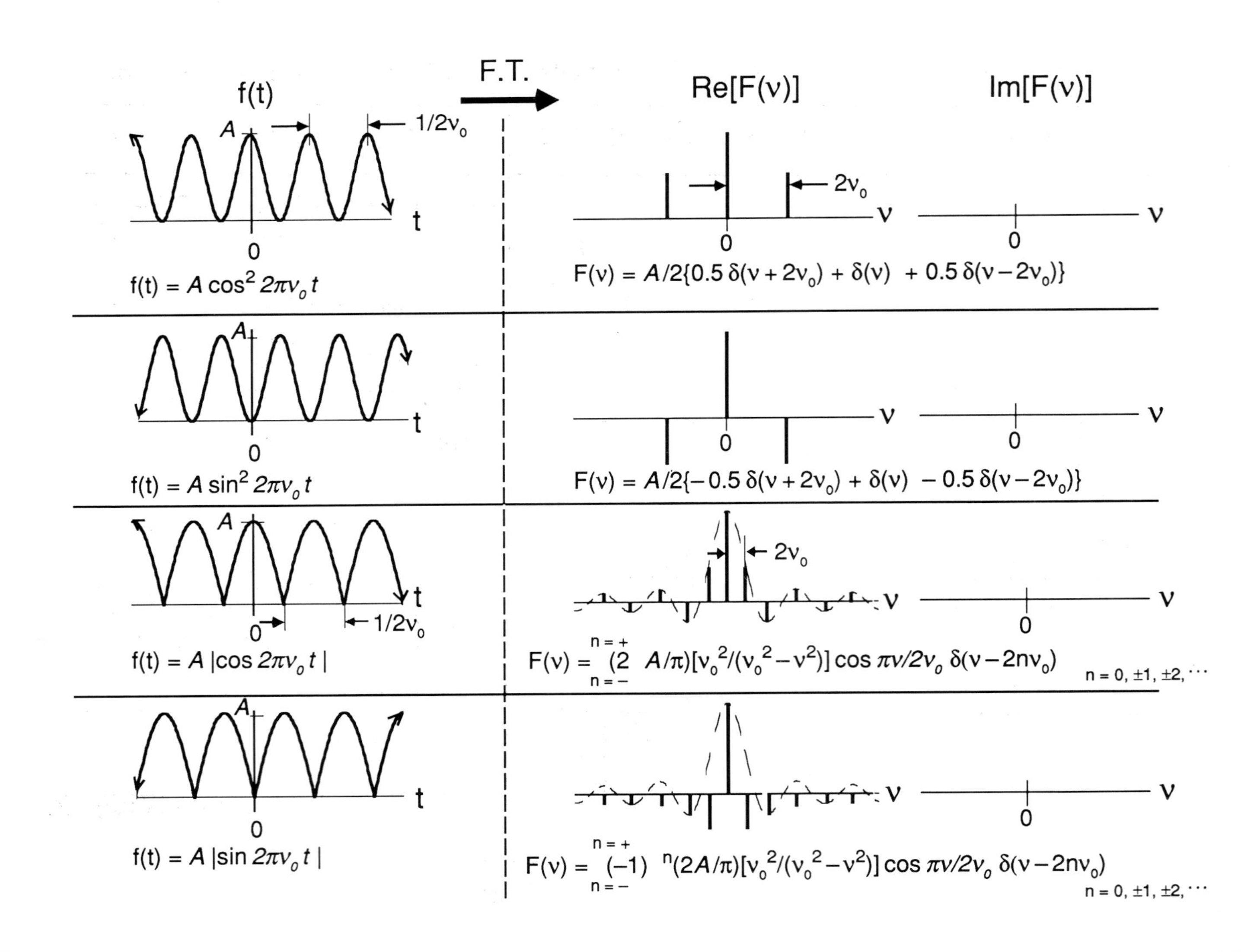
f(t)
F.T.
Re[F(ν)]
Im[F(ν)]
1/2ν0
2ν0
f(t) = A cos² 2πν0 t
F(ν) = A/2{0.5 δ(ν + 2ν0) + δ(ν) + 0.5 δ(ν − 2ν0)}
f(t) = A sin² 2πν0 t
F(ν) = A/2{−0.5 δ(ν + 2ν0) + δ(ν) − 0.5 δ(ν − 2ν0)}
f(t) = A |cos 2πν0 t|
F(ν) = (2 A/π)[ν0²/(ν0² − ν²)] cos πν/2ν0 δ(ν − 2nν0)
n = 0, ±1, ±2, ···
f(t) = A |sin 2πν0 t|
F(ν) = (−1)ⁿ(2A/π)[ν0²/(ν0² − ν²)] cos πν/2ν0 δ(ν − 2nν0)
n = 0, ±1, ±2, ···

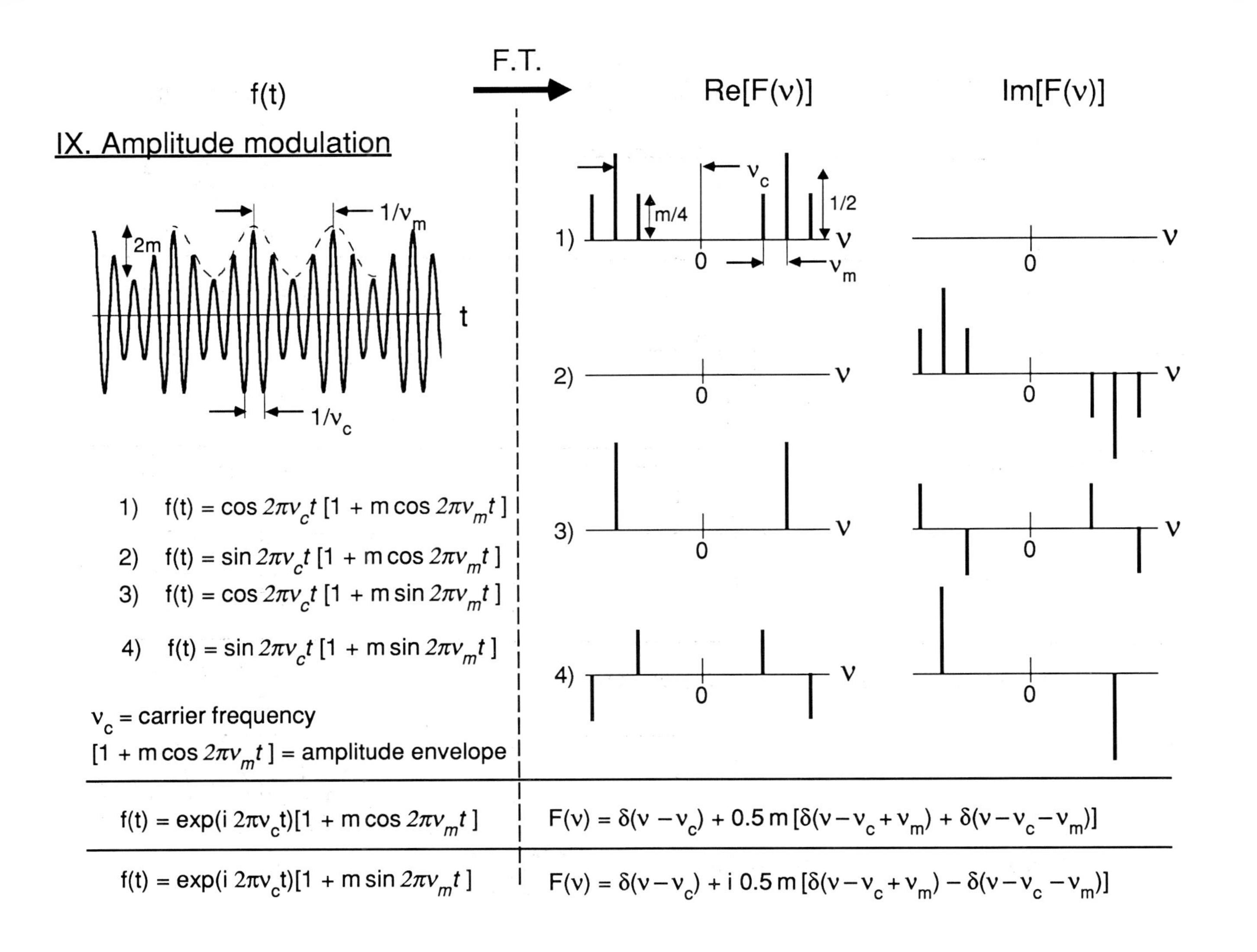

f(t)
F.T.
Re[F(ν)]
Im[F(ν)]
IX. Amplitude modulation
2m
1/ν_m
1/ν_c
t
m/4
ν_c
1/2
ν_m
ν
0
1)
2)
3)
4)
1) f(t) = cos 2πν_c t [1 + m cos 2πν_m t]
2) f(t) = sin 2πν_c t [1 + m cos 2πν_m t]
3) f(t) = cos 2πν_c t [1 + m sin 2πν_m t]
4) f(t) = sin 2πν_c t [1 + m sin 2πν_m t]
ν_c = carrier frequency
[1 + m cos 2πν_m t] = amplitude envelope
f(t) = exp(i 2πν_c t)[1 + m cos 2πν_m t]
F(ν) = δ(ν − ν_c) + 0.5 m [δ(ν − ν_c + ν_m) + δ(ν − ν_c − ν_m)]
f(t) = exp(i 2πν_c t)[1 + m sin 2πν_m t]
F(ν) = δ(ν − ν_c) + i 0.5 m [δ(ν − ν_c + ν_m) − δ(ν − ν_c − ν_m)]

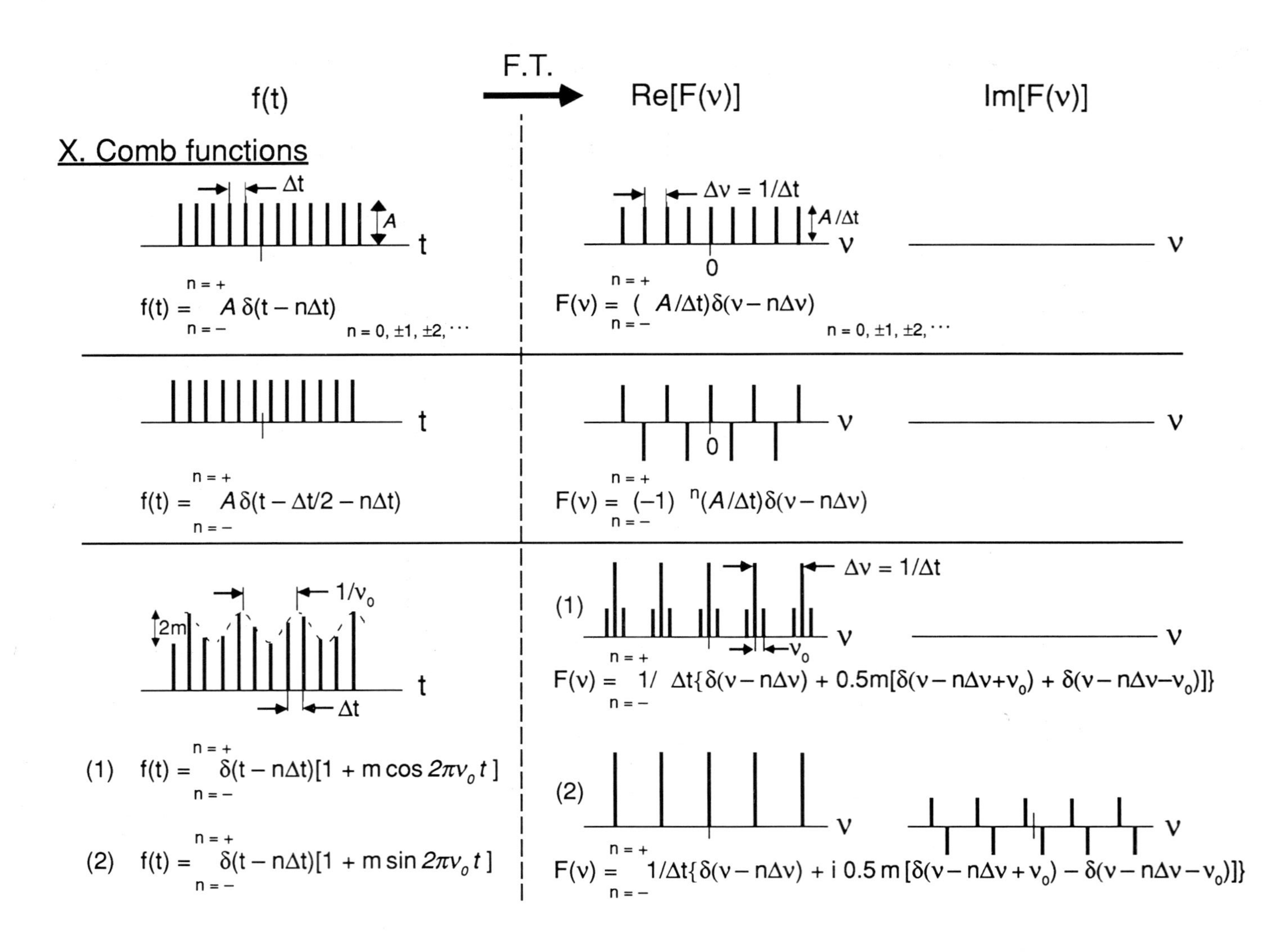

f(t)
F.T.
Re[F(ν)]
Im[F(ν)]
X. Comb functions
Δt
A
t
f(t) = A δ(t − nΔt)
n = +
n = −
n = 0, ±1, ±2, ···
Δν = 1/Δt
A/Δt
ν
0
F(ν) = (A/Δt)δ(ν − nΔν)
n = 0, ±1, ±2, ···
f(t) = A δ(t − Δt/2 − nΔt)
F(ν) = (−1) ⁿ(A/Δt)δ(ν − nΔν)
1/ν0
2m
(1)
ν0
F(ν) = 1/ Δt{δ(ν − nΔν) + 0.5m[δ(ν − nΔν+ν0) + δ(ν − nΔν−ν0)]}
(1) f(t) = δ(t − nΔt)[1 + m cos 2πν0 t]
(2) f(t) = δ(t − nΔt)[1 + m sin 2πν0 t]
(2)
F(ν) = 1/Δt{δ(ν − nΔν) + i 0.5 m [δ(ν − nΔν + ν0) − δ(ν − nΔν − ν0)]}

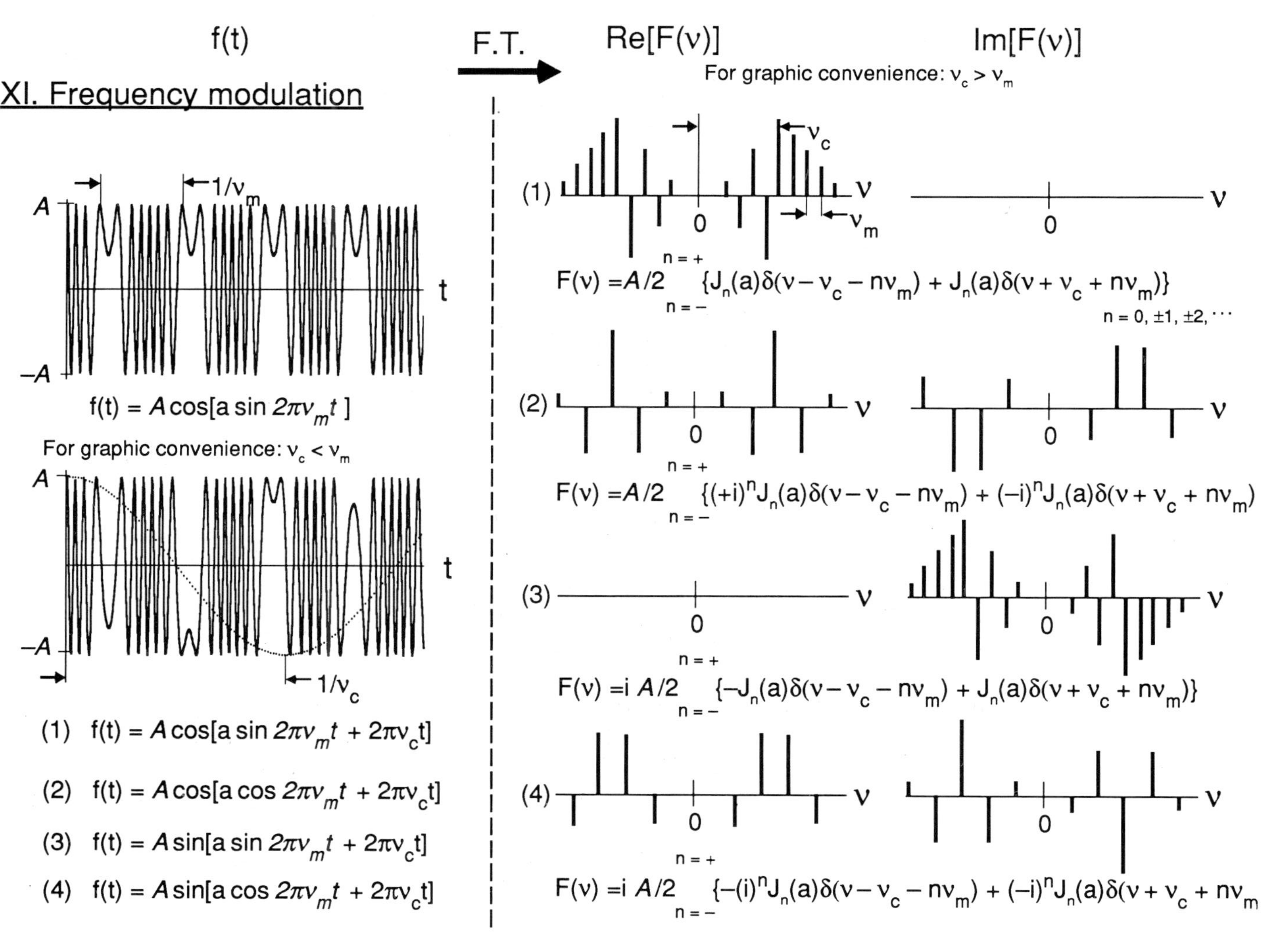
f(t)
F.T.
Re[F(ν)]
Im[F(ν)]
XI. Frequency modulation
For graphic convenience: $\nu_c > \nu_m$
$1/\nu_m$
A
−A
t
$f(t) = A\cos[a \sin 2\pi\nu_m t]$
For graphic convenience: $\nu_c < \nu_m$
$1/\nu_c$
(1) $f(t) = A\cos[a \sin 2\pi\nu_m t + 2\pi\nu_c t]$
(2) $f(t) = A\cos[a \cos 2\pi\nu_m t + 2\pi\nu_c t]$
(3) $f(t) = A\sin[a \sin 2\pi\nu_m t + 2\pi\nu_c t]$
(4) $f(t) = A\sin[a \cos 2\pi\nu_m t + 2\pi\nu_c t]$
ν_c
ν_m
ν
0
(1)
$F(\nu) = A/2 \sum_{n=-}^{n=+} \{J_n(a)\delta(\nu - \nu_c - n\nu_m) + J_n(a)\delta(\nu + \nu_c + n\nu_m)\}$
$n = 0, \pm1, \pm2, \cdots$
(2)
$F(\nu) = A/2 \sum_{n=-}^{n=+} \{(+i)^n J_n(a)\delta(\nu - \nu_c - n\nu_m) + (-i)^n J_n(a)\delta(\nu + \nu_c + n\nu_m)$
(3)
$F(\nu) = i\,A/2 \sum_{n=-}^{n=+} \{-J_n(a)\delta(\nu - \nu_c - n\nu_m) + J_n(a)\delta(\nu + \nu_c + n\nu_m)\}$
(4)
$F(\nu) = i\,A/2 \sum_{n=-}^{n=+} \{-(i)^n J_n(a)\delta(\nu - \nu_c - n\nu_m) + (-i)^n J_n(a)\delta(\nu + \nu_c + n\nu_m$

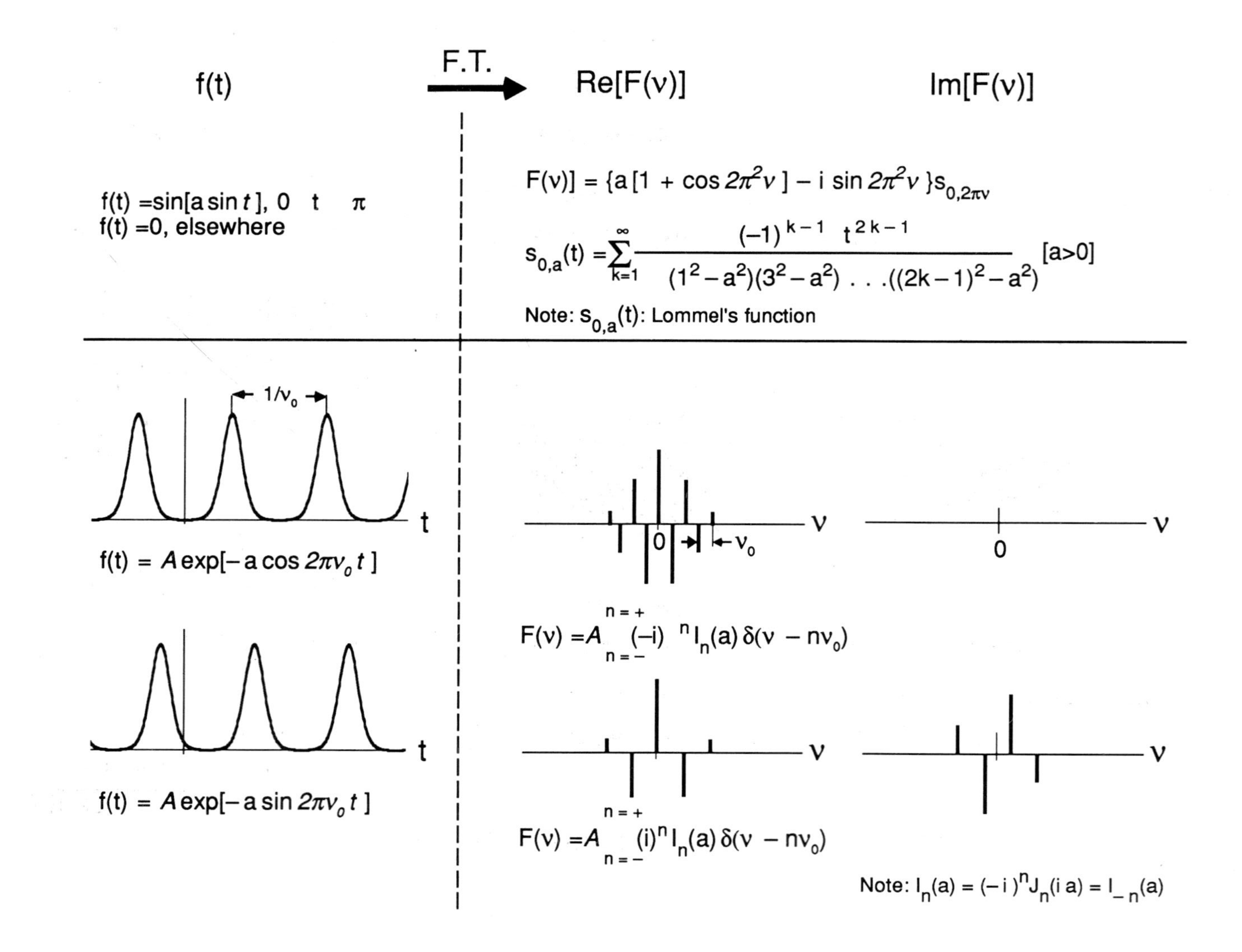

f(t)
F.T.
Re[F(ν)]
Im[F(ν)]
f(t) =sin[a sin t], 0 t π
f(t) =0, elsewhere
F(ν)] = {a[1 + cos 2π²ν] – i sin 2π²ν}s0,2πν
s0,a(t) = Σ (–1)^(k–1) t^(2k–1) / (1² – a²)(3² – a²) . . .((2k – 1)² – a²) [a>0]
Note: s0,a(t): Lommel's function
1/ν0
t
f(t) = A exp[– a cos 2πν0 t]
0
ν0
ν
F(ν) = A Σ (–i)^n In(a) δ(ν – nν0)
f(t) = A exp[– a sin 2πν0 t]
F(ν) = A Σ (i)^n In(a) δ(ν – nν0)
Note: In(a) = (– i)^n Jn(i a) = I–n(a)

f(t) $\xrightarrow{\text{F.T.}}$ Re[F(ν)] Im[F(ν)]

XII. Chirp function

Circular polarization:

$f^{+}(t) = A \exp[+i(2\pi\nu_1 t + (a/2)t^2)]$

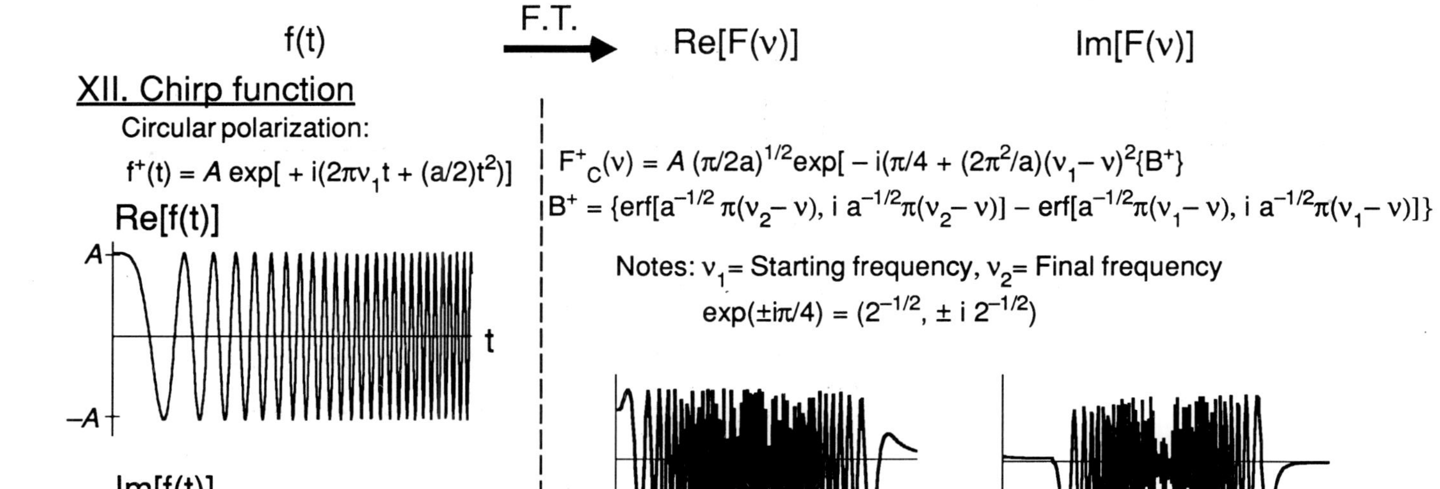

$f^{-}(t) = A \exp[-i(2\pi\nu_1 t + (a/2)t^2)]$

$F^{+}_{C}(\nu) = A\,(\pi/2a)^{1/2}\exp[-i(\pi/4 + (2\pi^2/a)(\nu_1 - \nu)^2\{B^{+}\}$

$B^{+} = \{\mathrm{erf}[a^{-1/2}\pi(\nu_2 - \nu), i\,a^{-1/2}\pi(\nu_2 - \nu)] - \mathrm{erf}[a^{-1/2}\pi(\nu_1 - \nu), i\,a^{-1/2}\pi(\nu_1 - \nu)]\}$

Notes: ν_1 = Starting frequency, ν_2 = Final frequency

$\exp(\pm i\pi/4) = (2^{-1/2}, \pm i\,2^{-1/2})$

$F^{-}_{C}(\nu) = A\,(\pi/2a)^{1/2}\exp[+i(\pi/4 + (2\pi^2/a)(\nu_1 - \nu)^2\{B^{-}\}$

$B^{-} = \{\mathrm{erf}[a^{-1/2}\pi(\nu_2 - \nu), -i\,a^{-1/2}\pi(\nu_2 - \nu)] - \mathrm{erf}[a^{-1/2}\pi(\nu_1 - \nu), -i\,a^{-1/2}\pi(\nu_1 - \nu)]\}$

Linear polarization:

$f(t) = A \cos[(2\pi\nu_1 t + (a/2)t^2)]$

$f(t) = A \sin[(2\pi\nu_1 t + (a/2)t^2)]$

Use of Euler equation

$F(\nu) = (1/2)[F^{+}_{C}(\nu) + F^{-}_{C}(-\nu)]$

$F(\nu) = (1/2i)[F^{+}_{C}(\nu) - F^{-}_{C}(-\nu)]$

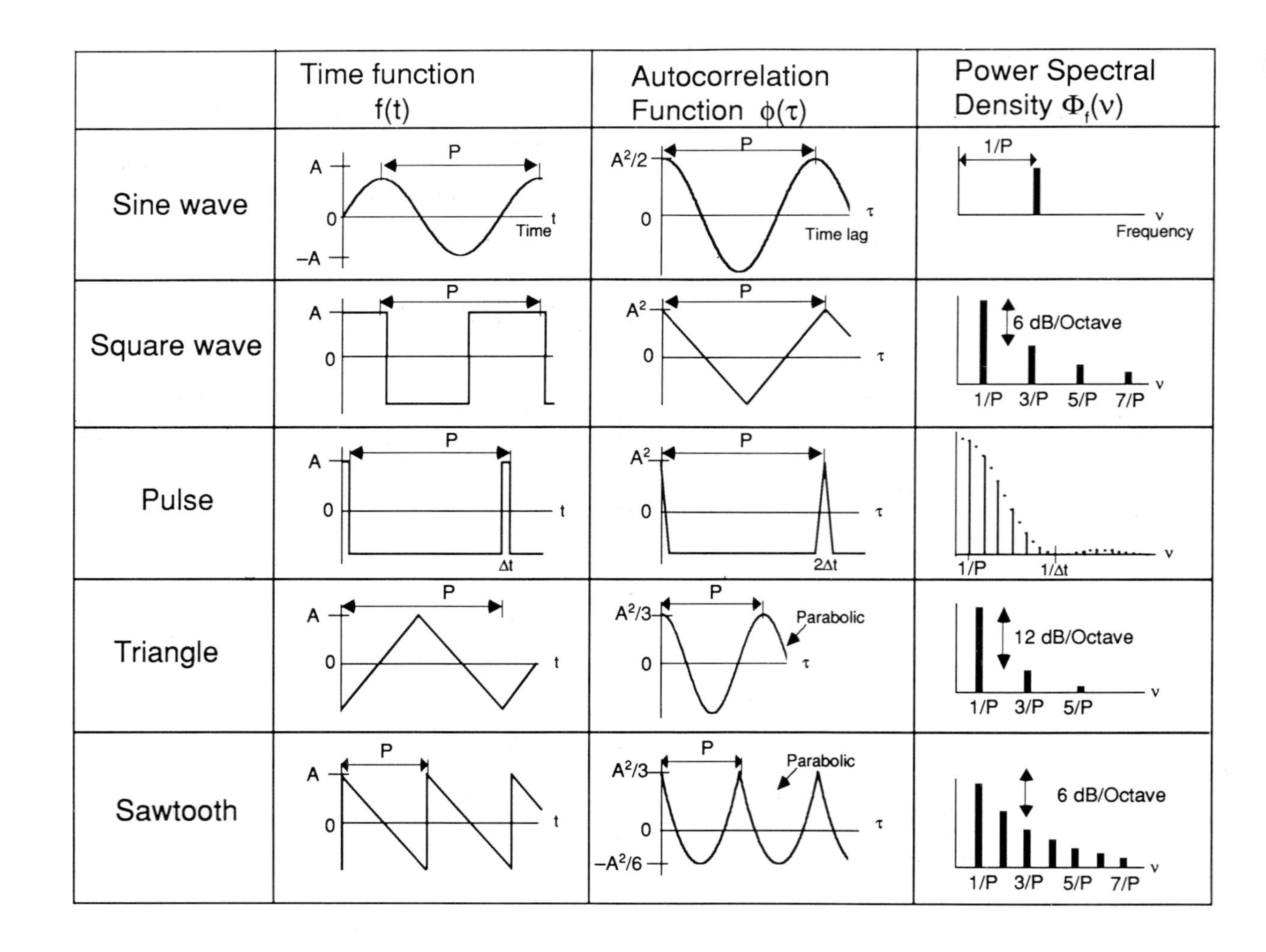

Time function
f(t)
Autocorrelation
Function φ(τ)
Power Spectral
Density Φf(ν)
Sine wave
Square wave
Pulse
Triangle
Sawtooth
A
0
–A
P
t
Time
A²/2
τ
Time lag
1/P
ν
Frequency
A²
6 dB/Octave
1/P 3/P 5/P 7/P
Δt
2Δt
1/Δt
A²/3
Parabolic
12 dB/Octave
1/P 3/P 5/P
–A²/6

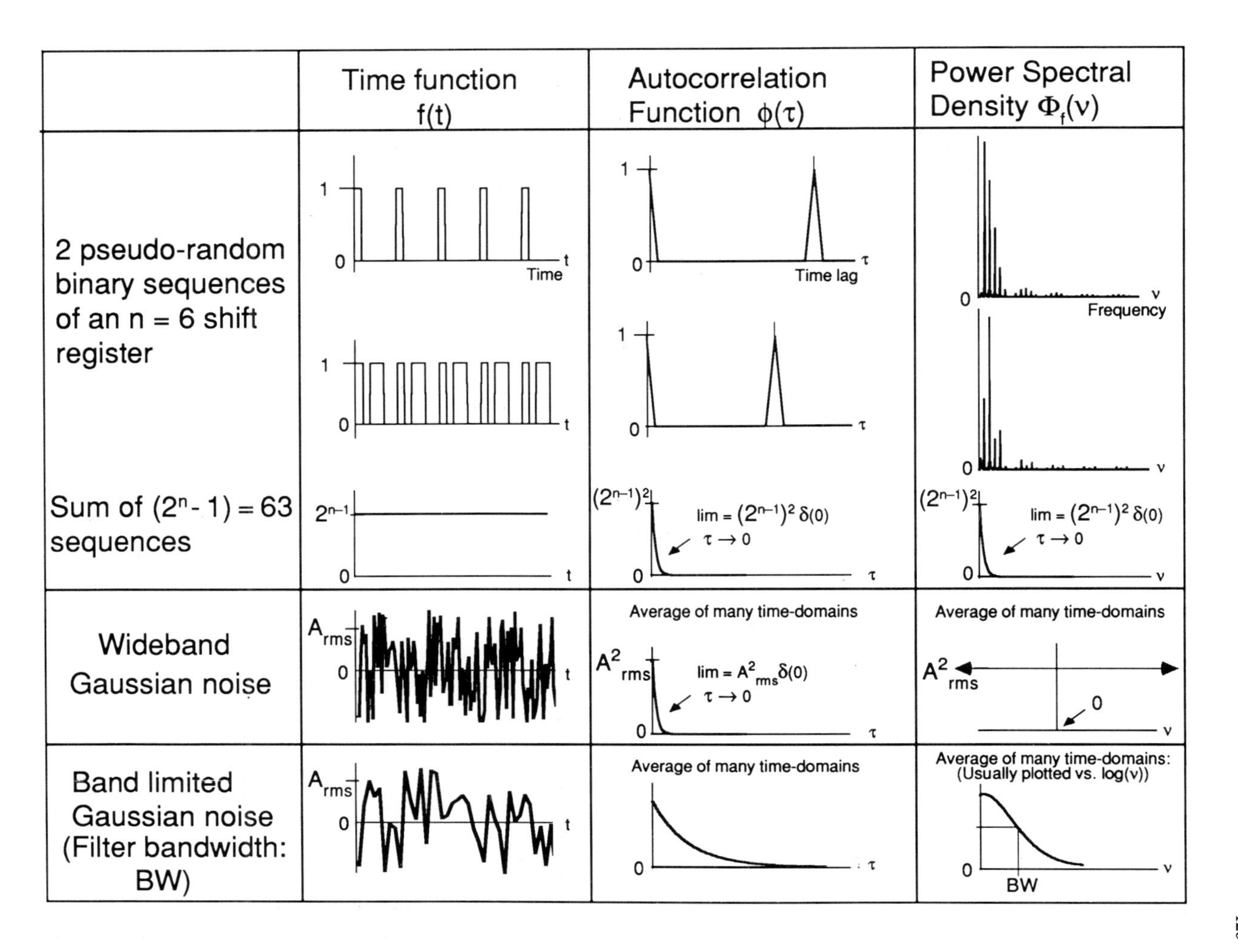

Time function f(t)
Autocorrelation Function φ(τ)
Power Spectral Density Φf(ν)
2 pseudo-random binary sequences of an n = 6 shift register
Time
Time lag
Frequency
Sum of (2n - 1) = 63 sequences
lim = (2n-1)2 δ(0)
τ → 0
Wideband Gaussian noise
Arms
Average of many time-domains
lim = A2rms δ(0)
A2rms
Band limited Gaussian noise (Filter bandwidth: BW)
Average of many time-domains: (Usually plotted vs. log(ν))
BW

Appendix E

Physical constants and units

Constant [a]	*Symbol*	*Value (SI units)*
Speed of light in vacuum	c	2.997925×10^{8} m s^{-1}
Elementary charge	e	1.602189×10^{-19} C 4.803250×10^{-10} esu (cgs units)
Avogadro's number	N_A	6.022045×10^{23} mol^{-1}
Atomic mass unit (Dalton)	u	1.660566×10^{-27} kg
Electron rest mass	m_e	9.109534×10^{-31} kg 5.48580×10^{-4} u
Proton rest mass	m_p	1.672648×10^{-27} kg 1.007276 u
Neutron rest mass	m_n	1.674954×10^{-27} kg 1.008665 u
Atomic masses	^{1}H ^{12}C ^{13}C ^{14}N ^{16}O ^{31}P	1.007825 u 12.00000 u 13.00336 u 14.00307 u 15.99452 u 30.97376 u
Faraday constant	$F = N_A\, e$	9.648456×10^{4} C mol^{-1}
Proton magnetic moment	μ_p	$1.4106171 \times 10^{-26}$ J T^{-1}
Proton magnetogyric ratio	γ_p	2.6751908×10^{8} s^{-1} T^{-1}
Molar gas constant	R	8.31441 J mol^{-1} K^{-1} 0.082057 l atm mol^{-1} K^{-1} 1.987 cal mol^{-1} K^{-1}
Boltzmann constant, $\frac{R}{N_A}$	k	1.380662×10^{-23} J K^{-1}
Planck constant	h $h/2\pi$	6.626176×10^{-34} J Hz^{-1} 1.054589×10^{-34} J s

[a] For a more complete list, see "*CRC Handbook of Chemistry and Physics*", CRC Press Inc, Boca Raton, FL, 67th Ed. 1986.

[b] For a discussion of S.I. units, see "Policy for NBS Usage of S.I. Units, *J. Chem. Educ.* **1971**, *48*, 569-572.

Conversion Factors

π	3.1415927
e	2.7182818
$\frac{\log_e x}{\log_{10} x}$	2.302585
1 eV	1.6022×10^{-19} J
1 cal	4.1868 J
1 Hz	6.6262×10^{-34} J
1 cm^{-1}	1.9865×10^{-23} J
1 K	1.3807×10^{-27} J
1 kg	8.9876×10^{-16} J
1 erg	10^{-7} J
1 D (debye)	3.3356×10^{-30} A m s
1 atm	1.01325×10^{5} N m^{-2}
1 torr = 1 mm Hg	133.32 N m^{-2}
1 lb	0.45359 kg
1 radian	57.29578 degrees
1 Angstrom	10^{-10} m = 10 nm
1 inch	0.0254 m

Greek Alphabet

α	Α	Alpha
β	Β	Beta
γ	Γ	Gamma
δ	Δ	Delta
ε	Ε	Epsilon
ζ	Ζ	Zeta
η	Η	Eta
θ	Θ	Theta
ι	Ι	Iota
κ	Κ	Kappa
λ	Λ	Lambda
μ	Μ	Mu
ν	Ν	Nu
ξ	Ξ	Xi
ο	Ο	Omicron
π	Π	Pi
ρ	Ρ	Rho
σ	Σ	Sigma
τ	Τ	Tau
υ	Υ	Upsilon
φ	Φ	Phi
χ	Χ	Chi
ψ	Ψ	Psi
ω	Ω	Omega

SI Prefixes

Submultiple	Prefix	Symbol	Multiple	Prefix	Symbol
10^{-1}	deci	d	10	deca	da
10^{-2}	centi	c	10^{2}	hecto	h
10^{-3}	milli	m	10^{3}	kilo	k
10^{-6}	micro	μ	10^{6}	mega	M
10^{-9}	nano	n	10^{9}	giga	G
10^{-12}	pico	p	10^{12}	tera	T
10^{-15}	femto	f	10^{15}	peca	P
10^{-18}	atto	a	10^{18}	exa	E

INDEX

A

B

C

D

E

F

G

H

I

J

K

L

M

N

O

P

Q

R

S

T

U

V

W

X

Y

Z

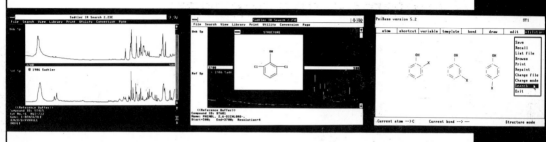

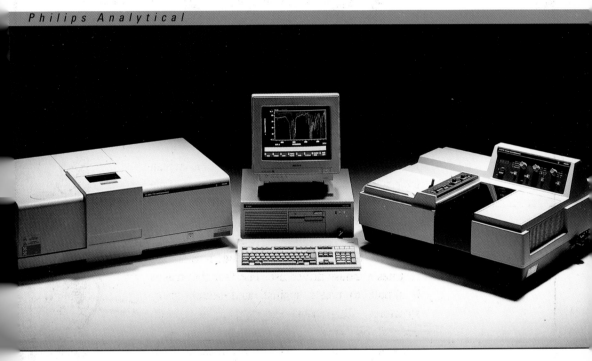